Keeping Peace in a Turbulent World

Keeping Peace in a Turbulent World:

Reflections from the Field

Edited by

Kawser Ahmed and Del H Khan

Cambridge
Scholars
Publishing

Keeping Peace in a Turbulent World: Reflections from the Field

Edited by Kawser Ahmed and Del H Khan

This book first published 2023

Cambridge Scholars Publishing

Lady Stephenson Library, Newcastle upon Tyne, NE6 2PA, UK

British Library Cataloguing in Publication Data
A catalogue record for this book is available from the British Library

ISBN (10): 1-5275-5110-5
ISBN (13): 978-1-5275-5110-7

To the fallen UN peacekeepers,

We dedicate this book to you. You sacrificed your lives in the line of duty to restore global peace and make this world a better place to live.

We hope that this book will help to raise awareness of the important work that UN peacekeepers do and the sacrifices they make. We also hope that it will inspire others to follow in your footsteps and work to make the world a more peaceful place.

Thank you for your service.

"Courage is not the absence of fear, but the triumph over it. The brave peacekeepers of our time stand as beacons of hope in the face of adversity."
—Nelson Mandela

"The pursuit of peace is a noble journey that transcends boundaries and unites humanity in its quest for a better future."
—Kofi Annan

"The brave peacekeepers who serve in the world's most dangerous places are the guardians of our collective security."
—Ban Ki-moon, former UN Secretary-General

"UN peacekeepers are the embodiment of the UN's commitment to peace and security. They are the front-line soldiers in the fight for a better world."
—Samantha Power, former US Ambassador to the UN

TABLE OF CONTENTS

LIST OF ILLUSTRATIONS

LIST OF TABLES

FOREWORD

United Nations peacekeepers perform multifaceted roles that are crucial in ensuring global peace and security. Indeed, the peacekeepers prevent conflict and ensure more rooms of manoeuver for diplomatic and political efforts, which signifies their great contributions to world peace and stability. In this connection, the involvement of Bangladesh Army in United Nations Peacekeeping Operations (UNPKOS) is a narrative of more than three decades. We are proud to earn the glorious top position as Troop Contributing Country (TCC). But, the journey of achieving such a noble reputation was not easy. Rather, it is the outcome of distinguished services, unwavering dedications and considerable sacrifices made by the peacekeepers who have left the warmth of their homes and families to uphold justice and order in the world's most perilous regions.

I am delighted to introduce the seminal work on peacekeeping titled 'Keeping Peace in a Turbulent World: Reflections from the Field' authored by the esteemed veterans. This volume is basically an anthology of narratives which reflects the lived experiences, enduring commitments, indomitable valour and great sacrifices made by the gallant men and women who have served under the Blue Helmet as United Nations Peacekeepers. It explores the multifaceted domain of peacekeeping from the vantage points of those who have been directly involved, especially the candid insights of the commanders offering invaluable observations on the conduct of various UNPKOS. The write-ups also serve as lenses through which the readers can comprehend the present challenges and envision the future role of the United Nations in an ever- changing global landscape. Thus, 'Keeping Peace in a Turbulent World: Reflections from the Field' stands as an important record of indomitable spirit, courage and dedication of UN peacekeepers.

I convey my felicitations to the authors and editors for their exemplary scholarship and dedication in bringing this enlightening volume to life. I am sanguine that the richness contained within its pages will not only educate but also palpably inspire succeeding generations to take up the sacred mantle of peacekeeping for a more peaceful, just and harmonious world.

May Allah bless us all.

Joy Bangla.

GENERAL SM SHAFIUDDIN AHMED
Chief of Army Staff
Bangladesh Army

PREFACE

Amidst a world rife with conflict and the unsettling unknown, the relentless quest for peace emerges as a resolute symbol of optimism and fortitude. The expedition to compile the volume "Keeping Peace in a Turbulent World" was impelled by an unwavering reverence for the often-uncelebrated champions who commit their lives to this noble undertaking – the valiant UN peacekeepers.

The pages ahead witness the culmination of numerous hours of collaborative effort, heartfelt dialogues, and meticulous curation. As both scholars and proponents of global harmony, we sensed an obligation to spotlight the narratives of Bangladeshi UN peacekeepers who have dared to venture into the most arduous and far-flung corners of our globe.

Within these subsequent pages, a portal opens into the existence of these extraordinary individuals, whose toils often transpire in the shadows, far removed from the glare of media coverage. Through their chronicles, we are afforded an intimate insight into the sacrifices, unwavering tenacity, and steadfast devotion that define the essence of peacekeeping. These anecdotes transcend geographical boundaries and cultural divides, resonating with the universal principles that underlie their mission.

However, this compendium transcends the confines of mere narration; it stands as a testament to the potency of collective endeavor and the shared resolve towards a world suffused with tranquility. Alongside the resonant voices of peacekeepers, we have interwoven the perspectives of former high-ranking architects of UN peacekeeping operations, thus enriching the narrative with a deeper comprehension of the tribulations encountered on the frontlines of conflict interception and resolution.

"Keeping Peace in a Turbulent World" is not merely a tribute to the gallant men and women who serve as custodians of peace but a clarion call for all those who harbor faith in the potentiality of a more harmonious global fraternity. We hope these narratives will kindle inspiration in readers across all walks of life – students, erudite, practitioners, and concerned global citizens – propelling them to unite with us in advocating for enduring serenity, empathy, and collaboration.

Our sincerest appreciation extends to all those who contributed to this endeavor – from the peacekeepers who candidly shared their experiences to the savants who graciously imparted their insights. This tome would have remained an unrealized ambition without unwavering commitment and support.

May the subsequent pages serve as a fount of inspiration, contemplation, and a poignant reminder that even amidst the tumult, the pursuit of peace perseveres as an aspiration deserving of our most earnest endeavor.

Kawser Ahmed, PhD
Del H Khan

ACKNOWLEDGMENTS

We would like to express our heartfelt gratitude to all those who have contributed to the publication of "Keeping Peace in a Turbulent World: Reflections from the Field." This book would not have come to fruition without the unwavering support, dedication, and collaboration of numerous individuals and organizations. Their collective efforts have made this publication possible, and for that we are truly grateful.

First, we would like to extend our deepest appreciation to the brave men and women who have served as UN peacekeepers from Canada, Bangladesh, and Argentina (all the contributors' countries) and shared their remarkable stories for this book. Your willingness to open up about your experiences and challenges is commendable. By sharing your narratives, you have not only enriched peacekeeping theories and practices but also touched the hearts of readers around the world. Your courage and commitment to peace serve as an inspiration to us all.

We would like to express our profound gratitude to the ex-UN force commanders from Norway and Bangladesh, who generously shared their memoirs, providing unique insights into the complexities of peacekeeping operations. Your contributions have added depth and perspective to this book, allowing readers to gain a deeper understanding of the leadership and decision-making processes involved in maintaining peace in challenging environments.

We would also like to acknowledge the subject matter experts (researchers, scholars, and officials) who contributed chapters on next-generation UN peacekeeping. Your expertise and forward-thinking ideas have broadened the scope of this book, exploring innovative approaches to tackle emerging challenges in the field. Your contributions will undoubtedly shape the future of peacekeeping operations.

A special mention goes to Niaz Mahmud Sakib (Bangladesh), whose exceptional translation skills brought the original Bengali stories to life in English. Your meticulous attention to detail and dedication to preserving the authenticity and essence of the narratives are deeply appreciated. Your contribution has enabled these stories to reach a wider international audience.

We express special appreciation to Wing Commander Raqib Mahmud, who has not only offered his stories as an author but has also generously provided his captivating photograph, which now graces the cover of this publication. Your unique position as both a storyteller and the source of the book's cover image has significantly enriched the overall influence of this project.

We extend our sincere thanks to Cambridge Scholars Publishing for believing in the value of this project and agreeing to publish "Keeping Peace in a Turbulent World." Your commitment to academic excellence and your support in sharing these powerful stories with the world are invaluable. We are grateful for the opportunity to collaborate with such a reputable publishing house.

Last but not least, through this project, CRRIC has fulfilled one of its research mandates, UNPKO, in order to contribute to the larger debate on bringing peace to this turbulent world. As of writing this book, the world is witnessing three intense armed conflicts: the Russian war on Ukraine, Azerbaijan's capture of Nagorno Karabakh, and the Hamas-Israel war in Gaza, resulting in millions of people being displaced and an unknown number of lives being lost. Thus, we all need to work together in order to bring peace by peaceful means.

Thank you all for your invaluable contributions, support, and belief in the power of sharing these stories. Together, we can make a difference in keeping peace in our turbulent world.

INTRODUCTION

Peacekeeping is not a job for soldiers, but only soldiers can do it.
—Dag Hammarskjold

In an era where the world is experiencing a multitude of conflicts and crises on multiple fronts, keeping the peace has become increasingly challenging. However, one might ask, what does 'peace' mean in the context of social unrest, political instability, or even an outright armed invasion of a nation-state against its neighbor? When discussing peace in the context of social unrest and political instability, it is useful to note that peace refers to a condition in which groups or nations resolve their conflicts in a non-violent way as well as strive to collectively achieve social, political, and economic goals. Peace is not only the absence of hostility (i.e., negative peace) but also 'positive peace', which denotes, "a more lasting peace that is built on sustainable investments in economic development and institutions as well as societal attitudes that foster peace" (Positive Peace, 2023). It should also be emphasized that conflicts are inevitable in societies, but it is possible to reduce their negative consequences and aim to achieve peaceful and non-violent resolutions.

One of the essential components of promoting peace among nations and its peoples is nurturing a 'culture of peace' (Daily Star, 2023). During the Cold War period, an Anglo-American approach to peace that largely focused on military might and hegemonic power dominated peace intervention discourse. However, as the understanding of peace has evolved, it is widely accepted that nurturing a culture of peace requires nations to contain their aggression and exhibit cooperative behavior in order to avoid conflicts and violence. The United Nations (UN) defines a "culture of peace" as a "set of values, attitudes, modes of behavior and ways of life that reject violence and prevent conflicts by tackling their root causes to solve problems through dialogue and negotiation among individuals, groups and nations." Since its founding over 60 years ago, the United Nations Educational Scientific and Cultural Organization (UNESCO) opined that, "since wars begin in the minds of men, it is in the minds of men that the defenses of peace must be constructed" (UN, 2023).

Additionally, Sommerfelt and Vambheim (2008) define a culture of peace as one that upholds tolerance among people, encourages the avoidance of situations where ethnoreligious or class differences would lead to conflicts and violence, promotes cooperative behavior, and fosters a willingness within citizens to contain their aggression. The UN defines a culture of peace as promoting non-violence through education, dialogue, cooperation, respect for human rights, and fundamental freedoms. Creating a culture of peace requires individuals and nation states to embrace values, attitudes, and traditions that respect life, promote non-violence through education and dialogue, support equality for men and women, and encourage all individuals to participate in the resolution of conflicts by peaceful means.

Since the end of the Second World War, the UN and other international organizations have been at the forefront of promoting and maintaining peace among nations. One of the most vital methods for reducing violence and advancing world peace and security for the past 75 years has been UN peacekeeping operations (UNPKOs). Amidst the chaos and uncertainty that often pervade regions torn apart by violence, UN peacekeeping missions stand as beacons of hope. Peacekeepers' tireless efforts to restore stability, protect civilians, and facilitate dialogue among parties have proven instrumental in resolving conflicts. Since its inception, over one million brave men and women peacekeepers across 71 missions worldwide have ventured into the centre of turmoil, navigated treacherous terrain and survived in volatile environments for the sake of world peace. The challenges they face are immense. From armed insurgencies and humanitarian crises to political uncertainties and cultural barriers, peacekeepers confront myriad obstacles in their missions to keep the peace.

The UN also runs special political missions (SPMs) that work on post-conflict peacebuilding, conflict prevention, and mediation in addition to peacekeeping operations (PKOs). The UN Security Council (UNSC) empowers SPMs to carry out a variety of tasks, such as assisting in political dialogue and reconciliation efforts, facilitating free and fair elections, keeping an eye out for human rights violations, organizing international development and humanitarian aid, and promoting the growth of strong rule of law institutions. While turning the page of history, we find the first UN mission was called the United Nations Truce Supervision Organization (UNTSO), whose objective was to monitor the Armistice Agreement between Israel and its Arab neighbors (UN). Since then, the UN has sent out more than 70 peacekeeping missions. Over the years, hundreds of thousands of military men from more than 120 countries have joined UN peacekeeping missions, along with tens of thousands of UN police and other civilians.

"Since 1948, over 2,400 peacekeepers have made the "supreme sacrifice" while serving in UN operations. Thus, the UN has suffered a historic average of 40 fatalities of uniformed and civilian peacekeepers a year".[1]

During the Cold War era, we witnessed several armed conflicts as colonial powers withdrew from their colonies and tensions rose between superpowers. The Suez Crisis is a case in point that formally launched the concept of modern-day PKOs. However, the nature of conflicts has evolved since then, becoming increasingly complex and protracted. In response to these challenges, the UN has developed peace-building missions as a means of promoting security and stability once conflict stabilization has occurred (Sens, 2004). As a result, complex peace-building missions that seek to eradicate the root causes of conflicts and advance lasting peace have increased in post-war environments along with the transition to capitalism.

First-generation PKOs were primarily deployed to monitor ceasefires between warring countries. Second-generation PKOs were deployed in the context of intrastate conflict and aimed to establish negative peace through disarmament, demobilization, and reintegration programs. The third generation of peacekeeping missions goes beyond traditional observer and monitoring tasks, incorporating complex multidimensional and integrative missions with more ambitious goals to promote comprehensive peace agreements and sustain peace in post-conflict periods. Finally, hybrid PKOs focus on working closely with local actors and building trust between conflicting parties, thereby facilitating reconciliation and peacebuilding efforts. Overall, the evolution of peacekeeping missions has been significant in keeping peace in a turbulent world.

United Nations Peace Mission Mandate data are useful for analyzing the types, goals, and particular tasks of 113 peace missions that took place between 1991 and 2020 to capture the trends of such a change (Hellmüller, Tan, & Bara, 2023). Scholars in this study considered three different periods to detect patterns throughout time: the post-Cold War period (1991– 2000), the War on Terror period (2001–2010), and the emerging multipolarity period (2011–2020). A transition in mission types away from PKOs and toward political missions is shown. The number of newly created UNPKOs has decreased; 36 new PKOs were established in the post-Cold War period, and this number dropped to 10 in the War on Terror period. In the recent

[1] Who Is Dying for Peace? An Analysis of UN Peacekeeping Fatalities, A. Walter Dorn, https://www.walterdorn.net/pdf/WhosDyingForPeace_Pkg-Fatalities_Scan_GPO-2008.pdf

multipolarity period, only six new PKOs were deployed, only one of which was established in the most recent five years. Since 2000, the number of political missions has consistently outnumbered PKOs. For SPMs and SE/SAs, the same is not true.

During the post-Cold War era, the dynamics of PKOs have also evolved. The UN Under-Secretary-General Boutros Boutros-Ghali's report "The Agenda for Peace" emphasized the role of peacekeeping missions in supporting the transformation of deficient national structures and capabilities as well as strengthening new democratic institutions. Moreover, since then, peacekeeping missions have expanded their mandates to include maintaining internal security and order, promoting institution-building and economic development, and ensuring humanitarian aid delivery. To effectively carry out their mandates, peacekeeping missions are required to work in coordination with a range of humanitarian and development actors.

The United Nations Department of Peacekeeping Operations (DPKO) 2008 Manual for Integrated Peacekeeping Missions[2] highlights this changed reality and emphasizes the importance of coordination between UN peacekeeping missions and a range of humanitarian and development actors involved in international crisis management. This evolution of peacekeeping missions has been driven by the belief that conditions have deteriorated to such an extent in post-war environments that only military intervention is likely to be successful in managing conflict.

The DPKO was established to give the UN a stronger institutional ability to respond to global emergencies. The Department was initially conceived in 1948 as a result of the foundation of the first UN peacekeeping missions, the UNMOGIP and the UNTSO. The UN Office of Special Political Affairs managed peacekeeping missions until the late 1980s. Early in the 1990s, the strategic framework for UNPKOs had changed from that of military surveillance to that of multifaceted initiatives created to assure the execution of comprehensive peace accords and to aid in establishing the groundwork for a lasting peace. Later, the current form of DPKO was established in 1992 as part of the UN secretariat's reform initiated by the newly appointed Secretary-General Boutros Boutros-Ghali. The broader range of tasks performed by peacekeepers now includes demining, human rights monitoring,

---

[2] United Nations. Department of Peacekeeping Operations. United Nations Peacekeeping Operations: Principles and Guidelines (Capstone Doctrine).
https://peacekeeping.un.org/sites/default/files/capstone_eng_0.pdf., chapter 5-1, p 53. August 2023.

electoral reform, humanitarian aid, security sector reform, and the disarmament, demobilization, and reintegration of former combatants. The Office of Planning and Support and the Office of Operations were created as two separate divisions of the DPKO in 1994 (Department of Peacekeeping Operations (DPKO)).

There is an assistant secretary-general in charge of each section. A Planning Division and a Field Administration and Logistics Division were established that comprised the Office of Planning and Support. The Africa Division, the Asia and Middle East Division, and the Europe and Latin America Division were the three divisions that made up the Office of Operations. The Situation Center and the Policy and Analysis Unit were established in 1993, while the Lessons Learned Unit was established in 1995. The Department of Field Support (DFS) was established in 2007 as a result of Secretary-General Ban Ki-Moon's desire to solve structural issues with the Department of PKOs, which included recruiting, logistical, and transportation issues. Through common functional and operational domains, close communication between the DPKO and DFS is maintained. The Office of Operations, the Office of the Rule of Law and Security Institutions (founded in 2007), the Office of Military Affairs, and the Policy Evaluation and Training Division were the DPKO's four primary offices as of 2012.

Under the UN Charter, Chapter VI (Pacific Settlement of Disputes)(United Nations, 2023a) allows member states of the Security Council to have the power to review and assess situations when opposing parties are in a dispute and struggling to engage in peaceful relations. Chapter VII (Action with Respect to Threats to the Peace, Breaches of the Peace, and Acts of Aggression) (United Nations, 2023b) allows the UN to take action more forcefully, using measures such as sanctions or military force. Chapter VIII also gives the UN the authority to organize regional peacekeeping arrangements. The traditional peacekeeping missions carried out by the UN serve as a way to observe the progress of ceasefire and peace agreements, especially in situations where formerly warring parties require assistance in transitioning towards peaceful coexistence.

The re-conceptualization of global security during the post-Cold War and War on Terror periods has gained widespread recognition, reflecting an emerging consensus among states, NGOs, and international organizations around the world. Several studies have also linked peacekeeping missions to humanitarian emergencies, which has provided greater legitimacy for interventions in domestic conflicts when human rights are threatened on a large scale. This gave rise to the concept of the Responsibility to Protect. As such, the nature of peacekeeping has evolved from traditional observer missions to multidimensional and integrative missions that involve a wide range of tasks, including peacebuilding, conflict prevention, and the promotion of human rights.

On so many instances, UN peacekeeping missions have acted as a crucial instrument for conflict resolution, de-escalation, and post-conflict peacebuilding. From the Suez Crisis to the Balkans, from Rwanda to East Timor, UN peacekeepers have played a pivotal role in mitigating violence, protecting civilians, and facilitating the restoration of normalcy. Thus, understanding the historical significance of UN peacekeeping allows us to appreciate its evolution and the lessons learned from past experiences. By delving into the historical context, we can identify the challenges faced, the successes achieved, and the areas that require further improvement. This understanding is essential for refining strategies, adapting to new complexities, and shaping the future of peacekeeping efforts.

The evolution of the literature on peacekeeping can be characterized as follows: initially, only a minimal number of studies were conducted during the Cold War. Then, in the 1990s, there was a surge of attention directed towards peacekeeping, which soon shifted to highlighting its shortcomings and unintended effects. Subsequently, systematic qualitative and quantitative investigations have emerged that have assessed the actual impact of peacekeeping through empirical evidence demonstrating that, despite its limitations, it is an immensely powerful policy instrument. Recent studies on peacekeeping have demonstrated an increased level of theoretical and methodological sophistication, giving way to more substantive discussions about its effectiveness. Among the issues being debated are the optimal organization or regional actor responsible for carrying out peacekeeping activities; whether utilizing force is effective; whether longer and more intrusive transitional administrations are effective; and how it affects other goals such as democratization beyond just establishing a stable peace. The latest research has also delved into nuanced analysis regarding local political actors and communities, highlighting their impact on this matter (Fortna & Howard, 2008). In conclusion, the literature on peacekeeping has evolved significantly over time.

This book focuses on UN-sponsored peace missions only, though other supranational organizations like NATO, the European Union (EU), and African Union (AU) carry out PKOs either independently or in collaboration with the UN. For example, as of today, as part of NATO's Kosovo Force, approximately 3,700 troops operate in Kosovo (NATO,

2023). NATO also actively takes part in securing the Mediterranean Sea (to expand the options available to counter the threat of international terrorism); in conducting a training mission in Iraq (the "NATO Mission Iraq" at the request of the Iraqi government and in coordination with the Global Coalition to Defeat ISIS) since 2018; and in providing air and sealift support for AU peacekeepers in 2005 and 2007. At the request of the AU, NATO is also offering the African Standby Force (ASF) concept support for capacity-building and expert training. The ASF is a component of the AU's attempts to build long-term peacekeeping capabilities.[3] The Common Security and Defence Policy of the EU gives the union the operational capability to send out both military operations and civilian missions. The scope of activities is outlined in the EU Treaty and includes tasks related to humanitarian aid and rescue, conflict prevention and peacekeeping, and the use of combat forces in crisis management, including efforts to promote peace, cooperative disarmament operations, military advice and support, and post-conflict stabilization (EU, 2018). The AU's Peace Support Operations Division (PSOD), also known as the African Standby Force Continental Planning Element, was established in 2004. Within the Peace and Security Department, PSOD responds to issues relating to Peace Support Operations (PSOs). It sets together policies, guidelines, agreements, frameworks, and processes to bring together regional peace and security institutions. All PSOs are authorized by the AU Policy Organs, including the Peace and Security Council and/or the Assembly of Heads of State and Government of the African Union (Assembly) (AU, 2023).

Today, PKOs are faced with new challenges that require them to adapt their strategies. Modern peacekeepers are now deployed in the midst of civil wars with active insurgencies, rebel organizations that recruit transnationally, and armed groups that employ terror tactics. Moreover, PKOs are also grappling with doctrinal changes and expansive mandates that sometimes make the mandates unachievable.

In this volume, we aim to shed light on the experiences and perspectives of UN peacekeepers, with an emphasis on the significant contributions made by Bangladeshi peacekeepers. Bangladesh, a country renowned for its commitment to peacekeeping, has consistently been one of the largest contributors of troops to UN missions. This book provides a platform for these peacekeepers to share their personal stories, insights, and lessons learned. Their firsthand experiences offer a glimpse into the realities of PKOs, exposing the human dimensions of this noble endeavor. By elevating

---

[3] Ibid

their voices, we not only celebrate their noble contributions but also offer valuable lessons and inspiration to future generations of peacekeepers.

This book has taken a novel approach to understanding the challenges of PKOs through the eyes of peacekeepers using storytelling. Being one of the world's oldest crafts, the power of storytelling reminds us of the sacrifices made by peacekeepers in foreign lands. Here, 37 peacekeepers from Bangladesh, Norway, Peru, and Argentina tell their stories of duty, responsibilities, and sacrifices, which inspire us to appreciate the crucial role played by PKOs in keeping peace in a turbulent world.

In this book, we have adopted a 'narrative inquiry approach' that enables us to explore and understand peacekeepers' lived experiences in UNPKOs. Through narrative inquiry, we seek to gather and analyze stories shared by individuals to gain insights into the ways in which we (the readers) can make meaning of their life experiences. Central to this methodology is the concept of meaning-making, which emphasizes how individuals construct and interpret personal narratives that shape their lives. The use of narrative inquiry in our research has been particularly useful in the exploration and understanding of sociocultural and institutional dimensions that peacekeepers witnessed during their tours of duty. Furthermore, this methodology is closely associated with the exploration of both lived experiences and cultural perspectives.

It is pertinent in this book to see how we used "new narrative research" (Casey, 1995), also known as narrative inquiry, which focuses directly on UN peacekeepers' lives and lived experiences. The shift to a teller's point of view has not only been observed in storytelling in the strictest linguistic sense in this book. For instance, some stories here do not have protagonists or climactic moments; instead, they present brief moments from the past that are thematically related (Kamil et al. 2001). While there are different interpretations of narrative inquiry, we agree that storytelling is at its core. In this book, the peacekeepers have shared significant events along an evolving timeline as a means of capturing meaningful experiences and insights.

Referring to Mead's lectures on 'mind, self, and society' at the University of Chicago in 1934, (Mead, 1934) narrative research on personal identity has coexisted alongside a history of relational-self thinking. The focus is primarily on the self within social contexts as we interact with others and how our experiences shape our understanding of ourselves. In this perspective, the 'peacekeeper' identity is more about relationships with people and the environment where they worked rather than individualism; it comprises narratives that reflect their daily accounts as they engage with the world around them (ibid, p, 243).

Narrative analysis examines how stories reflect the identities of storytellers (i.e., peacekeepers). These stories offer insights into unique social environments and are considered "documents of life". A person's experience is shaped by their social life, and so a story about them only makes sense in this context. Narratives can strengthen and rethink the relationships between individuals, communities, and their languages (Falconi & Graber, 2019).

The writing of the book involved a meticulous and rigorous approach to ensure the selection of high-quality narratives and expert insights. The compilation process began by reaching out to a diverse range of Bangladeshi and other peacekeepers, including those currently serving or retired from UN missions. Their stories were collected through interviews, written submissions, and collaborative discussions, allowing for a comprehensive representation of their experiences.

For the memoirs of ex-UN force commanders, a similar approach was employed, with in-depth interviews conducted to capture their reflections and insights. These interviews served as the foundation for crafting the memoirs, ensuring a balanced portrayal of their experiences and leadership journeys. The memoirs underwent a thorough review process to maintain accuracy and consistency.

In selecting expert contributions for the book's third section, researchers from various disciplines and backgrounds were invited to provide their perspectives. These experts were chosen based on their extensive knowledge, research, and practical experience in the field of peacekeeping. Their insights and recommendations were carefully evaluated to ensure relevance, originality, and alignment with the book's objectives.

Throughout the compilation process, rigorous selection criteria and quality control measures were implemented. The aim was to present a diverse range of voices and perspectives while upholding the highest standards of authenticity, credibility, and relevance. Each story, memoir, and expert contribution was thoroughly scrutinized to ensure its value and contribution to the overall narrative.

Preserving the authenticity of personal accounts while maintaining confidentiality and sensitivity was of utmost importance throughout the book's creation. Considering the potential risks and ethical considerations associated with sharing personal experiences from conflict zones, measures were taken to protect the identities and personal details of the individuals involved.

Pseudonyms or other forms of anonymization were used to safeguard the privacy and security of the peacekeepers and their counterparts. Where necessary, specific details that could compromise individuals' safety or violate confidentiality agreements were omitted or modified without compromising the essence of their stories or insights.

The book's contributors, including the peacekeepers, ex-UN force commanders, and research matter experts, were provided with clear guidelines and consent processes to ensure that their involvement was voluntary and conducted in an environment of trust and respect. The importance of informed consent, confidentiality, and respect for personal boundaries was emphasized at all stages of the compilation process.

By adhering to these principles, the book aims to present a truthful and authentic representation of the experiences and perspectives of the individuals involved in UN peacekeeping while upholding their confidentiality and ensuring their personal safety.

Although Canada is considered a pioneer peacekeeping country, the book offers a comparative analysis of research data from different nations' military organizations and, most importantly, the largest troop-contributing nation in the UN: Bangladesh (as of 2023). Canada's contribution to UNPKOs goes back decades since Pearson's ideas of using uniformed personnel with the consent of conflict parties were implemented as a PKO model. After all, Canada was a pioneer country in terms of troop contributions, and Canadian soldiers and leaders not only participated in UN peacekeeping missions but also helped shape the mandates and strategies employed in those missions. So far, Canada has participated in 12 UN PKOs, but its commitment to UNPKOs has been limited in recent years. Canada was also home of the premier UNPKO training center called Pearson's Peacekeeping Centre (1994–2013).

Recently, Canadian Peacekeeping Veterans Association, has published the inaugural edition of the anthology titled "PK75: Stories of Canadian Peacekeepers," accessible at https://www.cpva.ca/cpva-pk75-anthology. Additionally, for those seeking insights from a strategic military perspective in UNPKO, the works of Hanrahan, Maisonneuve, Caron, Cox, Crabbe, Elms, Goodspeed, Hale, Howard, Jorgensen, Labbe, Lightburn, Liston,

MacVicar, Pittfield, Ross, and Stevens are recommended for examination. Furthermore, this anthology boasts contributions from a group of esteemed Canadian academic scholars, presenting a promising resource for those with specific requirements in mind. Lastly, noteworthy is the fact that "PK75" has garnered opening paragraphs from distinguished Canadians, including General Wayne Eyre, the Chief of Defence Staff of Canada, who brings extensive experience as a veteran.

Despite Canada's historical commitment to UN PKOs, its recent limited pledges to engage in UNPKOs reflect a broader trend among Western democracies that increasingly favor more hybrid missions and operations outside of the traditional forms of peacekeeping authorized by the UN. This shift in focus represents a departure from the historical evolution of UN peacekeeping and poses challenges for the creation of a culture of peace.

Bangladesh is another country that has shown a strong commitment to UNPKOs and has continued to participate in various missions since 1988. So far, it has deployed 188, 558 peacekeepers, making it the largest contributor to UN peacekeeping missions.[4] It has emerged as a key player in UN peacekeeping efforts, having a full-fledged peacekeeping training center where peacekeeping participants get trained before their deployment. Despite the differences in historical involvement and current strategies, both Canada and Bangladesh have demonstrated a strong commitment to participating in UN peacekeeping missions. Two officials from the Bangladesh Army have contributed to this book, which allows readers to gain a comprehensive idea of Bangladesh's contributions in UN peacekeeping.

Bangladesh and Canada also lead a forum called the Women, Peace, and Security (WPS) Chiefs of Defense Network, which "exists to support member states' efforts to implement the WPS agenda and work to build capacities and increase opportunities for women at the local, national, and regional levels" (CHODS, 2023). Centring around UNSCR 1325, this platform widely covers women's participation in PKOs. However, as the world continues to face complex challenges and conflicts in various regions,

[4] In total, 188,558 Bangladeshi peacekeepers, including 2,728 female peacekeepers, have served in 63 UN operations in 40 states since 1988. Currently, 7,436 Bangladeshi peacekeepers, including 572 female peacekeepers, are participating in UN peace operations across 13 states. https://freshangleng.com/28641/bangladesh-s-benefits-from-usg-of-un-for-peace-operations--jean-pierre-lacroix-s-dhaka-visit

more countries must step up their commitment to UNPKOs and engage in the complex conditions of peacekeeping.

The book is divided into three parts. Part I contains true stories from Bangladeshi, Peru, and Argentinian peacekeepers. This section features a collection of personal narratives, recounting the experiences of the selected peacekeepers. Readers will encounter stories of courage, resilience, and dedication as they follow the journeys of these individuals in conflict-affected regions. The narratives touch upon diverse topics such as community engagement, gender perspectives, humanitarian assistance, and the emotional toll of peacekeeping.

Part II shares the memoirs of former-UN force commanders from Bangladesh, Norway, and Argentina. In this section, readers will gain insights into the policy realms and challenges faced by the former UN force commanders. Their memoirs provide a unique window into the decision-making processes, strategic considerations, and leadership challenges faced by those in command. The memoirs cover a wide range of topics, including negotiations with conflicting parties, collaboration with international partners, and the delicate balance between peacekeeping and peacebuilding. In Part III, three chapters capture critical conversations on next-generation UN peacekeeping from researchers/scholars. These contributions explore innovative approaches, technological advancements, and policy recommendations aimed at improving the effectiveness and adaptability of peacekeeping missions. Collectively, these three sections provide a comprehensive and multifaceted view of UN peacekeeping, from the personal experiences of frontline peacekeepers to the strategic perspectives of high-ranking commanders and the forward-looking insights of researchers.

PKOs can create a culture of peace by assisting with conflict resolution. Despite many challenges, recent studies have shown that PKOs play a vital role in maintaining international peace and security. Hultman et al. (2016) found that post-conflict states receiving PKOs were more likely to experience lasting peace than those without them. Furthermore, UNPKOs achieved this goal even under very hostile conditions where armed groups employ terrorist tactics or rebel organizations recruit transnationally.

In our quest to comprehend the complexities of peacekeeping, it is vital to recognize that no single perspective can capture the entirety of the field. PKOs are multifaceted and influenced by cultural, political, and contextual nuances unique to each region. By including diverse voices and experiences, we ensure a comprehensive understanding of the challenges, dilemmas, and achievements that peacekeepers encounter.

The objectives of our endeavour for this book are as follows.

1. To honor the courageous men and women who have served as UN peacekeepers. One of the primary objectives of this book is to pay tribute to the brave individuals who have selflessly served as UN peacekeepers. By sharing stories of the selected peacekeepers from Bangladesh, Peru, Norway, and Argentina and by acknowledging their contributions, we aim to recognize their courage, dedication, and sacrifices. Through their experiences, readers will gain a deeper appreciation for the personal and collective commitment required to uphold peace in turbulent environments.
2. To provide an insider's perspective on the challenges, triumphs, and lessons learned in the field. This book seeks to provide readers with an intimate and authentic glimpse into the world of UN peacekeeping. By sharing firsthand accounts, it offers an insider's perspective on the challenges faced, the triumphs achieved, and the lessons learned on the ground. Through these narratives, readers will gain a deeper understanding of the complex dynamics and realities faced by peacekeepers in diverse conflict zones.
3. To showcase the experiences and wisdom of former UN force commanders from Bangladesh. Their leadership roles and decision-making responsibilities provide valuable insights into the operational aspects of peacekeeping. By showcasing their experiences, strategies, and lessons learned, readers will gain valuable perspectives on the unique challenges faced by force commanders and the strategies employed to maintain peace and stability. This book also shares insights from female peacekeepers from the position of both leadership position and foot soldier.
4. To explore innovative approaches and ideas for the future of UN peacekeeping through contributions from subject matter experts. In addition to personal narratives, this book aims to explore innovative approaches and ideas for the future of UN peacekeeping. It includes contributions from subject matter experts who bring forward new perspectives, cutting-edge research, and creative solutions to the challenges faced by peacekeepers. By delving into these forward-looking ideas, readers will be exposed to diverse viewpoints and thought-provoking concepts that can shape the future of peacekeeping.

By achieving these objectives, this book makes a significant contribution to the broader discourse on peacekeeping. It promotes knowledge sharing by offering a comprehensive and multifaceted understanding of the field. Furthermore, by fostering understanding, this book encourages empathy and respect for the diverse experiences and perspectives of peacekeepers. It promotes a deeper appreciation for the complexities involved in maintaining peace in turbulent environments, cultivating a sense of shared responsibility and global solidarity in addressing conflicts. Through this broader discourse, the book aims to inspire individuals, policymakers, and organizations to actively engage in peacekeeping efforts and contribute to a more peaceful and stable world.

PKOs are not an end in themselves to stop hostilities, but rather a means to create favorable conditions for achieving lasting peace. Therefore, it is crucial to recognize the importance of PKOs in maintaining peace and stability globally and support their mandates with adequate resources. So far, 4,298 peacekeepers have lost their lives in the line of duty, reminding us that they gave their lives in a foreign land where they arrived to bring peace. Such sacrifices also remind us of the gravity and value of their mission to maintain peace, promote stability, and protect civilians in hostile environments (United Nations Peacekeeping, 2022).

As of the writing of this book, the war in Ukraine rages on. So far, we have witnessed the crippling UNSC and the lack of international consensus on how to address the conflict in the heart of Europe. This serves as a reminder of the continued importance and relevance of peace intervention in today's world, where conflicts persist even in regions outside the traditional conflict zones.

The situation in Ukraine highlights the continued need for effective peace intervention, whether in the form of PKOs or otherwise. The UNSC has a critical responsibility to work towards resolving conflicts and maintaining international peace and security by providing clear mandates, resources, and the political will to support PKOs. However, it remains to be seen how UNSC members come out of the imbroglio after the Ukraine war to bring about consensus in intervening in armed conflicts between nation states. Furthermore, member states must recognize the value of peace intervention through missions by providing adequate financial contributions, personnel support, and other necessary resources.

It is also crucial to acknowledge that PKOs cannot solve problems alone. Greater investments in diplomacy and conflict prevention techniques such as early warning systems should complement efforts aimed at building sustainable peace after violent conflict subsides. This approach focuses on addressing root causes while creating conditions conducive to successful PKOs.

Traversing the history of UN interventions teaches us one thing – a post-conflict society might relapse into conflict if deep-rooted inter-group issues are not resolved. Take, for example, the recent attack against K-FOR (a NATO-led mission that safeguards peace in Kosovo) peacekeepers in Serbia. Approximately 25 to 40 peacekeepers were left with "traumatic injuries involving fractures and burns" on 29 May 2023 when Serb demonstrators in the highly volatile region north of Kosovo assaulted them. They incurred injuries while safeguarding town halls against opposition from ethnic Serbs. Tensions had been brewing in this region for a while and recently as ethnic Albanian mayors took office in this northern part of Kosovo's Serb-majority area that ethnic Serbs boycotted (Squires, 2023). By this time, we know that violators of a rules-based order who undermine global peace might not be held accountable on a short-term basis by the UN or international community, as evidenced during the Ukraine war. Nonetheless, while peace may seem like an unattainable utopian aspiration, it remains a fundamental human need that underpins the dictum 'peace is possible'.

At the end, "Keeping Peace in a Turbulent World: Reflections from the Field" stands as a unique and invaluable addition to the existing literature on UN peacekeeping. While numerous publications have explored the subject of peacekeeping from various perspectives, this book's distinct contribution lies in its focus on the experiences and contributions of peacekeepers. Existing literature often offers a broad overview of UN peacekeeping efforts or focuses on specific missions or regions. However, this book fills a critical gap by providing a comprehensive account of the challenges faced by peacekeepers in diverse conflict zones. By highlighting their unique perspectives, the book brings a fresh and nuanced understanding to the field of peacekeeping. This book contributes to the growing body of knowledge on UN peacekeeping by shedding light on the experiences of Bangladeshi peacekeepers. Their narratives and insights offer a firsthand understanding of the realities faced by peacekeepers operating in complex environments. By capturing their diverse perspectives and reflecting on the lessons learned, the book provides valuable insights for future peacekeeping endeavors.

Furthermore, the memoirs of former-UN force commanders from Bangladesh, Norway, and Argentina offer a unique leadership perspective, providing readers with a deeper understanding of the strategic decision-making processes involved in PKOs. These accounts contribute to the evolving discourse on effective peacekeeping leadership and offer valuable lessons for current and future commanders. They serve as a valuable resource for policymakers, practitioners, scholars, and anyone interested in understanding and advancing peacekeeping efforts in today's complex global landscape.

In summary, although the effectiveness of UNPKOs has been a topic of debate for many years, their significance cannot be overstated. Collectively, these operations have helped conflict parties navigate the challenging path from violence to peace. As you venture into the subsequent sections of the book, readers will embark on a captivating journey through the real-life experiences of peacekeepers, the strategic perspectives of former-UN force commanders, and the visionary ideas of researchers. These captivating narratives promise to deepen our understanding of peacekeeping and shape our collective efforts to build a more peaceful world.

References

AU. (2023). Peace Support Operations. Retrieved from https://au.int/en/directorates/peace-support-operations

Casey, K. (1995). Chapter 5: The new narrative research in education. *Review of research in education, 21*(1), 211-253.

CHODS. (2023). Women, Peace & Security Chiefs of Defence Network. Retrieved from https://www.wpschods.com/

Daily Star. (2023). 'Culture of Peace': UN unanimously adopts Bangladesh's resolution. Retrieved from https://www.thedailystar.net/news/bangladesh/news/un-unanimously-adopts-bangladeshs-resolution-culture-peace-2914591

Department of Peacekeeping Operations (DPKO). *Summary of AG-050 Department of Peacekeeping Operations (DPKO), Office of the Under-Secretary-General (OUSG) (1992-present)*. Retrieved from https://search.archives.un.org/downloads/department-of-peacekeeping-operations-dpko-office-of-under-secretary-general-ousg-1992-present.pdf

EU. (2018). *EU Missions and Operations - As part of the EU's Common Security and Defence Policy (CSDP)*. Retrieved from https://www.eca.europa.eu/sites/eca-audit-defence/EN/Documents/EU%20Missions%20and%20Operations.pdf

Falconi, E., & Graber, K. (2019). *Storytelling as Narrative Practice: Ethnographic Approaches to the Tales We Tell*: Brill.

Fortna, V. P., & Howard, L. M. J. A. R. P. S. (2008). Pitfalls and prospects in the peacekeeping literature. *Annual Review of Political Science, 11*, 283-301.

Hellmüller, S., Tan, X.-Y. R., & Bara, C. (2023). What is in a Mandate? Introducing the UN Peace Mission Mandates Dataset. *%J Journal of Conflict Resolution*, 00220027231159830.

Hultman, L., Kathman, J. D., & Shannon, M. J. C. M. (2016). United Nations peacekeeping dynamics and the duration of post-civil conflict peace. *Conflict Management Peace Science, 33*(3), 231-249.

Mead, G. H. (1934). *Mind, self, and society* (Vol. 111): University of Chicago press Chicago.

NATO. (2023). Operations and missions: past and present. Retrieved from https://www.nato.int/cps/en/natohq/topics_52060.htm

Positive Peace. (2023). Our understanding of the term 'peace' has evolved significantly over the last 2,500 years. Retrieved from https://positivepeace.org/what-is-positive-peace

Sens, A. G. J. T. U. N. (2004). From peace-keeping to peace-building: The United Nations and the challenge of intrastate war. In R. M. Price & M. W. Zacher (Eds.), *The United Nations and Global Security* (pp. 141-160): Palgrave McMillan.

Sommerfelt, O. H., & Vambheim, V. (2008). 'The dream of the good'—a peace education project exploring the potential to educate for peace at an individual level. *Journal of Peace Education, 5*(1), 79-95.

Squires, N. (2023). Nato peacekeepers attacked by ethnic Serbs in Kosovo. Retrieved from https://www.telegraph.co.uk/world-news/2023/05/29/kosovo-attack-nato-peacekeepers-ethnic-serbs-zvecan/

UN. Our History. Retrieved from https://peacekeeping.un.org/en/our-history#:~:text=The%20mission's%20role%20was%20to,been%20deployed%20by%20the%20UN.

UN. (2023). Building A "CULTURE OF PEACE". Retrieved from https://internationaldayofpeace.org/culture-of-peace/#:~:text=As%20defined%20by%20the%20United,founding%20over%2060%20years%20ago%2C.

United Nations. (2023a). United Nations Charter, Chapter VI: Pacific Settlement of Disputes. Retrieved from https://www.un.org/en/about-us/un-charter/chapter-6

United Nations. (2023b). United Nations Charter, Chapter VII: Action with Respect to Threats to the Peace, Breaches of the Peace, and Acts of Aggression. Retrieved from https://www.un.org/en/about-us/un-charter/chapter-7

United Nations Peacekeeping. (2022). Fatalities. Retrieved from https://peacekeeping.un.org/en/fatalities

LIST OF ABBREVIATIONS

AAPTC - Association of Asia-Pacific Peace Operation Training Centers
ACLED - Armed Conflict Location & Event Data Project
APC - Armored Personnel Carrier
ASEAN - Association of Southeast Asian Nations
ASF - African Standby Force
AU - African Union
AU-led - African Union-led
BANBAT - Bangladesh Battalion
BANMED - Bangladesh Medical Contingent
BANSIG - Bangladesh Signal Unit
BN - Bangladesh Navy
BNCC - Bangladesh National Cadet Corps
C2 - Command and Control
CAX - Computer-Assisted Exercises
CFA - West African CFA franc
CIED - Counter Improvised Explosive Device
CIMIC - Civil-Military Cooperation
CORIMEC - Construction of Prefabricated Buildings and Living Camps
CPX - Command Post Exercises
CRSV - Conflict-Related Sexual Violence
DDR - Demobilization, Disarmament, and Rehabilitation
DFC - Deputy Force Commander
DOS - Department of Operational Support
DPKO - Department of Peacekeeping Operations
DR - Disaster Resilience
DR Congo - Democratic Republic of the Congo
DRR - Disaster Risk Reduction
ECOWAS - Economic Community of West African States
EIF - Elsie Initiative Funding
EU - European Union
FANCI - Ivorian National Armed Forces (Forces Armées Nationales de Côte d'Ivoire)
FARDC - Congolese army (Forces Armées de la République Démocratique du Congo)
FN - Forces Nouvelles
FTC - Full Training Capability

G5 Sahel - Group of Five for the Sahel
G5S-JF - G5 Sahel Joint Force
GI-TOC - Global Initiative against Transnational Organized Crime
GPS - Global Positioning System
GPOI - Global Peace Operation Initiative
HQ - Headquarters
ICRC - International Committee of Red Cross and Red Crescent
IDP - Internally Displaced Persons
IED - Improvised explosive devices
IEDD/EOD - Improvised Explosive Device Disposal / Explosive Ordnance Disposal
IAPTC - International Association of Peacekeeping Training Centers
ISR - Intelligence, Surveillance and Reconnaissance
ITS - Integrated Training Service
J&K - Jammu and Kashmir
K-FOR - Kosovo Force (NATO-led mission)
LA - Language Assistant (interpretation role in the military context)
LCM - Light Coordination Mechanism
LMG - Light Machine Gun
LRP - Long Range Patrol
MASH - Mobile Army Surgical Hospital
MG (R) - Major General (retired)
MILOB - Military Observer
MINUSCA - United Nations Multidimensional Integrated Stabilization Mission in the Central African Republic
MINUSMA - United Nations Multidimensional Integrated Stabilization Mission in Mali
MINUSCO - United Nations Organization Stabilization Mission in the Democratic Republic of Congo
MISAHEL - Mission for Mali and the Sahel
MoU - Memorandum of Understanding
MRAPs - Mine-Resistant Ambush Protected vehicles
NATO - North Atlantic Treaty Organization
NGO - Non-Governmental Organization
OC&S - Ordnance Centre and School
OPFOR - Opposing Forces
PCC - Police-Contributing Countries
PCRS - Peace Capability Readiness System
PDT - Pre-deployment Training
POC - Protection of Civilian Course
POTI - Peace Operation Training Institute

PR - Public Relations
PRIO - Peace Research Institute Oslo
QIP - Quick Impact Project
QRF - Quick Reaction Force
R2P - Responsibility to Protect
RPG - Rocket-Propelled Grenades
SEA - Sexual Exploitation and Abuse
SMG - Submachine Gun
SPM - Special Political Missions
SRSG - Special Representative of the Secretary-General
STM - Specialized Training Material
SUR - Statement of Unit Requirement
TCC - Troop-Contributing Countries
TRNC - Turkish Republic of Northern Cyprus
UGPS - Uniformed Gender Parity Strategy
UN - United Nations
UNAMID - United Nations-African Union Hybrid Operation in Darfur
UNESCO - United Nations Educational Scientific and Cultural Organization
UNDP - United Nations Development Programme
UNDPO - United Nations Department of Peace Operations
UNFICYP - United Nations Peacekeeping Force in Cyprus
UNHCR - United Nations High Commissioner for Refugees
UNHQ - United Nations Headquarters
UNMAS - United Nations Mine Action Service
UNMOGIP - United Nations Military Observer Group in India and Pakistan
UNPKO - UN peacekeeping operations
UNPOL - United Nations Police
UNPROFOR - United Nations Protection Force
UNSC - United Nations Security Council
UNSCR - United Nations Security Council Resolution
UNSG - United Nations Secretary-General
UNTSO - United Nations Truce Supervision Organization
US - United States
USARPAC - US Army Pacific Command
USINDOPACOM - United States Indo-Pacific Command
WFP - World Food Programme
WPSC - Women, Peace, and Security
ZOC - Zone of Control

United Nations Peacekeeping Mission Study Areas

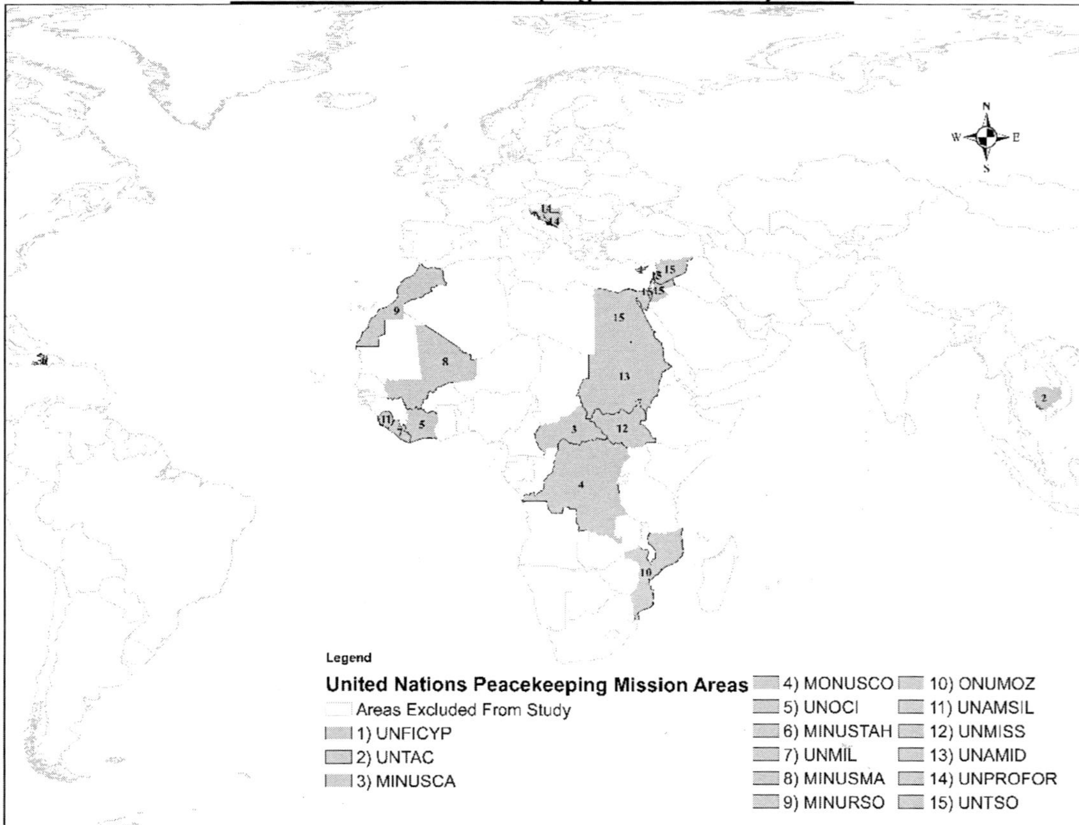

PART I

BEYOND THE BLUE HELMET: THE UNTOLD STORIES OF BANGLADESHI UN PEACEKEEPERS

This part tells gripping stories of Bangladeshi UN peacekeepers who risked their lives in the pursuit of peace across the globe. From their deployments in Africa to Asia, these brave soldiers share their personal experiences of serving under the UN flag and helping to keep the peace in some of the world's most challenging environments.

The Road to Timbuktu

Major Kazi Rafi

(United Nations Multidimensional Integrated Stabilization Mission in Mali - MINUSMA)

The November sun was merciless, its fiery rays scorching the earth to a crisp as our convoy traversed the desert region of Mali. By noon, temperatures soared to a searing 40 degrees, but as night descended the arid land was blessed with a mild 25 degrees. The Niger River flowed with renewed vigor and vitality, as if infused with new life, thanks to the bountiful rainfall in Bamako that year. The craggy hills, once barren and lifeless, were now adorned with lush greenery that stretched towards the sky. The African wild grass grew with reckless abandon, providing a feast for the goats, cows, and camels that roamed freely. But as with all things in the desert, this glory was fleeting, and come December, the once-glossy coats of these animals would wither away like dry grass without moisture.

The November morning air was crisp and cool as we set out on our journey from Bamako, the bustling capital of Mali. Our destination: the legendary city of Timbuktu, a once-thriving hub of African civilization that now lay shrouded in the Sahara Desert's unforgiving expanse. The journey ahead of us was daunting, spanning over 1,000 kilometers of rough and rugged terrain. But I was not alone in this perilous endeavor. Leading a contingent of 50 Bangladeshi soldiers, flanked by Togolese tankers for our protection, we were supported by an Egyptian helicopter that soared above us like a vigilant eagle.

We were not traveling light; we had a convoy of 37 colossal cargo transporters loaded with our essential supplies and victuals. These transporters were more common on the roads of African countries where the United Nations was carrying out its peace mission than buses for passenger transportation. The Bangladeshi military officers, Dr. Norman, a medicine specialist, and Major Sarwar, a communication officer, worked under my supervision. Major Sarwar was particularly vital in dealing with the unreliable mobile network, especially in the Sahel region skirting the border of Mauritania, where we needed clear communication during the operation.

Mali was a vast country, nine times bigger than Bangladesh, and its population of 15 million was widely scattered. Many people had migrated to the capital, Bamako, seeking refuge from the war's turbulence. But despite the population influx, the prevalence of AIDS was relatively low in the country, thanks to the predominance of Muslims, who made up nearly 90 percent of the population. As we made our way through the arid and rugged terrain, we were all acutely aware of the challenges ahead, but we pressed on, fueled by our determination to reach our goal, no matter what lay ahead.

The early morning sun beat down on us as we prepared to embark on our journey from Bamako. Kamara, our Malian interpreter, stood at attention, resplendent in his blue Haji Gamcha and beret, shielding his face from the harsh desert sun. Despite the sweltering heat, Kamara knew his role was crucial to our mission's success.

As we left the city behind, the landscape opened up before us. Mango orchards lined the sides of the road, and thorny bushes had sprouted greenery from the recent rainfall. In the distance, we could make out a few sparse trees, some with lush green leaves, and others with two large branches that hung like elephant trunks. Even Kamara was unaware of their name, but we had given them the nickname "Elephant Trees."

Our journey was not without its moments of levity, and we shared a moment of amusement when Kamara asked about the tree's name. With a smile, I told him they were called Elephant Trees, and Kamara promised to teach his people this new knowledge.

But beneath the humor, we knew the gravity of our mission. We were traveling to Timbuktu, an ancient city that was once the seat of one of Africa's largest civilizations. Our journey would take us over 1,000 kilometers of rugged terrain, with 50 Bangladeshi soldiers, Togolese tankers, and an Egyptian helicopter for protection. As the temperature continued to rise, we all knew the challenges ahead would be daunting.

As we trekked on, the Niger River flowed alongside us like a trusted ally. In the distance, a few rocky hills jutted out of the landscape like works of art, adding depth and diversity to the region's natural beauty. By the afternoon, we arrived in Segu, where the Segu Stadium stood majestic before us, it's perfectly manicured green grass a testament to the locals' unwavering care.

After a long day of traveling, we found a tranquil spot to rest and rejuvenate. As the sun dipped below the horizon, Noman and Sarwar set up a cozy area on the grass, complete with chairs and a table. We sat there, sipping tea under the gentle glow of the moon. As the night unfolded, our thoughts wandered to the wonders of Africa.

As we strolled along the path, we noticed the sharp thorns of the grass flowers standing guard over the delicate blooms. It seemed as though the thorns were silently protecting the beauty of the land, ensuring its safety from harm.

The next day, our journey took us deeper into the Sahara, towards the distant destination of 'San.' The landscape underwent a remarkable transformation before our eyes. The lush greenery gradually gave way to a more barren terrain. The once-guardian thorns now served as a reminder of the Sahara's unforgiving existence. We made camp at a stadium in 'San,' and as the sun began to set, I climbed to the top of the structure to take in the captivating yet desolate surroundings.

From the summit, a sense of eerie silence enveloped me, sending a shiver down my spine. But as night fell, the moonlight revealed an enchanting world that I couldn't ignore. The desert wind whispered secrets in my ear, and for a fleeting moment, I felt as if I were experiencing the first love of my life all over again.

Our arduous journey continued, and we ventured further into the heart of the Sahara. The horizon stretched endlessly with nothing but an expanse of thorny trees and beautiful yet isolated greenery. But as we pressed on, the number of trees diminished, exposing the stark reality of the desert. The once-green leaves vanished, leaving behind a landscape of leafless, thorny trees.

Finally, our expedition led us to a stadium in 'San' where we had no choice but to halt our journey. In the afternoon, driven by curiosity, I ascended to the highest point of the stadium. The deafening silence and a sense of solitude washed over me. Determined to witness the allure of the desert, I decided to stay until nightfall. And there, under the shimmering moonlight, the desert unveiled its enigmatic charm. Every detail became vivid, as if I were rediscovering the first love of my life.

As we journeyed down the dusty road, we passed by humble clay dwellings that looked as if they had been abandoned for ages. There was no porch or veranda, no living creature's shade, touch, or shadow. The only signs of life were the spiny branches scattered haphazardly around and the swaying trees that seemed to link us to the eternal flux of time in our precarious existence.

During our recess, I called my officers to join me, and we sat on camp stools while Noman purchased some delectable treats from a young woman who was selling her wares along the pavement. As we savored the delicious plump berries, almonds, watermelon, and papaya, Noman's voice brimmed with compassion. 'It was as though the watermelon was oozing nectar, and those plump berries! Never in my life had I tasted such ambrosial fruit. It was a rare delight to savor such bountiful berries in a deserted African country in November.'

Sarwar's voice quivered with anger as he spoke, 'Sir, they are letting foreigners plunder their country's wealth and resources.'

I posed a counter-question, 'And what can they do, tell me? If even a minor hindrance, like a mosquito, disrupts their looting of diamonds and uranium, they won't hesitate to deploy their arrows and warplanes.'

'You're right, sir,' Sarwar replied. 'Not long ago, a farmer was spreading manure in his field with a pickup truck near our super camp. The radar of the other contingent picked it up, and the media wove a tale of terror: an armed militant group had scouted the super camp for an imminent attack. Since then, two drones have been hovering over us like vultures, day and night. The escalating cost of safeguarding the super camp would have made the Mughal emperors envious. Yet, outside the camp, people starved, surviving on nothing more than roasted corn cobs at noon.'

Noman added, 'How times have changed! Eight hundred years ago, the king of this land, Mansa Musa, embarked on a lavish pilgrimage (Hajj) to Mecca with sixty substantial caravans laden with gold. Each camel carried sacks of gold on its back. Mali was the world's golden goose, producing half of its gold. Mali was the sole country producing such a vast quantity of gold. Legend has it that when Mansa Musa returned from Mecca, he lavishly distributed his gold from Egypt onwards. He was so generous that he caused the price of gold to plummet by half that year.' It was difficult to fathom that this land, once the envy of the world, now struggled to survive in the face of foreign exploitation and greed.

Meanwhile, Sarwar's typically cheerful countenance contorted with frustration. 'He lacked wisdom and prudence. He squandered his gold, mingling his black gold with the riches of Africa, tainting his oil and gold with the rest of the continent. The pirates of the Atlantic caught wind of his wealth and invaded Africa, pillaging the land.' We had halted to rest after covering a grueling 1.5 kilometers. Seeking shelter beneath a tree, I gazed out at the desolate horizon. While the influence of the Niger River had blessed San and Mopti with lush mango orchards, plum-berry groves, and fertile fields, the vast expanse before me evoked a profound sense of nostalgia.

My heart ached with pity for the impoverished African youth, clad in ragged and faded clothing. I couldn't help but think of a Malian girl whose beauty and grace could match that of any European girl, despite the fact that Europe had every resource at its disposal. As I envisioned her in my mind's eye, I was consumed by a profound sense of melancholy.

As we crossed the bridge over the Niger River, I couldn't help but marvel at the ingenuity of Mansur Musa, who had carved out canals from the river's veins, bringing life to the Diabali region and turning it into a lush land of green rice fields. It was a testament to the power of vision and hard work. But as we left Mopti behind and headed towards the Doyenja expedition, I couldn't help but wonder why they had lost all that glory.

As we journeyed on, we stumbled upon a hidden gem: a vast lake nestled under a craggy hill, adorned with countless white blossoms that floated like lotus flowers on the serene water. I ordered the convoy to halt and set forth with three guards for each officer. The closer we got to the lake, the more the arid sand gave way to moist earth, and sandy soil became normal soil. We soon spotted countless hoofprints on the sodden ground, evidence of nomadic herdsmen and their bovine and ovine charges quenching their thirst at this oasis. As they say, water is life. They must have been coming here regularly.

But my attention was soon drawn away by the mesmerizing sight of the desert's prickly trees, their jets of water reflecting the azure sky above. I ventured closer and discovered three trees laden with ripe, cool plums. Their sweetness was delectable. And as I moved further ahead, I was greeted by a breathtaking sight. From the small, leafy, thorny trees of yellow-blue and purple color emerged a swarm of butterflies, as if they had rained down from the heavens. It was like the final moments of Ursula from Gabriel Garcia Marquez's novel 'One Hundred Years of Solitude'. The beauty of this moment was like a magic spell that had enchanted me.

Marquez's vision was no fabrication. The rain of butterflies had indeed battered the forsaken city, covering every inch of the ground and sky. But here in Doyenja, the colorful trees with their thorny branches had become the butterflies' kingdom, a place where they could blend in with their own kind. I couldn't help but wonder if the butterflies had found their own kingdom by matching their environment with their own colors.

The scorching November sun mercilessly beat down on us as we arrived at Togor Camp in Doyenja. With regret, the operations officer informed us that there were no prefabricated air-conditioned rooms available for us officers. We bit our tongues, grateful for his kindness despite the unbearable heat that reached over 45 degrees. To survive, we hung our tents like swinging cradles under the sparse shade of a few trees in the camp. There, we spent our days brainstorming the safest routes to cross the treacherous desert ahead of us, sipping tea, and making small talk. But our patience wore thin as the days dragged on due to a lack of proper coordination.

As evening approached, the terrain was transformed into a magical wonderland. Our tent nestled next to the border fence, with a steep stone hill rising to the left. The moon shone bright in the African sky, inviting us to climb to the hilltop. We settled into our chairs and gazed out to the horizon, watching as people herded their cows and sheep back home. The moist air filled our lungs with freshness, stirring our hearts with passion and invigorating our spirits. But amidst the magic, a pang of nostalgia hit us, reminding us of our loved ones and the distant fields and rivers of our homeland. Home is where the heart is, but ours was far away.

The evening brought a respite from the heat, and we settled down on a mound to watch the large round moon rise. It was as if the moon wanted to eavesdrop on our hopes and fears, dreams and nightmares. We talked without purpose, conversing for the sake of conversation. We shared dinner under the moonlight, perched on that lofty mound like birds on a wire.

The next day, we set off towards Timbuktu, our colossal military caravan stirring up a storm of dust on the barren route. The Sahara loomed large, and the road was mostly devoid of tracks. We encountered few traces of civilian vehicles that had braved the desolate caliphate, but the path was mostly uncharted. Our progress was slow, with the recovery vehicle getting stuck in the sand time and again. The mammoth cargo and tanks trudged forward, but even they could only manage a speed of seven kilometers per hour. The camels and sheep with thick tails fled in terror, but we pushed on, determined to reach our destination. The desert was unforgiving, but we were adventurers, and we were not going to let it defeat us.

The barren wasteland lay in the grip of a notorious terrorist group – the merciless al-Qaeda. They ruled this land with an iron fist, and their savage ambush on a Togo convoy was a brutal testament to their power. But I wasn't one to cower in fear. I had come here for the 'Bound Clearing Drill,' a mission to reclaim this godforsaken place. And as I gazed at the towering mountains on the right, I knew this was going to be no easy feat.

Sweat dripped down my face as I waited for the tank escort to scout ahead. My trusty jeep roared beneath me, struggling to cool down in the scorching heat. The only sound was the deafening thump of a helicopter hovering above us. We were completely cut off from the world, with no mobile signal to rely on. But I was a one-man army, equipped with a GPS system, walkie-talkie, and ground-to-air communication, and I wasn't afraid to take on anything that came my way.

As I stepped out of my jeep to seek refuge under a thorny tree, I saw a lone shepherd guiding his flock of animals towards their destination. His peaceful life amidst this desolation seemed almost frozen in time. And as I gazed at the endless expanse of thorny shrubs and sandy plains, I couldn't help but feel a strange yearning to join him. How did he spend his nights in this mystery-laden land, with nothing but the moon to guide him? It was as if the moon itself was his teacher, bestowing upon him the wisdom of the ages.

And then I saw it – a seashell lying by my feet. It seemed to have a story to tell, and I couldn't help but pick it up. As I held it in my hand, the seashell seemed to come alive, whispering tales of the ocean and its mysteries. I realized that this barren wasteland was once the edge of the sea, and the seashell was a testament to the ocean's timeless power.

I whispered to Kamara, "Was this place ever under the sea?" And as we waited for the tank escort to return, I knew that the mission ahead would be as treacherous as the ocean itself. But I was ready for whatever lay ahead. The seashell had given me strength, and I knew that I could overcome any obstacle that came my way.

"The Atlantic was here, sir," Kamara spoke with a hint of wonder in his voice. "And not just here, but all across this region. According to a German historian, this land was once a thriving part of the great Atlantic some nine to ten thousand years ago."

My curiosity had led us on a perilous journey, and as the day turned to dusk we found ourselves only 97 kilometers into our trek. We had reached the treacherous Bambaramodite, where the moon hung low over the former seashore, now a barren desert. From our vantage point, I surveyed the flatlands on three sides, the moonlight casting a mysterious glow over the treeless horizon.

But our peaceful reverie was short-lived. At the foot of the mountain lay a small village of common folk and nearby an army camp belonging to the Malian government. Five machine gun posts circled the camp, evidence of the constant danger that plagued this region. Armed groups raided the camp three or four times a month, and the Malian army often fled, leaving their post behind.

With swift instructions, I directed our military to take charge, placing our vehicles in the center of the field and positioning tanks around them. Three mobile tents were erected on the veranda of the camp for the three of us, but water was scarce, making cooking at night a long and arduous task.

After dinner, as I lay down in my tent, the moonlight seeped through the mosquito net at the entrance, casting a soft glow on my bed. It was a rare privilege to bask in such pure, gentle light while the possibility of death loomed large. But in this land, death was an ever-present companion, a fact that became all too clear when we bid farewell to the Malian captain the next morning.

Upon our return, the camp looked unchanged, as if the deaths of the captain and his comrades were just another daily occurrence. But for me, their sacrifice would never be forgotten. It served as a stark reminder that in this adventure, death could come at any moment and those who survived were truly lucky.

The trees grew thicker as I pressed on towards Timbuktu, signaling that the Niger River was close at hand. The promise of water was palpable, and the flora responded accordingly, sporting a verdant coat. A village seemed to materialize before me, but it was a ghost town, lifeless and silent.

But then, the river came into view, and my breath caught in my throat. The Niger was a force of nature, a hungry lion with a mouth wide open, swallowing the land around it. The water sparkled in the light, and its reflection played with the lush grass, creating a wondrous scene that only those who had braved the desert's unforgiving heat could truly appreciate.

The river was swollen from heavy rains in the south and the flow of water from the northern Niger. On the riverbank, we encountered several civilian vehicles that had piled up, their presence a mystery since we hadn't seen them on the road. We estimated it would take three days to ferry our cargo across on two small boats, each barely able to accommodate two large trucks. We found a patch of land amid the wetland, where we shared our food with five families who had built their homes with wood and bamboo. As the sun set, the horizon became a floating canvas, and the sky's azure tint dissolved into obscurity. The colossal moon ascended like a blazing orb from the aquatic depths, a spectacle that left us speechless.

I dipped my feet in the water and was transported to a realm of matchless magnificence. But our journey was not over yet, and there was one last tale to tell – the story of a Malian maiden. But I'll leave it to your imagination to conclude this travelogue with that story.

As the sun beat down on the riverbank, the air buzzed with the excitement of the market on the other side of the river. Women hustled and bustled, loading their motorboats with goods for trade. But amidst the chaos, one girl stood out. She was singing, her voice carrying across the water in a language foreign to the author's ears.

Drawn in by her melodic voice and the sway of her hips, the author approached her, determined to discover the meaning behind her song. But when their interpreter, Kamara, informed the author that the girl was singing in Bambara, he was skeptical. Without hesitation, the author demanded that Kamara sing the same song to prove him wrong.

With hesitant laughter, Kamara attempted to replicate the girl's song, but it was the author who surprised everyone. He stepped forward and began to sing in a language that was not his own. The girl's eyes widened with surprise and delight as she listened to the author's rendition of her song.

In that moment, they connected, bridging the gap between their different cultures and languages. And as the author walked away, he couldn't help but feel a sense of joy and fulfillment that he had been able to share a moment of understanding with this young woman from Mali.

The tension in the air was palpable as Kamara gazed at me with disbelief. But I wasn't finished yet. With a sly grin, I handed the mobile device back to the African beauty and said, "Let me show you how Bengalis really dance." And with that, I began to move my body to the rhythm of the music.

The African maiden looked on in amazement as I spun and twirled, my hips swaying to the beat of the music. Kamara's eyes widened in shock as I effortlessly mimicked the dance moves of the heroes and heroines of my homeland. And as the music reached its climax, I threw my hands up into the air and let out a triumphant roar.

For a moment, there was silence. And then, slowly but surely, the crowd began to cheer. The African beauty clapped her hands in delight, and Kamara shook his head in wonder. “You are a true adventurer, my friend,” he said with a smile.

And with that, we set off into the night, our hearts full of adventure and our spirits soaring with the joy of discovery. As we walked, I couldn’t help but think about all the wonders that awaited us in this vast, wild world, and I knew that together, Kamara and I were unstoppable, for we were adventurers, explorers, and seekers of truth, and nothing would ever stand in our way.

SAVANNAH'S NIGHT WATCH

MAJOR DEL H KHAN

(UNITED NATIONS MISSION IN SOUTH SUDAN -UNMISS)

One

My senses came alive, a sudden jolt back to reality, as I spotted a distant flicker on the dark horizon, far ahead of our left-forward sentry. The African savannah, once a stage for safaris, had transformed into a turbulent battleground for our challenging peacekeeping mission, guarding a crippled UN helicopter. With urgency, I murmured into my radio, "Delta1 to Delta3, do you have eyes on that movement ahead? Can you confirm its nature, over?"

The sentry's response confirmed my suspicions, "Delta3 to Delta1, it appears to be a civilian, possibly inebriated. No threat observed, over." It was a drunken reveller, stumbling home from a late-night Christmas celebration. Alcohol had been his companion on the journey, and the chubby Mi-8 helicopter had borne witness to his inebriated exploits, a canvas for his haphazard artwork of vomit. Now, he wandered carelessly into the enveloping darkness, proclaiming "Merry Christmas" to the empty expanse.

Today was meant for celebration, but the tranquillity of my night had been shattered. I had spent the previous hours awake, watching movies and surfing the internet, believing there would be ample time to rest. However, my plans crumbled when my company commander's voice, unyielding and demanding, pierced the stillness over the telephone around 4:30 pm. "Are you prepared to lead a force protection patrol?" he asked.

My reply, delivered with a weary voice, was a simple "Yes, sir."

"Then hurry," he urged, "The chopper awaits you at the Juba airstrip."

With swift determination, I geared up and emerged from my quarters. Outside, my company commander sat in the driver's seat of his military jeep, its engine roaring with readiness. As I boarded, we embarked on our journey toward Juba, the South Sudanese capital, a temporary airport that signified the starting point of our perilous assignment.

En route, my commander, his gaze fixed on the road, delivered a briefing on the mission. “A Russian UN helicopter has crash-landed in the Lafon region of Eastern Equatoria.”

“Shot down?” My disbelief was evident in my voice. Given the prevailing security conditions in South Sudan, such an event would be exceedingly rare.

“No,” he corrected me, “The helicopter was carrying election observers for the upcoming referendum on South Sudan's independence. But upon reaching Lafon County, it suffered mechanical failure and couldn't take off.”

His words hung in the air, and I remained silent.

“Sunset is imminent,” he continued, “Your mission is to lead a force protection patrol, fly to Lafon, secure the damaged helicopter for the night. The helicopter that brought you here will return to Juba with the stranded election observers and the downed helicopter's crew. Tomorrow morning, a repair team will arrive at your location to fix the helicopter, and you'll return with them.”

As we passed through the iron gate of the Juba airstrip, allowing our jeep to proceed, he inquired, “Any doubts?”

“None, sir.” My response was steadfast as my eyes scoured the vicinity, seeking my team and the helicopter designated for our mission.

“Good,” he acknowledged as we reached the helicopter's side. Turning off the jeep's engine, he remarked, “In the daytime, a non-commissioned or junior officer would suffice for leadership. However, considering the remote location and the ominous shroud of night, an officer has been chosen.”

My team stood beside the helicopter, awaiting my arrival. Ten UN peacekeepers, their expressions a blend of frustration and resignation, had seen their holiday plans disrupted by unexpected circumstances. Nearby, two military observers from other nations displayed visible disquiet, questioning their involvement in this nocturnal mission.

Parting ways with my company commander, I assumed command from the second-in-command of the patrol. After a swift inspection of our weapons, ammunition, radios, and other supplies, my survey of the surroundings inevitably fell upon the Russian helicopter. Its design, meticulously engineered for the perils of nighttime flight, stood distinct. Heavier than the standard model, its rotors whirred in a deafening crescendo.

At last, the signal to board was given, and we embarked on our journey. But fate had one more twist in store for us - the pilot's announcement that two of us had to disembark due to the helicopter's overload with weapons and ammunition. Without hesitation, the two military observers disembarked, their parting words echoing in my thoughts as they disappeared into the obscurity, "There's nothing for observers to witness in the night."

As the helicopter descended toward Lafon, the cloak of darkness enveloped us. The pilot signaled, and the craft seemed to levitate, creating a cloud of dust upon landing. Swiftly, we disembarked, making way for the election observers and the crew of the incapacitated helicopter. The helicopter ascended towards Juba, leaving a dense trail of dust in its wake.

As the dust dispersed, the silhouette of the damaged helicopter emerged, idling under the starlit canopy. Swiftly, I retrieved my satellite phone and relayed our safe arrival in Lafon to our contingent headquarters in Juba. The night pressed in around us, and an unsettling sensation of being watched from the shadows seized me. The air was heavy with the scent of diesel, and the sound of the wind rustling through the grass was an eerie accompaniment, foreboding the challenges of the long night that lay ahead.

The Russian pilots secured the helicopter's doors before departing, leaving us isolated under a sky adorned with stars. The desolate, barren landscape extended as far as the eye could discern, punctuated by scattered bushes. The mountain range of Eastern Equatoria, a mere outline on the horizon, seemed worlds away. In the distance, a dim glow and the strains of distant music hinted at the Lafon village's existence. The village, a collection of scattered huts known locally as 'tukuls,' was illuminated by the Christmas festivities, as villagers gathered to celebrate.

Surveying my surroundings, an unshakable unease lingered. The darkness concealed our view and exposed us to potential threats from the mischievous and inebriated locals known to become belligerent under the night's shroud.

Peacekeeping was a distinct realm from warfare. Unlike warriors who concealed themselves in hostile territories, peacekeepers were obligated to reveal their positions. Yet, in the obscurity of night, our UN flag was no shield. If attacked, our assailants could claim ignorance, rendering the UN powerless and leaving only our lifeless bodies to return to our homelands, shrouded in national flags.

With meticulous precision, my team of eight established a security perimeter. An aura of invincibility descended upon us, fueled by the arsenal at our disposal. We had already coordinated with Bangladeshi Camp Torit, ensuring a three-hour window for emergency assistance. Yet, the true fear was surviving those hours if we were to come under attack. With no sanctuary to tend to the injured, a desperate predicament loomed.

My gaze inevitably settled upon the helicopter as a potential refuge. The Mi-8, I knew, was a formidable fortress, its walls impervious to bullets. However, the Russian pilots had sealed all the doors before departure, confining us to the wilderness with only the stars as our companions. As the night wore on, the uncertainty of what lay ahead intensified.

Sensing my thoughts, two of my team members directed their torches toward the helicopter, fumbling with the handles. Suddenly, one of the handles yielded, and the door began to inch open. In small, engraved letters, the word “Emergency” beckoned. Without hesitation, we swung the door wide, securing our heavy wireless equipment and other supplies inside.

As the midnight hour chimed and the moon ascended, bathing our surroundings in its luminous glow, I conducted one last inspection of our posts before settling inside the helicopter. Gazing through the window at the star-studded night sky, I marveled at constellations foreign to the skies of my homeland. The moon, too, appeared larger and more radiant, a phenomenon perhaps unique to equatorial regions. I took a sip from my water bottle, pondering whether I was on a UN mission or embarked on a surreal safari, as I watched the astral tides waltz across the savannah.

Two

Dag Hammarskjold once claimed, “Peacekeeping is not a job for soldiers, but only soldiers can do it.” However, as a young military officer, Hammarskjold's words appeared quite incomprehensible to me. The idea that a task might not be mine but that no one else could perform it seemed to defy logic. It was only after my first overseas mission as a UN peacekeeper that I truly comprehended the profound meaning behind Dag Hammarskjold's words.

It was a June morning in 2009 when my comrades bid me farewell, and I embarked on my maiden UN mission abroad. I joined a new unit, a composite infantry company that bore the name 'Additional Company,' an ensemble of 175 valiant soldiers. My honor swelled as I assumed command of the '9th Rifle Platoon' within this gallant company.

Our 'Additional Company' was destined to bolster the existing Bangladeshi infantry battalion (BANBAT-4) already stationed in South Sudan, becoming its fifth company. Our deployment was earmarked for mid-August, and in Dhaka Cantonment, we convened for the customary pre-deployment training interwoven with meticulous preparations. The procurement of equipment and the laborious chore of loading containers spanned a grueling three months. At last, the day materialized when our containers and vehicles embarked on their voyage to South Sudan, and the countdown to our departure via flights began, awaiting the arrival of the container ship at Port Sudan.

While our container-laden vessel reached Port Sudan as scheduled, the Government of Sudan, under President Bashir's leadership, wavered in accepting the addition of our 'Additional Company' to the existing UN force already deployed in South Sudan. This reluctance translated into a series of vexing flight delays, inflicting a harrowing period of uncertainty upon us. Fellow peacekeepers from other UN missions extended their empathy and support. Yet, we never foresaw becoming the 'Forgotten Company.' Our commander spared no effort to keep our spirits aloft, but the prevailing sense of abandonment lingered.

Ultimately, in the second week of February, we at last clutched our long-awaited boarding passes, bidding emotional farewells to our loved ones. The run-up to our deployment had been strenuous and emotionally draining for our families, preparing for another year or more of separation. They saw us off at the airport, and most of them returned home after midnight, embarking on the first night of this new chapter. To their surprise, late that night, a knock sounded at their doors, and they found us back at home. Just before boarding, we were informed of a 24-hour flight delay due to a lack of clearance from the Government of Sudan. While it was a heartwarming surprise for our families, it also offered a glimpse into the uncertainties that would accompany us in Sudan.

Nonetheless, the very next day, we set forth on our journey to Sudan. For most of us, it marked our inaugural overseas flight. On the plane, some playful soldiers concocted a rumor that we might have to return from Eden

Airport in Yemen. The rumor held such sway that even our ardent smokers abstained during the refueling stop in Eden!

As we touched down, the South Sudanese panorama unfurled before us, a mirage shimmering over the Juba runway and the tropical sun's embrace warming our spirits. We received a heartfelt welcome from Bangladeshi officers and BANBAT-4 personnel. The welcome tea gathering etched itself into our memories, a whirlwind of vital experiences, currency exchanges, and the exchange of mobile SIM cards among incoming and outgoing troops.

The night waned swiftly, and the very next morning, I, along with the 9th Platoon, boarded a Russian chopper, gleaming in white, destined for our next stop: Torit.

Torit, a UN camp nestled in Eastern Equatoria, held our affectionate monikers 'Tomping Site' or 'Team Site.' It was encircled by sprawling earthen walls, a meticulously arranged haven for UN staff and troops. The Kenneti River was our sole source of water, drawn each morning using water bowsers, and our weekly rations journeyed from Juba to us every Friday.

Our noble duties revolved around safeguarding UN personnel, assets, and operations, and ensuring the well-being of the local civilian population. Our sphere of responsibility spanned a staggering 82,542 square kilometers, a formidable task for our company-sized force.

Three

My experiences as a peacekeeper have been exceptionally diverse. The arduous journey of transforming a combatant into a peacekeeper commences long before our deployment to mission areas. This journey is delicate and intricately complex, yet undeniably essential.

In Lee Child's novel 'One Shot,' the protagonist, Jack Reacher, sheds light on why people join the military in the West. He mentions, “There are four types of people who join the military. For some, it’s a family trade. Others are patriots, eager to serve. Next, you have those who just need a job. Then there’s the kind who want the legal means of killing other people.” Individuals or soldiers motivated by such factors can be challenging to transform into effective peacekeepers.

However, the motivation of an oriental soldier diverges somewhat. Many Indians, as part of the British Indian Divisions, made supreme sacrifices during the 1st and 2nd World Wars. Intriguingly, during the Battle of Imphal-Kohima, Indian soldiers fought alongside both British and Japanese divisions, but their sacrifices were not for the British Queen or the 'Union Jack.' They made these sacrifices for their Regiments and Commanders. This strong tradition of sacrifice for the regimental spirit deeply resonates with our officers and soldiers, instilling profound loyalty and professionalism. This commitment is one of the reasons we excel as peacekeepers. To prevent any conflict between traditional soldierly principles and the ethos of peacekeeping, we provide careful induction training and emphasize the importance of peacekeeping. Moreover, due to disciplined regimentation, our officers and men maintain discipline and do not indulge in narcotics or unprofessional behavior.

With these advantages, Bangladeshi peacekeepers quickly earn the trust of the locals in host nations. Over time, we've learned that there is no better means of achieving success in peacekeeping than by establishing mutual trust. However, peacekeeping is always far more challenging than wars, and command becomes the acid test in such complex scenarios. Fortunately, the Bangladesh army conducts effective pre-deployment training, which has been instrumental in making Bangladesh one of the largest contributors to UN peacekeeping efforts.

During the initial weeks, our focus was on acclimating ourselves to the weather and familiarizing our troops with the terrain, threats, situation, SOP (Standard Operating Procedures), and ROE (Rules of Engagement). We worked diligently to ensure the motivation to serve effectively as peacekeepers, considering that soldiers, far from their families and parent regiments, may become fatigued and stressed due to the demanding nature of peacekeeping tasks unless consistently motivated.

Subcontinental soldiers are driven by unique causes that might not be familiar to others. Their family bonds are incredibly strong, and family is an integral part of their patriotism. 'Izzat'—a blend of self-respect, honor, and loyalty to one's family, commanders, and regiment or battalion—is an intangible and non-negotiable spirit that deeply resonates with subcontinental soldiers. Our soldiers can go to great lengths to keep their unit's 'izzat' free from any blemish. A good soldier serves according to the mission set by his superior. However, if the mission is well explained, and a soldier can find room for personal achievement, it serves as a true performance enhancer. We worked tirelessly to rekindle the regimental spirit among us to accomplish every peacekeeping assignment.

Manpower and logistics are typically sufficient during UN missions. Nevertheless, the Eastern Equatorial State, with its vast size, posed a challenge for a company-sized force like ours. Kapoeta was our farthest patrol destination, and we could hardly patrol another 200 miles beyond Kapoeta.

Except for a few miles of paved road in Juba, the capital of South Sudan, all the roads were composed of red gravel and were riddled with potholes. These roads, almost never-ending, cut through tropical grasslands dotted with shrubs and isolated trees, known as 'savanna.' We moved in slow convoys and spent our nights in tents. The UN Military Observers, the National Monitors from SPLA (South Sudan People's Liberation Army), and SAF (Sudanese Armed Forces) used to move with us during LRPs (Long-Range Patrols).

Kapoeta was divided into three areas: North, East, and South Kapoeta. These were remote counties where tribes like Toposas lived, far removed from modern civilization. Chukudum and Ikotos were neighboring counties, both mountainous. The roads through the mountain passes led to the famous Kidepo Valley. Lafon was a wide and open county, whereas Magwi, the most exotic county, bordered Uganda, and Kenya.

Torit served as the capital of EES (Eastern Equatorial State) and was the central hub of all lifelines. It was a lightly populated state, and the people were not as industrious. Liquor shops were abundant along the roads and near human settlements. Approximately 22 NGOs were operating in the area, and the SPLA was neither friendly nor hostile toward us. They had fought for a long time and were eager to secede from their North Sudanese counterparts as soon as possible.

Strangely, the South Sudanese economy was heavily reliant on cows, with the number of cows determining one's social status. A minimum of ten cows as dowry was required for a groom to marry his bride. Mothers held a significant role in the family, being the true heads of the households. South Sudan had been a source of slaves throughout history, and the society was still striving to eliminate slavery.

Approximately 60 tribes coexisted in South Sudan, each distinguishable by their unique body markings and tattoos. Most were born into animism but converted to Christianity every day. Nudity was common among the rugged Toposas of Kapoeta county and the cheerful Mundaris of Terekeka county.

It was genuinely surprising that, despite living along the banks of the Nile, the locals had limited knowledge of fishing with nets. The soil was incredibly fertile, yet the locals lacked enthusiasm for agriculture. In contrast, our soldiers, as hobbyists, managed to cultivate large pumpkins and papayas.

The ready-made garments and medical businesses were dominated by the Chinese and Indians. South Sudan represented a land of opportunities and a potentially lucrative market. However, Bangladesh's credibility was well-established through the actions of our valiant peacekeepers, and Bangladeshi peacekeepers were highly regarded in South Sudan for their humane behavior and ethical conduct with the locals. Like the Bangladeshi Peacekeepers of Torit who coexisted with the Dinkas and Toposas, Peacekeepers of Yambio were also welcomed by the local Azandes.

People often enjoy hearing stories of danger, excitement, and conflict, but the success of a peacekeeper lies in maintaining a conflict-free and peaceful environment. We, the Bangladeshi peacekeepers, managed to spend another peaceful year in South Sudan.

While we were far from our homes and families, the team sites became our new homes, and we became a family. It was a new and enriching experience of coexistence among UN members from different nations. We dined, played, and celebrated together within the team site to make our lives easier and more comfortable. Amidst challenging circumstances, we were truly united at our team site in Torit. After six months, my platoon and I were transferred to Juba as the Quick Reaction Force.

Four

The night passed without incident, and as the dawn broke, the recovery helicopter touched down in Lafon. Swiftly, the Russian technicians revived the damaged chopper, and it was evident that our mission had been a resounding success.

We boarded the helicopter for our flight back to Juba, but there was one stop left on our journey—Torit. As we landed on the helipad, I couldn't help but feel an irresistible pull to visit my cherished Team Site Torit. Stepping onto the familiar ground, I was met with a heartwarming sight: the Bangladeshi peacekeepers eagerly awaiting our arrival. They offered us chilled juice and extended a warm welcome that was like balm to my soul.

This unexpected reunion with my comrades and the heartfelt reception in Torit left me with a profound sense of contentment. As our helicopter lifted off once more, I felt a renewed sense of purpose. The world continued to turn, and another mission might be waiting just around the corner. But for now, I was filled with deep satisfaction, knowing that, amid chaos and uncertainty, we had found moments of connection and unity that made the journey worthwhile.

A Mission: Two Countries

Lieutenant Colonel Md. Shamsul Alam Siddique
(United Nations Mission in Ethiopia and Eritrea - UNMEE)

I could tell that I had to attend an interview at Force Headquarters Asmara, but I had no idea that it was going to take place in the blink of an eye. It was May 2007. I was employed as the military observer of the UN peacekeeping mission at the border area of Ethiopia called Humera. Referring to Humera as a "city" seemed unsuitable; it looked more like a suburb. Flowing on the outskirts was the Tekeja River, which had mercilessly split the Amharik tribe into two distinct groups. Other than the Amharik, the ethnic groups were mainly Tigrina, Tigre, and the Habsi. These tribes had their own mother tongue and followed their own customs and rituals.

Ethiopia and Eritrea engaged in a war in 1998 related to complications in determining their border. In 2000, the Algiers Agreement brought an end to the struggle between the two states. Unfortunately, within two years of war, both countries had lost 300,000 people in total. Moreover, Ethiopia and Eritrea are both poor countries, and after two years of fighting they had become economically and socially unstable. Under the Algiers Agreement, a temporary security zone was established within Eritrea, extending up to 25 kilometers from the international border. The United Nations Mission in Ethiopia and Eritrea (UNMEE) was tasked with monitoring this zone.

When I joined one of the UNMEE sites as a team member, I received a warm-hearted welcome from my fellow partners. After completing my three-day orientation, I became the team's admin and logistics officer. Major Asad of the Jordan Army Special Force happily taught me about the responsibilities of all the appointments. Though I was nervous about being the youngest on the team site, this issue didn't cause any trouble. On the other hand, to my greatest surprise, I quickly learned about the responsibility of all the appointments. I hope I didn't bore them with my repeated inquiries!

Our team leader had been Major Surendar Rohila, who was an Indian artillery officer. Perhaps he liked me a little more than others, as one day he said, 'The Force Headquarters wants names of volunteers for selecting a new team leader. You must apply.'

I disagreed as respectfully as possible. I said, 'Let some days pass, I'll see then.'

Looking at his face, I could tell he didn't like my answer. With a grumpy voice, he replied, 'Now or never.'

As ordered by my leader, I applied to the Force Headquarters. I was informed by the returned mail that the deputy force commander (DFC) of our mission would be visiting some team sites the next day and would return to Mission Headquarters Asmara. I was told to join them so that the trouble of sending an extra vehicle 500 kilometers far would be lessened.

I woke up before the sun rose, excited to do something new. I prepared myself and met my team, which consisted of Major Muio of the Bosnian Army, the patrol driver, Major Membe Haidra of the Gambian Army, the patrol leader, and a local interpreter named Negash. Our first destination was 180 kilometers away at the Shiraro team site. We were supposed to arrive there before the DFC.

The car galloped breathlessly. The rocky roads made it difficult to keep our balance. If the car was not in the correct gear or the driver was careless for even a millisecond, the car and its passengers would be turned into debris. We were fortunate to have Major Muio, an experienced driver. The cassette player was playing a Bosnian song. In fact, in almost all team sites in the mission, there was an unwritten rule: "The patrol driver gets to play their favorite song on the cassette player." This was mainly done to keep the driver alert during the ride on the dangerous roads. I could only understand a few Bosnian words, but the rhythm and speed of the song almost made us fly, and we reached the Shiraro team site just in time. What a roller coaster ride! I just hoped my bones were still in the same place as before we set out. Within minutes, we were informed via the radio that the DFC would be a little late.

Almost an hour and a half later, the DFC arrived at the Shiraro team site. Everyone seemed terrified. After a long time exchanging glances, Major Harun, the military assistant of the DFC, spoke up. He was the staff officer of the DFC.

While the car carrying the DFC was passing a village, they heard an explosion 50 kilometers away from the Shiraro team site. Everyone was shocked. The severity of the noise and the shock wave felt by the Bangladeshi armored general led him to understand that it was an anti-tank mine explosion. They had no security scout with them, so the only solution was to go check by themselves. In fact, it was their foremost responsibility to save the person in danger.

Without wasting a minute, they headed towards the source of the sound. They didn't have to go far. Upon seeing the crowd, the DFC stopped the car. Running towards it, they found a local truck completely smashed in the front. There was a large hole in the road as well. The victim of the explosion was the truck driver, whose body had been blown to pieces. Villagers were running here and there to gather every piece and put the body back together.

Due to the suddenness of the situation, the driver's assistant was terrified and unable to speak. Major Harun was given the responsibility of questioning the driver's assistant to gather information. Within a short time, the local police arrived.

After the driver's assistant gathered his senses, he explained what had happened. The previous day in the afternoon, they had come to this village carrying sand on the same road. After unloading the sand, they were supposed to return by evening. Unfortunately, the truck wouldn't start after they had finished unloading. The driver had no other option but to go to the city to bring a mechanic at night. Although the mechanic had fixed the car, they couldn't go back because they didn't have permission to drive on a road close to the border at night and so they spent the night in the village. Being tired from all these troubles, they woke up a bit late in the morning. Within minutes of crossing the village road, the explosion took place in the blink of an eye.

After hearing the account, the DFC glanced at his staff officer with a curious look. Major Harun was trying to sum everything up on his own. They eventually reached a conclusion – the truck had unintentionally been victimized. Moreover, very few cars travel on the border roads of the two war-stricken countries; therefore, the car carrying the DFC was supposed to be the first one. So, who was the first target? The DFC? "Alhamdulillah!" exclaimed the DFC as he realized he had been saved from the spectacularly tragic situation.

They had neither any security scout nor a weapon; a car with a UN monogram was the only support. Since the invisible mine setters had most likely targeted them, Major Harun told the DFC that staying there even for an extra minute would be unwise. But the patrol driver was so traumatized by seeing the scattered body parts of the truck driver that he became unable to drive. The DFC was an experienced driver, but due to the mission protocol, driving might leave a mark on his reputation. Moreover, Major Harun did not have a driver's license. So, the DFC comforted the driver, saying that he had to drive only a few kilometers to the Shiraro team site.

With no other choice left, the patrol driver started driving with shaky hands and somehow reached the Shiraro team site. When he arrived, he lifted his hands in the air and said that he couldn't drive any longer.

With the end of the story, Major Harun gave me a curious glance. I explained that I was an observer who had to drive along the border of the two countries and observe the border security situation. I also said that it would be a pleasure for me to drive a car of the DFC of my own nationality.

Having finished looking at the works in the Shiraro team site, we set off with our DFC. I was in the driver's seat, and our destination was the capital of Eritrea, Asmara, passing through Axum and the Ethiopian border. On the roadside, I caught sight of the debris of Axum Queen Sheba's palace. Queen Sheba was an ancient character whose description lies in Christian, Jewish, and Islamic texts.

My heart wished to take a glimpse of this incredible history, but my brain alerted me of how far we had to go and that we had to reach there before sunset. As expected, my wish to catch a glimpse of history remained incomplete. After completing our tedious journey, we reached Axum. Seeking permission from the DFC, I started playing my favorite song sung by the famous Bengali singer Hemanta Mukherjee on the cassette player. The lyrics go, "This time, I have set out to lose my way, for I have wandered too much in the hope of finding a straight path."

The tension in the car seemed to fade away. At last, the Kenyan driver spoke up, "I don't understand anything in your song. But it is quite melodious."

I asked, "Maybe you can drive now, right?"

Laughing out loud, he said, "You are a better driver."

Hearing his appreciation, I may have sped up a bit, and right then we bumped on a speed breaker. Right away, the DFC said, “Good drivers do not drive fast, they drive safe.”

Quite embarrassed, I slowed down from 90 km to 70 km. When we reached Asmara after traveling 500 km, the shepherd boys were taking their camels to their sheds. After I brought the DFC safely to his bungalow and having completed our arduous journey, I had a long, peaceful sleep. My interview would take place the next morning. God only knows what would happen!

A Mother's Tears

Lieutenant Colonel Md. Monirul Haque
(United Nations–African Union Hybrid Operation in Darfur - UNAMID)

In May 2016, a brave team of peacekeepers, including myself, were deployed to El-Jenina in Darfur, Sudan. Our mission was nothing short of monumental – to bring lasting peace to a region ravaged by conflict that was home to millions of vulnerable people.

El-Jenina, named after the Arabic word for "garden", was located near the border with Chad and was home to numerous refugee camps. The area was marked by a stark contrast between the arid Sudanese desert and the occasional patches of greenery. As a peacekeeper, I was reminded of my dual identity as a doctor and a protector, carrying both a stethoscope and a weapon while dressed in a Bangladesh army outfit and a blue UN helmet.

Sudan, the third-largest country in Africa, is bordered by many other countries, including Egypt, Ethiopia, Kenya, and Uganda. The political and social complexities of the region were staggering – Arabic was the predominant language, and Islam was the most widely practised religion. The northern part of Sudan is inhabited by people of Arab descent, while the southern part is predominantly Christian and black. The southern Christians staged a revolt in 1962, and a military coup followed in 1969. In 1972, the southern territory was granted autonomy following the "Addis Ababa" peace accord.

However, conflict flared up again in 1983 over the introduction of Sharia law in the south, escalating into a full-blown war under military dictatorship. In July 2011, South Sudan became an independent state following a referendum.

Despite the complexity of the political and social issues, our mission was clear – to bring peace and aid to the people of El-Jenina and the region. And so, with dry weather prevailing all around us, we set to work, a team of dedicated and fearless peacekeepers committed to our cause.

In the heart of Sudan lies a land of stunning natural beauty and abundant resources. Darfur, a western province four times the size of Bangladesh, boasts the world's third-largest uranium reserves, vast oil fields, and a thriving gum arabic industry.

Home to over 80 tribes, Darfur is a melting pot of cultures and traditions. While 27 tribes follow the Arabian culture, the rest have mixed cultural backgrounds, each bringing their unique flavour to this diverse region.

For centuries, Darfur was ruled by sultans, who maintained peace and order among the communities. But after Sudan gained independence in the 1960s, the central government took control, leading to chaos in Western Darfur. Rebels from the south and the people of Darfur blamed the government for the violence and instability that followed.

Despite the region's wealth of resources, Darfur remains plagued by conflict and violence. But we mustn't overlook the cultural and economic significance of this region. Gum arabic, the main ingredient in Coke and Pepsi, is primarily produced in Darfur, and its high quality has made it a valuable contributor to the global economy. Darfur may have its share of troubles, but its cultural richness and economic importance cannot be ignored.

Darfur was engulfed in turmoil and unrest. The conflict between non-Arabs and Arabs had brought with it nothing but despair, leading to a weak economy and an abandoned infrastructure. This was a region with no hope, a place where fear reigned supreme.

Then came a fateful day in February 2003 that pushed this region to the brink of disaster. It was a time when civil war broke out between the two groups, and the military dictator General Bashir did nothing to stop the chaos. He instead supported the Jan Zaid community (of Arab descent), who wreaked havoc on the non-Arabs. They committed mass murders, coercion, and tyranny, leading to the deaths of 300,000 people and the displacement of three million more to Chad.

Hopeless and dejected, Darfur cried out for help, and it finally arrived in the form of a joint UN and African Union peacekeeping mission known as Toghagwu in July 2007. Soldiers and peacekeepers from over 50 countries, including Bangladesh, were sent to this region of despair to restore peace and sanity.

Lieutenant Colonel Mizan, known as Commando Mizan, had a special calling. He was driven by his commitment to serve the distressed and provide free medical support to refugees in Darfur. It was this calling that led him to create a mobile hospital that accompanied the peacekeepers on their missions. This hospital provided free medical care and medicine to the underprivileged who lacked proper medical facilities. The joy and relief on the faces of those receiving care brought a sense of tranquillity to the peacekeepers. The memories of the contented faces of the refugees still touch the peacekeepers' hearts like raw silk from heaven.

The mobile hospital was a ray of hope in this dark land. It was a beacon of light for the sick and injured who had nowhere else to turn. It brought succour to the people of Darfur, who had been abandoned by the world. This hospital was a testament to the courage and determination of the peacekeepers and their commitment to making a difference in this troubled land.

The story of Darfur is one of courage, sacrifice, determination and of people who were willing to give everything to make a difference in the world. The story of the mobile hospital will always be remembered as a shining example of what can be achieved when people work together towards a common goal.

As we embarked on our mission to provide healthcare to those who needed it most, our focus was squarely on the university students of Darfur. In particular, we set our sights on tackling the prevalent diseases plaguing the region, such as HIV-AIDS and malaria. But we knew that to truly make a difference, we had to go beyond just treating the symptoms of these diseases; we had to reanimate the locals and equip them with the knowledge and tools they needed to combat these diseases themselves.

Our weeklong workshop, 'Train the Trainers,' was aimed at doing just that. We developed workshop modules and became facilitators for local health workers and schoolteachers, equipping them with the minimum knowledge of first aid treatment. And as we demonstrated how to save someone from a moribund state and bring them back to life, we knew we were making a true impact.

But our work didn't stop there. We distributed 20 pieces of first-aid equipment, including sphygmomanometers, to our 50 trainees, with the hope that they would go on to train others and spread our efforts even further. And we didn't forget the school students of the refugee camps either, distributing educational materials and sports equipment to them.

As we worked tirelessly to make a difference, the sight of the Bangladeshi flag flying high on foreign soil filled us with pride as patriots. The UN recognized Bangladesh's contributions to the cause, and the people of Darfur appreciated the initiatives of Bangladeshi peacekeepers.

But as felt proud as we felt of what we had accomplished, nothing compared to the impact we had on the lives of the people we helped. Despite the limitations, we were able to provide healthcare to the refugee camps in Darfur, offering free visits, medical services, medicines, tests for malaria, and prescriptions for all other diseases.

I still remember the old woman who came crawling towards my table at a gathering of thousands of patients. She had lost both her legs in the 'Darfur War' and had been leading a difficult life ever since. But as I looked into her eyes, I saw the impact we had made on her life, and I felt a sense of kinship with her. At that moment in Darfur, I found a mother – a word that transcends all boundaries of caste, tribe, religion, and language. It is the sweetest and most powerful word in the world, a word like no other. And it reminded me of the true importance of our work and the significant impact it has on people's lives.

On that day, we set out on a mission to distribute much-needed supplies to those who desperately needed them. It was a hot and dusty day in Darfur, and as we arrived a crowd of people had already gathered, eagerly waiting for us to arrive. We had sewing machines and wheelchairs to distribute, but little did I know that this day would be etched in my memory forever.

As we were giving out the supplies, I noticed an elderly woman in desperate need of a wheelchair. I immediately went to my commanding officer, Lieutenant Colonel Mizanur Rahman, to help her; however, he told us that we had already given out all the wheelchairs, and my heart sank. I couldn't bear the thought of leaving her without the aid she so desperately needed.

So, I pleaded with my commander to go to Khartoum and purchase a wheelchair for the woman. Without hesitation, we set out on a mission to find her small mud hut in the scorching heat.

When we arrived, we were met with a heart-wrenching sight. The woman and her family lived in a small space that was no bigger than 20 by 10 feet. There were 14 of them, including her daughter, daughter-in-law, and grandchildren. Despite the lack of space, their smiles were contagious, and their hospitality was overwhelming.

As I presented her daughter with the sewing machine and wheelchair, the woman's eyes welled up with tears of joy. I could feel her emotions and see the gratitude in her eyes, even though we couldn't speak the same language. It was a moment of pure fulfilment, and I realized that life truly has meaning and purpose.

As we made our way back to camp, I thought about my own mother and how much she means to me. I realized that a mother's love transcends all boundaries, including language, culture, and geography. When the Sudanese mother embraced me with love and gratitude, I knew we shared a bond that went beyond our differences.

To all the mothers of the world, I extend my love and gratitude. May you always be surrounded by love and happiness, just like the evergreen leaves!

Tongue-Tied Triumph: Bengali Takes Center Stage in Sierra Leone

Major Md. Asadullah
(United Nations Mission in Sierra Leone -UNAMSIL)

Ah, behold the winter of 2001! A gloomy morning shrouded in fog unveils a mysterious enclave disguised by mango trees. It's almost as if a secret army brigade has found solace in this covert abode, concealed from prying eyes.

In this peculiar space, one can spot a makeshift structure, a massive room that seems to have been salvaged from an abandoned car garage or workshop. Through the branches of the mango trees, the dense fog seeps into the peacekeepers' hideaway, as if nature itself conspires to join their clandestine operation.

Inside, tables are meticulously arranged, each surrounded by chairs occupied by officers who exude an air of importance. Their collective focus is fixated on the center, their intensity akin to individuals huddling around a hearth for warmth in the face of a chilly world.

Meanwhile, soldiers scurry about, either engaging in rigorous exercises under the watchful eye of their superiors or burying themselves in their studies. It is here, my friends, that the headquarters of Bangladesh Signal Unit 2 (BANSIG-2) resides. Their grand mission: to reform and prepare the unit for deployment to Sierra Leone as proud contributors to the UN peacekeeping mission.

This unit, comprised of approximately 800 troops, was bid farewell by their esteemed leader, Colonel Asif, and his deputy, Lieutenant Col. Moin, accompanied by a group of seasoned officers in their golden years. In a delightful throwback to ancient war expeditions, the unit even included barbers, tailors, cobblers, and sweepers.

Among the ensemble, the chosen adjutant, Major Sarkar Nazrul Islam, was missing, his absence created confusion in the unit. Alas, he had been transferred to a division of the armed forces half a mile away, rendering it nigh impossible for him to join his comrades on the eve of their maiden flight.

An army unit without an adjutant is akin to a herd of cows without a shepherd. Thus, a substitute adjutant was deemed necessary, a proxy to shoulder the weighty responsibilities. The initial burden fell upon Senior Officer Miyaji Sir, but, alas, he hailed from Savar Cantonment and was burdened with the laborious task of relocating his residence. The demands of city life, you see. Consequently, he was swiftly excused from duty on the second day of assuming his post, for he had acquaintances to tend to – the necessary ties that bind in the relentless urban landscape.

Ah, my dear readers, allow me to regale you with the tale of my presence in the illustrious Bogra cantonment, situated across the meandering Yamuna River. At that moment, I had no place to call home. They saw me, a young and carefree soul, devoid of any permanent dwelling or familial entanglements, and deemed me the perfect candidate to serve as the proxy adjutant. Little did they know that I possessed no personal or familial complexities – or if I did, I certainly never voiced them to anyone, ever.

I must admit, I did hold a certain charm in the eyes of my seniors, who saw me as the epitome of "polite suavity" within our unit. With an unwavering resolve to endure tribulations and afflictions silently, I was a paragon of quiet resilience, much like the steadfast earth itself. However, amidst all the accolades, what I needed most was a roof over my head, an urgent and pressing matter that loomed over me like a shadow.

You see, I had always been a tad rotund since childhood, save for a brief period of being underweight due to multiple bouts of malaria in 1988. The concept of "weight" held unparalleled significance in the armed forces of Bangladesh and beyond. Thus, I shouldered the duty of a proxy adjutant for BANSIG-2, sacrificing everything else, be it finding suitable lodgings or even shedding a few pounds. Oh, and let's not forget that BANSIG-1 was simultaneously carrying out their duties in Sierra Leone at the time.

Now, picture this: The sun rising from the west and setting in the east not once, but twice. During this extraordinary time, Colonel Siddique, the captain of BANSIG-1, sent an email to Colonel Asif proclaiming that BANSIG-1 had been chosen for the illustrious "wait-lease" program destined for the grand “Sierra Leone Mission.” With great difficulty, they managed to transport their radio sets and antiquated signal equipment to Sierra Leone.

Upon arrival, the UN conducted a thorough inspection of BANSIG-1. After their on-site observations, the UN found it challenging to support the mission with the type of equipment we had. However, they didn't want the mission to go unsupported. So, a difficult decision was made – the entire unit would be repatriated while we awaited the arrival of a signal battalion from another country. During this interim period, BANSIG-1 offered services to the UN, performing other tasks assigned to us.

How could we, the members of BANSIG at all levels, resist the temptation to turn back the hands of time? We seized the opportunity with fervor, accomplishing the task within a mere 15 days. And thus, like clockwork, every Thursday became a weekly ritual where all military and non-military personnel of the United Nations were invited to the enchanting shores of the Atlantic, a paradise known as Goderich.

Ah, the tantalizing aroma of Turkish chicken kebabs, like a savory sizzle dancing on the Atlantic breeze. The taste of Bosnian bread exploded with flavors, teasing the palate of all who dared to indulge. Oh, how our hospitality overwhelmed the BANSIG-1 headquarters in Goderich, just a stone’s throw away from Freetown, the capital of Sierra Leone! It transformed into a festive haven, a place where UN members could gather, indulge, and enjoy.

Meanwhile, in the Magburaka area of central Sierra Leone, the Bangladesh sector showcased their remarkable performance. Their Nigerian and Kenyan counterparts, however, struggled to find their footing and establish a robust presence in the same operational territory. What set the Bangladesh sector apart? It was truly a wonder! Their genuine love and hospitality toward the people of Sierra Leone effortlessly fortified their position and endeared them to the locals. It was no surprise that the Bangladesh army quickly became the darling of the community. Their unwavering commitment to the well-being of the people they served was a shining testament to their compassion and dedication.

On a sun-drenched morning, as golden rays illuminated the world below, the United Nations Force Commander, General O’pandey, made a decisive move. Despite the deficiencies in modern equipment, he resolved that BANSIG-1 would continue operating within Sierra Leone. They would rely solely on UN-provided resources, strictly adhering to allocations. Thus, an unbroken line of communication was established between BANSIG-1 and the United Nations force stakeholders, with BANSIG-1 acting as the flawless communication wing, ensuring ceaseless contact. They put in tireless efforts to maintain seamless communication within the UN's internal bodies.

However, amidst our successes, we faced certain difficulties. One notable challenge was the proficiency of our operators and soldiers in English. The linguistic barrier proved embarrassing at times and posed challenges in difficult situations. Inaccurate interpretations of English by foreign UN stakeholders led to operators communicating with the wrong individuals. This raised concerns about potential internal conflicts among different operating bodies of the United Nations.

Our captain, Colonel Asif, highlighted the importance of two lines at the bottom of that lengthy email, deeming them a priority and adopting them as his own. The very next day, he announced to the unit court that all members of BANSIG-2 would undergo a compulsory month-long English education program. And to top it off, there would be an exam upon completion of the training. Those who failed the test would not be considered for the mission, you see. A testament to the importance of linguistic prowess in the face of UN operations!

Oh, the illustrious Ziaur Rashid Safdar, who was bestowed with the noble task of planning the training program. With his boundless creativity, he crafted not just one, but many curriculums all at once. How he managed to swiftly combine them into a single curriculum was truly awe-inspiring (at least, for me). I was genuinely amazed to uncover the intricate correlations between them. The training took place at a vast ground on the west side of the Army Signal Brigade, where groups of soldiers ran thrilling drills according to Zia Sir's plan. Like an orderly swarm of ants, they rotated every 30 minutes, mastering various skills and knowledge within the narrow span of a month. From physical fitness to weight loss, firing skills to driving skills, and whatnot, Zia Sir didn't forget to include even the most challenging subjects like communication and "mutual intelligibility" in the curriculum. After each period, the soldiers shuffled from one training shed to another, creating a relentless frenzy akin to the childhood game of 'roundabout release run.'

Even we officers jumped in with great enthusiasm to teach English to the soldiers. And after a month of systematic instruction, the day of reckoning arrived—the test results! Oh, they shall never be forgotten! Only ten soldiers out of approximately 800 passed. Colonel Asif stood dumbfounded, utterly flabbergasted. But within a day or two, it became clear that Colonel Asif wasn't the type to wallow in despair. He knew all too well that a battlefield never grants an easy victory and the journey to a war zone and conquering it is never as smooth as a parade. So, with a battalion of 700 English-challenged soldiers, he set off for Sierra Leone on the anticipated date.

As I reflect upon the scene that unfolded before my eyes, an indescribable euphoria engulfed me upon our arrival in Sierra Leone. A sight of extraordinary joy welcomed us, filling my heart with immense delight. I discovered that our soldiers, not only amongst themselves but also with their foreign counterparts and the locals of Sierra Leone, conversed in Bengali. It seemed as if Bengali had become the vernacular tongue as they exchanged stories, anecdotes, and emotions with one another. Laughter and amusement filled the air, except for the leader himself, who insisted on speaking English even with his runner. One fateful night, a fierce hurricane from the Atlantic Ocean swept in, blowing the tin roof off his abode.

In the dead of night, we sprang into action to come to his aid. When we reached him, we found him pointing his finger at his runner and shouting, "Do not go to that side!" Oh, the quirks of language barriers!

Zia Sir, the newly appointed head of the Signal Company in the illustrious 'Canama Sector,' assumed his commanding role. He was accompanied by his deputy, the charismatic and experienced officer Syed Kamruzzaman Sir. With exceptional leadership and command abilities, Kamruzzaman Sir possessed a towering presence and a strong personality. His imposing physical stature was matched only by his unwavering sense of duty and dedication to his profession.

Once, during a company visit, Zia Sir and his deputy encountered a young man in the company suffering from cerebral malaria, the severe form of malaria prevalent in that region. Zia Sir inquired, "How long have you been suffering from such a fever?"

"For five days," replied the young man. Zia Sir then asked if he had consulted a doctor, to which the boy replied that he hadn't sought medical attention because he had failed his English exam back in his home country. Ah, the perils of English exams determining access to healthcare!

Ah, the doctor in charge of the Canama sector, a foreigner among us. Apart from the English language struggle, nothing could hinder our path. In fact, whenever any Bangladeshi patrol hit the streets or showcased their might, the children, teenagers, young adults, and even the elderly of Sierra Leone would shout “Bangla-Bangla!” in unison, their voices faltering and flickering through the Sierra Leonean air, leaving us in awe. Their outpouring of love overwhelmed us, and we couldn’t resist the urge to share our food with them whenever the opportunity presented itself. It seemed like the logical thing to do, you know?

The United Nations provided us with an abundance of food. Unfortunately, a significant portion of it didn’t quite meet the discerning palates of Bengalis, except for the occasional delight of ice cream, honey, and juicy grapes. Nearly a year passed, and the peace process was on the brink of conclusion. We emerged victorious, capturing the hearts of Sierra Leoneans, and not through the language called “English” but through the universal language of the heart called “love.”

Imagine that, resolving conflicts in a country at odds with our mother tongue! What an extraordinary achievement, my dear friends! As a testament to our triumph, the Sierra Leonean government proclaimed Bengali as their second official language, leaving the world utterly astounded. Sierra Leoneans paid homage to “Bangladesh” and its “Bangladeshi peacekeepers.”

Colonel Asif was brimming with joy, and so were we. The countdown to our departure was in full swing. It was time to wrap things up and bid farewell to Sierra Leone. The Savar Cantonment was bustling with activity as BANSIG-3 regrouped to replace us, the valiant BANSIG-2.

In a state of elation, Colonel Asif, the commander-in-chief of BANSIG-2, sat down to pen a letter to the commander of BANSIG-3. With a mind filled with ecstasy, he wrote, “We faced countless challenges here and conquered all the odds, but none quite like the ‘language barrier.’ Yes, nothing could hinder our progress, except for the initial hiccup of the ‘language.’ But guess what? We did it! We overcame it! We tamed and subdued that barrier with our very own mother tongue, Bengali!”

So, my dear commander, whoever you choose to bring in now for the job, rest assured that language is no longer a barrier here. All they will need is expertise in ‘DDR’ (Disarmament, Demobilization, and Reintegration).

BEYOND THE CALL OF DUTY

LIEUTENANT COLONEL MUHAMMAD LUTFUL HAQUE

(UNITED NATIONS MISSION IN IVORY COAST -UNOCI)

In the quietude of that November morning, as the crisp air enveloped Yakuba Camp, I found myself engrossed in a meeting with Captain Shafiqur Rahman - our camp commander. The room resonated with the hum of logistical matters, for I held the mantle of chief logistics officer in our contingent. It was my duty to oversee the meticulous coordination of supplies and inspect the various camps and posts under our watchful gaze. Five months had slipped by since our arrival in Ivory Coast, and routine had settled upon us like a familiar companion.

Our conversation flowed seamlessly, dwelling on the mundanities of logistics, when an abrupt intrusion shattered the tranquility. A signalman, his face etched with urgency, burst into the room, bearing tidings from Zenuoula. The government forces had launched an audacious air strike, unleashing a torrent of devastation upon the rebel forces' stronghold in Bouake. The missive from our contingent headquarters reverberated in our ears, urging us to heighten our vigilance and closely monitor the unfolding situation. For the past five months, a rare tranquility had caressed our mission area, with gunfire becoming unfamiliar to our ears.

As per the meticulously crafted plan, I was slated to embark on a journey to Zenuoula via Daloua on that very day. Yakuba Camp nestled in the Zone of Control (ZOC), a neutral expanse bridging the territories of government and rebels. Daloua and Zenuoula, on the other hand, remained firmly within the government's grasp. With this knowledge, I embarked upon the road, a sense of security tugging at the edges of my consciousness, deeming the government-controlled territory a sanctuary from perils.

The wheels of my vehicle churned the dusty road as I ventured forth, leaving Yakuba Camp behind. The hours danced on, and as I arrived in Daloua, whispers of anxiety reached my ears carried by the winds of uncertainty. The rebel forces, their spirits stung by the air strike's audacity, teetered on the precipice of retaliation. Bouake, their heartland, simmered with a fervor that threatened to ignite the country.

Through the crackling radio waves, I engaged in conversation with Colonel Mohammad Iqbal, the commanding officer of our contingent. From the confines of Daloua, I painted a vivid picture of the tense and volatile atmosphere that had gripped the land. It was imperative, I argued, for a senior figure to be present in Seguela, a territory firmly under rebel control. Proposing myself as the harbinger of presence, I awaited his response, aware of the weight of my words in the delicate balance of our mission.

Days had blurred into a flurry of activities since our arrival. Camp construction, coordination with mission headquarters, the collection of UN materials, and forging connections with the locals had become the familiar choreography of our lives. Yet, the air strike in Bouake injected an electric current into our tranquil existence. It was as if the heavens had whispered of an impending storm, the anticipation of something grand, something monumental. An exhilarating frisson coursed through my veins as I embarked on the journey from Daloua to Seguela, bidding farewell to the comfort of routine.

Ivory Coast, a nation steeped in the warmth of independence since 1958, had fallen into the clutches of political turmoil. Four decades of stability crumbled beneath the weight of military coups and counter-coups. The year 1999 marked the descent into chaos, with a fierce tug-of-war for power tearing the fabric of the nation. It was a law on voting rights and citizenship in 2000 that ignited the flames of a full-fledged civil war, leaving the land scarred and divided.

The factions were clear-cut, each staking its claim. In the north, the rebels held sway, their grip firm and unyielding. The south, a domain controlled by the government, stood as a bastion of power, encompassing the capital and the vital ports that fueled the nation's commerce. A slender ribbon of neutrality, the ZOC, snaked its way between the warring sides, a demarcation that dared to keep them at bay. This neutral zone, a refuge from the clash of arms, became an enclave barred to both factions. It was within this sacred space that peacekeepers stood as sentinels, their posts peppered along the roads, guardians of neutrality.

Yet, the forbidden allure of the neutral zone proved irresistible to the restless spirits of conflict. Like fleeting shadows, the factions would often breach its hallowed boundaries, sowing chaos and discord wherever the opportunity beckoned. The rebel army, lacking in officers, had partitioned their territory into military regions entrusted to the leadership of sergeants and corporals. And it was within one of these strategic enclaves that our destinies intertwined in the shifting sands of war.

Our contingent's dominion stretched vertically through the heart of the country, dividing it with purpose. Like a complex tapestry woven with intention, our five companies found their places in this intricate landscape. Two stood sentinel in the south, their allegiance tied to the government's embrace. Two others nestled within the neutral zone, a delicate balance teetering on the edge of uncertainty. And one lone company dared to tread the rebel-held territory, its presence a beacon of resilience.

French, the language that danced upon the lips of Ivory Coast, held dominion over communication. To venture beyond the confines of our camp without the knowledge of this linguistic tapestry was to wander aimlessly through a labyrinth of incomprehension. There was a tangible weight to the necessity of understanding, a tether that bound us to the boundaries we called home.

From the vantage point of Daloa, my journey unfurled cautiously, its path tracing the lines of government-controlled lands. The roads of Ivory Coast, like the veins of a living creature, pulsed along the meridian, carving paths that stretched both northward and southward. As I passed through the checkpoints, eyes scanning the faces of those who stood guard, I searched for a flicker of emotion, a telltale sign of the imminent. But their demeanor held no secrets, their masks firmly in place. There was no grand campaign being orchestrated, no momentous operation poised to unfold. And in the recesses of my being, a pang of disappointment took root.

Zedosi, the sanctuary that housed Major Tapan's Banalfa encampment, beckoned me to linger. Together, we pondered the possibilities, the hypothetical scenarios that could unravel if the tides of fate turned against us. With the afternoon sun casting elongated shadows, I resumed my journey, retracing the steps of those who had tread these trails before me. Yet, as the northern zone, shrouded in rebel control, greeted my arrival, a somber transformation painted the landscape. Sparse settlements dotted the road, devoid of the usual bustle of activity. A weighty silence hung in the air, mingling with the scent of uncertainty. Ivory Coast, a land with a gentle population density, unfolded its secrets before me. Once teeming with wildlife, its forests had succumbed to the greed of the past, the echoes of colonial exploits resonating through the ages. The unpaved roads that sliced through these ancient woods bore witness to the scars of progress.

And then, in the distance, a mere seven or eight kilometers from Seguela, a scene unfolded that tore through the fabric of the mundane. From the depths of the forest trail on the left emerged a jeep, its velocity betraying urgency. Mounted upon the open pickup that trailed behind, a rocket launcher stood poised, its presence an undeniable proclamation. It took but a heartbeat to recognize the rebel convoy and its enigmatic leader, destined for the highway and the embrace of Seguela. With conviction, I veered off the road, halting my car by the wayside, beckoning fate to dance in the space between us.

As the rebel convoy drew nearer along the highway, I raised my hand, a signal commanding them to halt. To my surprise, the procession screeched to a stop before me, and from the confines of the jeep emerged none other than Kone Zakaria, the deputy commander of the rebel zone. His presence commanded reverence, surrounded by a handful of security personnel who shielded him with unwavering loyalty. Though our previous encounter had been brief and unremarkable, the weight of his reputation preceded him. Zakaria, a man of around 45 years, possessed a modest stature, his lean figure reminiscent of a refined bodybuilder. Behind his eyes and etched upon his face, one could discern the sharpness of an agile mind. At first glance, he exuded warmth and affability, but whispers of his merciless nature lurked in the shadows. Yet, such rumors were just whispers, perhaps nothing more than figments of imagination. Despite his humble origins as a sergeant, Zakaria had honed his military prowess through training in various lands. Language barriers notwithstanding, he communicated with us in fragmented English and expressive gestures.

After exchanging the obligatory pleasantries and formalities, I sought Zakaria's insight on the unfolding situation; however, his responses danced on the fringes of evasion, leaving me with the impression that he feigned ignorance. He skillfully skirted around the subject, refusing to acknowledge the repercussions of the Bouake air raid that lingered like unspoken shadows. Although I expressed my desire to maintain contact amidst the shifting tides, his silence on the matter was telling. And so, bidding him farewell, I left him with my intention to keep in touch.

It was later, through my interpreter, that I pieced together fragments of the conversation between Zakaria and his associates. It became clear that Zakaria had taken refuge in his clandestine hideout, our encounter with him nothing more than a fortuitous accident. He had no intention of exposing himself within the confines of the public headquarters, vanishing into the realm of secrecy. The promise of future contact, it seemed, was an empty gesture, a ruse to deceive.

Upon reaching Seguela, whispers of armed rebels in the Zenuoula region swirled through the air, painting a portrait of tension and uncertainty. Their presence manifested in a meticulously organized manner, as if rehearsing for an imminent act of aggression against Zenuoula. The insurgent force consisted predominantly of militia members, uneducated souls hailing from impoverished and marginalized regions. These individuals, lacking stability, proper attire, and regular income, had sought refuge in the militia, their hope for a brighter future. The vast expanse of the land became their shelter, their encampments nestled deep within remote areas and ancient jungles. Rarely did they venture into the public eye, save for the sporadic checkpoints that dotted the highways. They remained elusive, shrouded in mystery, their presence felt but seldom seen by the local populace.

The insurgents, like vultures, seized every opportunity to ravage the city. Prior to our arrival in this mission area, they had left a trail of destruction in their wake, pillaging urban centers throughout Ivory Coast. Zenuoula, situated on the southern border of the neutral zone, stood as a city smaller in both population and area compared to our average sub-district headquarters. The rebels had previously descended upon it, unleashing chaos upon public and private assets. At gunpoint, they brazenly looted government properties, their motives devoid of any semblance of reason or consideration.

As night descended, casting its dark shroud upon the sky, Colonel Iqbal took to the radio, reaching out to Kone Zakaria. He urged the rebel leader to rein in his forces that lurked ominously around Zenuoula, brandishing their arms as a show of force. The warning carried a weighty message: any attempt by the rebels to assault or linger near Zenuoula would not go unanswered. The UN forces would not remain idle if provoked. Upholding the mandate bestowed upon us, we possessed the authority to employ maximum force to restore peace and safeguard the lives and property of the innocent. We stood prepared, resolute and determined for any step that needed to be taken.

Zenuoula's vulnerability threatened not only our contingent headquarters but also the numerous camps and checkpoints scattered throughout the area. Within the confines of the city, a company of government forces held their ground, while several additional companies and platoons were stationed in the surrounding regions. Although outnumbered, the rebels were ill-equipped, their arms feeble in comparison to the well-trained and well-armed government troops. An assault on Zenuoula would undoubtedly result in catastrophic loss of life and property. The rebels may have been greater in number, but they were weak in arms, whereas the opposing government faction possessed advanced weaponry. If an attack upon Zenuoula truly occurred, the consequences would unfold in unimaginably horrific ways.

Under the cover of nocturnal darkness, I embarked on a search for Zakaria. It became evident that his main office lay empty and deserted, devoid of any trace of his presence. The rebel checkpoints held no knowledge or inclination to divulge his whereabouts. Finding myself at a dead end, I made my way back to our checkpoint on the highway, discovering that Zakaria had never crossed its path. A flicker of hope ignited within me, offering a brief respite. I retraced my steps, returning to the city, resolute in my pursuit of Zakaria. Alas, my efforts proved futile, and I returned to the company base empty-handed. Word of my fruitless search reached the chief, who promptly contacted me directly through the crackling radio waves.

The chief imparted crucial information he had received from the locals, who had ventured forth to our headquarters with news of the swelling insurgent numbers on the opposite bank of the northward-flowing river from Zenuoula. He further revealed a disheartening alliance in the making—the city's rogues were joining forces with the rebels, driven solely by their insatiable greed for plunder. The chief implored me to redouble my efforts in locating Zakaria, emphasizing the dire need to appease or pacify the rebels through his intermediary.

Once again, I embarked on a relentless pursuit of Zakaria, traversing the city's shadowed streets with only a glimmer of hope lighting my way. As I wandered through the urban maze, following faint trails and subtle signs, my determination led me to a rebel checkpoint nestled deep within the heart of the city.

There, by some stroke of fortune, I encountered one of Zakaria's unwavering companions, a constant presence by his side. Pleading with him earnestly, I implored him to stay put and remain vigilant until I returned with news of our encounter. Reluctantly, he vanished into the obscurity of the night, leaving me alone with my thoughts.

Darkness cloaked the city, its streets devoid of light except for a few pockets that emerged after dusk. It was all part of the rebels' plan, shrouding their intentions in secrecy, concealing their movements within the veil of night.

With the overall state of affairs as my guide, I couldn't shake the conviction that Zakaria held a prominent role and was perhaps even leading the impending assault on Zenuoula. Walking through the dimly lit obscurity, I pondered the commander's role in the escalating conflict.

The consequences of an attack on Zenuoula weighed heavily on my mind. Lives would be lost, and the ripple of violence would extend beyond the city's borders. The rebels' belligerence would not be contained, spreading to the surrounding settlements and territories. Inevitably, the government forces and their allies would rise in response, escalating the situation into an uncontrollable spiral.

After a brief delay, Zakaria's assistant returned, signaling us to follow his lead. We found ourselves in a modest colony, a cluster of low-cost single-story dwellings designed for the impoverished. The night swallowed all light, leaving no trace of human presence.

We entered an unassuming house nearby, its interior radiating with brightness amidst the prevailing darkness. Passing through the living room, I caught sight of Zakaria seated at a table in the inner chamber surrounded by a group of individuals. As my interpreter and I entered, several figures rose and exited the room, leaving a crowded space behind.

The chamber was cramped, accommodating 10 to 12 people, some seated while others stood against the walls. Clad in military attire, their aura exuded a sense of intensity, a readiness for the imminent conflict. Their intentions were palpable, their purpose clear to us all.

As I contemplated how to initiate a dialogue with Zakaria, how to convey our message, ultimatum, or threat, I chose to begin with flattering words: "It is truly admirable how swiftly you have prepared for war and gathered yourself," I began, acknowledging his tactical prowess. "It seems that the enemy will face formidable resistance."

After searching for the elusive commander for hours, I expressed my astonishment. "Although I failed to locate you, it appears you have been staying close to our camp," I remarked, emphasizing the irony. "Considering your strategic advantage, it would be unwise to take hasty actions and jeopardize your assets. Exercise patience and trust in the United Nations."

With measured words, I hoped to appeal to Zakaria's rationality, to deter him from plunging further into the depths of conflict. But whether my plea would find its mark in his hardened heart, only time would reveal.

I had been the sole voice in the room, my words echoing into the silence as Zakaria and his council listened intently, withholding any response or feedback. It was clear that they were unwilling to divulge their plans or reveal their true intentions. They stood firm, unwavering in their position, refusing to yield. My attempts to soothe their anger with my bland, dry words seemed futile, as if our efforts to pacify them were mere whispers lost in the vast wilderness.

After making more remarks in that vein, I yearned to hear Zakaria's opinion. In essence, he conveyed that they would retaliate, but not through aggression or deception. Their course of action would be decided after consulting with their fellow fighters. Curiosity compelled me to inquire about the timeframe. "How long will it take?" I asked. He replied, "Perhaps two or three days." Eager for further details, I inquired about their upcoming discussions with the fighters. "When will they commence?" I questioned.

His response came swiftly, "The next day."

Determined to keep a watchful eye on him and ensure his security, I proposed to accompany him during those crucial exchanges. I knew not what thoughts swirled in Zakaria's mind before he acquiesced, but he agreed. This proposal of mine was not born of profound insight, but it granted me the opportunity to observe the unfolding situation and the enigmatic figure of Zakaria. The time and location for our meeting the next day were settled.

Leaving Zakaria's place behind, I headed straight for our post checkpoint, where I briefed the JCO in charge. I emphasized that under no circumstances should Zakaria be allowed to breach or cross our checkpoint. I implored them to notify us via radio if Zakaria made any such attempt and to buy us as much time as possible to stall his progress. Laden with doubts and apprehension, I returned to our camp.

I reported everything to the commander, not concealing my lack of trust in Zakaria's words. I made it clear that I doubted his sincerity. As I voiced my concerns, a nagging question haunted me during the night – Why had I taken this gamble? It was not within my assigned duties, and any impulsive decision by Zakaria or his militia could inflict irreparable harm upon me.

The night passed in a turbulent blend of doubt, excitement, and fear swirling within me.

By 9 am, we arrived at Zakaria's hideout. To ensure our safety, we had a contingent of soldiers accompanying us in a pickup truck. However, Zakaria was nowhere to be seen upon our arrival. Yet, before long, he emerged with a fleet of four to five cars, oblivious to the trap we had set. Could we have whisked him away to a secret location? Perhaps. But it seemed unnecessary, for fate had something else in store for us.

With Zakaria's sudden arrival, our journey to an unknown destination commenced without delay. His whereabouts, as well as those of his fellow rebels, remained shrouded in mystery to me. I knew little about his revolutionaries, and my ignorance left me with an air of intrigue. For the first two hours, Zakaria wandered aimlessly through the bustling main thoroughfare and the adjacent alleys where their checkpoints and small hideouts were scattered.

Curiosity failed to stir within me as I observed his purposeless meandering. It seemed as if he was merely passing the time, deliberately avoiding the actual location or the gathering of his fervent followers. Meanwhile, the crackling sounds of irregular gunfire began to reach the ears of the locals, a testament to the rebels' advance across the river towards Zenuoula. Informed by the radio, I approached Zakaria and relayed the impending assault by our forces. The air would soon be filled with the relentless hail of bullets from automatic weapons and the thunderous explosions from mortars and rocket launchers positioned near Zenuoula. Our APCs had taken strategic positions along the riverbank. I urged him to order his men to retreat immediately, lest they face grave danger and suffer needless casualties. Failure to heed my warning would result in the unnecessary loss of life and chaos for both sides.

Unbeknownst to me at the time, the operation to secure Zenuoula and repel the insurgent assault had already commenced. Colonel Iqbal had devised the plan the previous night, naming it 'Operation Zenuoula.' By the morning of November 7th, the battalion had set up the mortar and skillfully marked their targets. As the rebels attempted to breach the city by crossing the river around 10–11 am, they were met with a ferocious onslaught of fire from automatic weapons, mortars, and rocket launchers.

Shortly after noon, Zakaria led us down a path veiled by dense forest. The road was relatively smooth, its firm surface facilitating easy passage. Dust clouds billowed behind us as Zakaria's convoy raced ahead. After walking for about 30–40 minutes, we arrived at a secluded spot nestled amidst the trees—a long-abandoned school building. Inside, I beheld a gathering of approximately two thousand rebel militiamen fiercely loyal to Zakaria. Only a few were clad in military uniforms, while the majority donned civilian attire. Most sported tracksuit pants and T-shirts, with the iconic image of Che Guevara emblazoned across their chests. Each rebel bore arms; some brandished bazookas, while others carried heavy automatic weapons, their ammunition belts draped like cross-belts across their torsos.

Zakaria's presence ignited a fervor among the rebels, their excitement tangible in the air. Chants and slogans reverberated through the space, creating a spectacle both remarkable and strangely intense. After a few moments of fervent commotion, Zakaria finally began to address his eager audience. Every word that escaped his lips was met with thunderous applause and cheers from the rebels, their loyalty and admiration on full display.

Surrounded by his entourage of rebel fighters, Zakaria became an inaccessible figure. We dared not venture too close, fearing that they might seize our weapons, rendering us defenseless. Moreover, if they discovered the bullets lodged in their knees were a consequence of the Zenuoula attack, our true intentions would be laid bare, and we would find ourselves in a precarious position.

I turned to my interpreter, seeking clarification on the content of Zakaria's impassioned speech. According to my interpreter, Zakaria conveyed the following to his fervent followers: "Our victory is within reach. We shall not sit at the negotiation table, for when they attack us, we must respond with weapons! An attack with weapons shall be met with an equally forceful response. Inqilab, Zindabad! Long live the revolution!" The atmosphere brimmed with excitement, enveloping us amidst the dense jungle. Although our exact location remained unknown, we surmised that we were somewhere in close proximity to Zenuoula. Later, we discovered that we stood in a place called Janjra, approximately 10–12 kilometers north of Zenuoula City. From there, the rebels could reach Zenuoula in less than two hours.

Upon observing the rebels' fervor and aggression, it became evident that they were on the verge of heading towards Zenuoula at any moment. Preventing their advance seemed like a distant hope; all I could do was witness the unfolding atrocities. Amidst the tumultuous chaos and the frenzy of the crowd, an idea sparked within me. I swiftly relayed a message via radio to the commander expressing my lack of control over Zakaria. Only the force commander possessed the authority to disarm Zakaria. I requested the captain to provide me with the force commander's satellite phone number, enabling me to facilitate a direct conversation between them. Each rebel leader possessed one or more satellite phones (Inmarsat) for communication amongst themselves and possibly with their foreign mentors or advisors.

Though the captain had managed to withstand the initial shock with minimal casualties, he remained deeply concerned that a large-scale attack would result in a bloodbath. In a remarkably short time, he furnished me with the force commander's phone number. Approaching Zakaria, I informed him that the force commander wished to speak with him. He appeared somewhat surprised but handed me his phone. Dialing the force commander's number, I introduced myself and passed the phone to Zakaria, providing a brief summary of the events that had transpired over the past twenty hours. The assembled crowd fell into silence, their eyes fixed upon us. Zakaria and the force commander engaged in a lengthy conversation conducted in French.

Although I couldn't comprehend the unfamiliar language or hear the force commander's voice from my position, Zakaria's gestures and expressions suggested that the dialogue progressed favorably. Once their conversation concluded, the force commander addressed me once more, expressing gratitude for my involvement in the situation. He thanked me for keeping a watchful eye on Zakaria and informed me that Zakaria had agreed to retreat from his current position and return with his men, relinquishing any further pursuit. However, he requested that I remain with Zakaria until their retreat. I assured the force commander and concluded our exchange.

Zakaria, once again, addressed his rebel militia fighters. Although I relied on my interpreter to understand his words, it became apparent that Zakaria had received assurances regarding several of their demands from the force commander that his militia perceived as their triumph.

It felt like a cathartic resolution to a nerve-wracking ordeal, a gratifying end to the tense excitement that had consumed us. I relayed every detail to the commander, who insisted that I remain in close proximity to Zakaria at all times, never to part ways until their complete retreat. When I inquired about Zenuoula, he informed me that our display of force had inflicted significant damage, forcing the advancing rebel group to withdraw.

As evening descended upon us, I found little appetite after the thrilling events of the day. I consumed a meager meal of dry food and fruits that I had carried with me. Zakaria embarked on his journey back, spending some additional time with his militia. We trailed behind, reaching Seguola as dusk painted the sky. A profound sense of peace washed over me, permeating my body and soul. The grueling psychological warfare had finally come to an end. Regardless of who emerged as the victor or the vanquished in this war, it was undeniable that we had averted a major humanitarian catastrophe.

There remains a slight satisfaction within me, knowing that my prompt actions, immediate decisions, and courage played a part in averting such a calamity. The risks I took at that time extended beyond my responsibilities, but they were also inevitable. In those grave moments, my conscience led me down the arduous path, allowing me to glimpse humanity and peace with newfound clarity.

In recognition of my contributions to 'Operation Zenuoula,' Colonel Iqbal forwarded a citation to the force commander. His recommendation likely echoed along these lines:

> "The officer played an indispensable role in the operation of Zenuoula, employing strategic and diplomatic skills. Lieutenant Colonel Lutful Haque demonstrated remarkable prudence and leveraged his extensive knowledge gained through service in the Bangladesh Army. Despite facing grave danger, he approached his duties with unwavering courage and composure. Entrusted with the task of communicating with Zakaria, under immense pressure from 'Anusir,' he successfully compelled the rebel leader to withdraw his forces. The COM zone commander personally arrived at Janjra to order the retreat. The ability to restrain and monitor the rebel leader will forever stand as a testament to the diplomatic success of BANBAT.
>
> Even in dire circumstances, Lieutenant Colonel Lutfullah never hesitated to follow the orders of the contingent commander, despite the risk of death or being taken hostage by UNMEEF. He maintained a composed demeanor in every situation, fulfilling his professional duties with unwavering responsibility and dedication."

Throughout my 28-year military career, I had faced numerous challenges. From confronting the Shanti Bahini in the rugged terrains of the Chittagong (also known as Chottogram) hill tracts to thwarting smugglers on the border, I had even weathered the storm of public unrest during times of military rule. All of these encounters were part of my duty and responsibility. However, the incident that unfolded in Ivory Coast, as recounted earlier, transcended the ordinary. It was a test that went "beyond the call of duty."

I often ponder if it was my involvement in that pivotal moment that forged a connection with Zakaria. In the days that followed, he sought me out on several occasions. The gifts he bestowed upon me serve as constant reminders of that fateful day.

As the memories settle, I reflect on the journey that brought me here and the challenges, risks, and sacrifices made in service to my country. Each chapter in my military career led me to that extraordinary moment in Ivory Coast, where duty took on a whole new meaning.

And so, with the echoes of that event still resonating, this chapter comes to an end. The tale of a soldier's unwavering commitment, is etched forever in the annals of time.

ECHOES OF AFRICA

LIEUTENANT COLONEL MD ASHIQUE HASAN ULLAH

(UNITED NATIONS MISSION IN THE DEMOCRATIC REPUBLIC OF THE CONGO -MONUSCO)

George Moboyu was more than just an interpreter to me – he was a symbol of the incredible African continent who embodied the struggles and aspirations of its marginalized and impoverished communities. Looking into his piercing gaze, I could see the pain and hardships of those living in the midst of indescribable suffering and deprivation. Despite the stress, anxiety, and immense suffering he had endured, Moboyu always greeted me with a genuine, friendly demeanor, addressing me with, "Hey Chief," which never failed to lift my spirits. His playful banter and warm smile were a source of joy, even amidst the grief that sometimes weighed heavy on my mind.

I still vividly remember the day we first met at Mbandaka airport. From a distance, Moboyu greeted me with an unreserved and hearty smile that immediately put me at ease. It was clear that I was in good hands under his benevolent care. Although our initial relationship was limited to the formalities of an interpreter and chief, Moboyu's impact on me and my perception of the world was transformative.

Over time, a remarkable bond formed between us, and Moboyu became one of my closest friends. He evolved from being my interpreter to my French instructor and eventually became an indispensable companion with whom I could converse for hours on end. During our conversations, Moboyu often spoke like a philosopher, expounding on African society, history, tradition, and culture. His ability to convey weighty matters in a facile and spontaneous manner amazed me.

Before the Congolese war, Moboyu had a good life. He had been gainfully employed by an American company, and his family had been happy and fulfilled. But the brutal civil war changed everything. The company was unable to withstand the economic onslaught, and they folded, leaving Congo. Moboyu and his family were left in poverty and penury. Forced to flee Kinshasa due to the tyranny of rebel groups and government forces, he sought refuge in Bakanda.

Through all his hardships, Moboyu remained resilient. His spirit and determination inspired me, and I am forever grateful for the friendship we shared. George Moboyu will always be a symbol of the strength and perseverance of the African people and a reminder of the power of human connection.

Moboyu's Journey

Moboyu's journey is one of perseverance, hope, and selflessness in the midst of war and hardship. After graduating from a prestigious university in Kinshasa, Moboyu began his career only to have it nearly destroyed by civil war. Eventually, he landed a job as a UN interpreter, which allowed him to provide for his family and care for his late sister's children.

Through conversations with Moboyu, his exceptional education and awareness became evident, surpassing that of the average Congolese individual. He inspired me to learn more about the exploitation of the Congolese people and their unspoken agony.

Despite the aid I tried to provide, Moboyu felt a sense of penance for relying on my assistance. Later, Moboyu announced that he was going to be a father again, and months later he visited me with the joyful news that he had been blessed with a baby girl.

But what made the moment even more special was the revelation that Moboyu had named his daughter after me. Despite facing resistance from hospital staff and priests, Moboyu persevered and successfully named his daughter Ashique George.

I was deeply touched by this act of pure generosity and selflessness, as it was such a rare occurrence in my life. The experience taught me the joy and satisfaction of winning someone's heart, and I expressed my deepest gratitude and admiration for Moboyu.

In the end, I arranged to bring gifts for Moboyu's daughter and expressed my intention to visit his family. Moboyu's journey shows that even in the darkest times, there is always hope, perseverance, and selflessness that can light the way forward.

A Visit to Moboyu's Family

As I arrived, George was already waiting for us in front of his house. He greeted us warmly, expressing his excitement for our visit. George's wife appeared soon after with their baby daughter, a little angel from Africa with delicate features, curly hair, and small hands and feet. We all gathered in the courtyard, where our differences in language, nationality, and culture disappeared in the face of love and compassion. As I held the baby, I asked George to promise me that he would take her to Bangladesh one day when she grows up and make her listen to my stories.

Later that evening, many of George's neighbors came to meet us, excited to see the Original Ashique, or Captain Ashique, as the sun set over the African horizon. We enjoyed a light supper together before leaving for the camp with our gifts. I expressed my gratitude to George and his wife before retiring for the night.

As I lay in bed that night, I couldn't help but think of George's daughter, reflecting on my own child back home. George was a man who found joy in the simplest things, like a child, and his amazement at the diversity of physical appearances in Bangladesh made me smile.

Over breakfast one morning, we watched a TV program about agriculture and farming practices in rural Bangladesh that caught George's attention. He was surprised to see a farmer on the screen who looked like his Congolese brethren and asked me how it was possible to have so many different people of different colors in the same country. I explained that Bangladesh's diversity was due to centuries of migration and the confluence of different ethnic groups, leading to a rich tapestry of ethnicities, cultures, and physical features.

Our conversation led to a discussion of the shared burdens of pain and suffering that exist in both our countries. I shared with George that I have seen the same shades of anguish and affliction in the eyes of both his fellow citizens and mine, despite the distance and cultural differences.

George then shared his dreams of finding a job with his former American company or teaching in college. He dreamed of moving his family to a new home and leaving behind their dilapidated building. As I listened to George, I couldn't help but think of the black wood tree with a circumference of four centuries he had spoken of earlier that represented the Congolese people's strength and endurance in the face of natural disasters and colonialism.

My time with George and his family left a lasting impression on me. The universality of love and compassion transcends all barriers of language, nationality, and culture, bringing us together as one human family.

I am grateful for the opportunity to have met George Moboyu and his family. They have taught me so much about the power of hope, perseverance, and love. I will never forget my time with them, and I am proud to call them my friends.

In the Line of Fire

Major Md. Hasan Shahriar

(United Nations Mission in the Democratic Republic of the Congo -MONUSCO)

11:00 AM. Level Two Hospital, Democratic Republic of Congo.

The room was an eerie sight, bathed in blinding white, its walls, floor, and ceiling all painted in the same sterile hue. Six beds, adorned with tangled wires and draped in thin white sheets, stood like sentinels in the desolate chamber. On the bedside table, a note accompanied a gleaming fruit basket bearing the inscription: "With the compliments of Force Commander, Manusque." The air hung heavy with the suffocating blend of disinfectant, medicines, and air freshener, assaulting my senses. The anticipation of the horrors that awaited me in this hellhole gripped my very soul.

As Major Asif, clad in a hospital gown and nursing a fractured leg, hobbled into the immaculate white room of Level Two Hospital, he felt an ominous shiver run down his spine. He had been relocated to this room in preparation for the force commander's visit, and it had been meticulously arranged for the occasion. Concealed beneath a pristine white bandage wrapped around his knee lay a small wound that oozed a subtle trickle of blood, resembling a dark mark painted on a child's forehead to ward off evil.

It all began yesterday. In response to intelligence on a possible militia attack, Asif had ventured outside the camp with a patrol team on a mission to assert dominance and showcase their force. They had been closing in on the Supreme, a significant building, when the incident occurred.

"Aaah...!" Asif winced, his thigh and knee throbbing with agony. Cold beads of sweat formed on his forehead as he contemplated the daunting prospect of reliving the nightmare. The force commander was en route to see him, and despite having recounted the harrowing ordeal seventeen times already, it served as the only distraction from the excruciating pain in his leg.

His daily routine typically commenced with waking up before the break of dawn, just in time for the Fajr prayer. Following the prayer, he would embark on a jog around the army camp, culminating in a well-deserved breakfast. Yesterday was no exception as he meticulously adhered to his ritual.

Breakfast had been consumed, yet his steaming coffee remained untouched when the news reached him. The tantalizing aroma of the hot brew wafted through the air, teasing his senses. Reports indicated a possible militia movement near the 'Supreme' approximately 24 kilometers southwest of the camp involving two militant sections.

In the aftermath, the community liaison assistant delivered a chilling report of their heinous deeds. Each village they encountered along their path had fallen victim to their merciless looting. The destitute villagers lost their meager possessions – cows, chickens, goats, anything of value that caught the marauders' eyes. A tragic fate befell an elderly woman who had refused to surrender her belongings willingly, meeting her demise from a bullet wound. Time was of the essence, and the search for the culprits demanded swift action and an ounce of luck. If they moved without delay, there was a glimmer of hope to teach these militia groups a lasting lesson. However, a formidable challenge stood in their way – the assailants' ability to blend in seamlessly with the innocent villagers, shedding their weapons in the dense jungle. Who would dare to expose them? The journey ahead promised to be an arduous struggle.

Major Asif, consumed by the memories of that fateful day, relived the harrowing episode in his mind. The vivid recollection unfolded before his eyes as though it were transpiring in the present. With agility honed through rigorous training at the Bangladesh Military Academy, Major Asif sprang into action, changing his attire in a matter of seconds. Just as he prepared to move, a commanding voice crackled through the walkie-talkie ordering him to join an emergency patrol. A standby unit always stood ready to confront any situation with swift precision. Major Asif and his comrades, UN peacekeepers bound by duty, leaped into motion. The resounding thumps of their boots echoed through the air accompanied by urgent commands pressing them to move faster, fix their equipment, and assemble at the briefing point with haste.

Leading the patrol that day was Senior Warrant Officer Morshed, an indomitable commando from the quick reinforcement platoon. Having previously served in the Sierra Leone mission, he returned to the Democratic

Republic of Congo after a seven-year hiatus, older but no less agile. Tales of his legendary feat, single-handedly vanquishing a group of 17 Shanti-Rokkhi Bahini militants in Bangladesh's treacherous hill tracks, reverberated among the soldiers, earning him their admiration. From a fresh graduate of the Army Commando Course, Morshed had risen through the ranks, overcoming countless challenges to attain the esteemed position of senior warrant officer just days before.

Senior Warrant Officer Morshed maintained a strict demeanor, demanding perfection in every task, yet he possessed the gift of lightening the mood when off-duty, regaling the soldiers with jokes and laughter. Wherever he went, life and vitality followed in his wake.

Familiar with Senior Warrant Officer Morshed's ways, Major Asif knew that he was not one to raise an unnecessary fuss. The urgency in his voice over the walkie-talkie hinted at something extraordinary or an imminent crisis. Morshed swiftly rallied everyone, forming a disciplined line at the briefing point, assuming command in the absence of the patrol leader. In the face of emergencies, he possessed the uncanny ability to act swiftly and calmly, free from the grip of panic.

Each soldier meticulously checked their weapons, ensuring every component was in perfect order – barrels, magazines, ammunition, safety catches, and locks. "Is everything in order?" Morshed inquired.

In unison the nine soldiers replied, "Yes, sir."

"Listen up, everyone," Major Asif commanded, his tone laced with urgency and determination. "We're about to embark on a mission of utmost importance. The specifics of our destination and objectives remain unclear, but we'll receive our orders from the patrol commander. Regardless, we must maintain a calm and composed mindset. Understood?" His piercing gaze scanned the faces of his comrades, demanding their unwavering attention.

"If we encounter any pauses or require deployment along the route or within the operational zone, Lance Corporal Aftab, you will advance on the left flank, securing a position with optimal cover. Soldier Majid, you will be his LMG II gunner. Clear?"

Major Asif's voice resonated with authority as Aftab snapped to attention and responded, "Yes, sir!"

"Excellent. Now, Lance Corporal Amjad, you will do the same on the right flank, providing cover for Aftab's right side. Soldier Asad, you will act as our marksman with a rifle," Major Asif continued, outlining the strategic plan of securing cover from all directions. Senior Warrant Officer Morshed nodded approvingly, his instructions reinforcing the importance of protecting civilians and executing precise rescue operations. They had rehearsed these maneuvers countless times, their proficiency earning the highest score in the previous operational efficiency inspection.

Donning a blue helmet, black sunglasses, and a bulletproof jacket over his uniform, Major Asif exuded an aura of preparedness. His jacket pockets held three extra magazines, gloves clung tightly to his hands, while elbow and knee guards offered an extra layer of protection. Gripping an SMG in his hand, he appeared ready to face the horrors of war. In this volatile and unpredictable land, where the situation could erupt at any moment, Major Asif understood the necessity of constant vigilance and readiness.

Adjacent to the camp stood the briefing point, adorned with a meticulously crafted model of the critical areas. Soil, sand, stone fragments, and wood formed a replica of the terrain, revealing the entire landscape—roads, villages, culverts, and more. Villages were labeled on small sheets with vibrant colors, while pebbles marked the paths, painted red to denote danger. Each soldier had traversed these routes so frequently since their arrival that every location was etched in their memories.

The drivers, well-acquainted with the treacherous twists and precipitous drops of the mountainous roads, still attended the briefing sessions, participating in the detailed mapping and strategic planning. Everyone received their instructions, reinforcing the mental imprint of the destination and its intricacies.

Having briefed his team on the necessary guidelines and outlined the mission plan, Major Asif raised his hands in prayer, invoking a time-honored tradition of the army. "O Sovereign of the worlds!" his voice reverberated, seeking assistance, courage, and divine collaboration in the face of impending perils. "Deliver us from calamities, guide us in making the right choices, and grant us a safe return to our camp. Amen." A resounding chorus of "Amen" echoed through the air, binding them together in shared determination.

The car stood at the center, positioned meticulously. In the darkness, venturing out without an armored personnel carrier was a risk nobody dared to take; however, during the daytime patrol, they had no choice but to rely on lighter vehicles like pickups. Many areas were inaccessible to armored carriers, so they had prepared the pickups as best they could for protection. The empty space on the side of the pickup had been reinforced with a plain steel sheet, firmly welded into place. The canopy remained open, while a makeshift mount had been fashioned at the back of the cabin to hold a formidable light machine gun. A vigilant soldier stood perpetually on guard behind the weapon, prepared for any eventuality.

As dusk settled, the Congolese army, known as FARDC, joined their ranks. Clad in berets, ropes cinched tightly around their waists, AK-47s cradled in their hands, and magazines taped together, they exuded an air of readiness. Ammunition belts crisscrossed their chests and shoulders, some wearing boots while others opted for cades. Lieutenant Kambato, a battle-hardened veteran of 45 years, assumed command of the Congolese contingent. His intimate knowledge of the territory made every corner his domain, with not a single inch escaping his grasp. Accompanying him were seven seasoned FARDC members prepared to face any obstacle. After the patrol, a few packets of biscuits would be sufficient to satiate their hunger.

"Hello, Delta Two calling Delta Nine, over," Major Asif's voice pierced the airwaves, ensuring all was in order before venturing into the battlefield to eliminate the forces of evil.

"Delta Nine receiving. Send your message, over."

"Delta Two to Delta Nine-Sub, all systems green. Are you ready?"

"Delta Nine to Delta Two, everything is in order, sir. We are ready to proceed, sir."

"Good. Move..."

The engine roared to life, the vehicle trembling with anticipation. With a single command, "move," the entire team advanced in unison, maintaining their formation with disciplined precision.

Their path led them to a right turn from the camp, following a mountain road for approximately two kilometers until they reached Badjangga Bazaar. A short distance ahead, a three-way intersection awaited. Taking the left fork, they pressed on for another two kilometers until they reached a Y junction. The straight path led to a camp named Bukiringi, while the rightward direction veered toward Supreme. It was this very route that the marauding militia group had been using to unleash havoc and advance in their ruthless raids.

With his weapon slung across his shoulder, Major Asif led the patrol, his vehicle taking point. His eyes remained keenly vigilant, scanning the surroundings on both sides. Prepared for any eventuality, he exuded an air of authority and readiness. Behind him, Senior Warrant Officer Morshed followed in the pickup truck, maintaining a safe distance. The LMG men in each vehicle were on high alert, their senses honed to detect danger.

All seemed ordinary at first. The sun bathed the landscape in radiant light, the air crisp with the purity of nature. In the distance, unfamiliar birds chirped intermittently, adding a touch of serenity to the atmosphere. Men and women traversed the picturesque mountain road, the women gracefully balancing baskets atop their heads or clutching bundles of cassava and other goods. Meanwhile, the men navigated the rough stone road astride their motorcycles. On one side of the road, a peculiarly dense crowd gathered, leading Major Asif to wonder if someone had passed away. Rumor had it that they celebrated death in these parts – perhaps that was the cause. Suddenly, a voice cut through the air, shattering the tranquility. It was Lieutenant Kambato of FARDC shouting from the middle vehicle, 'Militia! Militia!'

Almost instantly, the same chilling word reverberated through their walkie-talkie sets, echoing the grave danger that lay ahead. A surge of electricity coursed through Major Asif's veins. He barked his command, 'Halt the vehicles! Now! Quickly! Move!' Though they had been briefed not to stop the vehicles in the middle of the road, the drivers skillfully sought cover before coming to a haLieutenant Driver Imam demonstrated his competence by skillfully parking the car behind a colossal stone at the roadside, adhering to the briefing with precision.

Before Major Asif could fully comprehend the situation, Lieutenant Kambato leaped out of the car and sprinted away. It became evident that someone had dashed out of the crowd, gasping for breath. It took no time for Major Asif to realize that the man was a member of the militia group. But why had Lieutenant Kambato fled? Was he a scout for the militia, alerting them to the presence of the patrol? Perhaps he had run off to inform the others, leaving them vulnerable. Maybe Kambato's true allegiance would be exposed and he would pay the price for his treachery.

“Suddenly, the sound of gunfire pierced the air!” Major Asif recounted, his voice filled with intensity.

Alas, it had begun! Had they fired upon Kambato? Or had Kambato already unleashed his own hail of bullets upon them? Major Asif remained uncertain.

“But my thoughts were shattered by the sound of more gunfire. And again. And again!” Major Asif continued, his words painting a vivid picture of the unfolding chaos.

War erupted, transforming the once bright and serene day into a battlefield of ferocious encounters. The idyllic spot morphed into a theater of violence and bloodshed. Major Asif’s suspicions had been confirmed: Kambato was indeed a member of the militia group. They had likely been pillaging along the road, sending him ahead as a scout to assess the situation and provide a preview of the impending danger. The moment Lieutenant Kambato had cried out, “militia, militia,” he had fled, setting the wheels of conflict in motion.

However, the situation had taken a dire turn. The ominous sound of bullets drew closer, a relentless march toward their position. Major Asif swiftly relayed instructions through the crackling walkie-talkie set, commanding everyone to assume their positions. With nine years of military training under his belt, he reminded them to follow the principles of dash, dine, crawl, observe, and fire. In such a treacherous moment, it was crucial to remain as composed as ice and act with precision. Too often, panic gripped individuals caught within the crossfire, transforming them into unwitting victims of battle shock. In such cases, those stricken by battle shock became more perilous than the enemy they faced.

Abandoning their vehicles, the team sought refuge wherever they could, crouching on the roadside, hiding amid the bushes, or taking cover behind stones and mounds of soil. They awaited their chance to spot the enemy and unleash their fury. The sound of bullets grew ominously near, slicing through the air with a haunting hiss. Some rounds ricocheted off the body of the pickup truck, thudding against the metal. Suddenly, a voice erupted from the side, exclaiming, “Sir, I’ve got them! I’ve got them! I see them, sir! Militia! Militia! Armed militia! They’re advancing and firing relentlessly, sir!”

Major Asif pressed the walkie-talkie set to his lips and bellowed, “Fire!” His voice was drowned out by the deafening chorus of gunfire as the

BANBAT members opened fire, unleashing a symphony of bullets that reverberated through the air.

"Delta line for Delta Two, send your message! Over, sir!" Senior Warrant Officer Morshed's voice trembled with excitement.

"Delta Two for Delta line-sub. I'm keeping the militia pinned down with suppressive fire. You lead your section and maneuver behind the hill to flank the militia's position. Brace yourselves for a counterattack, then advance and eliminate the target with coordinated fire and movement. Understood?"

"Affirmative, sir."

"Good. Report to me once you're in position. If you fail to do so, signal me with a red flare from your pistol. Now, move! Move swiftly!"

"Roger that, sir."

Following the tactical plan, Senior Warrant Officer Morshed and his team maneuvered into position for the counterattack. In an instant, the situation descended into chaos, the deafening roar of gunfire engulfing the surroundings. Bullets wove a sinister tapestry of war, leaving empty casings strewn about like macabre confetti. Spent cartridges littered the ground, ejected from their chambers. Some of the scorching-hot shells landed on hands, but amidst the chaos, who had time to care? The air grew thick with smoke, shrouding the area in a haze of battle. The acrid scent of gunpowder, mingling with adrenaline, became an intoxicating elixir, fueling their resolve. Meanwhile, Senior Warrant Officer Morshed led his team in a stealthy flanking maneuver, preparing to launch a relentless counterattack from the rear.

The tempest of bullets raged on for ten relentless minutes before abruptly falling silent. The whistling menace of the projectiles haunted their eardrums for an additional five minutes. Then, as if swallowed by the shadows, the enemy's gunfire ceased. They must have retreated, cunningly using sporadic shots to provide a smokescreen for their escape. In an instant, the air settled, but Major Asif and his comrades dared not lift their heads. What if their foes lay in wait, ready to spring an ambush? What if they had devised a new strategy, a nefarious plot waiting to ensnare them?

A small structure materialized from the dense thicket ahead, shrouded in uncertainty. It was difficult to discern whether it belonged to the militia or an innocent civilian. The audacity of the militia to expose themselves seemed unlikely, yet the world had become a playground for sinister surprises in those troubled times.

In the blink of an eye, Major Asif recognized the figure – Lieutenant Kamvato! He had survived the onslaught! But why had he charged through the hail of bullets like a madman? Just as the thought consumed him, a bush rustled fifty feet away. Someone emerged, brandishing a rifle with lethal intent. Swiftly taking aim, the target in their sights was none other than Lieutenant Kamvato, who was vulnerable and exposed. He teetered on the precipice of escape, his fate hanging by a thread. Oblivious to the imminent danger, Kamvato pressed on, flashing a smile that betrayed his innocence.

Major Asif's frantic gestures conveyed a desperate warning as he attempted to steer Kamvato clear of impending doom. "Move! Get out of the way! You'll be shot!" he yelled, his words pleading with urgency. But his impassioned pleas fell upon deaf ears.

In a moment charged with electricity, Major Asif felt an overwhelming surge to protect Kamvato. Casting caution aside, he abandoned his cover, lunging toward his comrade and pulling him down onto the unforgiving earth. Another soldier swiftly followed suit, joining the harrowing rescue. A gunshot pierced the air, followed by an eerie silence. Lieutenant Kamvato and Major Asif lay in pain, their bodies throbbing from the impact. The weight of Major Asif's muscular frame had taken its toll on the aging Kamvato, who at 45 struggled to bear the burden. Slowly, they separated and moved to opposite sides, each gasping for breath. For several moments, time stood frozen in stillness.

Finally, the trio rose, their senses heightened but mired in a state of shock. Lieutenant Kamvato's uniform was marred by the stains of blood, the lower part drenched in crimson. Yet a faint smile still clung to his lips. Major Asif, too, summoned the strength to stand, but his body betrayed him, collapsing like a toddler learning to walk. Private Hasanur, the valiant soldier who had sprinted alongside Kamvato in the rescue, lifted Major Asif upright, only to witness his commander falter once more. As Hasanur steadied Major Asif, he noticed a telltale blotch of blood on his knee and couldn't help but grin. "Sir, you've been shot!" he exclaimed. Asif's ire rose at the sight of Hasanur's smile, infuriated by the audacity of the scoundrel's laughter. Yet, the soldier's next words melted away his anger, leaving only gratitude in their wake.

Soldier Hasanur's heartfelt words echoed with a mix of remorse and devotion. "Sir, I thought you had taken a bullet to the chest, and I felt like a wretched criminal," he confessed. Innocently, he continued, "Sir, I am your runner. I wondered if I could shield you with my own chest. If it pierced me, it wouldn't have pierced you. But, sir, no need to worry. You were shot in the leg. You'll be fine, sir. As long as I breathe, nothing will harm you, sir." Hasanur's words were accompanied by a genuine, wholehearted smile, pure and untainted.

Major Asif, overcome with exhaustion and a growing sense of drowsiness, found solace in the profound serenity that Hasanur's selfless love had brought him. A surge of laughter swelled within him, but his weariness took hold, and he slumped onto Hasanur's sturdy shoulder. Darkness enveloped his consciousness, yet a smile of profound gratitude remained etched on his face—a testament to the satisfaction he feLieutenant

This is the way of the peacekeepers, relinquishing their own lives and willingly sacrificing themselves for others. Major Asif, Senior Warrant Officer Morshed, and soldier Hasanur, like valiant warriors, embrace the essence of selflessness every day as they spearhead peace in a land plagued by turmoil and unrest. They are the courageous sons of Bengal, who have planted seeds of love and compassion on foreign soil. And perhaps that is why, in Congo, people of all ages extend their right hand, show a thumbs-up, and joyfully exclaim, "Bangla Rafiki!"—meaning "Bangladesh friend! Best friend! Bangladeshi friend! Our friend!"

BEYOND THE SRANG HILLS: AN UNFORGETTABLE ENCOUNTER IN CAMBODIA

BRIGADIER GENERAL MD BAYEZID SARWAR
(UNITED NATIONS TRANSITIONAL AUTHORITY IN CAMBODIA - UNTAC)

I was transferred from the town of Siem Reap to Tram Khanar. Siem Reap, a small yet captivating town, is home to the world-renowned Angkor temples in the northwest of Cambodia. On the other hand, Tram Khanar is a village and communication center located in Kampong Speu province approximately 35 kilometers south of the capital city, Phnom Penh. As a United Nations Military Observer, I began to appreciate the peacekeeping work at the grassroots level in the heart of rural Cambodia.

In the distance, the serene Srang Hills stood tall and still. Every time I stepped onto the balcony of my living quarters to relax and breathe in the fresh air, the picturesque Srang overwhelmed me with its majestic beauty. A delightful friendship had already blossomed between me and the lush hills, one of the most captivating features of rural Cambodia.

One morning, I received a call. It was Major David Basely, an American officer from the sector headquarters in Kampong Speu. He informed me about an upcoming "helicopter reconnaissance" mission. He confirmed the details over the wireless set. Although we had been stationed in Tram Khanar for about three months, we still needed a comprehensive understanding of the entire area in sector 9 (west), particularly the remote and rugged terrain that was difficult to access.

It was a sunny morning in October 1992 when Cameroonian military observer Major Ferdinand, Cambodian interpreter Cho, Marsila, the Election Official from the neighboring Basat District, and I boarded a combat-colored helicopter made in Russia from Phnom Penh's Pochentong Airport. The military observer team from the sector, including Major David

Basle and Captain Jackson, had already departed from Kampong Speu to join us on the same mission.

Two young Russian Air Force veterans who had served in the Afghan War piloted the chopper. After a while, the colossal Russian grasshopper took off amidst the deafening noise of the engines running at full throttle.

When the chopper started hovering with its rotary wings fluttering to gain altitude, it looked like a big bird against the stunningly beautiful backdrop of blue sky. Soon, the city of Phnom Penh was visible down below. In the distance the bright spire of Wat-Nam temple, Prince Sihanouk's gilded palace, and the museum. We could also see the twist and bands of the Mekong and Tonle Sap rivers.

The chopper then changed direction towards the west of the airport, following the number 3 national highway, and started flying over the area we were responsible for looking after. The panoramic view of the landscape changed quickly. To the west of National Highway 3 lay the long-expanded greenery of the Srang Hills, a large wilderness area that occupies areas from both Kampung Speu and Takeo provinces.

These rolling hills were the undeclared battle zone of the Phnom Penh Government Forces and Communist Khmer Rouge guerrillas. As the chopper was following the NH 3, I enjoyed the bird's eye view of the greenery of the wild, hilly terrain and the beauty of rural Cambodia.

I found an incredible similarity between the hilly terrain of the Chittagong hill tracts of Bangladesh and the area in question. Undulating green hills were dotted with scattered habitations, and an eerie silence was suspended over the long stretch of the sylvan shadow, broken only by the occasional chirping of the colorful flit of birds – all seemed to have been woven neatly on a rich tapestry of nature. Cambodia thus started unfolding her indescribable natural beauty.

Sitting next to the chopper's window, I was imbibing the exquisite beauty of rural Cambodia. While the natural sublimity of Cambodia enthralled me, British Captain Jackson was busy gossiping with his female compatriot Marcella. I found Cambodia as appealing as a beautiful garden full of amazing flowers. But why did some people or parties want to destroy such a wonderful place?

"We are now flying over the area that the Khmer Rouge controls," said our young interpreter, Cho. Major Beasly requested the pilot to fly low. At the aiguille of the hill in the far distance, a yellowish Buddhist temple peeked through, and a few houses could also be seen below. We wondered whether it was a Khmer Rouge base.

And that was it…! All of a sudden, an unexpected incident took place – as we looked down, we saw a group of Khmer Rouge armed guerrillas clad in drabby olive clothing pointing AK-47 rifles at our helicopter.

Without a moment's delay, they started firing. All my excitement turned to horror. My thrilling experiences faded within seconds as the gunfire intensified. Had a bullet hit our helicopter, the incident could have easily led us to the valley of death.

"Oh God! Oh my God!" Marcella screamed out of fear. I took my face away from the window out of an instinctive sense of self-preservation.

"Speed up! Speed up! Speed up, man!" Major Beasly yelled at the pilots. What a stifling situation. It was nearly taking our breath away. We were confused as to whether it was only the mere trigger-happy Khmer Rouge guerillas or a well-planned attempt to harass UN peacekeepers.

Thanks to the Almighty, we somehow got away from such a terrifying incident unscathed. Luckily, the sensitive parts of the helicopter, such as the fuel tanks, engines, or rotors, had not been hit. And since the chopper was gliding at a high altitude and accelerating quickly, it flew into the airspace of Takeo province within no time, crossing the risky hilly hunting grounds below. The key decision soon came… The mission for that day was aborted.

After a while, our chopper landed on the field of the Tunisian battalion at Tram Khanar. The dust and debris being kicked up by the whirling motion of the rotor blades created a pall of obscurity that temporarily blinded our vision. The swirling clouds of dust settled as the rotors stopped, and gradually everything came to a standstill.

Both the Russian pilots quickly got to business assessing the damage. What they found was a damaged landing skid and dented main body window. UN election official Harold was on the other side, waiting with a car for Marcella. When Marcella went to step down, one of the Russian pilots asked, "Are you okay, Madam?" Nodding her head and waving her hands frequently to assure the man that she was all right, she got into the car.

As I reached my accommodations, freshened up and gathered my calm, the day's turbulence loomed large in my mind. What dreadful consequences those bullets could have inflicted on all of us – the thought haunted me.

It is customary to give a verbal 'OK report' to the sector monitor team every afternoon over the wireless set fixed and fastened with the car. Unlike other usual days in Tram Khanar, we reported about the Khmer Rouge firing incident. Colonel Keefer (French), the chief observer of the sector, ordered an investigation. UNTAC HQ (the United Nations' transitional authority in Cambodia) was preparing for a formal condemnation.

Just after sending the 'OK report', when we left for my room, I found a media car waiting in front of my residence. Journalists from different media outlets were there. A few followed us on their way back to Phnom Penh from Takeo province.

Our colleague Major Likaboth, in his usual English-French, was briefing them about that day's incident. I added that it might not have been a planned firing. With a cold beer in hand, the British journalist Kelvin Barrington of The Cambodia Times enquired and then started drafting a report with the title "Khmer Rouge fires at UN helicopter."

After getting settled, I contemplated the risky life of peacekeepers and the peace of Cambodia. Sakun Tha, one of my Cambodian friends from Phnom Penh, through a letter, informed me that 'Bon Om Touk' – the traditional water festival of Cambodia (the festival of the reversing current) would soon take place on the 7th of November (1992) in Phnom Penh.

On the forthcoming full-moon night, the Phnom Penh people would gather at the confluence of the two rivers. On a surprising note, that day, the Tonle Sap River would start flowing in the exact opposite direction. No other river in the world is as magical as Tonle Sap, which flows in the opposite direction on a full-moon night.

October 23, 1992 marked the first anniversary of the Paris Peace Agreement. Mainly due to the lack of cooperation, hostile attitude, and violence on the part of the Khmer Rouge, the peace accord could not be implemented the way it was expected. The accord was still languishing in the dark of uncertainty and was yet to come to the light of day.

Although it was not completely far-fetched, I could see that durable peace in Cambodia was a long way off. The event of firing at the UN chopper in the Srang Hills only consolidated my assumption.

I have a hobby of reading, and I like to read anything that interests me before I go to sleep. That night, I reached out to a few selected poems of the Irish poet W.B. Yeats. From 'The Second Coming,' I began reading....

> Things fall apart; the center cannot hold.
> Mere anarchy is loosed upon the world,
> The best lack all conviction, while the worst
> Are full of passionate intensity.
> Surely some revelation is at hand.
> Surely the second coming is at hand.

Coincidentally, with the theme of 'The Second Coming' by Yeats, I found an amazing allegorical similarity between the theme of the poem and that of present Cambodian situation. Two decades of civil war, genocidal Khmer Rouge rule, heavy bombing by the United States, and destructive foreign policies pursued by the neighboring countries and foreign superpowers had combined to lead Cambodia to the brink of destruction. Almost everything had fallen apart in this country.

Only divine blessings could have saved the country, and it seemed as if the peacekeepers of UNTAC resurrected for the second time, like the great Jesus, to save Cambodia from all miseries.

At present (at the time of the writing), approximately 16,000 peacekeepers from 32 different countries are serving in Cambodia to establish peace. Those peacekeepers have been carrying out their duties in cities-villages, hilly areas and jungles to serve the cause of peace at great risk. Along with other peacekeepers of international communities, we, the peacekeepers of the Bangladesh contingent, are also there.

UNTAC (1992) is the largest UN peacekeeping mission up to this point. Also, the 'Bangladesh peacekeeping contingent' deployed in Cambodia was the largest peacekeeping contingent sent abroad.

Bangladesh and Cambodia share a profound historical connection. Over a thousand years ago, the illustrious Khmer civilization, also known as the Angkor civilization, emerged as a powerful force in Cambodia, both culturally and politically. This civilization reached the pinnacle of architectural development, sculptural beauty, and innovation. They were responsible for constructing the magnificent Angkor Wat, the largest religious temple in the world. History bears witness to the significant contribution of our region, South Asia, in the development of this great civilization. Consequently, our commitment to establishing peace in Cambodia surpassed that of other countries' peacekeeping efforts. This was the prevailing sentiment among us, at least.

On my arm, I proudly wore the national flag of my country alongside a blue armband bearing the United Nations emblem. Our mission was clear—we had to succeed. Yes, we had to bring peace to Cambodia and shield it from the watchful eyes of evil. Gradually, I managed to overcome the haunting memories of the Khmer Rouge attack. Another day in Cambodia unfolded, filled with a mix of romance, anxiety, and a hint of frustration, and, ultimately, ended with hope.

Lost in Time and Africa: A Female Peacekeeper's Journey

Captain Sharmin Haque
(United Nations Mission in the Central African Republic -MINUSCA)

"Good morning, officer! We will be landing shortly. Please fasten your seat belt Lieutenant" The Ethiopian air hostess delivered her briefing with a charming smile as I stretched and moved my body, attempting to shake off the fatigue accumulated during the long flight. Her radiant smile seemed to alleviate my jet lag, at least momentarily. The temporal discrepancy disoriented me, and I found myself wondering about the time. Being in the air prevented me from truly grasping whether we had crossed into a different time zone or remained within our own.

According to Bangladesh's account of the time zone, it should have been dusk back home, but here the sun shone brightly. The Ethiopian flight attendants, paragons of virtue, displayed both beauty and grace. Despite their ebony complexions, they possessed stunning beauty. Throughout the grueling 13-hour flight, they attended to our needs diligently. However, there seemed to be some confusion with the meal schedule. Why else would they offer us lunch packets for breakfast and snack packets during dinner? Nonetheless, we expressed our gratitude and devoured whatever they provided. As the proverb goes, beggars can't be choosers. Whenever the food trolley passed by, a mischievous smile played on my lips, acknowledging the whimsical assortment of beverages and dishes.

"What is this?" I would inquire.

"Fanta. Would you like some?" she responded, eager to help me choose and serve.

"No. What's that?" I pointed to another juice can.

"Pineapple juice. Would you like to try it?"

"Who would have thought of trying these... Does anyone? Please, bring something solid so I can evaluate it a bit... I mean, taste it... hehe!"

"You're still immature, officer," the African beauty laughed with her eyes closed. I, too, found happiness in the obedient act of sipping the pineapple juice, feeling like an amenable girl. After all, 13 hours was a long stretch of time, and I needed something, even if it was just pineapple juice.

We, the 210 valiant members of the Bangladesh Army, boarded the plane united. Our destination: Kagabandoro in the Central African Republic, where we would join the UN peacekeeping mission. With a sense of duty and honor, we embarked on this mission, fully aware that we would serve as ambassadors of our nation. As I blinked and rubbed my eyes, I glanced out of the window at the small airport. It was Bangui, the capital of the Central African Republic, a city teeming with life and energy. It served as a hub for international flights, attracting diverse cultures and languages. Our final destination lay in Kagabandoro, a remote town located an hour and a half away by air. While Bangui airport was modest in size, its beauty and orderliness were captivating. Stringent rules prohibited photography, depriving us of the opportunity to capture countless images of that exotic place.

As I made my way to the door, stifling a yawn and carrying a hefty backpack, a gust of wind infused with dust and sunlight greeted me. A group of soldiers from the Bangladesh contingent awaited us outside, offering warm greetings and camaraderie. It felt as though I had not left my country. In reality, we were there to replace them, as per the government's decree, as they would soon be returning home while we took their place for the next year. On November 10, 2017, at 12 noon local time, the restless girl, born and raised in a small town in Bangladesh, set foot in Africa for the first time. It was a continent as vast as it was diverse, as rich as it was impoverished, and as beautiful as it was harsh. She stood ready to confront the challenges and embrace the opportunities that awaited her in this land of striking contrasts.

I blinked my eyes futilely, unable to discern whether it was dawn or dusk. The pale whitish sky pierced through the window, deepening the mystery that engulfed me. In the distance, the piercing cry of a rooster jolted me from my daze, bringing back memories of waking up to the rooster's call at my grandfather's home. Glancing at my mobile phone, I checked the time: 4:50 in the morning. I was never an early riser, certainly not this early. Puzzling it out in my mind, I realized the cause of my confusion. Bangladesh was five hours ahead of Central Africa, making it 9:50 in the morning back home. The early bird may catch the worm, but not in this case. While my biological clock had awakened on time, my natural clock had fallen behind. There was nothing to be done but restlessly wait in bed until the sun cast its natural light. As soon as daylight broke, I sprang into action, preparing myself with all my belongings. I had arrived in Bangui from Bangladesh just the previous day, and today, or rather tomorrow, I was set to head to Kagabandoro.

After battling jet lag and succumbing to a deep slumber, I woke up feeling as though I barely had any time to contemplate whether I was in my homeland or a foreign land. However, as I sat down for breakfast, a pang of nostalgia struck my heart, a hint of homesickness perhaps. I wasn't usually one to long for home so soon, but the craving for homemade food had crept in. Unfortunately, the offerings before me hardly resembled the comforts of home. A doughy round item, something between bread and paratha, sat before me, accompanied by poorly boiled frozen vegetables with an overwhelming amount of peas and an abysmal taste. A fried egg was served as well. "There is no smoke without fire," the proverb echoed in my mind as I contemplated the debate over plastic eggs. But could this be a plastic egg? After consuming it, no one would harbor any doubts about its artificial nature. Never could I have imagined that a fried egg could lack color, aroma, taste, and flavor. With a bitter taste lingering in my mouth, I got into the car, feeling stuffy-nosed, listless, and utterly bland.

The airport was a mere five-minute drive from the Bangladesh Special Forces camp. I attempted to steal glances out of the car window, trying to absorb the beauty of arid Africa as much as possible. The entire road leading to the airport was unpaved, and the red soil dust billowed up, offering glimpses of Africa's untamed wilderness stretching far and wide. There were only a few people on the road, each going about their own business. No crowds like back home, no roadside drains, no clutter signaling drain repairs or excavations. No tea stalls, no peddlers. Truly, there was nothing reminiscent of Bangladesh. Ah, Bangladesh! Nostalgia surged within me once again.

A security officer intercepted our car at the airport entrance, rambling on in the local French language, of which I understood nothing. Our Bangladeshi soldier retorted in Bangla, bearing a strong Chapainawabganj accent, asserting our desire to proceed inside. Conversations flowed in their respective tongues, leaving a sense that no one truly understood one another, yet they allowed us to pass regardless. Eventually, our path was cleared, and it confirmed my belief that gestures were the primal language of the world. With a congenial posture, I positioned myself before the immigration counter, surveying the surroundings. The people I encountered were bizarrely diverse, with peculiar attire and even more eccentric hairstyles.

Our interpreter was bustling about, desperately trying to convey our purpose and intentions to those who didn't understand English, while we struggled with our limited knowledge of French. Communication relied solely on our interpreter's skills. He had no respite and was constantly on the move. Suddenly, he vanished like the wind, leaving me with a stack of passports in hand, instructing me to handle the immigration counter. Standing before me was a tall and broad African officer speaking rapidly in French after inspecting my passport. I chuckled, unable to comprehend his words. With my lack of knowledge in French, I resorted to friendly gestures as my only means of communication. However, the African gentleman seemed uninterested and unpersuaded by my positive and cheerful demeanor. He signaled and articulated something else, leaving me no choice but to pay closer attention.

"Aaato..."

I hesitated, unsure of how to respond. He continued walking towards the restricted area of their office. "Aato" sounded like a Bengali word. It closely resembled "Asho," a word children sometimes mutter as "Aato" in Bengali, meaning "come." Hearing a word from my native tongue was like a balm to my ears, and it brought a sigh of relief. I trailed behind him, but as he turned around, he uttered, "Aatooooo," his face displaying a mix of wonderment and surprise. I quickened my pace to keep up with him. He moved a little further, pivoted again, and looked at me with a hint of astonishment, saying, "Aaato...." His voice carried a tinge of irritation. I was puzzled. He was beckoning me, and I was obediently following. I glanced at the African officer and then at the gate of the terminal. What was the hitch? It felt as if I had become trapped in an endless sea. After covering some more distance, he gently reprimanded me, "Aaatoo." I felt like a fish out of water.

Just at that moment, the interpreter appeared out of nowhere like a knight in shining armor. He approached and inquired, "Sir, where are you headed?"

"This fellow said, 'Aaato...,' so I'm doing what he asked. I am following him. But now he's getting angry again; I don't understand. Can you find out what's going on?" I asked the interpreter, seeking clarification.

"Sir, what are you doing? 'Aaato' means 'hold on' or 'wait.' He was asking you to wait there. You are trespassing into their restricted area now. Our helicopter isn't even that way, sir," the interpreter tried to explain, highlighting my mistaken path.

"Oh, I see." Finally, I grasped why the officer's eyes had blazed with frustration. A sheepish grin spread across my face as I retraced my steps, rectifying my error.

The clock was ticking, and the helicopter was scheduled to take off at eight in the morning. Time was running out, and we had to act swiftly. As the first light of day broke, the helicopter ascended into the sky, its propellers generating a deafening roar. Despite the noise, some of us managed to doze off, finding solace in uninterrupted slumber that not even the helicopter's propellers could disturb. Those who remained awake yawned constantly. Exhausted and restless, we had already endured a thirteen-hour flight and now an additional seemingly endless hour and a half in the air, longing for the journey to conclude.

I peered down repeatedly, scanning the landscape beneath me. Endless stretches of trees and forests greeted my gaze, with no signs of human habitation in sight. I marveled at the unfamiliar terrain I had ventured into, surprised by the corner of the world I had arrived in.

Soon, the lush scenery transformed into a desolate, barren expanse resembling a wasteland. The trees and bushes vanished, leaving behind only thorny shrubs. Suddenly, tin shacks emerged into view, signaled by the Pakistani aircrew as our destination for the next year. These shacks, known as CORIMECS, would be our living quarters for the foreseeable future. The moment of touchdown approached swiftly.

After landing, we made our way on foot to the officers' mess. It resembled a modest army camp, reminiscent of those back in Bangladesh. Rows of CORIMEC sheds stood closely together, providing our simple accommodations.

The CORIMEC sheds, constructed mostly of tin, appeared decent from the outside but were far from cozy. They transformed into scorching furnaces under the midday sun and turned into chilly freezers during the biting cold of the night. Each night reminded us, like the famous Ned Stark warned, that "Winter is coming," even in the desert. As the saying goes, appearances can be deceptive.

However, the real challenge came with the beds. Oh yes, wrestling with the beds presented a greater ordeal. They were wooden contraptions that would tilt slightly to one side with the slightest movement, producing a resounding thump that made it seem as if the ceiling might collapse.

Four female officers, including myself, were assigned to a partitioned CORIMEC. Each room contained a few flimsy pieces of furniture that I managed to squeeze and arrange to my liking before finally leaping onto the bed.

I had approached Major Salma, our anesthesiologist, with a playful request the night before. "Salma, sir, I think it would be best if nobody wakes me up for PT tomorrow morning. I'm hoping to skip it and feign illness due to jet lag. The jet lag has already taken its toll on me, and I need a few days to recover." However, before I could finish my sentence, Colonel Amina's voice boomed from the neighboring CORIMEC.

"Sharmin?" she called out, addressing me.

"Yes, sir!" I promptly replied.

"You shall attend PT in the morning. I expect you to be there on time. After all, you're my adjutant!" Her tone was commanding yet gentle this time.

"Yes, sir!" I had no choice but to agree and repeat my response.

She added, "Whether others go or not, the adjutant must be present. Do you understand?"

"Yes, sir!"

"And remind me in the morning – I need to assign someone to exchange some dollars for the local currency, Shefa (the currency of Central Africa)."

"Yes, sir," I replied for the final time before succumbing to the depths of slumber.

"Okay, fine! Go to sleep then!" Colonel Amina concluded.

With a greeting of "Assalamu Alaikum, sir," the conversation came to an end that night. As they say, honesty is the best policy, and I learned that the hard way on that day.

The tin walls of the CORIMEC were as thin as paper, making everything easily audible, even the faintest whisper. As an additional duty, I had taken on the role of an adjutant. Instead of receiving orders over the phone as before, I now received covert mandates and directions. Blasting loud music on our speakers became out of the question (having to work as an adjutant for the time being as an additional duty, where earlier orders came only by telephone, now orders began to come from the other side of the thin tin walls of this CORIMEC without a telephone. Thus, we stopped playing loud music on the speakers). We would stifle our giggles and laughter, trying to contain our amusement before it burst out. There was no more freedom to relax and flash a smile upon returning from the office because every sound was detectable from the neighboring CORIMEC. What a predicament!

Communicating on our mobile phones was even more challenging. The Wi-Fi was weak, and without raising our voices, nothing could be discerned on the other end. But if I raised my voice even slightly, the entire Bangladesh contingent would overhear without even trying to eavesdrop. What a catastrophe! As a result, we all unwittingly uncovered each other's hidden secrets within a matter of days. We soon discovered that even the most serious-faced officers of the Bangladesh army would use endearing terms like "Babu" and "Baby" during their private moments. We, the junior officers, were amazed.

Our respected seniors, too, used those words! Oh my! Loose lips sink ships, and that's precisely what was happening. It turned out that the romantic spirit dwelled within every man's heart, and those words were not uttered in vain. These amusing incidents became a regular occurrence in the dining room. It seemed that everyone had mastered the art of teasing and pulling each other's leg. Major Moazzem, our surgical specialist, particularly excelled in playful banter with remarkable candor. The members of BANMED 3, who had stayed behind to hand over their responsibilities to us, did their part perfectly, and now it was our turn. As the new BANMED 4 team, we embraced our work with great enthusiasm and motivation. Still waters run deep, and perhaps we were about to dive even deeper into that realm.

I was preparing my instruments to extract a tooth from a Pakistani soldier when I was interrupted by a question. The inquiry came from Senior Warrant Officer Farooq, my assistant. He handled various tasks like filling, scaling, and maintaining the operating room. He was more skilled at jumping around than demonstrating loyalty. His robust health seemed excessive for his own good, but it didn't dampen his flamboyant nature. He would eagerly dive into any task, almost like curiosity killing the cat.

After we relocated to Africa, we noticed a remarkable change in him. He underwent a positive transformation. He started praying regularly, and the beautiful sound of the Azan from the nearby Pakistani contingent's mosque might have led him there, as he donned his traditional Punjabi outfit for the Maghrib prayer. At first, we couldn't understand why he wouldn't complete his prayers in our own mosque but rather went to the Pakistani mosque. However, due to the playful banter and his attire, including his thick beard, he soon earned the nickname 'Darwish Baba' among the soldiers.

"Farooq, sir, what did you tell the patient?" I asked.

"Yes, sir. I wanted to know if his anesthesia was properly administered," he replied. I was momentarily taken aback by his response. It had been nearly three weeks since we had arrived in Africa, and we had already started seeing patients after overcoming the challenges of the journey. And that's what led us into trouble!

When in Rome, do as the Romans do.

We found ourselves surrounded by one Nigerian, one Rwandan and three Pakistani units. Naturally, most of our patients were Pakistanis. The trouble was, they hardly spoke or understood English. I had picked up some Urdu from watching Hindi movies, but I couldn't speak it fluently to save my life. Every time I needed to communicate with a patient, I would break into a cold sweat. I tried to converse with them using a mix of broken English, Bengali, and Hindi, but I couldn't grasp the connection between anesthesia and drunkenness. I asked, "Farooq Sahib, what could possibly link anesthesia and drunkenness?"

"Why, you said it yourself the other day when you switched from English to Urdu and asked if he felt heavy, implying that it was burdensome, didn't you?" He was reiterating the same sentiment now.

I responded, "That day, he used the Urdu phrase 'Nasha ho gayi keya...'"

"That translates to getting high in Urdu; how did you interpret it as drunkenness?" (In Urdu, the same word can be used to express love or fascination.)

"It's essentially the same thing. When you get high, you become intoxicated. If you get intoxicated, you'll be drunk. It's all interconnected, isn't it, sir?" he insisted.

Deciding to let go of the discussion, I immersed myself in my work. Extracting the patient's teeth required precision and speed. Firstly, the patient struggled to follow my instructions due to the language barrier. Secondly, Farooq Sahib was a constant nuisance.

"Ei je rogi, ha karenge. Ha karenge to ha karenge... Continue ha karenge. Ebar thuthu felenge, katha nehi bolenge, only dorod lage to bolenge..." Farooq Sahib communicated with the patient using a mix of Urdu and Bengali pidgin. (Keep your mouth wide open, now spit, don't speak, only tell if it hurts.)

"Farooq Sahib, what is dorod?" I asked. (The word should have been pronounced as 'DARD.')

"Oh, Allah! How can you not know dorod means 'pain'! Yes, in Hindi, dard means pain." He exclaimed, emphasizing the mispronunciation of dorod. My laughter disrupted my work as I chuckled intermittently. I couldn't help but laugh. Nevertheless, I continued to extract his teeth. Suddenly, a dreadful sound reached my ears, causing me to rush outside, where I witnessed a plume of smoke rising from one of the Pakistani units. It didn't take long to understand the cause—a militant faction had greeted us in Africa with a mortar attack. What a splendid way for Kagabandaro inhabitants to welcome their guests! I pondered as I contemplated returning to the room; however, three more Pakistani patients awaited me, leaving me no choice but to attend to them.

From the tin abode of CORIMEC, I could clearly hear Farooq Sahib's voice, "Ei j patient, ajke scaling nehi karenge, pore korenge, tomar unite bomb porsenge, giye dekhenge ar doya korenge." (Dear patients, we will not perform scaling today. We will do it later. Your unit is under bombardment. I will go there, inspect, and pray for you guys.)

I sighed deeply in my mind and prayed, placing my hands on the door handle, "Ya, Allah, spare me, spare us from Farooq Sahib's Urdu, please." I had no opportunity to resume my work as I entered the room. Before I could even start, I was forced to crouch on the floor amidst the sound of bullets. Then came a relentless barrage of gunfire that lasted for a considerable duration. The contingent commander instructed us over the walkie-talkie to don bulletproof jackets and helmets until the situation was under control. Every time I saw my 14-kg bulletproof vest, I felt uncomfortable. What a merciless twist of fate; I had to wear that jacket that day. Allah, how was all of this written in our destiny? I tried to recall the date as I took cover and sat down, blending into the surroundings; it was December 9, 2017.

Was that day a grand occasion or any other remarkable event? I highly doubted it; however, had I perished on that day, it might have been a memorable one for my parents—the day they lost their beloved child. Sometimes, I couldn't help but wonder. If I had died, if I had been killed that day, would anyone shed tears for me in my homeland? Would anyone write a heartfelt eulogy in my honor? Would anyone exclaim, "Oh, such a young woman sacrificed her life for the country, alas!" Or would I have disappeared with a brief headline reading 'Soldier's Demise in Central Africa' tucked away on a trivial page of the newspaper?

As mid-December approached, Africa began to grow colder. It wasn't a bone-chilling cold, but rather a pleasant one that called for a sweater in the mornings and evenings. Winter carried an inherent gloominess, a unique melancholy that being in a foreign land seemed to amplify. On Facebook, I saw people getting married in Bangladesh. A deep sigh escaped from within me. Ah! I was far away while the festivities of my homeland were in full swing. People roamed in groups, enjoying themselves, while my mood turned sour. I wondered how much fun I would have had in Bangladesh during winter! But regardless of my state of mind, when one is away from their country, the affection for it intensifies to such an extent that it becomes difficult to grasp if one hasn't experienced it. Bengali TV channels played throughout the day in the common area. By then, we had listened to almost all the songs on Gaan Bangla channel. A song by an obscure band, 'Gausul Azam Maiz Bhandar School khuilase' (The School Has Opened), had become popular among the listeners. As this song played, even the serious-faced officers could be seen humming and nodding along. It seemed that everyone knew the song.

In the evenings, our Bengali movie sessions would commence. At that time, it would be late at night in Bangladesh. Why Bengali channels chose to air such melodious songs during those late hours, only Allah knew. However, these Bengali movies soon became a tremendous source of entertainment for us. Despite the exaggerated acting, mismatched dialogue, and unrealistic settings, we thoroughly enjoyed watching Bengali films. Pori Moni, Bobby, and Bubly had managed to captivate some of our hearts, an aspect I would never have discovered if I hadn't started watching Bengali movies.

Meanwhile, our pathologist, Major Fahad Sir, had established himself as a true connoisseur of Bengali cinema. If we ever failed to recognize any new actors or actresses from Dhallywood, all we had to do was ask him once. Sir would eagerly share all the details, and we absorbed them with great fascination.

Not every day unfolded with tranquility. There were moments when clashes erupted at the most inconvenient times, and critical patients arrived with wounds from bullets, large scissors, or swords. Everyone rushed to witness the scene. Sometimes, the echoes of rocket launchers, mortars, or rifles pierced the air, serving as a stark reminder that we were far from the familiar shores of Bangladesh. We were on a mission, a peacekeeping mission. During our time there, we managed to pick up some French. Our interpreter, Habib, painstakingly taught us a few words and phrases. For instance, "Je m'appelle Sharmin" meant "My name is Sharmin," "Camo ça va" meant "How are you," and "saba bein" meant "I am fine." We often playfully referred to our poor Habib as "Je m'appelle Habib," relinquishing his Bengali name in favor of a French one. Despite our occasional jests and lightheartedness about the French language, it served as our sole means of communication with the locals. We held respect for all languages but learning French within a limited timeframe proved challenging. Nevertheless, we had to acquire at least a functional understanding to navigate our daily lives there. God forbid, if ever faced with the barrel of a gun, I would have had to utter at least this much: "Je suis casque bleu, ne me tuez pas" meaning, "I am a peacekeeper, don't kill me."

While many compatriots back home frowned, scoffed, or sneered at the mere mention of the army or equated our mission with monetary gains, there could have been a valiant army soldier who prematurely sacrificed their life in the pursuit of peace on a peacekeeping mission. Some of us grappled with the haunting terror of mortality in the war zone. And there were innocent souls like myself who slumbered with a bulletproof vest and helmet, serenaded by the symphony of gunfire. Who else cherished our memories? Who would have spared a thought for us in our homeland at that very moment? Yet, nestled in a remote corner of Africa, thousands of miles away from Bangladesh, I felt an overwhelming surge of pride to proclaim, "Josui Kaske Bera, Josui Kaske Bera from the Bangladesh Army and proud to be so." For I understood that the army was not merely a profession, but a way of life—noble yet perilous. It wasn't a bed of roses, but a path paved with thorns, as arduous as it was rewarding. And fortune favored the brave.

SERGEANT MARVIN

ADDITIONAL DEPUTY INSPECTOR GENERAL MD. FEROZ AL MOZAHID KHAN (UNITED NATIONS MISSION IN LIBERIA - UNMIL)

The first rays of the sun had not yet kissed the earth when an uproar broke out beyond the confines of our lodgings. It wasn't even morning yet! The sound of the commotion, coupled with the piercing hullabaloo, had roused me from my slumber, engulfing me with an ominous feeling of dread. With bated breath, I tried to comprehend the gravity of the situation, while my companion Mazhar also stirred from his peaceful repose. Any sense of calm eluded us as we stepped outside the home, as the cacophony of human voices reverberated through the vicinity. Sound sleep in a foreign land would be rare – we knew this beforehand. Could it be possible that the rebel insurgents, equipped with their lethal armaments, had come down the street? The previous day had been marred by mass shootings and a series of violent clashes between the insurgent MODEL (Movement of Democracy in Liberia) rebels and the common folk at the 'Buchanan'.

There were many armed rebel groups in Monrovia, so there were valid reasons to be afraid. Finally, what I learned from the landlord was that a thief had crept into a house next door. The custom there was to beat a thief to death if he got caught, with no involvement of the law or court. Who would move to court with such an issue in Liberia at this time? Who would ring the bell? Everybody used to take a share in the diabolic pleasure of beating a thief. Nowadays, cruelty here was a bit more unrestrained than in other regions of the globe. However, since the United Nations Peacekeeping Mission had commenced there, the public had been urged to not take the law into their own hands. Larceny had always been there. In every region, theft is a known term. But here, theft was happening at a large scale and at a burgeoning rate. The mission officials had become targets of thieves as passports and money were being frequently stolen.

Thieves are inherently reckless in nature. Nevertheless, in such impotent socio-economic conditions, preventing crimes like theft was no doubt a daunting task.

There was no arrangement or guideline, and to be honest the conditions had led to a situation and setting where people did not know how they would survive, how they would be able to make ends meet, or how they would feed their children. My notion was that such events would increase in number despite the danger of direct encounters with death. Morality had succumbed to hunger, and the tumultuous commotion eventually quieted down and dissipated into a hush.

Slowly, dawn broke in Monrovia, the capital of Liberia. The day marked the end of my six-day vacation. I reclined in indolence. On every Sunday morning, I heard the sacred chime of the church, and on this Sunday the bell resounded from every direction. Sunday signifies a day of rest, a holiday. The road was empty, deserted like the Sahara. The shops were closed. Only the church's ding-dong sound kept reverberating. Monrovia abounded with churches whose names I did not know. The maiden who assisted us with our domestic chores, Lily, went to attend church every Sabbath as she used to take leave early that day from work at the place we resided.

"Does visiting the church require money, Lily?" I asked Lily curiously one day.

"Whoever has whatever ability, gives it," Lily replied.

Before reciting the sermon on Friday's Jumma prayer, a donation box is sent out and is left open to pass from one to another until the end of the lines of ummahs who came to pray. That's how it works in our country. That money gets spent on the mosque's maintenance or development or on some religious initiative.

The people there lived in fear of God and invoked His name often, as the belief in God's justice was prevalent in them. Yet, they did not follow the common norms of society. There were no such established societal norms, such as the obligation of marriage to have children. They used to have children without getting married, as the law there allowed them to wed whomever they wanted except those prohibited from marrying in their societal norms, and they could marry as many girls as they wished. In brief, marriage was not mandatory to have children, and one could marry a number of girls as per their ability.

The old men were more prone to taking advantage of this and marrying young girls who had no choice. Poverty only exacerbated this system. I had heard that the former president, Charles Taylor, was a notorious example of this. That's why everything was going on! He had one wife and twenty mistresses (girlfriends) and had fathered dozens of children with them. When he lost his power, he fled to Nigeria with his women and mistresses and went into hiding.

Near our abode, there was a wall with a cartoon on it. It showed a schoolgirl standing half-naked in front of her teacher, who was leering at her. The cartoon vividly portrayed how the teacher was only there to enjoy the girl and indicated that such problems existed there, with the artist or those beneath the drawing wanting to spread the message. Yes, the painting had a message, as below it 'Stop abusing girls' was written in English.

The prolonged war had wrought a humanitarian crisis of epic proportions, causing established societal norms to crumble, shattering time-honored social mores, and leaving in its wake a society in disarray. The once-venerated tradition of marriage, founded on the noble pursuit of procreation, had all but lost its relevance. In Liberia, women had borne the brunt of the conflict, with their children being raised in impoverished circumstances by grandparents with little means to provide for them. The grandparents too were poor. In a cruel twist of fate, no one seemed to be willing to shoulder the responsibility for the welfare of these hapless children, leaving their mothers feeling powerless and desperate. The helpless mothers could do nothing more than see their children grow up that way. As the world around them carried on, the plight of these Liberian families remained largely unrecognized and misunderstood. Those who are used to the "live together" culture in the developed world could never fully comprehend the severity of the condition there, especially the difficult circumstances faced by families. Day by day, the situation grew ever direr, to the point where the very institution of marriage was in danger of collapsing. I was afraid that the young would grow up without knowledge of their fathers, and a nation of fatherless children would emerge. One day, they would become fathers, too! Had this once-proud society been brought to its knees, robbed of its very heritage and identity?

"Dear Father," I queried of a learned church elder following Sunday's church service, "Can a society truly be considered respectable if it lacks the presence of a strong paternal identity?" I noted the congregation's attentiveness to his words and continued, "It seems that the public readily accepts and embraces the teachings and proclamations that come from figures of authority such as yourself."

"We do all to enlighten and make all aware of the canon, religious edict, or decrees, but whether they act upon or adhere to the principles or not depends on the individual's own moral compass," replied the father thoughtfully.

The police force in Liberia had been introduced after the end of the war, and the recruitment went on under the supervision of the police commissioner of the United Nations Mission in Liberia.

That police force commenced with the motto to work to uphold democratic values and to preserve the honor of the nation as well as to respect the legitimate rights and desires of the general public and be representative of all groups. To this end, rigorous campaigns were underway to ensure the participation of all communities and the inclusion of only those fit, educated, decent, deserving, and worthy of the position. Eligible candidates were encouraged to join the police force.

In Liberia, there was a Jamaican police sergeant named Marvin who was a part of my team. Marvin had a dark complexion, was tall and lean, and exuded youthful energy. Her youthfulness blossomed in the curves of her body, and she had a sprightly, buoyant, naive, lively, and easygoing personality. She was provided with an official apartment in Kingston, the capital of Jamaica.

With the purpose of promoting police recruitment, we visited schools and colleges often and occasionally universities. We talked with both boys and girls. Learners from all levels, regardless of whether it was a school, college, or university, had children to take care of. Most of them came from an impoverished background, yet their age helped them keep their youthfulness and exuberance intact.

Despite the challenging tenure and ongoing issues to deal with in their country, those young people had never ceased to display the joy of juvenility, and this always overshadowed the challenges they faced. Witnessing their vibrant countenance of juvenility was an inspiring experience, as they refused to let their circumstances dampen their enthusiasm for life. Dimming all the negativity, they moved ahead and hummed the melodies of life.

One day, I found Marvin broody and crestfallen. She looked upset.

“Marvin, are you feeling down?” I asked, leaning back in my office chair.

“Yes, I feel awful,” she replied.

“If you don’t mind my asking, why are you feeling so upset?”

“It’s because of my little son. I haven’t seen him for a long time.”

“Take some time off. Go home right away. His dad must be there. Your son is definitely doing well,” I said.

“Actually, he doesn’t have his father. His father is not with him, I mean. He is in America right now as an immigrant! Other than that, we were not married, so his dad never felt that responsibility for his son. In the hopes of a better life, he settled in America,” Marvin explained.

“How many siblings do you have?” I asked, breaking the taboo of personal subjects. It is not polite to ask about personal matters, but at that moment, I seemed not to care about that fact as much as I desired for Marvin to come out of her gloom.

“We are three sisters. No brothers,” she answered.

“Why didn’t you get married?”

“I am smart,” Marvin drolled.

“But your sisters are married?”

“Yes, but they all live in the village.”

“So, your sisters are not as smart as you, huh? Are they?”

Marvin felt a deep pain inside that appeared visible in her eyes. She said something to herself and then asked me, “Tell me, who will marry me now? I am already 34,” her heart-wrenching query echoed. “All men now tend to marry young girls. Nowadays, it seems that young maidens are the only ones who are deemed suitable for matrimony,” she added.

“In the blissful days of our relationship, everything seemed fine. I was blind in love and didn’t care that much about marriage. The notion of marriage never occurred to me. However, now I know that tying the knot is a matter of great reverence and responsibility. Living with the family is a lot of fun, you know; that joy is unparalleled. The contentment of my sisters, who are all happily married, is a testament to this fact. My sisters are very happy. My first love left me for America while I had another child with my second partner, and he also left me unmarried. Yes, I am that smart, and, therefore, I have such a life now."

Marvin burst into unrestrained tears upon saying those words. No words could comfort her. I, too, could not find the words to console her, even for a moment.

'Bailey' lived exactly opposite to my flat in Monrovia. He had his wife and children but lived alone as he often slept with his girlfriends in that flat and kept his wife and children away in a different flat in the same city. Bailey was educated and worked for the 'World Food Program'. To him, too, the term 'family' was not much more than just a word, as it signified nothing for him.

The importance of education cannot be understated, for it serves as the bedrock upon which civilization, humanity, and peace are established in society. Every society requires individuals who possess a strong sense of conscience and a deep commitment to the well-being of others. Such individuals are needed now more than ever, particularly in Liberia, where the need for conscientiousness seemed acute.

If only I could pull together my scattered thoughts like a gorgeous string of flower beads!

Days abroad were filled with longing, while at other times they were filled with memories. I wondered what Anindya was doing then. And his mother! My wife! Anindya was probably playing with his small hands and feet. Perhaps Anindya had forgotten his father. At the end of the daily chores, during the evening respite, 'Tani' (my wife) may have been thinking of me and all about the delightful merriment and gripes of my small world and my family! It is always the wife and children who illuminate the atrium of an expatriate.

In the languorous hours of our holiday afternoon, Belal and Rahman took me to Silver Beach, not far from Monrovia city. As we strolled along the shoreline, the 'Kino' boats were upended and scattered about as if they had been uprooted by a forceful wave but were kept adorned, maintaining a beautiful line. It appeared as though the fishmongers had vanished, leaving us the tranquil beach all to ourselves. The salty ocean breeze tousled our hair, imbuing us with a sense of serenity. As we ambled across the warm sand, we came upon an abode ensconced amidst the coconut groves. A young boy came out of the dwelling and approached us with a friendly grin.

Knowing that we had come to buy fish, Suho assured us in his Afrikaans English, "Today, the fishermen will all be out at sea, and tomorrow morning, they will return to the shore with the fish they have caught throughout the night. If you can come tomorrow morning, I will take you there with me. Don't worry."

Knowing that we had come to buy fish, Suho assured us in his Afrikaans English, "Today, the fishermen will all be out at sea, and tomorrow morning, they will return to the shore with the fish they have caught throughout the night. If you can come tomorrow morning, I will take you there with me. Don't worry."

As the sun sets over the deep waters of the sea, we returned to our flat in Monrovia in the evening. I was fascinated by the sound of the waves, the crashing white waves, the birds soaring through the sky, and the tiny crabs scurrying across the sand. The wind rustled the bushes and swayed the coconut leaves, just as I swayed. I spent a lot of time on the shores of the Atlantic.

As Rahman gazed upon the "Kino's", he couldn't help but marvel at the ancient and timeless beauty that emanated from them. "Look at the 'Kino's. Africa seems to be a thousand years old," Rahman mused aloud.

"Why is that so?" I inquired curiously.

"Just take a moment to contemplate the remarkable craftsmanship of the locals who fashioned boats out of a single tree. Such a feat could only have been accomplished centuries ago, yet it still endures to this day in the same fashion. Can you not feel the weight of time that has passed, embedded within each of these Kino's?"

That's right. The whole tree! Yes, if the tree from which the Kino was crafted had been slightly bent, the vessel would have inherited that same curvature.

The Kinos seemed to never sink. The highest, those that could not get capsized even against the strongest stream, these water vessels with a deck full of fish could turn upside down yet remain afloat. It was possible that kinos were more essential than sampans, small boats, or schooners. Who could know for sure!

The local fishermen were known to catch shrimp on Saturdays. I didn't see shrimp in the market on any other days of the week. The beef was scarce, with availability limited to two days a week; however, fish remained a daily staple. Despite this, it didn't seem like many fish were being caught lately, or perhaps there were simply no buyers at a good rate. As a result, the fishermen resorted to catching and counting whatever fish they could to make ends meet.

The day before, I had heard Nelson Mandela, the former statesman of South Africa, on the radio. He would not contest the presidential election for a second term. He would voluntarily relinquish power. He was known as "Madiba" in his own country and by then all over the world. He had spent 27 years in jail. His wife, Winnie Mandela, who did so much for him, also had to leave him in the end. At that time, he was campaigning for the African National Congress (ANC) party, supporting his successor, Thabo Mbeki, from his stronghold in Soweto. He urged all to vote for the ANC with his elevated voice. This immensely influential figure devoted his life to the world. I felt extremely happy while listening to his voice, as hearing the voice of one of the greatest minds of our time was a great pleasure.

We mourn for 'Madiba', as he is with us no more.

Throughout his life, he was unable to devote his attention to his family, as he spent too much of his existence confined within the walls of a prison cell. Upon his eventual release, he demonstrated a remarkable magnanimity in sharing power with those who had once imprisoned him, seeking to establish a lasting legacy of black equality within his country. Despite the many years he spent incarcerated, he remained attuned to the great events that shaped the world around him, never hesitating to offer his opinion in a manner that was both unflinching and fearless. When the prospect of a US invasion of Iraq loomed large, he did not hesitate to condemn it.

Liberia was a country in West Africa that had never been governed by any particular method or policy. Honestly speaking, the people were not well-prepared enough to enjoy the resources they owned, of which there was no shortage. Farsightedness and determination were attributes that could have helped them take possession in their own palm, yet no leader in this country seemed to possess these qualities. The UN missions had been there before. Perhaps our mission was the end of the UN endeavor in Liberia. If the wheel of fate could not have turned around now, at least I personally foresaw no light of hope in the near future for the people of that country, which would slowly sink into the abyss.

Flying High: The Mission to Alimba

Wing Commander Raqib Mahmud
(United Nations Mission in the Democratic Republic of the Congo - MONUSCO)

Air Force peacekeepers rarely get the chance to execute their duties in a camp area, especially without any specific reason. I was no exception. However, whenever the opportunity arose to set up a camp for a week in the militia-controlled area of 'Alimba' in DR Congo and serve as a military air operator on a UN mission, I did not miss it.

The story of Air Force peacekeepers in the UN peacekeeping mission in DR Congo is one of courage, sacrifice, and dedication to restoring peace in a troubled region. For more than six decades, these peacekeepers have faced numerous challenges and dangers, yet they have persevered in their mission to protect civilians, neutralize armed groups, and support political processes.

Their journey began with the United Nations Operation in the Congo (ONUC), the first UN peacekeeping mission to involve air combat. Its mandate was to prevent a civil war, and to achieve this, ONUC had to use force. But the path to peace was fraught with danger and controversy, from the assassination of Prime Minister Patrice Lumumba to the secession of Katanga province, from the interventions by foreign mercenaries to the involvement of Cold War powers.

Despite the obstacles, ONUC achieved some stability before ending in June 1964. Still, the work was far from over. Since then, other UN peacekeeping missions have been deployed, including MONUC, MONUSCO, and MONUA. Throughout these missions, Air Force peacekeepers have played a vital role, providing air support, transportation, reconnaissance, and medical evacuation to other peacekeepers and civilians.

These peacekeepers faced all kinds of challenges, from dangerous armed groups to poor infrastructure and complex terrain. But they didn't let that stop them from achieving incredible success in their mission. They restored stability in some of the most troubled areas of the country, neutralized armed groups, and helped deliver much-needed aid to millions of people affected by conflict, displacement, hunger, and disease.

Of course, it wasn't easy. The Bangladeshi Air Force peacekeepers faced attacks from armed groups and had to work with limited resources. But they persevered, and their efforts paid off. They received UN medals for their outstanding performance and professionalism, and their all-female Formed Police Unit provided security at Monusco Airport in Kinshasa.

These peacekeepers weren't just focused on maintaining security, either. They engaged in community policing, raised awareness about human rights, and promoted gender equality. They were shining examples of bravery, resilience, and commitment to peace, and their sacrifices will never be forgotten.

The Bangladesh aviation contingent played a crucial role in this mission, with six MI series helicopters and one C-130 cargo aircraft. Alongside peacekeepers from other countries, they were responsible for maintaining peace throughout the entire airspace of DR Congo. Thanks to their efforts, the people of DR Congo were able to find stability and hope and to progress in the midst of conflict and suffering.

There was much discussion and speculation in the mission area on the concept of the 'Rapid Deployable Battalion', which was stirring up controversy. A quick reaction force (QRF) was always ready to be deployed anywhere at any moment, regardless of the area type, whether remote or outlying.

Luckily, there had never been any occasion that required the deployment of the QRF. But it is worth noting that even if such a situation had arisen, how would everything have worked out since such a situational context had never been practised? Thus, it was decided that the QRF would fly on a chopper to an outlying area called 'Alimba' in DR Congo for a week-long tenure and then return to base on a chopper.

I was tasked with the responsibility of planning the heli-lifting operation of the exercise since I was the military air operator of the mission. The calculation I had made was that it would take at least 24 sorties of 14 support helicopters to transport a company of peacekeepers to Alimba with a minimum of one week's supplies. But there was no place in Bunya that could accommodate such a number of helicopters at once. Moreover, there were thousands of other inconveniences, such as fixing refuelling plans and air-marking maintenance areas.

Nonetheless, one of the major considerations we had in our plan was to utilize the available resources in an excellent manner to bring out the best output so that we could defend both ends as effectively as possible. The plan included six Bangladeshi MI series choppers, three Russian aviation choppers, one Uruguayan Bell 212, and two Ukrainian MI series choppers. Two Ukrainian attack helicopters were also part of the plan.

The shrill militia groups were spread out, covering DR Congo at every end. The initial plan was that when the support helicopters carried the provisions, the attack helicopters would provide security during flight movements, takeoff, and landing. Despite not having had a moment's slumber, I planned where those helicopters would land, who would board, what supplies would be loaded, which chopper would move first, what the sequence would look like after each sortie, where the choppers would refuel, and which one would return to Alimba first after forays.

The last stage of planning included the timeframe and territorial array in which any helicopter would go on maintenance operations (for seven days after being deployed in Alimba). Above all, these were the blueprints of our operational schemes, flight schedules, and security arrangements for that seven-day exercise.

On the morning of the scheduled day, the QRF consisting of Guatemalan Special Forces and Uruguayan commando forces boarded the aircraft with meager provisions and arms, and we took off to inaugurate the exercise operation. For the first time in the mission's history, the deputy force commander accompanied us. Two Air Force peacekeepers were tasked with directing the heli-lifting operation, with one commanding the flights and assisting us from Bunia. The safe landing and takeoff of helicopters in the operation area of Alimba were solely my responsibility. By the grace of God, to everyone's surprise, we managed to safely land the QRF at Alimba within two hours.

After landing, the first challenge I faced was regarding the use of the toilet. The toilet consisted of a big hole and a blue-colored polythene awning around the hole, leading me to feel that the less the consumption of food, the better. A friendly Moroccan colleague who had an immense interest in the nitty-gritty details of aviation accompanied me, and we both decided to maintain a different diet than usual, given the condition of the toilet.

While we were discussing the action plan for the following day, night patrol (area disbursement), and duty delegation, food was served in the form of MRI (Meal Ready to Eat), which was half-cooked food inside a polythene bag. I was not good at heating the food, and my awkward look made the Uruguayan commander laugh. He then earnestly explained to me how to heat the food. As the heavy rain thrashed the lands and threatened to blow away our tents, we held onto the tent poles for dear life. The soldiers guarding the outpost caught my attention with their unwavering stance and staggering professionalism, even as armed militias patrolled the area.

Earlier, the two commanders had jokingly asked if I wanted to join the night patrol, but I was advised against it due to a lack of extra bulletproof jackets and weapons. As the patrol set out without me, I stayed behind to enjoy camp life.

Upon their return, I listened in awe to the stories of their thrilling experiences. I couldn't help but notice that the special forces members were never without their weapons, even when they slept. They kept their arms in a manner that allowed them to fire at any moment, ready to safeguard themselves and others.

But it wasn't just their preparedness that impressed me. Every morning, I watched in amazement as they quickly organized themselves and made all necessary arrangements with incredible efficiency. And they didn't stop there – every day, they neatly arranged and tidied up the entire camp, laying out all equipment according to the order of helicopters and loading points.

Thanks to their efforts, we accomplished the first-ever rapid deployable QRF heli-lifting operation in record time in the history of the DR Congo mission. In just two hours, we, 154 peacekeepers, returned safely in 24 sorties with 12 tons of cargo. Ukrainian attack helicopters circled overhead, ensuring our safety as we flew off to Bunia one by one, carrying soldiers and supplies.

As I watched the Ukrainian assault choppers diligently patrol the sky and the ground soldiers patrol the vulnerable areas below, I felt a sense of awe and respect for the professionalism and dedication of these soldiers. It was an unparalleled operational experience and one that I will never forget.

Sahara's Rapture

Captain Eman Nur Shamsuddin
(United Nations Mission for the Referendum in Western Sahara -MINURSO)

Solemn midnight, unperturbed. We were in the western part of the shoreless Sahara Desert at the uninhabited Tifariti UN camp. I was making my bed to have a tight night's sleep. At that moment, we were a group of eight in total. My medical assistant and I were the only Bangladeshis in the group; the rest were military observers from different countries.

No matter how scorching the heat in the Sahara was during the daytime, the area abruptly became shivering cold at night. The wind started howling and blowing, startling the UN containers at regular intervals. At the initial stage of the mission, there had been sleepless nights. But after getting used to it, the howling wind and my tight night's sleep emerged hand in hand. There was no dozing, and siestas were rare; consequently, slumbering became easy.

I woke up with fluttering eyes, twitching, and took a few steps towards the door to open it after hearing a knock. The imperious Croatian officer Zelco was shouting behind the door, "Doctor, wake up! Wake up!"

"Wait, brother! You are breaking my door!" I replied hurriedly, afraid of the big guy breaking my door.

As soon as I opened the door, Zelco informed me that a civilian patient would visit the camp and was waiting to see me. But the team sight (the team responsible for outpost security) was awaiting the commander's approval to let the patient in. Therefore, the commander asked for my opinion on whether I was willing to treat the patient or not!

I swiftly rushed to the Nepalese commander and asked to allow the patient in. By then, most people had rarely heard about the disputed Western Sahara, as the region that had been the center of controversy for so long was still unknown to many.

In short and simple terms, there were two discordantly belligerent groups. One of them was ruling Morocco, armed with heavy resources and a larger population, while their libertarian counterpart, Polisario, occupied a small part of the Sahara.

United Nations observers were in the middle of this conflict as mediators. The mission's aim was to set the stage for a referendum by implementing a peace accord.

Of the eleven team sites, few were in Moroccan territory, and even fewer were in Polisario territory. Each team site had 15 to 20 military observers, and observers from 34 different countries were involved in the mission. Consequently, at each team site, there was a mixture of officers from seven to eight countries.

At each team site (three in total), a doctor and a medical assistant served as representatives of the Bangladesh Medical Unit. Our responsibilities included ensuring healthcare for military observers, maintaining the cleanliness of the team site, providing healthcare coverage for other team sites during heli-recce (helicopter reconnaissance), and being responsible for medivac or Casevac.

Meanwhile, members of the Bangladesh Army's medical corps also provided medical assistance to the vulnerable locals. The Bangladesh Medical Unit had served alongside the Swiss Medical Unit, Korean Medical Unit, and Malaysian Medical Unit for seven years and had a good reputation.

It was difficult to understand the nature of the injuries or the medical complexities suffered by the patients. The camp cook, who was a local, was the only person who could help us interpret the locals. Perhaps due to his long-time service in the UN camp, he had some knowledge of English, and he always tried his best with his rudimentary vocabulary. As part of his volunteering effort, he tried his utmost to interpret the details of the accident.

At that moment, all the information we had was "A lot of blood! Critical condition!" The local hospital doctor was struggling to handle the patient's condition and asked for help from UN doctors.

So, I asked our medical assistant to get ready. Even with all the preparation and life-saving medicines, predicting the extent of the challenge of treating a major injury in the heart of the Sahara was dicey.

The Polisario fighters were barred from entering Moroccan territory due to an embargo; therefore, the only option was to move to Algeria, which was home to the Polisario's largest refugee camp, for treatment.

The distance by air was about three hours, and the only way to communicate was through the UN satellite phone.

Time flew as there were no roads around, and the desert was the only reality there!

We were waiting at the gate for the patient to arrive, and all that appeared was darkness all around. Only the dark horizon momentarily peeked through before our eyes. But we were waiting there, hoping that a car with the patient might arrive soon. Meanwhile, the whole camp had woken up. Other officers accompanied me at intervals, most of them my seniors. Among them were the veteran observers who had already completed five missions, who repeatedly asked me if I needed anything.

All I requested was for them to stay with me. With a gentle smile, I asked them, "Brother, please accompany me."

From a distance, the glaring headlights caught my attention. Yes, it was the patient! A vehicle with the patient!

Right after we had taken the patient to the camp, it didn't take long to realize that their condition was much more critical than expected, though the local doctor had done his best. The incident that added insult to injury for the patient was when a jeep tire burst, striking him directly in the jaw, breaking his nose and injuring his hands. The result was terrible, and blood gushed down his face. Examining the patient's face seemed as difficult as nailing jelly to a tree. Immediately after providing the patient with first aid, I called the headquarters.

"Sir, I need Casevac immediately." Receiving a call for Casevac from me at two o'clock at night would flabbergast anyone, and the headquarters trembled after receiving such an unusual call.

But the UN's Casevac (helicopter) was at least an hour and a half away from the camp. Moreover, there was confusion surrounding the patient's identity and related issues, whether he was a Polisario fighter or simply a civilian, and why the local hospital was not interested in treating him.

I was concerned and assessed the challenges of flying to the nearby Algerian Polisario refugee camp, safely conferring the patient to responsible camp healthcare, dealing with queries from UN authorities, and many more. All these thoughts and calculations were driving me crazy.

Amidst these international complications and limitations, I also started empathizing with my patient. Setting aside all the calculations, our medical assistant and I began trying to reduce the bleeding.

The patient soon started vomiting blood, and his panicked mother began wailing in Arabic. With my mix of pidgin, Creole, and attempts at lingua franca using Bengali, English, and some broken Arabic, I tried my utmost to soothe her.

The headquarters informed me that I must, "Pass the night somehow! The chopper will fly in the morning."

If only the night had passed as swiftly as the winds of the Sahara! But it was not to be!

Our dragoman cook was out of the camp at that moment. I grabbed some food from the store and offered it to the patient's mother. She pointed back to her son. She was not going to eat until her son recovered. Above all, she was a mother! I didn't insist and left the room after I adjusted the saline drop limit.

When would this night end? How long would it take for the night to pass? Yes! The night went by!

Come morning, the patient's mother seemed anxious and fidgety. When would the helicopter arrive and rescue her son from such misery? When would her son recover...?

I tried to communicate with the HQ. Decisions from their end kept changing every hour. Her impatience grew at an increasing rate.

She angrily expressed her despondency and desperation, this time in Arabic. Our camp cook translated...

She complained that we were not helping her son escape death just because they were poor. Determination took over me. This time, I became more desperate and assured her of her son's treatment in Algeria.

Being unbiased, I calmly and clearly informed the headquarters over the phone that the patient would soon die in a UN camp without proper treatment if they didn't send helicopter evacuation support as soon as possible. I also reminded them of the responsibility for the consequences if anything bad happened.

After the tense morning went by, the scorching sun greeted us. Henceforth, my practice of polemics and the pleading of a mother ended after a helicopter reached our camp after 12 long hours.

However, by then, my patient was as stable as expected, and he, a young man (likely my junior), could barely open his puffy eyes. The bleeding on his face was gradually subsiding, but his features were indiscernible due to the severe injuries and damage.

After all the necessary arrangements were made, all the camp observers gathered to witness and be a part of the emergency heli-evacuation.

The helicopter was to perform a touch-and-go maneuver. The patient had to be placed inside the chopper within a very short span while the rotor was running. The only person accompanying the patient to Algeria was me. I left the spot with my patient after the last-minute evacuation briefing.

But yes, on board, the patient was fortunate to have his mother with him.

As soon as the Casevac chopper touched the ground, the pilot signaled me to board. I, along with my patient and his mother, got inside and took off...

That was the first-ever helicopter ride for both the mother and son. The weary mother sat holding her son's hands. In the harsh, scorching mid-afternoon heat, the helicopter bounced with a jolt and departed for the Algerian refugee camp two and a half hours away.

My duty would be over as soon as I reached the hospital in Algeria. The young man held his mother's hand for a long time and uttered a few slurred Arabic words in the meantime.

At one point, he extended his hand toward me. His mom beckoned me with a grateful smile!

I tightly held his hand and whispered in a slurred tone in Bengali, "Bhoy Nai, Ami Achi" ... (Translated as: "Hey, young man, nothing to be afraid of, I am still here! I will be...")

The language had never been a barrier for me to reach his heart, and the young man grasped those words of solace, reassurance, spirit, and trust. A faint smile appeared on his broken jaw!

Ah! What a heart-touching smile! Delightful, graceful, and cold as frost on a windowpane! Like a sealed envelope!

Bloodshed at Aveba

Major Md. Tarek Mahmud Sarker
(United Nations Mission in the Democratic Republic of the Congo - MONUSCO)

The Democratic Republic of Congo (DRC) stood as a troubled land marred by relentless internal conflicts and political upheaval. In response to this urgent call for peace, the United Nations dispatched its experienced peacekeeping force, casting a ray of hope amidst the chaos. Our mission, baptized under the name "MONUSCO," was a testament to the international community's unwavering commitment to bring solace to the DRC's war-torn soil.

Within the intricate tapestry of MONUSCO, the Bangladesh Battalion (BANBAT 1/11) emerged as a formidable force entrusted with the weighty responsibility of preserving and disseminating tranquility throughout the revered Ituri district nestled within the Oriental Province. Our battalion's nerve center lay firmly within the hallowed confines of the Andromo camp in Bunia, serving as a bastion of strength and resilience. Yet, our reach extended far beyond, like a constellation of outposts speckling the region, each bearing its own unique moniker: Bogoro, Aveba, Bukiringi, Marabo, and Commanda.

As the commander of the esteemed support company, my days found their purpose within the hallowed halls of the battalion headquarters, where we orchestrated our operations and fostered seamless harmony with our fellow encampments. However, the ebb and flow of fate occasionally bestowed upon me a rare opportunity – a chance to temporarily assume the mantle of leadership at the camps whose commanders ventured on well-deserved leaves. It was during these profound junctures that I would embark on an immersive odyssey spanning more than a month's tenure within each camp. This serendipitous journey allowed me to immerse myself in the local terrain, as I absorbed its secrets and unraveled the enigmatic dynamics of the conflict that gripped our mission in the heart of the DRC.

Our mission, fraught with danger and uncertainty, demanded unwavering resolve and steadfast determination. But amid the trials and tribulations that awaited us in the DRC's unforgiving landscape, we stood united, resolute in our pursuit of peace.

One such occasion arose when Major Mahi, the esteemed camp commander of Aveba, was due for a well-deserved leave, and I found myself stepping into his shoes as the acting commander of Aveba. It was a daunting yet pivotal responsibility that afforded me the rare opportunity to stand on the frontlines, bearing witness to the valor and unwavering dedication of our soldiers while actively contributing to the region's peacekeeping endeavors.

During my journeys to the Bukiringi camp, I had the privilege of visiting the Abhebha camp on multiple occasions. Perched atop a modest mountain, the camp offered a captivating panoramic vista of the surrounding landscape. However, as members of the peacekeeping force, our eyes were not solely fixated on the enchanting beauty that unfolded before us; rather, they keenly sought out the hidden perils that lay beneath. Abhebha camp possessed a vantage point that allowed UN peacekeeping officials to skillfully survey the terrain with a discerning eye.

The presence of militias in the area long preceded the United Nations' intervention, evident in their sporadic bursts of gunfire followed by swift retreats. The camp occasionally responded with retaliatory fire, while on other occasions it served as a vigilant observer, silently documenting the movements of the militias and their hurried attempts to evade capture. They meticulously surveilled the area without initiating hostile actions.

Before embarking on his leave, Major Mahi took the time to brief me on the camp's overall condition and the weight of responsibility that rested upon my shoulders. He remarked, "I leave you in a challenging situation, where your engagement will likely be essential."

Abheba camp presented itself as a captivating spectacle surrounded by incredible natural beauty that was truly mesmerizing. However, the familiar PREFAB CORIMEC 1 structures commonly seen in such settings had been replaced by more humble tin sheds repurposed from previous units. For our soldiers, a number of tents also provided shelter. Though enduring the hardships of living in those tents for an entire year was undoubtedly challenging, the camp boasted an exceptional supply of pure, untainted water. Unlike other camps, where water had to be laboriously fetched from distant sources using water browsers or used sparingly, there, water was conveniently pumped from a natural cascade using a submersible pump. It was then stored in a vast tank before being distributed to both the soldiers' quarters and the cookhouse with the aid of a small pump. Thus, unlike many camps, there was no need for an extensive workforce dedicated solely to managing and supervising the water supply, which brought a sense of respite.

Furthermore, the encampment featured an expansive field that doubled as a helipad, inviting the local Congolese people and army personnel to partake in friendly games of football. The encampment was embraced by the surrounding mountains, elevating its natural magnificence and transforming it into an awe-inspiring sanctuary.

The mission in Congo presented us with an unequivocal challenge that proved to be a formidable test of courage for all those involved. On February 25, 2005, Captain Shaheed Ashraf and nine other brave peacekeepers lost their lives while fulfilling their duty in the crucial mission of establishing peace.

Our relationship with the people of Congo was amicable, but a substantial language barrier separated us. That was the only difference between us, as everything else seemed to be progressing smoothly. The locals were fluent in their local language, Swahili, and had some knowledge of French. However, they struggled to comprehend English, while we found ourselves unable to communicate even the simplest phrases without relying on the French term "Pesho" as a substitute for "good morning." This language barrier posed a significant obstacle, necessitating the assistance of a French interpreter to facilitate communication.

Most of the interpreters available to us during the Congo mission were Congolese; however, doubts lingered regarding their ability to accurately convey our intentions with the necessary finesse and accuracy, despite our efforts to eloquently express our thoughts. It was disheartening to discover that our carefully chosen words, even when delivered with utmost care, failed to generate the desired level of understanding among the local population.

In contrast, during my previous mission in Sierra Leone in 2002, there had been no need for interpreters. We would pause on the roadside, stop our vehicles, and engage with the people in Krio. We actively listened to their concerns and grievances, fostering a genuine connection. This simple yet profound act of empathy left a lasting impression on both those we encountered and ourselves, enveloping us in warmth, fondness, and positivity.

However, our interactions with the Congolese felt more akin to those of foreign visitors in an unfamiliar land, as we were separated from the native inhabitants by a vast cultural divide. Nonetheless, we made great efforts to establish meaningful lines of communication with the assistance of our interpreters, and the results were somewhat positive. While we achieved some degree of success in our endeavor, we remained keenly aware of an intangible sense of earnestness that accompanied these exchanges. It undoubtedly stood as the most significant challenge of our Congo mission.

We had a distinct advantage while serving in Sierra Leone as there we had the full leverage of language, which allowed us to seamlessly connect with the locals and establish a rapport, a feat not easily accomplished in Congo. It was not possible to get along with the local Congolese as quickly as the locals in Sierra Leone. It always took a bit longer to adjust in Congo. The most fun was with the kids in Sierra Leone. Those local kids were the highlight of my trip, as they would gleefully exclaim, “Give me chop-chop,” an endearing phrase in Bengali that translates to “give me food.” Conversely, such pleas for sustenance were conspicuously absent in Congo, where communication often necessitated the presence of an interpreter.

However, undeterred by the linguistic obstacle, with great patience, we tried to minimize the divide between us and the Congolese. Our goodwill extended to buying footballs for the community, constructing a clubhouse there, gifting them televisions, and even procuring sewing machines for the women.

The communication infrastructure throughout Congo was notably lacking, as few roads were built/paved through the rugged mountains and cars often became bogged down in the muddy pathways, particularly during rainy weather. However, we utilized our armored personnel carriers (APCs) for transportation, as the rescue vehicle provided some relief from the fear of being stranded. When smaller vehicles became stuck, we had no choice but to rely on the APCs to pull them out of the mire. As someone who had personally experienced the anxiety of being trapped on these treacherous roads during patrols, I can attest to the challenges they posed. At times, when we encountered local trucks that were stuck, we made it a point to assist them in getting back on the road, recognizing that their struggles could easily impede our own journey.

Our time at the Ebola camp and Bukiringi of Congo was marked by a high degree of stress, as the camps were situated in areas controlled by different militia groups. Gunfire and the sounds of militia audaciousness were an almost daily occurrence, with Congolese army personnel and UN peacekeepers on one end and the militia on the other, all engaged in a game of cat-and-mouse.

In the middle of the night, we would often hear the unmistakable sound of militia gunfire kilometers away from us. It was difficult to discern whether they were engaged in looting or fighting amongst themselves, but the constant threat of violence meant we had to remain vigilant at all times, even as we struggled to get the sleep we needed to function.

On occasion, we were forced to undertake "stand-to"[1] twice in a single night in order to be ready to negotiate any potential attacks. Such cases of 'stand to' and 'false alarms' occurred for at least three to four days every week.

This lack of rest inevitably took a toll on us, leaving us feeling drained and depleted in the days that followed. Living through such unrestful occurrences day after day, I developed a terrible sensitivity to the sound of gunshots. Our experiences in the Congo were a true test of our physical and mental resilience, requiring us to remain steadfast in the face of adversity.

One night, I was jolted awake from my deep slumber at two o'clock, startled by two gunshot-like sounds that echoed through the silent air. I got up, half-asleep, wearing just my shorts and a cozy t-shirt, but in a state of groggy confusion, I hastily adorned myself with my uniform, grabbed my bag full of bullets, shouldered it and the SMG weapon in hand, and then headed towards the perimeter bank. Yet, to my consternation, it suddenly occurred to me that there was no stand-to position of our soldiers, gadding, nor the crackle of gunfire.

Doubt crept into my mind as to whether I had made any miscalculation. One of our sergeants came up to me and asked, "Sir, you are with your SMG! Is there anything wrong, sir?"

I replied, "There were sounds of gunfire. I am wondering why you guys are still not in the stand-to position! Why so!?"

---

[1] 'Stand-to' is a military term meaning soldiers remain ready and on guard to take on enemy attack on their positions.

But to my surprise, he countered, “No sir” and reassured furthermore, “Sir, there was no fire.”

“What about the sounds I heard then? I heard the sounds,” I insisted, my confusion mounting.

Yet the sergeant calmly replied, “Sir, I have been awake since 11 o’clock. I have not heard any sounds. Perhaps, you may have misheard something?”

He told me that he had been awake since 11 PM and had not encountered any sounds. And I may have had heard it wrong.”

“I woke up because of those sounds, and I came here,” I retorted. “And now you’re saying I heard it wrong. Bewildering! simply!”

I moved and talked, and the sergeant accompanied me as I was going to every sentry post. There were six sentry posts in total with twelve soldiers on duty, two at each post. I thought that since I had woken up, I would return to my room after having conversations with the soldiers on duty. After meeting and talking to the sentries of a post, I was moving toward another when an unfurled tin body caught my attention.

I inquired to the sergeant about why the tin part was lying open in such a way. He said, “The wind blew it over from our prayer room’s veranda to here. And the howling wind is still whistling, sir! But not with the great power it possessed earlier on. And sir, that tin body indeed made a distinct sound when it fell onto the ground that might have reminded you of the sound of gunfire in your sleep. So, that’s what happened, undoubtedly, sir!”

The following day, I shared the story of the incident with three officers at the breakfast table. Their laughter knew no bounds. Captain Siraz told me that Company-2 IC Captain Rafi had a lot to tell about ‘firing’ and ‘stand to’. He had stories that were enlightening upon listening.

He added, “Captain Rafi was then at the Bukiringi camp. And that Bukiringi camp was then under the command of Aveba and operated with all the logistics support from the Aveba camp. Everyone at their camp had experienced the feeling of being woken up by unprecedented noises multiple times and taking a stand to position at night.”

The medical officer of our camp, Dr. Lieutenant. Colonel Nasir, was my senior. I asked him if there was a way to get away from such experiences of pseudo gunfire in the middle of the night. "This, too, shall pass." I reminded myself of his words while standing before the mirror.

What he suggested was to invest more time in physical training games and rigorous workouts, as a tired body tends to fall asleep early and deeply. "And then sounds might not affect the slumber that way. One more thing, if any firing really happens to be the case at midnight, you will be receiving the control call, and also the duty NCO and duty JCO will wake you up," he added.

I replied, "Well, saying this might be easy for you, but doing it is a daunting task for me. You asked me not to get tense, but please, you speak well. Nonetheless, henceforth, for it is only tension that holds me back from slumber."

"We officers lived in a tin room there. And at night, in a tin room, there had been peculiar noises, sometimes many distinct noises. The whistling wind brought forth a symphony of sounds – patuspatus, phatas phatas – reminiscent of firing from a distance.

Two of the four officers in the camp used to stay up and vigilant throughout the night while the other two slept early and woke up at the crack of dawn. These two pairs of officers exemplified two divergent habits, and as the company commander I always recognized the value of such variance and honestly and liked that approach. Half of the night passed guarded automatically, and the same happened early every morning.

As we were at the Aveba camp, the looming threat of militia attacks was never far from our minds. In such a climate of uncertainty, remaining vigilant was not just a matter of habit; it was a necessity. And so all our men remained ever watchful to protect the Aveba camp from any sort of attack, regardless of any militia group's threat or other external threats. In doing so, they safeguarded not only themselves but the entire camp.

My recollection of my inaugural night at the Aveba camp remains indelibly etched in my mind. It was a night cloaked in the quietude of darkness, and my companion and I had retired to our quarters to unwind and enjoy TV shows. I still remember the first night at the Aveba camp. At around 130, we were all awake and enjoying our time.

However, our peaceful reverie was abruptly disrupted by a raucous sound that did not seem too far away and might have been emanating from close quarters. Our eyes then gazed in shock at luminous, dense smoke billowing out from the light machine gun (LMG) post.

Reacting with swift alacrity, I shouted and asked to stop firing first. Inquiring into the root cause of the calamity, I approached the LMG officer and queried about the cause of the fire. I needed an answer from our soldiers, so I asked, “Why are you guys firing? What happened?”

He declared, “Sir, we have been attacked from that direction. Enemies engaged in firing, sir!” I probed further, inquiring if he had observed any flashes in the vicinity of the fire. To this, he assented.

I then expounded that he must have been privy to the sound of the conflagration and had inadvertently caused the flames. Then, I inquired again! “It’s okay that you responded when fired upon” but when I asked about the number of rounds fired, it was discovered that the soldier had unleashed an astounding 75 rounds.

I inquired further, “How come you shot 75 rounds, whereas the possible militia personnel had only discharged a mere two! How is this possible? I am really expecting an answer!” The incongruity of this fact was confounding, and I could not help but ponder the circumstances that could have led to such an astonishingly lopsided discrepancy.

A profound sense of anger gripped me. In the wake of a trying situation, I engaged in a discourse with the contingent commander, Colonel Iftekhar. That discussion propelled some ideas, as with remarkable composure and calmness Colonel Iftekhar uttered the words, “This is nothing but the battlefield stress. Your soldiers fired extra rounds in a state of acute anxiety and discharged a surplus of ammunition. Hey, listen, nothing to be panicked about. Rather, let us try to elevate the soldiers’ motivation and introduce refreshment training. If possible, we should continue it frequently.”

The following day, the training with the manifesto slogan “One enemy, one bullet” commenced to alleviate the mounting pressures and anxieties that existed in the mission area. Surprisingly, it reduced the stress of our soldiers. In the program, each day there was physical training in the morning and ‘game-on’ in the evening. It was mandatory for camps to instate morning physical training and evening games as a means of reducing excessive stress. Also, to maintain caloric intake and a healthy weight, there is nothing better than regular physical training and games.

Furthermore, Bangladeshi peacekeepers possessed a special reputation in United Nations missions. Hence, to keep that honor and prestigious reputation, Bangladeshi peacekeepers engaged in a variety of grueling refreshment training that gradually became an integral aspect of their daily camp life.

After the incident of that midnight firing and the rebounding effort from our soldier, sometimes in the middle of the night, I would wake up. Just a few words or mere sounds would make my ears perk up. The very next day, I went to the liaison officer of the Congolese Army to figure out the root cause of the last night's firing. He replied that he learned that the militia had glided into the village about 700 yards away from the camp to loot, and before taking entry to the area they fired a few rounds to scare everyone.

"It was only a week ago that we raided, beat off, and chased them away out of the area and arrested two of them (the militants). How did they come to have the sheer audacity to be here again?" I asked him.

"Those whom your team beat off had been replaced by a new group of militias to carry out acts of terror in the area," he informed me. One afternoon, three local Congolese wanted to meet me at the Aveba camp. They were local NGO employees. One of them knew good English and shared with me his unique background of being an employee in Sierra Leone while his family resided in Uganda. And his children studied there in Uganda. Interestingly enough, I noticed a striking similarity among the Congolese people; many of them had family members residing in Uganda due to security concerns.

Congolese UN staff, especially interpreters, kept their families in Uganda for security reasons. At one point in the conversation, they informed us that they were there for a reason and had weighty information on a grave matter that required our attention.

"You guys patrol the area day and night, and the show of force never stops, yet a few militias (malicious individuals) had been carrying out acts of terror such as robbing and looting in the gaps between the patrols (replacement time). They are going to different areas at different times. Even in the last week, they attacked three villages in three days," added the one whose English was good.

Indeed, to make matters worse, they were attacking different villages at different times, with three such attacks occurring in the last week alone.

"Would you please tell me the time?" I asked.

"Eight to ten," they informed.

"How long has this been going on?" I enquired further.

"Three or four such occurrences have taken place in the past week," they said. I kept the names jotted down in my notes.

That day, I rescheduled our patrol time to 8 o'clock instead of 7 o'clock. I had one hour saved, as my plan was to move out for patrolling with the patrol team at 9.00, and I kept it a secret.

Gradually, I would be buying time; otherwise, there would have not been any privacy to the plan. I informed all earlier on that night that I myself would be joining as the patrol commander.

At 8 that night, the Second-in-Command of the patrol, Subedar Munir, approached and proclaimed, "Sir, the patrol is ready and primed for action."

I said, "Ask the patrol to take respite from their APC and indulge in some television enjoyment. I shall tarry for a moment before embarking on the patrol."

Our Congolese interpreter had taken the afternoon off with the plan to return to the camp in the evening. However, at 9, when I asked Sergeant Munir to have all the patrol members mount the vehicle, he informed me that the interpreter had not shown up. He was yet to arrive.

I became quite upset and felt a surge of anger because without the interpreter, it would have nearly been impossible to move with the patrol. So, I asked Sergeant Munir to wait for a little more.

"Let him arrive. He (the interpreter) might be arriving soon," I added.

After thirty minutes had passed and he still had not arrived, I turned towards the clock with a perturbed voice and instructed, "No more waiting. Let's go."

With two APCs, a jeep, and a pickup, we set off, Bangladeshi peacekeepers on the jeep and APC, and eight Congolese Army soldiers in the pickup. Our operation progressed.

We bumped along the path toward the camp's gate when suddenly the interpreter appeared, hobbling towards us, out of breath. I let out a burst of English profanity in his direction. He apologized, and as we continued he explained the reason for his delay: "Sorry for being late! Sir, the reason I am late is," with frequent fleeting breathing, he said, "Looting is going on."

I relented a bit, and I asked him (the interpreter) to get on the jeep, as I and wanted to know whether he could be able to help us by showing us the way to that village.

"Sure, sir. Let's go," he said.

Within half an hour, we reached the village and found two men and three girls injured as militias had left them thrashed after a flogging with sticks. I ordered to treat the injured with quick first aid. When we talked to the village locals, it was revealed that they had been looting only 10 minutes before we arrived, and they had been chased out by the Congolese Army stationed in the neighboring village. We were about to get into our vehicle, preparing to go down that same road where the Congolese Army ran after the militias and scared them off; right at that moment, however, heavy widespread firing started targeting our convoy. I shouted to my patrol, "Lie down. Crawl and take cover, track down the flash of their arms, and fire back. Rebound back, fire back!"

My runner and I took cover in a mound of dirt, seeking refuge in a small vegetable garden behind a house by the roadside. From our vantage point, I could clearly observe the flashes of gunfire from the militia positioned roughly 500 yards downhill. Shooting downhill was always challenging. At that moment, Subedar Munir, my second-in-command, asked over the walkie-talkie, "Shall we engage from the APC, sir?"

"Fire towards the direction of their weapon's muzzle flash," I instructed. Luckily, our APC was equipped with tracer rounds. We immediately halted our small arms fire as the potent burst of 12.7 MG rounds began from the APC.

Two minutes into our concerted effort, Captain Parvez from the camp radioed, requesting a temporary ceasefire on the APC firing. "Sir, please suspend the APC fire for a moment. The tracer rounds are dangerously close to the base camp."

Without hesitation, I ordered everyone to cease firing from the APC and stay clear of it. We were left with no choice but to rely on our instincts and shoot in the dark night, fully aware of the potential risks of collateral damage.

In the midst of the engagement, some Congolese Army members who were accompanying us managed to capture two militants, both armed only with sticks. This information became crucial, as earlier, when we arrived in the village, we had learned from the villagers that only 5–6 militiamen were equipped with AK-47s, while the others, around 10–12, carried only sticks.

While these events unfolded, I communicated with my contingent commander, Colonel Iftekhar, through the wireless set. I sought his advice on whether I should pursue the militia or not. Though my intention was to apprehend or completely dislodge them, I always prioritized arresting the militiamen first.

Maintaining his calm and composed demeanor, Colonel Iftekhar suggested that I bring two more APCs and an officer to the location. He also recommended further interrogation of the locals to gather more information before making any decisive moves.

It became evident once the entire situation was assessed that pursuing the militia would not have been worthwhile or beneficial in any way. The safety of the patrol was naturally a significant concern for the contingent commander.

Within moments, Captain Siraj appeared from nowhere with two APCs. "Brother, the battle is over. Let's get back to the camp after a patrol," I said to Captain Siraj. We then spent another hour around and returned to the Congolese Army camp after we were done with the last patrol of the night.

Let me describe now the two militiamen we arrested. After we arrested them, they were sent to the company headquarters of the local Congolese Army. That was the rule. As there were no Congolese police in the village, the accused held captive were to be handed over to the Congolese Army camp. Police activity was being carried out by them. I went to their camp and found that they had tied the two militants with four pegs ground roughly in two holes vis-a-vis, with only underwear on their bodies, and their hands and feet spread apart like animals.

I tried to remind the company commander of the Congolese Army through the interpreter of how the two prisoners should be treated according to the Geneva Convention. The company commander seemed not quite happy with my words as he looked agitated and upset. He got crazed up in a manner and started yelling at the interpreter. I enquired of the interpreter about why the company commander was behaving that way.

The interpreter replied, “Sir, he is saying that he knows very well what to do and what not to do. It’s better for the UN not to whine about it. We are not expecting their instruction here. They better be silent.”

To avoid trouble and any chance of misunderstanding, I pretended not to have a grasp of the Congolese Army commander’s words. Then, I asked my interpreter to “Tell the company commander that the reason for his anger was advisable as militias committed heinous crimes over time, ran rampage there, looted. However angry, interrogate them with patience and do the necessary to bring them to the battalion headquarters or their brigade headquarters after the inquiry.”

On the way back, I asked the interpreter about the company commander’s plans for the prisoners. To my dismay, the interpreter shared a disturbing account of how the militias would be brutally treated during the interrogation. If they were members of minor militia groups, they would be segregated and subjected to consecutive bayonet strikes in the chest until their death, after which they would be thrown into a pit and buried. If the captives were deemed highly dangerous or leading members of a militia group, they might be taken to the battalion headquarters.

I immediately expressed my dissatisfaction to our operations officer, stating that this treatment was unfair. He informed me that prisoners captured by peacekeepers are usually handed over to the Congolese police. However, in the absence of police, they should be transferred to the Congolese Army and there was no further recourse available to us.

Feeling helpless, I reported the matter to the military observers as instructed, acknowledging the situation for their further inquiry. I realized that there was little else I could do, as no established judicial system existed in the region. The concept of human rights seemed to exist only in words or dreams, not in reality. Thoughts of whether showing mercy or empathy to those who held power over us and fired bullets at us would make any difference crossed my mind, but it seemed pointless.

As I untied the laces of my boots upon returning to the camp, exhausted from the intense operation, my phone rang. It was the operations officer once again. I patiently recounted every detail of the incident to him, explaining the intricacies for an hour. He urgently requested a detailed report, so I hurried to the operations room, meticulously transcribed the entire situation, typed it up, and faxed the comprehensive report to Battalion HQ. It was already late, and I finally sat down to have dinner at 2 o'clock.

The following day, as the sun called the earth to awaken, I found myself up at six in the morning. I quickly dressed in my uniform and headed straight to the site where the confrontation had taken place the previous night. There, I noticed a patch of grass and leaves stained with blood, hinting that our shots in the dark may have injured a militia member during the scuffle. However, the presence of wild animals in the area made it uncertain. It could have been the result of an animal encounter as well. How could we be sure?

As the cycle of bloodshed persisted, the peacekeepers serving in the contingent counted down the days of their designated year of service. Their yearning for home grew stronger, and they eagerly anticipated reuniting with their wives, children, and loved ones. Yet, the game of bloodshed continued.

Despite the inherent dangers of their work, a glimmer of joy and anticipation sparkled in their eyes as they prepared to return home. Meanwhile, in my home country, another contingent was being prepared, ready to face new challenges in the game of bloodshed known as war. Our soldiers embraced each new challenge with unwavering determination and a renewed sense of purpose.

The rotation of old and new soldiers continued, with peacekeepers remaining focused solely on the task at hand: the establishment of peace. New games and new challenges were part of the cycle of bloodshed. Even as I write these words in 2018, the game continues in the DR Congo. Despite the challenges, we persevere. We are committed to the cause of peace, the betterment of the Congolese people, and upholding the principles of the United Nations. It is an honor to serve alongside my fellow soldiers while striving to bring hope and stability to a troubled land.

CHUKUDUM

MAJOR MD. ABDUR RAHIM
(UNITED NATIONS MISSION IN SOUTH SUDAN - UNMISS)

The acquaintance between the head of office, Mrs. Winnie Babihuga, and Mr. Captain was a long and pleasant one. Their acquaintance dated back to when Mrs. Winnie served as the head of the office in Juba, South Sudan, and the captain as the liaison officer under the sector commander there.

BANBAT-5 celebrated Pahela Baishakh (the Bengali New Year ceremony) once in Juba. On one such occasion of Pahela Boishakh, Battalion 5 organized a program on resplendent dawn in the month of Baishakh where the young officers of the battalion had showcased their creativity and artistic talents through a captivating display of drama, songs, poems, jokes and jests representing their country. Their creativity and vibrancy left a profound impression of their ingenuity and spontaneity on all the international staff based in Sudan. Such creative fervour and enthusiasm made Mrs. Winnie regard the young officers as a delightful bunch.

Contemporarily, Colonel Sanaul, a kind, gentle, amiable, and highly professional person, was the sector commander in Juba. He was a remarkable figure in his own right. He displayed an exceptional level of professionalism and commitment to his fellow officers. He always treated his subordinates as his own family. His handling of various political matters during his tenure as sector commander was exemplary and earned him the admiration of many. Sudan would be segregated into two different parts, two different Sudans: one would be North Sudan, and the other would get revealed as a different independent state, South Sudan. The latter's revolutionary military force had shared an understanding with the Bangladesh sector and enjoyed a cordial, open-armed relationship. The situation was gradually escalating towards normalcy. Was peace imminent? Was that land on the verge of meeting the most cherished entity – peace?

Those South Sudanese people knew that a new country was to be born and revealed on the map of the world, and they were by then mentally prepared to be part of a new country, a new Sudan.

The captain went on leave. Prior to embarking on the mission, he got married, but as his call to the mission manifested, he could not even manage the time to take his wife home to fulfill the culture-based marriage ritual, and he left the homeland to go on the mission of establishing peace in Sudanese lands. Ah! Sigh! Who could see a newlywed bride crying that way? Her tears were a storm of emotions, raging and pounding against the shores of her heart as she watched her husband leave for his mission, a river of sorrow flowing endlessly down her cheeks and carrying away any hope of seeing her beloved husband soon.

Major Jahid, the senior officer of the battalion, used to throw his facetious pokes at him, capitalizing on the weeping bride. After almost eight long months as the sector commander's liaison officer, it was then my turn at the Torit team site.

The team site was beautiful, as such serene surroundings prevailed there. A BANBAT-5 company was induced there on the site. Torit was the capital of the Eastern Equatoria State. Captain was weighed/assigned there with the onus of coordinating with the civil counterparts and effectively liaising between PR and leading operations.

Major Kabir was a nice fun-loving man with an intriguing personality who never failed to leave the audience amazed and spellbound with his awe-inspiring magic tricks. It was then the time to go on a field mission at 'Chukudum' as part of the regular operations.

Chukudum was an incredibly beautiful region of Budi County. The breathtaking natural allure of that area was almost surreal and beyond belief unless witnessed with one's own eyes and experienced firsthand.

That was the land of the Didinga tribe as the majority of the people there belonged to Didinga. During the Second Sudanese Civil War (1983–2002), this region suffered from extensive damage and destruction, sustaining devastating casualties. The war was still going on, and its impact was evident as the people there had not seen any substantial progress in their lives. As we delved a little deeper into Chukudum, it felt as though we had stepped into a primitive, timeless, ancient society. The clothing and food of the locals were reminiscent of a bygone era. Their traditional attire and the foods they consumed remained an emblem of their age-old customs and practices, lost to the passage of time. But they nourished every bit of their customs and culture.

All preparations were underway for the cars to leave at 7 o'clock in the morning. Military observers, UN police, and the civilian component (international civilian officials) were all readily present. There, we had Captain Choy also, the Cambodian officer of the electoral division who loved to cook and feed everyone at the camp, and he was a great friend to all.

German Air Force Captain Roni, a military observer, was also a good friend of the captain. Ronnie joined us that day as one of our tour companions.

It was going to be a great time. As a language assistant, Hooton always traveled close to Captain Ronnie and used to call him 'The Chief'.

The convoy left for Chukudum. That journey would take roughly six to seven hours. En route, I met the local army commander of Sudan, Lieutenant Colonel, to discuss the situation in Sudan. So, yes, we had a meeting on the journey. Despite the commander's assurance that everything was under control, one could not help but wonder if such a statement held any truth.

However, when he was asked about the existing situation around Sudan, he assured us, saying, "They had everything under their control".

During the journey, UN staff and observers accepted the captain's hospitality and had lunch together (hosted by the captain). However, Ronnie's ears began to burn after having an excessive amount of spicy food. Despite such discomfort, he struggled but maintained a nice smile on his face in an effort to please a friend.

A group of peacekeepers was quickly moving towards a conflict zone to convey the 'message of peace' through the crooked, meandering terrain.[1] Conversing with individuals from diverse areas and backgrounds had always been a delightful experience, as it was a pleasure talking to new people, which proved to be the beauty of cultural exchange.

At the end of the day's activities, as the evening ended, a get-together culminated in a sumptuous dinner party. Following the feast, Ronnie had to depart from the camp and find lodgings in a nearby hotel in Chukudum. On the way, he was stopped at a local army checkpoint at gunpoint and was instructed to go back. Ronnie turned back to the captain in fear. Later, the

---

[1] Teams of peacekeepers continue to deliver a message of peace through South Sudan's crooked friend road. It is a pleasure to be able to talk to new people in a diverse environment during this journey.

captain took him in his car to the hotel. Poor Ronnie was emotional and thanked the captain.

It was a call to return after the end of the three days Long Range Patrol (LRP). The captain's car was the first to go. Suddenly, they spied a local army pickup truck approaching from the opposite direction. The truck was sending a warning that any United Nations vehicle that ventured forth would be met with great peril.

A group of rebels, armed and ready, were poised to launch an attack at any moment at any provocation and were not far ahead, only two hundred yards away.

One of their commanders was killed, while several others were reported to be in dire straits. Only thirty soldiers were with the captain's LRP, who were just returning to escort Ronnie. The situation was promptly relayed to the company commander, who, given the fact that representatives from all United Nations CompoNet (protocol) were present, deemed it unwise to proceed.

The military observers, including the captain, headed toward the direction of the skirmish as they were keen to find out what on earth had taken place there. It was then that the phone rang and Major Kabir advised, "Return to the camp as soon as possible and join back again only when the situation turns back to normalcy." The entire group retraced their steps to the old camp, waiting for the situation to improve before they returned again.

So, yes, then, the whole party returned to the old camp. The new camp setting was set up/installed again. The tent of Major Nizam, the Malaysian observer, had been set up near the captain's tent. Then, it was only time to wait.

When everything was settled, everyone might move back again. But the security situation was deteriorating quickly. Who knew if the wait would be longer or not?

We all went to meet the Budir county commissioner. He was an innocent-looking man who appeared quite helpless at the time. He said, "You are like our guests. But the situation is not under our control. They are killing my innocent villagers one by one."

There was nobody left to understand the fact that all of us had to wait/stay there for at least a few more days. But then we were taunted by logistical challenges! We were running out of rations near the end. Oil was taken out of the car of the UN officials to fill the BANBAT vehicle. Our lifeline of communication, the batteries of mobile phones, were also running out. That is, the entire caravan/company traveling together was isolated from the world. The only hope that remained was the satellite phone.

Finally, after another two days passed, the helicopter arrived with provisions. The company commander was preparing to send an APC to receive them, but the situation was too heated at that moment to get a green signal.

When the day bids adieu and evening comes, time tends not to pass quickly as it takes a breath on the cozy couch of the dark horizon and stars.

The poor civilian staff started panicking.

A soldier's friend in times of crisis is their weapon. A soldier is very confident with his weapon in hand when confronting the enemy. After midnight, the warrant officer Omar suddenly woke the captain up and said, "Sir, a number of lights are visible in the distance. The lights seem to be moving toward us slowly. Please have your sub-machine gun (SMG) ready."

The captain held his SMG close to his hand. The surroundings were also alerted, especially on-duty officers. But the problem was how many days the mere 30 armed soldiers would fight against an unknown enemy front without knowledge of their number or ability. And they were also unlikely to receive any support at Chukudum, a village far from the team site of 'Torit'.

However, the soldiers in the battalion were both high in morale and prepared to deal with any nighttime ambushes. All of them were morally strong enough to face off against any enemy and defend themselves. Most of the soldiers were from East Bengal and were brave and loyal. At times, whenever the situation seemed to get a little more favorable, the very next moment the news of the killing of five people in a nearby village stunned us.

Six days passed that way. Meanwhile, I built good relations with a few of the locals of that area. I thought sports at that moment would fit. Senior Warrant Officer Hamid once used to play for the Army team, and we had some other good players, too. We invited the locals to a friendly game, but after playing with them, Mr. Hamid was so astounded by the speed, movement, accuracy, and enthusiasm they had for the game that he did not hesitate to describe them (the locals) as similar to our national team!

A rebellious mindset grew among all of us. We then all agreed that we could not stay there any longer. We decided that we had to prepare to retreat. The families of everyone involved were worried because it had been a long time since the last contact. Moreover, that was a mission area.

In fact, a soldier always looks up to the commander on the battlefield because the morale of the soldier depends on the morale of the commander. Since that area was still under the control of the rebels and we had to move with 12 vehicles using that route, more preparation was a necessity. The captain briefed everyone, "My car will be in the forefront. You guys can rely on me. You put your status on me. Do you have that trust? We are going to make no wrong! No slippery movement! Right! No mistake! Right! Are you guys confident enough to show them what you are capable of? No issues! The situation will be dealt with accordingly. Let's see what's next, then!"

The roaring, raving, flamboyant army made the grenadier captain proud. In fact, the situation was such that there needed to be a way to know what the higher authorities were thinking about the ongoing problem due to the communication complexities. All our vehicles moved in the direction of 'Torit.' As we all moved along, panic grew amongst the civil officials. Their dismay seemed never-fading like an unending fire. Their only hope turned out to be the veteran captain, a man with seven years of service experience.

When one is in grave danger or on a journey or mission where there exists the possibility for death, what does he/she think to ask for first? Right! The loved ones! Family! It's always the family one tends to have on their mind in times of danger. BANBAT soldiers and civil officials, too, were no exception. They yearned to reach out to their loved ones, but alas the cruel grip of circumstance rendered all means of communication inaccessible.

In a bid to gain a better understanding of the situation, we cautiously stopped at least a kilometre behind the scene of the action. We then surveyed that terrain. According to reliable sources in that area, the conditions were not yet ripe for any semblance of communication and were suitable for a grand movement. In such a dire and trying moment, Captain Ronnie, a close ally and confidant, provided the captain with invaluable counsel from the sidelines. The language assistant, Hootuni, also by then alerted the UN stakeholders (especially the captain) of the possibility of an attack on the United Nations convoy.

Yet, we knew that we had to press on from there regardless. BANBAT's soldiers mentally prepared themselves for any eventuality. Our indomitable soldiers of the battalion had steeled themselves to face whatever peril lay ahead and any unprecedented intervention.

Every driver and second-seat occupant of the convoy was called and briefed on the situation. Military observers and civil staff trusted the captain. The rest of the vehicles followed the captain's car. All the drivers were briefed that the vehicle should not stop no matter what happens. Sitting in the front seat of the jeep, the captain loaded his SMG and put the barrel of the weapon out on the mirror top. Alamgir, the runner's soldier, seemed more motivated then, and confident Alamgir boasted, "Sir, I will not let anything touch you. They have to go through me first before anything happens to you. I'll have your back, sir! You can have this much trust in me now, sir!"

At that moment, the captain remembered Napoleon Bonaparte's famous quote, "The first value in a soldier is the endurance of fatigue, and courage is only the second virtue".

The flag of peace flitted high on the impetuous Bangladeshi UN peacekeepers' vehicles that crossed the perilous area under complete militia control at a dauntless, overwhelming speed. A convoy with only 30 soldiers was ready to doom the enemy, but the groups of armed hyenas could not manage enough courage to fight out a convoy with a sheer 30.

In the meantime, no chirpings of terrorists reached their ears.

As soon as the convoy reached the camp, Major Kabir embraced the captain like he had met his youngest brother after a long time. A few moments away, the head of the office, Winnie, rushed to meet us all at the tea party.

She shared her concern with us. "I was breathless when learning to know about the situation you guys were in! It was like my heart stopped beating."

The biggest concern of mine was none but my son-like, ever-smiling Captain Rahim. It is with great conviction that I can affirm the resolute strength and courage of the soldiers of Bangladesh, who I am confident possess the fortitude to overcome any challenge that may come their way, the way of peace. As such, it brings me immense relief to have it reported that all the UN personnel, police officers, and civilian staff were safely escorted to their homes (She was talking about 'The Torit Team Site' - the place that I still call home).

Thus, all of us were greeted with an overwhelming sense of elation and jubilation there. These emotions rapt all of us who made it to the Torit team site that day.

KIDAL

MAJOR SHAHRINA BINTE ANWAR LIBRA

(UNITED NATIONS MULTIDIMENSIONAL INTEGRATED STABILIZATION MISSION IN MALI - MINUSMA)

January 2017. Setting afternoon. A late-sinking sun walking towards twilight.

Winter's chill lingers, making the warmth of the sun a welcome respite. The tinge of winter stayed yet, and the sunshine felt perfect, full of warmth. A crow perched on a nearby coconut tree and punctuated the air with a sense of familiarity, as it had been consistently cawing for several days.

At such a moment, the cell phone on the table started ringing (as if it had sprung to life). It was a logistic recce and survey (LRS).

"We've got an LRS and will go to Mali," one of my course-mates informed me.

Footnote: Recce refers to a 'Reconnaissance' mission.

The following day, an official letter revealed that our team of 15 would be moving to or deployed at different spots throughout the country of Mali. My destination was 'Kidal', a place nestled on the fringes of a desert contiguous of the Sahara.

En route, as our flight had a stop in Turkey, a layover there offered a fleeting yet enchanting glimpse of Istanbul. It was a fascinating experience indeed. Before long, we touched down in Bamako, the vibrant capital of Mali. As I stepped onto the African soil, I couldn't believe my eyes as I had truly understood that I left my home continent behind miles away. A sense of awe washed over me, and the adventure began.

After it had gotten a bit late at night and the darkness and silence deepened, I went straight to the 'Bangla House.' Within a day or two days later, our journey toward the Kidal Super Camp commenced. In those days, I had only read about the Sahara Desert in books. But it was time – I was going to experience the Sahara with my own eyes! Witnessing the magnificence of the African landscape, especially of a different continent than mine, was truly special. As the first Bangladeshi woman to set foot in the heart of Kidal, an unparalleled sense of elation surged through my being, filling me with an indescribable rapture.

The time following my return to the homeland after the completion of the LRS seemed to vanish in a haze as it flew by in a blur, leaving me bewildered. The relentless demands of daily life had blurred the lines between day and night. Then, the moment appeared for bidding adieu, a final goodbye. I was departing from the city I had always known that was so close to my heart to journey to a distant land. The familiar, cherished faces would be absent from my life for an entire year. The vibrant energy of this city would be but a memory. I would no longer be able to experience all of it for at least a year. The hustle and bustle of this city would no longer stir my heart. To be honest, it still surely would, deep inside. However, I ventured forth, leaving my own people, the people with a brownish complexion, to embrace the land of blacks.

With a heavy heart, I donned my military uniform and boarded the venerable white pigeon-like old-fashioned United Nations aircraft, a symbol of peace, which I called the "white dove".

Tears threatened to escape from my heart and strain down my face, but I fortified my resolve. That day, we readied ourselves with a solemn commitment to peacekeeping. Strength and courage accompanied me on this journey, bolstered by my fellow colleagues on the mission.

May 2017.

We found ourselves touching down at the airport in Gao, a conspicuous city situated 1200 kilometers away from the capital of 'Bamako.' It was hard to believe that I was on a mission and it had started…

The unrelenting heat of the desert besieged us at every move, leaving us torpid, leaden, and exhausted. The harsh heat of the sun and its keen rays visibly bleared the high-ranking officers who showed up to welcome us at the airport. The heat was too intense to open my eyes. Even through my aviator sunglasses, I could barely see except the perplexed frowning mirages in that dreary deserted arid landscape.

During the stopover, how could we miss visiting/staying over at the BANBAT and BANSIGNAL camp! Seeking respite, we sojourned at the

BANBAT and BANSIGNAL camps, but the wait seemed interminable as if an eternal abidance – 'The everlasting wait'. Our days in the BANSIGNAL camps unfolded akin to those of nomads, burdened by the meager possessions of our hand luggage.

As we had no UN identity cards issued after our names by then, it was as if the names on the PAX list were floating on a cloud and would float there forever. Going to Kidal would be a reality soon that I understood very well and early, while our main luggage would travel directly to Kidal from Bamako.

Our journey to Kidal remained elusive. This predicament persisted for a week. As we finally made it to the Kidal Super Camp, the holy month of Ramadan commenced, and construction work began in full swing and earnest. The unyielding fervent sun above our heads all the time and the work for the mock camp's development and establishment conspired to make life insufferable, a plight further exacerbated by an acute water shortage. Soldiers faced an unpredictable water supply, with water for bathing available one day only to vanish the very next day; or sometimes it would last for two days and then we would experience scarcity for the following few days.

June 2017.

We found ourselves living at that remote, secluded super camp for roughly a month. Time passed swiftly, and 15 holy days of the month of Ramadan had already elapsed. The date was June 8, 2017 on the calendar; the clock showed 6:10 PM.

I was seated in the field mess and was eagerly awaiting Iftar. In a mere 15 minutes, the call to prayer was about to resonate through the air. But what resonated was the sound of an explosion from what might have been a cannonball or grenade. When that sudden sharp crack pierced the silence, it took me no time to realize that the camp had been attacked.

Instantly, all the walkie-talkies pealed out with the alarm "Back storm". We made haste to the nearest bunkers as the cacophony of gunfire enveloped our surroundings with thunderous explosions with no respite. The whizzing of bullets whistled overhead, striking the ground perilously close to the bunkers we took shelter in. The crackling sounds were clear, with the shells falling somewhere nearby. Then, a series of harrowing explosions shook the very foundations of our existence. Blast after blast, everything was happening quickly, and we had to beat the time to rejoinder appropriately.

Just thinking about it sends shivers down my spine. The impact of the sprinter attack had destroyed Sir's office. I had offered my gratitude to Allah, the Lord of the Worlds. In that terrifying attack, six people had been injured in Kidol Super Camp, and a military vehicle and several barracks had been destroyed, but luckily all the members of our contingent had remained unharmed. There have been many attacks on the camp since then, but the unbearable memory of that day still haunts me.

It had been quite some time since Iftar time had passed. Despite having fasted for nearly 15 hours, the situation did not allow me to break my fast, though I had dutifully not missed performing my prayers.

8:20 PM, Night!

Yes, it was 20 minutes past 8 pm then, as per the hands of the clock. I had just returned from the perimeter defense and placed my weapons next to the dining table after I removed my BP helmet. The tables were adorned with plates in lines for Iftaar.

We were awaiting the arrival of the officers, who entered the field mess in procession, but, sadly, no one could break their fast that day in time. Hunger had long since dissipated from our minds. Lieutenant Colonel Mehedi, the esteemed G-Force logistics chief, made his entrance, bringing with him the news that his sector's office was on fire, and by then the office might have been engulfed in flames.

Footnote: Iftaar - The Muslim ritual for breaking fast.

I flinched in shock. Anxiety gripped me, and I rose to my feet.

I recalled the conference from the previous day, which had necessitated breaking my fast in his office. Yes, the same sector office that was set fire. The day before, I clearly have it in mind, during the Iftaar break, we were so immersed in following our religious rituals and speaking our thoughts together with our colleagues.

Had the conference occurred that day, or had the attack transpired the day before, just thinking about the possibility sent shivers down my spine! The outcome would have been beyond imagination. His office was decimated like a sieve due to the continuous sprinter coup.

I offered my heartfelt gratitude to Allah, the Lord of the Worlds, right then for our deliverance from harm. In that harrowing assault, six individuals sustained bloodcurdling injuries within the very confines of the Kidal Super Camp and a military vehicle and multiple barracks were obliterated, yet all the members of our contingent miraculously remained unscathed. The camp has since been subjected to numerous attacks, but the haunting memory of that Iftar moment and that whole night lingers with me on until today, a specter that refuses to be exorcised.

During the 10-month mission, we faced mortar shelling and complex IED attacks. Our main camp, occupied by Bangladeshi Engineers (Construction), was located approximately 1600 km away from the capital. Another camp, called Agulhaq, was situated around 300 km from Kidal. Both the Kidal and Agulhaq camps relied on logistics provided from Gao Super Camp, which was approximately 450 km away. We would gather two weeks' worth of supplies at a time, and these logistics would be transported by road. The journey along the 450-km route typically took four to five days, assuming the convoy remained unharmed. Unfortunately, we rarely had the opportunity to enjoy fresh supplies, unlike the contingents stationed in Gao or Bamako. Take bread as an example; a loaf of bread usually lasts for four to five days. And that is why we asked to be given frozen bread.

40 to 60 per cent of the fruits and vegetables on every turn of supplies returned rotten to us. Despite being in the middle of the hostile weather, extreme security risks, and challenging surroundings, our engineer soldiers couldn't hope for any fresh food. However, our soldiers, too, were inexorable.

They had cultivated their own vegetable gardens within the encampment territory with a variety of golden crops, green vegetables like calabash, gourd, cauliflower, pumpkin, Malabar spinach, red spinach, eggplant, green chili, lettuce, and whatnot! At the heart of the desert, our soldiers weaved green while those vegetable gardens kept serving the contingents with the supply of fresh foods.

In that place of hardship, we were confronted by the acute scarcity of water. We felt every moment of struggle of the water crisis. With no other viable sources of water, we had to rely solely on the water supply the United Nations provided for our daily needs. Astonishingly, four thousand precious litres were allocated every day by the UN, which served as the vital lifeline for all our basic activities ranging from quenching our thirst, ablutions, preparing meals, cleaning clothes and bodies alike, and ensuring the cleanliness of the camp.

We never received any bottled or alternative drinking water sources only for drinking purposes amidst such dire circumstances. Our soldiers tried to save every drop of water. They did everything possible to economise their usage of the precious available resources at every turn. Even during ablutions, they remained utmost vigilant of the imminent crisis if even a single drop was wasted.

Standing near the butt-end of the Neem tree, they performed their ablutions lest their efforts affect the reserve of water.

Many of our soldiers had previously left for Gao and Bamako on different assignments, and every time they returned they came back with small neem saplings in their handbags. So, they were no doubt the guardians of those Neem trees and stood watchful vigil around them lest harm reached the Neem, such is their incredible value.

Under wastewater treatment and management, even the bathroom water was being recycled to use for the vegetable gardens.

Ah! How dear those beautiful vegetable gardens were! Only the engineering contingent could achieve such a feat! Only they could make such a situation possible, and they did not disappoint.

Sergeant Faruk devoted the majority of his time to nurturing his cherished vegetable garden. He tended to the plants with the same affection and care one would give to their own children. The UN provided much but never fresh green chilies; instead, the UN offered spicy capsicum always, which never was enough to appease a Bengali's palate.

Faruk's joy knew no bounds when the chili plants of the garden started yielding fresh green peppers. What on earth could restrain his mirth? Finding small pleasures amidst adversity was the essence of life always. The mere joy of such sheer moments was life truly – the definition of life in a nutshell.

Between March 2017 and March 2018, ration supplies at Kidal Super Camp ceased for 80 days. The supply also remained halted for a stretch from October 2 to October 15, 2017 as well as a long tenure from January 8 to 23, 2018.

As we were left with no way to store extra supplies, the absence of supplementary ration storage rendered the rationless days insufferable.

The anguish experienced was beyond the power of words to convey. Security risks severely limited the opportunity to purchase from local markets. In the barren land where locality had been a rarity, reality left little in the way of eking out an existence.

Nevertheless, we consistently maintained a stockpile of rice and dry food sufficient to cover at least two months (in the event of an emergency). The trouble persisted in regard to fresh rations. And at Kidal's market, onions and garlic, too, happened not to be found regularly but only occasionally.

Usually, oil and potatoes could be found in limited amounts on usual market days that could sustain us for a duration of 10 to 15 days. However, finding mutton at a remarkably low price in the market was easy. The mutton price only seemed affordable. We had to rely on combat ration packs for the remaining days.

August 17, 2017

When the ration truck arrived at the camp, its arrival was no less than Eid day for the soldiers. The company sergeant major himself was on the ground for the unloading of rations. The diligent logistics clerk checked all the ration delivery notes bit by bit and meticulously documented the supplies.

The sun overhead felt like a gargantuan burning furnace, with temperatures soaring to at least 50 degrees Celsius, if not higher. So what! Despite such heat, there were no signs of exhaustion on anyone's face. Nobody showed up with any sign of fatigue. Regrettably, a significant portion of the rations reached us either rotten or spoiled. And we had no way but to reject and return 40 kg of tomatoes, 20 kg of oranges, and 35 kg of grapes.

I stood there to observe the ration supply work (pound by pound) as the sky ominously darkened. A dust storm commenced and started showing its frenzies as the wind's severity progressively intensified.

Due to the urgency of the situation, the ration trucks were hastily unloaded. However, the camp faced a shortage of space, resulting in the containers being stacked on top of each other. Unfortunately, some of the containers had already fallen due to the impact of the dust storm. Fortunately, we had safely taken shelter in the bunker and assumed our positions. It is worth noting that, like other MINUSMA contingents, we were unable to establish our camp using CORIMEC (CO.RI.MEC., which refers to the construction of prefabricated buildings and living camps) due to security risks. Instead, we resided in secure underground bunkers.

Our valiant soldiers never compromised guarding their posts like vigilant sentinels, even as the heavy dust storm raged around. The storm's ferocity had somewhat abated as the noon left to make way for the afternoon. Grounded containers were lifted up back into place with the aid of a forklift and excavator. The sheer intensity and destructive power of a desert storm in the heart of the desert we would never have known if we had not come to the Sahara.

Our struggle for survival had not only involved evading mines and repelling surprise attacks on our camp but also contending with the unyielding hostility of nature itself. The following morning, the company quartermaster sergeant (CQMS) had reported a concerning discovery: water was dripping from the onions in the supplies received the previous day, indicating they would spoil soon.

"Sir, the supplies that were delivered to us yesterday had been leaking water, especially from the onions. And, sir, onions get rotten so quickly" said the CQMS.

I had instructed that they be reduced to wet pulps, pounded, and preserved with the utmost care. If all had gone according to plan, the next shipment of supplies would have arrived in 15–20 days. These provisions used to be transported via the Gao-Anfis-Office with the Kidal logistics convoy.

Typically, BANBAT peacekeepers escorted that convoy throughout the journey and as it traversed the treacherous terrain. The perilous and inhospitable path was strewn with various improvised explosive devices and mines, and the ever-present threat of surprise attacks had always been there. In September of the previous year, a logistics convoy, while returning via the Gao-Anfis-Kidal route, fell prey to a mine deadfall, claiming the lives of four peacekeepers. My course-mate, Major Zadid, had right then been immediately sent to the hospital and subsequently confined to a bed in a Singapore hospital; his condition was critical. With serious injuries, he was still bedridden.

March 2018

We were nearing the end of our mission. There was a six-hour time difference between Kidal and Bangladesh, so staying in touch with our home country was not always possible. There, a few months ago, Carpenter Junayed's wife passed away. Junayed could not even help her with a sip of water in her final moments. Soldier Harun never got to see the face of his newborn, as he left for an eternal journey just two days after the birth. The father never got to meet his child. It went to the land of no return two days after birth, never seeing the father.

News of loved ones' deaths from nearly 10,000 kilometers away cast a shadow of grief over the harsh desert. Yet, we continued carrying out our responsibilities with the utmost dedication. Life goes on despite everything.

That night was a 'super moon night.' The full moon with a huge halo around it in the desert was quite a sight. Perhaps a soldier thought of his beloved back home staring at the moon, lying in his bunk bed while his pillow turned wet with tears. It was then late at night. Reports of enemy fire were creeping in from the distant posts. Mike One (duty post) was alarmed with the news of firing and every time they received news of a new camp attack.

Our camp also got attacked. The Kidal Super Camp in the heart of the desert and our base would become a battlefield once more.

War was coming, and we had to dress up for the battle. Completing the write-up was a luxury. Leaving my write-up, the way it needed to be left, I put on my bulletproof jacket, placed my helmet on my head, and put my arms in a position to attack at any moment. On my notepad, I wrote, "Might return soon and write again, filling the white pages with black letters, piecing together fragments of our experiences if I come out alive again."

Mozambique's Theft Paranoia

Brigadier General Bodrul Millat
(UN peacekeeping mission in Mozambique - ONUMOZ)

"Did he flee in that direction? Did he go that way? Did you see him go silently? Didn't you do anything?" I was so surprised that I nearly shrieked. "Yes, I saw! What was I supposed to do? What could I do?" he replied in broken English, equally surprised. I was dumbfounded, gazing at him with my hand on my head, staring at him in shock. What a den of thieves. What an anarchical state.

Let's recap the incident that had taken place just 10–15 minutes earlier. I had my microbus parked right there, and I left it there after locking it to buy bread from the bakery alongside the road.

Mozambique was rife with theft, which we had been cautioned about many times. Many had already fallen prey to robbers, and so we stayed extra vigilant. That's why I stood by the bakery on the side of the road only 10–12 yards away from the car, ordering bread and keeping an eye on the car almost all the time. But at some point, perhaps when I was paying for the bread, it was a matter of a few moments, the thieves broke the glass lock of the back window of my microbus and sprinted away with my briefcase that had been on the passenger seat!

On the sidewalk were a few shops, and quite a few people were there. Everything happened right in front of their eyes, and the crowd did nothing except act as mere spectators. Nobody intervened. They just supported their fellow countrymen. Yes, it might have been the consequence of having the backs of their compatriot or, in other words, the thief of their own nationality; otherwise, they would have made even a minimal effort. The thief got away with what he wanted because he had that backing! How marvelous! There was no money or anything precious in the briefcase, and they would not have benefited much from their theft. But in that briefcase, I had critical documents like my passport, national ID and UN ID card as well as other vital papers. And this alone was enough to make me sit in the middle of the road (stranded).

The terrifying concern I had was what would happen to my passport. What would I do without it? How would I make it to my home in a month as I was supposed to be on vacation the following month? How would I travel without a passport?

Moreover, there was no Bangladeshi consulate or embassy in Mozambique. So, if I had to get another passport, I would have had to go to the neighboring country of Zimbabwe, where there was an embassy of Bangladesh. But how would I make it to Zimbabwe without that passport? This was enough to leave me highly anxious.

The people of a country ravaged by war face extreme poverty; the poverty and despair of a war-torn country can impact the mindset of its people and drive them to the highest level of desperation, and Mozambique was a striking example of such consequences. There was no assurance of having two meals a day, no employment opportunities, no means of earning a decent income, no work to make two cents. In such circumstances, what could they do? They resorted to stealing, looting or robbing as their daily routine. And they did not target their fellow Mozambicans, who had nothing to offer them or who may have been spared out of a sense of affection or love towards their fellow countrymen. And no one intervened if they harmed others. Nobody stopped them. Their main prey was us, the military members working for the UN, civilians, or various volunteer and human rights workers employed by the United Nations.

Mozambique stretched for about six hundred kilometers across two thousand kilometers along the east coast of Africa. It was a vast country in East Africa torn apart by war. In 1994, peacekeeping forces from Bangladesh and four other countries were deployed in three regions of this country to maintain peace and stability. The northern region was the responsibility of the Bangladesh sector, which had its headquarters in Nampula, the regional capital. The Bangladesh Armed Forces constituted eighty percent of the sector staff. A complete battalion there was diligently carrying out their duties. For the strategic benefit of safeguarding the security of that region, the members were scattered in different positions in different instrumental cities but in a planned chain.

I was an officer at the sector headquarters. I earned a handsome salary, enough to rent a duplex. I rented the upper floor of a splendid duplex house with two other colleagues. It had three bedrooms, a spacious living and dining area, a TV, a fridge, an oven, an iron, and even all the crockeries and furniture. It was a lovely little duplex. So even though it cost 600 dollars a month (200 dollars for each of us), I felt lucky. But there was a catch: it had only one toilet. That was another weird thing.

Now I am involved in real estate business. Often, I have to deal with all kinds of clients and their peculiar demands like "You will have to prepare three attached bathrooms and one common one." They even seem not to care about the sub-vent bath that already existed in the plan, which means in a 3–4-room house, they want one bathroom for each person living in the house as if they all would have to fulfil their natural functions at once. They do not easily understand that having one less bathroom means having more living space, less maintenance, and lower costs (avoiding unnecessary expenses).

I try to reason with them, using examples from Europe and even Mozambique, but who pays any heed? These customers are not the type that can be easily convinced. They have their own ideas of what a good house should be. A friend of mine who is also in the business once joked, "It's not about the convenience or inconvenience of the client or what the client wants or needs; it's about what the owner/payer makes. Your loss means the landowner's profit." Is that so? Was he right? I wonder. When I built my own house, I decided to have one less bathroom than I could afford. I did not want to invest unnecessarily in an extra toilet.

I wanted to have bigger rooms and more comfort, so my rooms became more spacious. I thought I had made a wise choice. But then something happened that made me question everything. It all started with an incident of theft…

Anyway, let's put this matter aside for now and get back to the theft incident.

Although I knew it was futile, I chased after the thief who had fled with my briefcase along with two or three young men from there with me. Even after crossing the city limits and going a long way through a village, we never caught a glimpse of him. One of the youths spoke some English, so I used him as an interpreter to inquire among the villagers if they had seen anyone carrying a briefcase like mine. I also offered a reward of one lakh meticais (the currency of Mozambique) to anyone who could help me locate him. But it was in vain. They were unwilling to betray their 'brother' for such a 'despicable' deed for one lakh meticais. The reader might assume that one lakh meticais were equivalent to one lakh taka in Bangladesh. In fact, however, this was only worth about five thousand rupees. Disappointed, I returned to the police station and filed a GD. I even made an announcement to the people near the bakery with the assistance of the police that I would drop the case and pay the reward money if the thief returned my passport and two ID cards that were inside the briefcase and kept everything else (a Walkman, sunglasses, some Mozambican currency). But nothing came of it.

The language was one of our major challenges in Mozambique. The people of that country spoke Portuguese, a language completely foreign to us, and they had no grasp of English either. Some of the enthusiastic junior officers attempted to learn Portuguese from books, but their speech was incomprehensible to us and probably to the ones they addressed as well! (only God knew what they meant by what they said, and even whether the person they asked understood or not). We were scared lest they misinterpret their words and utter something offensive by mistake! I remembered a scene from a movie where the hero's friend claimed to know French as well as English. He tried to ask in French, "Can your daughter help us find our way?" But what he said implied, "Can your daughter come and stay with us?" The man fetched a gun immediately from his house and ran after them. So, we were in such a predicament. There were very few who could communicate in English. So whenever we had an important task, such as signing a lease agreement or a house rent agreement, we made sure we had our UN interpreter from our headquarters.

However, my housemate Mahbub once went to the market and told me, "Sir, see if this language works or not?" He asked the shopkeeper in clear Bengali, "Brother, how much is a kilo of tomatoes?" The shopkeeper understood the language correctly and replied, "Two mil meticais" (two thousand meticais). It was a remarkable occurrence. Actually, it was not so remarkable since Mahbub conveyed what he said with his gestures, and the shopkeeper grasped it correctly. Anyway, that was how we managed. We were going on like this.

But this was not my sole encounter with theft in Mozambique during my eight-and-a-half-month mission (we were the last group of this mission). Yes, that was not the only such experience. The thieves did not single me out. Almost everyone in Nampula had faced these robbers in some way or another. House burglary and car part thefts were everyday affairs. One even fell victim to mugging! In Beira (the second largest city in Mozambique after Maputo, the capital of Mozambique, about a thousand kilometers south of Nampula), a military observer found his Land Cruiser SUV standing on bricks the following day! The thieves left the body there only and stripped off the four wheels, glass, mirror, and all the fittings inside! The officer had to pay a hefty fine in dollars to the UN.

My second experience occurred in my absence. But that time, I did not incur much loss; rather, the thief did. After much trouble, I managed to arrange a temporary (six-month) passport by contacting the Bangladesh embassy in Zimbabwe, obtaining a letter from the sector commander, and sending another officer. When I returned from the country, I saw that my room had been tidied up as usual. I was astonished that neither of my two housemates knew anything about the incident (of cleaning up). Because I had locked my room when I left for a 21-day vacation, neither I nor my two friends knew exactly when the theft took place.

As soon as I opened the door of the room, I stood there stunned. The mattress on my bed and the pillows were cut and torn, and the whole room was littered with feathers, springs, and pieces of cloth. All the clothes that had been in the closet were on the floor. There were some books also scattered on the floor. In a word, it was a terrible situation. But the most horrific thing was that the whole room was stained with blood as if a murder had taken place in this room!

The blood had dried up and turned black, but blood stains were all over the furniture, wardrobe, table, and floor. We three did a lot of research, and what we could find out through our investigation was something like this.

Before I explain, let me share one more detail. Although Mozambique was a war-torn country and once a Portuguese colony, it retained purely European culture. For instance, all shops would close at six in the evening except for bars, restaurants, and nightclubs. But even though the shops were closed, all the lights inside stayed lit up. Through the huge glass windows, one could see all sorts of fancy electronics, clothes, shoes, etc. There was no grill behind the glass! It would have been easy to steal those goods with a little effort. But in that very country that had become famous for burglary and theft, nobody would steal those goods. Why? Because of three particular reasons.

The first one, in my opinion, was that they would not want to be the reason for the infliction of harm on their 'kinfolk', the second one was reverence for the law, and the third one, which was actually the primary one, was the dread of death. A few shops had enjoyed a system or arrangement of guards based on the area. And our area was also one that had guards who had permission to shoot down any thief. The guards did not wield sticks or shotguns but AK-47s! And I heard that if someone attempted to steal, they would shoot without hesitation, and there would be no trial for that.

That was why none of our house's windows had any grills along with the glass windows. One window of my room faced a huge veranda in front of the house. The thief or thieves (because we had found at least two kinds of footprints in the blood: one barefoot and one with shoes) had shattered the glass of that window to enter my room. While entering, one or more thieves had wounded their hand or foot. Then why was nothing stolen? The outcome of our 'investigation' was that there was nothing to steal. They had not come to take my clothes or belongings, and there were no electronics or anything fancy or precious in the room. They had only come to take one thing – dollars!

That was why we had no faith in the banks in Mozambique in addition to the possibility of war erupting at any moment. Moreover, the banks there were in a critically miserable condition. None of us kept any money in the bank. We would send some money home through someone who traveled to the country and keep the rest hidden in our room. The thieves were aware of that. But they were unaware that we would never stash money where they would search for it. Moreover, they overlooked one thing: this room's occupant had gone on vacation, and he had simply taken all his money with him!! Wasn't it supposed to be that way?

So where did we used to hide our money? We would conceal it in a place that, if the thieves learned of it now, they would undoubtedly pull out all their hair in anguish. We would keep our money inside some old Bangladeshi papers and magazines that came with the books. Moreover, so that I didn't mistakenly give it to someone else or discard it, I would write in Bengali on the cover, 'There is money inside!' Furthermore, so that it didn't appear like anything different, I would write something else in Bengali on the cover of the other magazines, such as 'Today's weather is wonderful', 'Vaidi' (a Bengali expression), 'Brother, how are you doing', etc.

And that time, I didn't bother to go to the police anymore as I had witnessed their competence before. We knew their ability. Besides, I had not suffered much loss; on the other hand, at least one of them had been severely wounded. If anything, I felt a little sorry for them!

The last incident was like something out of a thriller; I acted like a Bangla movie hero and performed all the Dhishum Dhishum acting (the sounds from fighting scenes of 1990s Bengali movies).

Among my two housemates, Apu was my coursemate, and Mahbub (not their real names, but they are irrelevant to the story) was our one-course junior but a close friend nonetheless. As we had similar mindsets, we enjoyed and endured all the difficult days of Mozambique and made it through without any family quarrels, mental anguish, or unrest. Now, let me get to the point. One day, Mahbub noticed his belly fat protruding slightly, so he started jogging every morning. I realized my belly was even more noticeable a few days later, so I joined him. However, he would leave for the jogging when it was still dark outside, while I would follow a bit later. We could never synchronize our schedules. We would only catch a glimpse of each other when I was heading out and he was returning. As for my coursemate, he was a lover of comfort and leisure. He was not into jogging. As you will soon discover, he was also a notorious Kumbhakarna (a character of Indian mythology).

The main predicament revolved around the sole key to the house. Whenever Mahbub ventured out each day, he would leave the key in the front lawn, cleverly concealed within a bell flower bush that we were all familiar with. However, to our dismay, it became evident that the thieves had also learned of the key's hiding place. Our house, though a duplex, posed an additional challenge with its external staircase leading to the upper floor. Upon entering the main gate, a small vestibule stood before the stairs, providing access to the upper level. The ground floor had a separate entrance. The property owner had rented out both floors while reserving the basement for his own family. To enter our portion of the house, one could open the door facing the staircase from the inside without a key, but a key was essential for entry from the outside. Therefore, when someone departed, we would diligently place the key in our "secret" spot. Since our morning runs occurred at different times, the first person to leave would take the key to ensure easy access upon their return. Without the key, it would prove challenging for them to enter the house if they were to return earlier than expected after their run.

I opened the door from inside and left it open. Then, I went jogging. I met Mahbub after a short while as we used to have a crossing. He said he had finished his exercise and was heading to the vegetable market. I returned home after running for about half an hour. I was doing some free-hand exercises on the pavement close to our house when my eyes wandered to the second floor's balcony. I spotted a young black man. We had a local young man who worked in our house, Anthony, who never came to work in the morning. I was a bit puzzled, as I couldn't tell if the young man was Anthony or not. Then, I saw another young local man join him on the balcony. He turned out to be looking nervous as he made direct eye contact with me. I had no more difficulties understanding that they were thieves, but I didn't know what to do. I was struggling to figure out if I had anything to do right at that moment. I shouted as loud as I could, calling my friends' names at the house. But it was to no avail, as my course-mate was a descendant of Kumbhakarna, as I had mentioned before. He was not one to break his slumber so easily.

At my scream, our lady on the ground floor and even two girls from the next house came out, but who would wake up my friend! The two thieves seemed clueless about what to do either. They knew for sure that I was one belonging to that house. Anyway, I did not do a brave thing but rather something stupid. After not getting any help from my friends (the one sleeping and the other one in the market), I thought I had to catch the two thieves anyway without gauging any potential risk. And I acted on it. All I knew was that I had to seize them.

I picked up the key from the Bela bushes and opened the outside door, only to see two young African men of 20–22 years old running down the stairs. I didn't feel like my brain was working properly at that moment. I knew that fighting those giant figures would be a huge risk. Yet I stood in the middle of the stairs blocking them with both arms and yelled. They too stopped for a moment, clueless as to how they would make their escape safely. The two young men paused for a second, then came down with a fierce scream and pounced on me. Consequently, I lost my balance but dragged them both to the outside lawn using their momentum.

I was trying to hold both thieves with my hands, but one of them slipped away and escaped in an instant. I ignored him (the one that escaped) as it would have been a pointless to chase after the first one while the second one made an escape. So, I clutched the other one tightly. I noticed that many people were enjoying the spectacle from the balconies or lawns of the surrounding houses (they might have been making fun of my situation). I pleaded with them for help, but none of them showed any interest. And my two friends were nowhere to be seen either!

Suddenly, the second one also wriggled out of my grip and ran in the opposite direction of the previous one. I chased after him without hesitation.

He was not too far ahead of me. He was running just a few feet in front of me. But then, some people on the side of the road came to my aid after hearing my cries and seeing me run. They caught the boy and restrained him. I joined them and tied his hands behind his back with a rope brought by one of those who came to my aid. Then, I brought him near our house.

Just then, Apu's voice rang out from the balcony of the second floor. "What happened? What is all this fuss about? Why is there such a crowd of people?" Apu was asking, twinkling his eyes.

Mahbub had returned as well. He inquired, "Sir, what's the story?" After explaining everything to them, Apu went to fetch our car. Oh, by the way, we had a brand-new microbus provided by the United Nations. We three used to use that microbus, but we were afraid of the car or its parts being stolen, so we never parked it near our house or in the garage. Instead, we left it with the other government vehicles in front of the office every day and walked back home. There was a squad of the Bangladesh Battalion that guarded the office and the cars.

Then, Mahbub, I, and a few locals went to the police station. Apu drove the car. After handing over the thief to the police and filing a GD, it was almost afternoon when we returned to our abode. We had already informed the office about the whole incident over the walkie-talkie and told them we would be late for work.

That was the conclusion of the episode of my encounter with the third burglar. But since he had taken our keys and might return with his accomplices and break into all the rooms with ease, we replaced the five locks of our house that very day. I didn't want to risk making new keys. We incurred many expenses. Among all the losses, this incident of theft and our time was our net loss!

But when we returned home and sat down to review the incident and the aftermath, I was terrified, and my limbs were stiff. What if they had had a gun or a knife! And what if he had shot me dead on that stairway? Mahbub added more fuel to the fire and said, "If one of their knives grazed or had a scratch on you, then you're doomed with AIDS!" As if dying there from AIDS was more normal than dying from a stab wound! Actually, before and during our stay in Mozambique, we had heard so much about AIDS that we were constantly afraid. I was even reluctant to shake hands with the locals because one of our 'expert' colleagues had once said, "Imagine his hand is bleeding and your hand is also wounded somehow or you nicked it while shaving in the morning, then that's a bad contact, meaning AIDS!" We even kept our own shaving razors in our room every morning after using them and locked them up so Anthony could never get his hands on them, even though we had no idea if he had AIDS or not!

I realized when I pondered over the incident again that it was a mistake to move to the stairway and to attempt to confront them or get involved in a scuffle with them. Fortunately, they were also scared like me or even moreso. What if they had grabbed me and thrashed me instead of running away? Or what if they had assaulted me with a weapon! During my eight and a half months in Mozambique, I had many opportunities to explore beautiful places, endure terrible and painful experiences, face linguistic challenges (including being a victim of linguistic harassment), teach the Bengali language to local children and teenagers and organize a Bengali music program with their participation and purchase a satellite dish to watch the World Cup football match (1994) with shared financial contributions (there was no other option at that time) and many other events of joy and sorrow that could fill many other stories. Still, I feel deep in my heart that these three encounters with the thieves left the greatest marks. Besides, I have already written quite a bit on those topics and so have others. Therefore, this time, I wrote something different – 'The story of theft paranoia in Mozambique'.

PYGMY DANCE

LIEUTENANT COLONEL NASIR UDDIN AHMOD
(UNITED NATIONS MISSION IN THE DEMOCRATIC REPUBLIC OF THE CONGO -MONUSCO)

Participating in a United Nations mission bestowed upon me the extraordinary privilege of immersing myself in the vast and awe-inspiring African continent. Among its magnificent lands, the Democratic Republic of Congo emerged as a profound imprint etched upon the deepest recesses of my heart. It stood as a testament to nature's grandeur, boasting colossal mountains that pierced the heavens, lush forests teeming with life, meandering rivers that told tales of ancient civilizations, and sprawling lakes that mirrored the sky's expanse. Everywhere I turned, unparalleled natural beauty enveloped me, filling me with a profound sense of wonder. The verdant landscape was an orchestra of resplendent greens extending as far as the eye could fathom, as if the very essence of Congo's territory was rooted in glorious emerald hues.

To this day, the memories of that staggering country, adorned with incomparable natural wonders, continue to reverberate through the depths of my being. Our journey led us to none other than the enchanting realm of Ituri Province, where our base camp was nestled amidst breathtaking surroundings. Bounded by the majestic Ituri River on one side and the expansive Lake Albert on the other, it was a haven immersed in untamed beauty, a place where the spirits of exploration thrived.

Diminutive compared to its neighboring provinces, the verdant province of Ituri sprawled across an area roughly half the size of Bangladesh, its borders brushing against Uganda and Sudan. Here, the weather remained steadfast throughout the year, offering a gentle embrace even on the coldest of nights. Wrapped snugly in a light quilt or blanket, slumber became a cherished companion, cocooning me in comfort.

In this ethereal landscape, I came face to face with the true meaning of the word "temperate." The chilly winds, borne from the surrounding hills, possessed a tranquil bite that permeated my very being. They whispered stories of resilience and secrets of the land, forging a connection between myself and this captivating terrain. Indeed, the weather here remained consistent, with occasional subtle variations in temperature that only added to the region's allure.

Never before had I encountered a region so perfectly attuned to the concept of a "temperate atmosphere." It cocooned me in an ambiance of unparalleled bliss, evoking a profound sense of comfort that defied description. In a single word, it was "comfy."

This place unfolded as a veritable paradise of natural splendor, a sanctuary where nature's artistry flourished. Those who had ventured to Bunya, Congo before me had proclaimed a singular longing for the captivating weather that would persist long after departing. The warm caress of sun-kissed mornings, the gentle breezes that carried whispers of forgotten tales, and the cool evenings that provided respite from the day's endeavors were all treasured memories I held close to my heart. Oh, how I yearned to capture those fleeting moments once more, to relive the joy that Bunya bestowed upon me. It was a perpetual source of comfort and delight forever etched in the tapestry of my adventurous soul.

Bunya, the capital city of Ituri Province, stood before me, its bustling streets and towering buildings a stark contrast to the dense forests that hid the elusive Pygmies in the depths of Congo's Ituri and Kivu regions. Since my childhood, the tales of these enigmatic people had captivated my imagination, fueling a longing to witness their extraordinary lifestyle firsthand.

The Pygmies, known as the world's most diminutive tribal community, thrived in the remote reaches of Ituri and Kivu. Although the exact number of Pygmies remained uncertain, an unreliable source hinted at a population of approximately 300,000 in Congo. These remarkable individuals sustained themselves through hunting and gathering in the unforgiving jungle, leading lives of simplicity and harmony with nature. Encounters with them were exceedingly rare, as they moved with the grace of shadows through their domain.

Yet, within the vast Congolese territory, other tribes lurked that were infamous for their dark and gruesome practices. The Lenku community, known for their cannibalistic tendencies, had long preyed upon these harmless creatures—the Pygmies. These gentle souls fell victim to the Lenku's insatiable hunger, their flesh providing sustenance to the savage tribe.

My yearning to meet the Pygmies burned with an intensity matched only by the legends that surrounded them. Finally, the opportunity arose, and I seized it without hesitation. A medical campaign was planned in Pygmipara, the heartland of the Pygmies, where I would not only witness their existence but also provide them with essential medical services, food, and clothing.

Lieutenant Colonel Hasina, Captain Kamrul (then a major), and I formed the medical team entrusted with this noble mission. Accompanied by numerous military and civilian personnel, including the esteemed chief of the mission, we embarked on our journey to the land of the Pygmies.

Our adventure began in the quaint city of Bunya, our base camp situated two hundred kilometers away from the Pygmy settlements. Knowing the arduous nature of the task that lay ahead, we had made meticulous preparations, aware that establishing a camp in such a remote location would prove no easy feat.

To ensure a smooth operation, some of us had ventured ahead, leaving Bunya the previous day to set up a temporary encampment in Mambasa, the closest point to the Pygmy land. Mambasa stood on the edge of the world's second-largest rainforest, a place of ethereal beauty and untamed wilderness. As we soared above the landscape in a helicopter, the vastness of the pristine jungle unfolded before our eyes, leaving us in awe.

This untamed wilderness rivaled only the Amazon rainforest in its sheer expanse and magnificence. It would take us nearly 50 minutes to navigate even a small portion of this grandiose forest, where colossal, unnamed trees towered above us. The lush green canopy concealed the secrets of the forest, leaving us with a sense of wonder and intrigue. Amidst it all, a mysterious river snaked its way, its murky waters adding an enigmatic air to the already bewitching surroundings.

In the heart of the verdant Congolese jungles, a land of mystique and hidden treasures, the rivers flowed with a breathtaking, golden hue. Whispers among the locals spoke of rivers infused with the essence of gold, their waters shimmering with a lustrous, ethereal glow.

Legends were woven of daring individuals who had mastered the art of extracting gold from the loamy depths of these sacred riverbeds, casting their lines like humble fishermen in search of a bountiful catch.

Yet, such abundance of wealth did not go unnoticed, for various militia groups sought to control these sacred jungles, where the precious metal flowed. Even the locals, who had been intertwined with these jungles for generations, were often denied access to these sacred places, which were jealously guarded by those who sought to profit from the golden veins that coursed through the land.

As the eve of the medical campaign approached, I arrived in Mambasa by helicopter, greeted by the sprawling expanse of an abandoned coffee factory transformed into our makeshift campsite. Surrounded by the lush greenery of the forest, the camp stood as a testament to our determination to provide aid and support to the marginalized Pygmy tribe.

However, the beauty of our surroundings did not mask the dangers that lurked within the untamed wilderness. The dense forest was a realm teeming with venomous snakes, scorpions, and spiders, their presence a constant threat to the safety of those who called the camp home. Among them, the fearsome black mamba reigned supreme, a deadly serpent whose venom could claim a life within half an hour of its strike.

Vigilance became our shield, and we remained ever alert, knowing that a single misstep could lead to perilous consequences. Despite our best efforts, the infamous Mr. Black Mamba managed to infiltrate the camp, his presence instilling an aura of danger and uncertainty. However, through the sharp eyes and quick reflexes of one of our warrant officers, disaster was averted, and the camp remained secure, sparing the lives of our soldiers.

In Africa, swarms of mosquitoes buzzed incessantly, bearing the mark of death with their venomous sting. These relentless creatures posed yet another threat to the peacekeepers, who valiantly fought against both natural and man-made adversaries in their quest to maintain order and survival. Warding off these relentless insects required a constant battle, a fight as arduous as fending off a camouflaged militia group.

The path that lay ahead was a long stretch of unpaved, muddy terrain devoid of proper road infrastructure. In fact, the entire Ituri Province lacked macadamized roads, save for a few kilometers of rough tarcrete within the capital, Bunya. The journey ahead would test our resilience and perseverance as we pushed through the treacherous terrain, determined to reach the Pygmies and fulfill our mission.

And so, with the golden rivers and unforgiving wilderness as our backdrop, our expedition continued, braving the perils of nature and the greed that tainted these sacred lands. Adventure awaited at every turn as we ventured deeper into the heart of the Congo, ready to embrace the unknown and forge a path through the untamed jungles of our quest.

The air was thick with the deafening symphony of roaring lorries, buses, and trucks, their massive frames rumbling along the muddy pathway, leaving a trail of dust and debris in their wake. During the relentless downpours of the rainy season, this once-passable road transformed into a treacherous quagmire, swallowing vehicles whole and rendering them immobile. These muddy trenches became the bane of travelers, causing blockades that lasted for days, as the vehicles struggled to break free from the clutches of the feculent mire.

Despite the challenges of reaching our remote camp, tucked away in the distant reaches, it teemed with a vibrant energy. A vast field dominated its center, where the grimness of daily life gave way to the camaraderie and thrill of cricket matches. Players from neighboring camps would occasionally venture to this hub, seeking friendly competition amidst the backdrop of our challenging existence. Yet, the serenity of our camp was shattered by a devastating storm, leaving a trail of destruction in its wake. Tents were torn asunder, their flimsy frames no match for the ferocity of the elements. Days of turmoil followed as we picked up the pieces, rebuilding and revitalizing our camp from the ruins left by consecutive storms. Amidst the chaos, we found solace in the PREFABs, our prefabricated accommodations that became our sanctuary in those trying times, providing shelter for the valiant peacekeepers who served in that tumultuous area.

In the early hours of morning, our spirits soared as we embarked on our journey towards Pygmy Para, the mythical land of the Pygmies. The excitement was palpable as we crammed ourselves into jeeps and pick-up trucks, donning our helmets and bulletproof jackets. The weight of the bulletproof jacket draped upon our bodies brought forth a strange mix of trepidation and inspiration, a reminder of the risks we willingly embraced in the pursuit of our mission.

The fervor of patriotism coursed through our veins, igniting an unwavering dedication to our homeland and a burning desire to contribute to its betterment. It was this deep-rooted affection for our country that propelled us to join the United Nations peacekeeping operations, where we became active participants in the noble endeavor of promoting global peace and stability. Every moment spent in direct service of this cause was cherished, for our hearts swelled with an unwavering love for our nation and an unyielding commitment to serve with utmost devotion.

With each passing day, the weight of the bulletproof jacket became a familiar burden, a constant reminder of the risks we willingly embraced as defenders of global peace. It draped over our shoulders, accompanying us on our journey, instilling within us a distinct sense of honor and pride. Armed with our SMGs, we ventured thousands of kilometers away from the embrace of our beloved families and friends, braving the inherent dangers of our duty to arrive at the camp.

The looming threat of militia attacks hung over us like toxic insects, ready to strike at any moment. We understood the importance of vigilance to ensure we were always armored and prepared for swift evacuation, defensive actions, or a swift counterattack should circumstances demand it.

Squeezed into the car, our anticipation grew as we prepared to welcome our chief guest, the contingent commander, and many others who were set to arrive by helicopter for the inauguration ceremony near Pygmy Para. A team had been dispatched ahead of our departure to secure the safety of the helipad, leaving no room for complacency.

Leaving the safety of the Mambosa Camp behind, we embarked on the journey towards Pygmy Para, fully aware of the need for meticulous preparation. Every detail had to be taken care of, from arranging panels and adorning the ceremony dais to organizing the proceedings. The local authorities had already been informed of our imminent arrival, the word about our noble cause spreading like wildfire.

The pathway that unfolded before us was wide and capacious, a majestic trail paved with soil that meandered through the lush wilderness. On either side, dense forests and towering mountains stood as sentinels, their outstretched branches welcoming us with open arms. As we traversed the route, scratched runnels and troughs accompanied our journey, their murmuring waters providing a rhythmic symphony to our ears.

The fragrance of nature in full bloom filled the air, an intoxicating blend of the raw wilderness and the essence of nearby villages. Sungrass and dilapidated bamboo grooves adorned the Choth Bazar, their intricate patterns and hues seamlessly blending with the natural landscape. After every few kilometers, we encountered dense forests that beckoned us to explore their depths, their chorus of chirring crickets serenading us along the way. The air buzzed with the vibrant hum of insects, and a kaleidoscope of colorful butterflies danced in the sunlight, transforming the surroundings into a living, breathing tapestry. Occasionally, an eerie atmosphere would cast its ghostly veil upon the forest, adding an otherworldly touch to the natural marvels that surrounded us.

Amidst this breathtaking scenery, we would often come across herds of cows grazing peacefully on the mountain slopes. These majestic creatures boasted massive horns seemingly larger than their own heads. In this land, cows were abundant and affordable, their sturdy frames a testament to the richness of the region. Acquiring a robust and sinewy cow would not require more than a couple of hundred dollars, evidence of the simplicity and abundance of life in these parts.

As we continued our journey deeper into the wilderness, we reached the Ipulu reserve forest, a place of natural beauty and untamed power. The Ipulu River roared by, its torrential waters rushing past with a fierce determination, as if challenging us to test our mettle. Alongside the river, the Ipulu stream danced playfully on the rocks, adding a touch of whimsy to the landscape. With each step we took, we could sense the stories embedded in the soil, tales spanning thousands of years.

To our surprise, we came across a bungalow nestled within the forest. It was there that we encountered Miss Rozy, a Scottish woman in her fifties who had tirelessly dedicated herself to preserving this natural wonder. Despite being a foreigner in this land, she had made it her own, immersing herself in its essence day after day. Her unwavering commitment and passion for conservation were inspiring, evoking awe in all who crossed her path. It was through her efforts, along with the support of the local community, that this breathtaking paradise remained untouched and unspoiled, ensuring that future generations could revel in its beauty.

While Miss Rozy stood out from the locals and didn't fully assimilate into their culture, her tireless work continued undeterred. Her focus was on research and preservation, particularly on an exceedingly rare animal species known as the Okapi. This elusive creature, found only in the Ituru region closest to Pygmy Land, was a remarkable hybrid of a zebra and a giraffe. It possessed a unique body structure, resembling a zebra with its stripe-banded back legs and buttocks, while sporting an elongated neck and head reminiscent of a giraffe.

Filled with anticipation, we ventured away from the well-trodden pathway and delved deeper into the lush and verdant forests of Ituru, leaving the bungalow behind. The forest welcomed us with open arms, its vast expanse beckoning us to explore. As we proceeded, the sound of helicopter rotors filled the air, signaling the arrival of our team members, the chief, and other invited guests who had traveled to the area by air. Tents were pitched within the forest, creating a temporary basecamp amidst the bosom of nature.

The gathering was graced by all 20–25 Pygmy families, each radiating a sense of unity and camaraderie. The air was filled with a palpable togetherness as the families arrived, their presence illuminating the occasion with their unique charm and grace. Despite living a simple life, their countenances were filled with deep and abiding joy, reflecting the contentment that comes from embracing a life connected to nature.

Addressing the area as the Pygmy Forest may have seemed like an exaggeration, as it consisted of no more than twenty to twenty-five humble abodes made from leaves. Each house was tiny, barely accommodating one person, similar to our own tents. Sleeping in those miniature houses required careful maneuvering to avoid discomfort.

Amongst the dwellings, I could identify their leader, a man in his sixties with a face adorned with a full white beard and weathered skin. His forehead carried the marks of a life well-lived, while his tough demeanor was accentuated by bushy eyebrows. His hair, disheveled and unwashed, gave testament to the simplicity of their lifestyle. He wore torn and dirty clothing, with half-torn and used jeans shorts that struggled to stay on his waist, often requiring him to hold them up with one hand.

Amongst the Pygmy community, their leader stood out as a black man with a height of 4 feet, a tattered hem, and a bent nose. He held a light purply plastic mug with a yellowish stain, which he used to sip tea repeatedly. To catch a glimpse inside his home, I had to bend my neck downward, where I discovered a fired-up mud/clay stove providing warmth and protection from the cold, insects, and wild animals at night.

Inside, there was an earthen basin covered with a polythene sheet on the muddy floor where he slept, finding contentment in his simple dwelling. There were no beds, comfortable couches, stools, shelves, pillows, blankets, quilts, or mattresses. The absence of extra clothing, lights, lamps, or lanterns highlighted the minimalistic nature of their lives.

Many of the Pygmies lived on tree branches, using the trees as their homes and shelters. They kindled fires and cooked their food with utmost simplicity, using a clay pot as their primary vessel. Leaves, possibly gathered from the wild, were boiled in the clay pot and served as their main source of sustenance. My observation was interrupted by a man who appeared with cassava in his hands, a scrumptious vegetable that served as a delicacy for these resourceful inhabitants. He ate the boiled cassava with utmost concentration, savoring its delicious taste.

Palm oil trees were abundant in the forest, providing an essential resource for the Pygmies. They used the fruits of these trees as vegetables, either boiled or crushed, roasted, and mixed with other ingredients. These resilient people thrived on the ingenuity of using the resources available to them. I observed a woman skillfully pounding and grinding fresh green vegetables with a wooden pestle in a large wooden bowl, showcasing the staple foods of their diet.

As we embarked on our mission to provide medical care, we encountered patients with ailments that were not severe in nature. Most common were bodily discomfort, localized pain, and malnourishment. While we didn't come across any complicated diseases, the challenge of communication arose as we struggled to understand their language. However, we bridged the gap through the language of the heart, expressing empathy and hope to those in need. Our team included interpreters, including a local, who helped us understand the Pygmies by translating their language into English. With their assistance, we were able to provide effective medical services to the Pygmy families.

As we embarked on our journey back to Mambasa camp, Miss Rozy and the contingent commander distributed clothing, food, and sports equipment to the Pygmy camp, bringing immense joy to the people. The Pygmy land buzzed with excitement and elation, grateful for the generosity and support they received.

However, as we flew above the forest, the expected descent didn't come after 15 minutes. Instead, we found ourselves still in the air after nearly 30 minutes had passed. The passage of time felt distorted, and a sense of unease and trepidation gripped my heart. There was a sudden jolt, as if a gust of wind had swept through the helicopter, causing everyone to notice the change. Silent understanding filled the air, and though no words were spoken, it was clear that something had gone wrong.

The helicopter continued to circle in the sky of Ituru for another ten minutes, leaving us uncertain as to whether we would ever reach our intended destination. I voiced my concerns, asking if we were going back directly to Bunya, but the camp commander emphatically assured us that we were indeed heading to the camp in Ituru, not Bunya.

As we questioned why it was taking so much time, speculating if there was a reconnaissance mission underway, the commander reassured us that was not the case either. The mystery of the prolonged journey lingered, and I wondered if we had done something wrong.

Then, in a sudden and unexpected moment, the helicopter swooped down, landing on the helipad nestled within the heart of the dense forest. Another strong jolt rippled through me, intensifying the sense of anticipation and uncertainty about what awaited us upon reaching the camp.

After the emergency landing in the dense forest due to the helicopter's technical difficulties and dwindling fuel, we were filled with exasperation and relief. The situation was precarious, and it was thanks to the bravery and decision-making of our wing commander, Khayer, that we were able to land safely, as he put his own life at risk in the process. It was a moment that reminded us of the danger we had narrowly escaped and the gratitude we felt for being alive.

A year later, Colonel Hasina contacted me to share that some junior officers had sent her birthday wishes and a beautiful cake. However, she clarified that it wasn't her birthday, but rather a celebration of the second anniversary of surviving the helicopter emergency. It was a lighthearted moment combining fun and gratitude as we reflected on the perilous incident we had overcome and the miraculous way we had been saved.

While some returned to the camp riding on another helicopter from Bunya, I remained in Mambasa. The grounded helicopter stayed amidst the lush foliage of the forest for the next seven consecutive days, waiting for repairs or retrieval. It served as a constant reminder of the ordeal we had endured and the importance of valuing each day we were given.

GUNPOWDER AND GLORY

MAJOR MEHEDI HASAN
(UNITED NATIONS MISSION IN IVORY COAST - UNOCI)

On February 25, 2011, in Abobo Abidjan, Ivory Coast, around 11 pm, we received orders to move to Abobo, a territory controlled by President Gbagbo and known for its unrest and tension. The Togo Battalion's force commander had been stationed there with a heavily armed force patrol since morning, and it was now our turn to replace them.

We departed for Abobo at 11:40 pm, accompanied by four armored personnel carriers (APCs) and one belonging to the Jordan police. Captain Asif was also with us. As we crossed Sebroko and proceeded further, we encountered a checkpoint manned by soldiers from the Forces de Défense et de Sécurité (FDS). Despite exchanging words with them, they refused to let us pass the checkpoint, leading to polemic arguments.

In an attempt to defuse the situation, our acting force commander and his convoy considered getting down from the vehicles to assess the situation and seek a peaceful resolution. After a brief conversation with the FDS soldiers, the commander realized that there was no time to waste and ordered us to "Just go through and take control of Abobo..." In those moments, his words seemed to embolden us, and the soldiers at the checkpoint yielded and cleared the way, showing signs of fear.

Having the advantage of the Jordanian police's familiarity with the terrain, they guided us through the deserted surroundings, pointing out the way. The roads were marked by the remnants of burned tires, and at regular intervals we encountered small signs of barricades in the eerie silence.

Amidst this unsettling environment, my attention was drawn to a somber object lying in our path. Upon closer inspection, I realized with a sense of horror that it was a corpse. As I surveyed the scene, I discovered more bodies scattered along the road, emitting a strong and pungent odor. It was evident that some of them had been there for at least a day. With caution, we moved past the corpses, aware of the possibility of danger, and continued our journey.

In the distance, we noticed a flickering light on the horizon, indicating a blazing inferno likely originating from a nearby dwelling. Recalling our steps, I estimated that we had traversed almost twenty kilometers from our encampment and had entered the infamous and chaotic region known as PK 18. The situation was tense and filled with uncertainty as we pressed forward into the heart of Abobo.

Despite the victory of Alassane Ouattara in the presidential election held in October and November 2010, the political situation in Ivory Coast remained unstable. The incumbent president, Laurent Gbagbo, refused to relinquish power, leading to a deepening crisis and deteriorating security conditions.

In response to the escalating unrest, the United Nations launched the United Nations Operation in Côte d'Ivoire (UNOCI) on April 4, 2004. The mission aimed to promote peace and stability in Ivory Coast, which had been plagued by political and social tensions for years.

It is against this backdrop of political instability and conflict that our experiences in Abobo, Abidjan took place. As members of the military, we were deployed to replace the Togo Battalion in the territory controlled by President Gbagbo. The situation was tense, and our duty involved maintaining vigilance and guarding against potential adversaries.

Ivory Coast, located on the Atlantic Ocean coast in the southwest of the African continent, has a rich history. It was a French colony from 1893 until it gained independence in 1960. However, political instability and internal conflicts have marred the country's progress since 1999.

The long-awaited national election, originally scheduled for 2005, faced numerous delays and setbacks. It was eventually held in 2010, with Alassane Ouattara emerging as the winner. However, Gbagbo's refusal to step down created a power struggle, plunging the country into a state of uncertainty and triggering a civil war.

The events and circumstances surrounding the election and the subsequent refusal to transfer power by Gbagbo had a profound impact on the security situation in Ivory Coast. As members of the military, we found ourselves in the midst of this turbulent period, carrying out our duties in Abobo while tensions ran high and uncertainty loomed.

It is important to note that the political landscape and security situation in Ivory Coast are complex and dynamic, and the events described here provide a glimpse into a specific moment in time. The country continues to work towards stability and reconciliation, but the path to lasting peace remains challenging.

Former President Laura Bagbo's brazen refusal to relinquish power plunged the country into a dark abyss of political instability and escalating security concerns. Ivory Coast's future hung in the balance, its once-promising political stability now a distant dream.

Abidjan, the pulsating heart of the nation's commerce, took center stage despite Yamoussoukro being the official capital. This bustling metropolis became the epicenter of Ivorian administration, housing the prestigious ministries and the grand Presidential Palace itself. But there was no designated national palace for the president to call his own. Disregarding this inconvenient truth, Bagbo audaciously occupied the grand edifice as if it were his personal residence. Meanwhile, the internationally recognized President Alasane Ouattara found refuge in the heavily fortified Golf Hotel, its very structure shielded by the United Nations' protective canopy.

Close by, the UNOCI course headquarters loomed, nestled in the proximity of the Sebrocco hotel. This establishment, not far from the presidential palace, stood adjacent to the Gbagbo-supported FDS army camp. A dangerous combination of power, loyalty, and volatility simmered in the shadows, threatening to ignite a powder keg of violence and chaos.

On January 21, 2011, the 'Mechanized Infantry Company-4' of BANBAT 3/19 completed its deployment to Abidjan, the commercial capital of Ivory Coast. As the vice-captain of the company, the weight of responsibility rested heavily upon your shoulders. With a deep sense of honor, commitment, and dedication, you took charge of the operation, inheriting the mantle from the preceding company.

Your mission was fraught with peril, and the onus upon your shoulders was significant. Operational support had to be provided to the BAN Support Company, the unwavering force responsible for securing the heavily fortified Force Headquarters at Sebroko. Manning the automatic weapons' positions for the local defense of Sebroko, your team became the guardians of this crucial stronghold. Deploying rocket launchers and grenade-firing rifles, you held the line, ensuring the security of the main gate and the safety of all within.

But your responsibilities extended beyond the confines of Sebroko. Vigilant patrol of the surrounding areas became paramount, as you steadfastly safeguarded the lives and well-being of innocent civilians. No adversity would go unchallenged, no danger left unmet. The safety and security of the people were non-negotiable, and you embraced your duty with unwavering resolve.

Obstacles littered your path, but never once did panic cloud your judgment. The army's rigorous training had instilled in you an indomitable spirit, preparing you for the multitude of challenges that lay ahead. Courage flowed through your veins, buoyed by the unwavering support of senior officials, comrades, and cherished soldiers who stood shoulder to shoulder in times of need.

The weight of responsibility failed to diminish your morale; instead, it steeled your determination. In the face of adversity, you marched forward with unwavering courage, resolute bravery, and a firm commitment to fulfill your duties. Never once did you look back, for your devotion to the cause propelled you relentlessly forward.

As the nights grew darker and the stakes higher, the fate of Ivory Coast teetered on a precipice. It was within this crucible of chaos and uncertainty that you and your comrades stood as a beacon of hope, staunch in your mission to restore peace and stability to a nation in turmoil. The echoes of your unwavering resolve reverberated through the corridors of power, and the people found solace in your dedication.

The education bestowed upon us by the army served as a consistent guide, fueling our courage and determination in the face of trials and challenges. With each hardship, we defied the odds, fueled by the resilience instilled within us by our seniors. Shoulder to shoulder, comrades and colleagues united, we remained vigilant, committed, and ready to execute any duty handed down from the Force Headquarters. Our diligence in fulfilling our responsibilities was evident, even amidst the countless obstacles we encountered.

Reflecting on our deployment to Abidjan, it became clear that our collective efforts and the army's relentless pursuit of excellence were unmatched. It was a profound honor to have served our country with commitment and selflessness.

February 28, 2011, Abidjan, Ivory Coast. The day brimmed with duty and purpose. My mission: to transport the mortar section of Battalion 1 from

Sébroko to the position held by the Togolese Battalion. As dawn's first light cast its hesitant glow, we set forth. Three APCs and a pickup truck comprised our modest convoy, slicing through the deserted roads like a swift blade.

The uncertainty that cloaked our path weighed heavily upon us, but with resolute determination, we ventured into the unknown. Time crawled with an eternal slowness until, finally, we arrived unscathed at the Togolese Battalion, delivering our cargo without incident.

The return journey, however, promised to be arduous. The sun had climbed high in the sky, and Abobo's streets stirred with awakening life. Vehicles swerved and retreated as we approached, a pandemonium of chaos hinting at the troubles that awaited us. Undeterred, we navigated through the tumult, blaring our horns, weaving through the tangled mess of vehicles.

In the distance, plumes of smoke pierced the sky, a haunting testament to homes ablaze and the brewing unrest. Nevertheless, we pressed on, our sense of purpose guiding us through the tempest. With firm determination, we returned to camp unscathed, our resolve unbroken.

March 2, 2011. Pargola Hotel, Abidjan, Ivory Coast.

4:30 am. The morning air hung heavy with stillness as Captain Rafi and Captain Asif, their minds sharp and focused, embarked on a daring mission to Sebroko—nine stranded Ukrainian helicopter pilots awaited their rescue. With three APCs and one jeep at their disposal, they set forth, determined to retrieve their comrades from deep in hostile territory.

Leaving the safety of the Pargola Hotel, situated close to the airport and just beyond the Lycorn base, they ventured into the unknown. Their destination was reached uneventfully, but the return journey would be marked by treacherous obstacles. An armed militia, unwavering in their intent to impede their progress, crossed their path.

In a heartbeat, our APCs roared to life, retaliating against the assailants' aggression. Bullets whizzed through the air as the exchange of fire intensified. Amidst the chaos, two bullets found their mark, striking our interpreter, Ruhul, and Lance Corporal Hakim. Yet, through a stroke of luck, the bullets merely grazed their fingers, failing to penetrate their flesh.

Though danger loomed, the skirmish resulted in only minor injuries. The ricocheting bullets, potent enough to leave their mark, could not vanquish the indomitable spirit that burned within Ruhul and Lance Corporal Hakim. Their bravery and resilience shone through, a testament to the intense dedication of those who served in the face of peril.

In the tumultuous tale of our experiences, through trials and challenges, we emerged unbroken. The shadows of uncertainty and danger failed to extinguish the flickering flame of our resolve. Shoulder to shoulder, we stood as guardians of peace and protectors of the innocent, undeterred by the perils that lay in wait.

The chaos of the assault was met with firm resilience and a unity of purpose. The enemy's intentions to overrun our camp were swiftly met with an unyielding defense. Our APCs, strategically positioned and armed to the teeth, unleashed a torrent of firepower upon the assailants.

The distinctive roar of the 14.5-mm and 7.62-mm machine guns resonated through the air, a symphony of defiance against those who sought to sow terror. With each well-aimed shot, we pushed the enemy back, their ranks faltering under the relentless barrage.

At the forefront of the battle, Captain Asif, a beacon of courage, emerged as a guiding force. With unflinching determination, he unleashed a devastating counterattack, striking fear into the hearts of the enemy. His actions served as an inspiration to all, a testament to the unrelenting spirit of the Bangladeshi peacekeepers.

March 30, 2011 marked a pivotal moment in our journey. We proved our mettle, defending our camp and our comrades with indomitable will. The assault may have been unexpected, but our preparation and resolve ensured that the enemy's plans crumbled.

As the dust settled and the echoes of gunfire faded, we stood victorious. The strength of our unity and the dedication of every soldier who stood shoulder to shoulder fortified our position. The camp remained secure, a testament to our solid commitment to peace and the safety of those under our protection.

In the heart of Sebroko, Abidjan, Ivory Coast, we had triumphed against adversity, leaving behind a resounding message that the pursuit of peace will always prevail over the forces of chaos and violence.

The events of March 31, 2011 in Patu/Kokodi, Abidjan, Ivory Coast tested our resolve and proved our unwavering commitment to our mission. Our team, entrusted with the critical task of escorting esteemed UN personnel, encountered unforeseen hostility that put our mettle to the test.

As the convoy moved through the Patu area, hostile fire erupted, threatening the safety of those under our protection. However, under the courageous leadership of Captain Rafi, we stood firm in our duty to protect and serve. Captain Rafi's swift and decisive response, returning fire with great force, forced the enemy to retreat, ensuring the safe return of all those in the convoy to the camp unharmed.

Undeterred by the previous encounter, we embarked on another crucial mission – escorting the SRSG to the radio-television station in Kokodi. However, upon assessing the surroundings, the SRSG deemed it unsafe to proceed and immediately ordered the escort patrol to regress to the camp.

As we made our way back, only a short distance from the radio station, the relentless militia launched a brazen ambush. Acting swiftly, our three APCs formed a protective shield around the SRSG's vehicle, and we retaliated with a fierce and resolute counterattack. Our heavy machinegun fire quelled their aggression, preventing the enemy from advancing and taking advantage of the situation. Despite the enemy's bullets falling around us, we successfully made it back to the base with the SRSG unharmed.

Our steadfastness and determination in the face of adversity thwarted the enemy's futile attempts to cause harm. The Force Headquarters recognized our valorous efforts and honored us with the prestigious 'Force Commander's Certificate' as a testament to the strength of our collective spirit and commitment to our mission.

March 31, 2011 stands as a shining example of our resilience, bravery, and dedication to peacekeeping. We overcame the challenges thrown at us and emerged victorious, ensuring the safety and well-being of those we were entrusted to protect.

On April 1, 2011 in Patou, Abidjan, Ivory Coast, Major Khabir and his show of force returned from patrolling the vicinity of Abidjan. Despite being attacked again on their way back, the force responded with precision and trapped the enemy within their own ambush. The patrol successfully returned to the camp unharmed, showcasing their constant awareness, resolute aim, and fearlessness in the face of worsening security conditions.

In the midst of the prevailing dire circumstances in Patou, Captain Rafi and his team of three APCs valiantly rescued a dozen officers, including the charge d'affaires of the Nigerian Embassy, from the treacherous Patou regime. This act of bravery occurred at mid-day, around 1 o'clock, and earned us admiration for our courage in the face of the complex challenges in Patou.

Our activities on Abidjan.net, the online platform, received significant attention and promotion, rightfully acknowledging our vital role in the operation. It is indeed a source of pride to have contributed significantly to such a complicated and challenging mission.

During the chaos, Captain Rafi identified an enemy position and swiftly ordered a counterattack. As a result, at least six enemy casualties were inflicted. Simultaneously, an APC of the Jordanian police returning to Sebroko after a patrol was targeted with RPGs by the Gbagbo Force (FDS). Fortunately, the APCs sustained minimal damage as the rockets failed to strike accurately.

On April 2, 2011 in Sebroko, Abidjan, at 5 o'clock in the morning, our peaceful slumber was abruptly disrupted by the deafening sounds of heavy artillery fire. The alarming situation intensified as the firing continued throughout the day, echoing with an intimidating force that shook our camp situated near the FDS base.

Despite being peacekeepers, our team had no choice but to adopt a defensive stance and respond to the attacks. Positioned within our APC, we retaliated against the gunfire. The situation was dire, and amidst the chaos the bullets flying overhead and the smell of gunpowder filled the air, creating a surreal atmosphere.

Captain Rafi displayed remarkable acumen by accurately identifying the enemy's position. He swiftly ordered retaliatory fire, and we followed his command, resulting in at least six enemy casualties. Our counterattack proved decisive in the face of the ongoing hostilities.

On April 3, 2011 in Sebroko, Abidjan, the sound of gunshots persisted throughout the day, keeping everyone on high alert. The exchange of gunfire occurred intermittently, and our defensive position was occasionally challenged. At around 3 in the afternoon, the enemies targeted our APC, and we retaliated in response. At least four enemies were shot dead by that time.

Throughout the day, news circulated among us that the UN's MI-24 helicopter would launch an assault on the Gbagbo base and the supporting militias. However, due to adverse weather conditions, the news remained speculative and did not materialize into an actual operation.

On April 5, 2011 in Sebroko, Kareena Port, the morning began with intense fighting as there was a fierce exchange of ammunition between us and the Gbagbo-backed militias. Following a directive issued the previous night, we took refuge in the basement of the Sebroko Complex around 7 in the morning, anticipating an assault by a UN MI-20 helicopter on the FDI's base adjacent to ours.

However, at 8:30 am, we received information that only civilians would seek shelter in the basement, while military personnel were instructed to hold defensive positions. We immediately returned to our camp and prepared ourselves to defend against any malicious intentions. By 11 am, we were ready to respond with force if necessary with our APC on standby.

The enemies attempted to target our APC, but their efforts resulted in minimal casualties. Neither side suffered any significant harm. Tension and anticipation continued to escalate as the day progressed.

Finally, around 5:30 pm, the long-awaited attack on the FDS base commenced. The MI-20 helicopter unleashed a hail of fiery bullets and missile attacks, illuminating the sky with explosive bursts.

On April 6 and 7, 2011, our team was stationed at the Novotel Hotel in Abidjan, Ivory Coast, providing protocol and security to the international journalists residing there. Despite the tumultuous circumstances of insurgency, the Bangladeshi peacekeepers, including our team, went above and beyond to ensure the safety of the journalists.

During those 48 hours of unrest, we had to navigate a total of seven roadblocks while maintaining security in the area surrounding the hotel. Our dedication and commitment did not go unnoticed, and the international journalists expressed their gratitude for our unwavering protection and support.

As a result of our efforts, we were honored with the "Testimonial Certificate of the Force Commander." This recognition served as a testament to our commitment to duty and our devotion to maintaining peace and security in Abidjan, Ivory Coast.

On April 21, 2011 in Abidjan, Ivory Coast, news rapidly spread that Laurent Gbagbo, the former president, had been arrested earlier that morning. At precisely 2 am, the confirmation arrived that Gbagbo had indeed been apprehended and taken into custody. The atmosphere was filled with collective jubilation as everyone expressed their uncontained joy, and the belief that the war was finally over ignited hope for a new beginning. However, only time would reveal the true outcome, as Africa was a realm of uncertainty where unforeseen conflicts could emerge at any moment. Thus, the future remained unclear, known only to God.

The values of "The Rebel," a poem by the rebel poet Nazrul, were deeply ingrained in us during our training at the Bangladesh Military Academy. The poem begins with the lines:

> O hero, pronounce—
> Pronounce high is my head,
> Beholding my head, even the Himalayas lowers its peak.
> The ever-elevated head, O hero, pronounce your greatness.

With these verses resonating within us, we embarked on a journey across thousands of miles to the African continent, far from our homeland, proudly holding our heads high. As a generation born after the war of independence, we were unacquainted with the scars of conflict. However, during our UN peacekeeping missions, we witnessed the aggression of war and actively worked to oppose it. We were unswerving in our commitment to peace, as it was the only path we knew.

Our duty was to maintain peace, and we approached this responsibility with profound sincerity. At the end of the day, this was what truly mattered. Entrusted with the flag of the United Nations, we, the Bangladeshis, were proud to bear its weight. We were not merely peacekeepers; we were proud Bangladeshi peacekeepers. I, too, share in this sense of pride. What more could a peacekeeper ask for in life?

Before concluding and sharing an incident, let me say one last thing. After my journey to Ivory Coast, I took to Facebook, and the words I wrote resonated with many. And so, today, I conclude with those very words: "You know why I am proud? I am proud because as I move around the streets of Africa, people shout 'Bangla! Bangla!' I hear their affectionate and warm shouts, 'Bangla... Bangla...' reaching me from afar. Yes, I am a proud peacekeeper and proud to be Bangladeshi."

TREASURE HUNT

MAJOR JAMIL AHMED KHAN

(UNITED NATIONS MISSION IN SIERRA LEONE - UNAMSIL)

As I leaped over the tall grass, I unexpectedly found myself plunging into knee-deep water. Although there was no clay or mud beneath, the water soaked my uniform, leaving me thoroughly wet. Being the first to encounter this watery spot, I had the opportunity to warn everyone about it. However, before I could do so, the Pakistani doctor caught up to me and unintentionally became a victim himself. The rest of the team proceeded cautiously, taking extra care. Even the bulky EOD team leader, known for his strength, revealed his fear of water, much to our surprise.

Located in the vast wilderness of the Komba district, Jui Town was our last encounter with civilization. After scouting the area, I selected this field of tall grass as a suitable helipad during the reconnaissance operation. Accompanied by two Ukrainian helicopter pilots and two crew members, we failed to notice the water concealed beneath the grasses during our aerial survey. Even the pilot remained unaware of it until he skillfully landed. It was an oversight on our part.

Our team consisted of 15 members. I had conducted the solo reconnaissance before devising the entire operation. Although we thoroughly briefed the helipad procedures after finalizing the preliminary plan, the other 14 team members still regarded me with doubt and curiosity. Ahead lay a dense jungle, raising concerns about our route and how long we would have to trek through the dark forest. The Pakistani doctor, a man in his fifties, appeared the most helpless and uncomfortable in his drenched uniform.

In Liberia, my perception of what comprised a town was challenged. Five or six small houses scattered in the jungle or by the road constituted a town, while a larger area in the city was also considered a town. Jui Town, nestled in the thick jungle of Komba, was relatively significant by that standard, resembling a large village in our country, complete with shops and markets. The road ahead was arduous. Even as an experienced officer who had worked in the hilly region of Chittagong, it was difficult to estimate how long it would take to walk seven kilometers in the dense forest and mountainous terrain. The true distance covered in the jungle and mountains was challenging to comprehend.

The presence of the doctor had the potential to slow down our progress on the challenging path. Additionally, I had estimated that it would take us six to seven hours to reach our target location, Safulla Town, which meant we would have to begin our cache search operation in the jungle by late afternoon.

I discussed the matter with the two Pakistani captains, Niyar and Shahrukh. I expressed my concerns about Major Kashif's ability to handle the demanding road and long journey. I asked them if they would mind if I left Major Kashif behind and proceeded with our plan. Both Niyar and Shahrukh assured me that they had no issue with it, as long as I was comfortable with the decision. I then had to convey the news to Major Kashif.

Major Kashif was a courteous and humble officer. I respectfully informed him, "Sir, I kindly request that you do not accompany us any further. The road ahead is uncertain, and the nights are unpredictable. If it would not cause you any official trouble, you may return to Tabmanberg using this helicopter."

He met my gaze, smiled gently, and gratefully responded, "But you will have to report it that way then."

I assured him, "That's not a problem, sir."

During our conversation, Imam Yusuf Sarlif, my main source, became an important and popular figure in Jui Town. It was considered prestigious for anyone in the jungle town to board a UN helicopter, and, naturally, he had gained prominence.

He was surrounded by a group of ten to fifteen individuals of various ages, each wielding a large machete similar to the ones used by our hill tractors. I understood that they would be our guides, clearing the way through the jungle ahead.

Bringing a non-UN local civilian on a UN helicopter for the operation was not an easy task. However, there was no way to exclude him. Three factors allowed me to bend the usual UN rules in this situation. First was the significance of the intelligence before the upcoming national election after a prolonged war. Second was the influence of my senior Russian counterpart,

Milber. And third was the presence of Bangladeshi staff officer Squadron Leader Rashed in Air-Ops. We recognized the importance of Imam Yusuf Sarlif's presence at every stage of transforming primary information into actionable and effective intelligence, from the initial report to the reconnaissance report to the final plan. I knew, believed, and was convinced that the main operation would be a struggle without him and would never succeed. Furthermore, my experience in the Chittagong hill tracts supported my decision to keep the primary source with me, providing the necessary confidence at that crucial moment.

Before delving into the main story, I feel it is necessary to provide a summary of the preceding events to address any potential questions or uncertainties. It marked the beginning of my career in UN missions, which had spanned over a decade at that point. On April 9, 2005, I arrived in Liberia as a UN military observer. Following a week of orientation training, obtaining a driver's license, and completing other formalities, I was assigned to Military Observer Sector 2 headquarters, located approximately 120 kilometers from the capital, Monrovia. I was thrilled about my posting. My superior, Colonel Sergei Spak, a distinguished Russian officer standing at 6 feet 3 inches, arrived at the Bangladeshi Milon House in Monrovia on the morning of April 22. He greeted me warmly in fluent English, debunking the assumption that Russians were not proficient in the language.

By four o'clock in the afternoon, I had reached my designated workplace. The accommodation arranged for me was conveniently situated just a short distance away from the office, around 20–25 yards. Adjacent to the headquarters, there was also an observer team site where Major Tofayel, a fellow Bangladeshi, served as the operations officer. I was delighted to discover that my room was next to his. We both belonged to the artillery branch and shared a cordial relationship.

As a recent computer engineering graduate from MIST, my background impressed the UN personnel, leading them to assign me the role of sector communications officer. Within a week, I became a military information officer (MIO), and before the month came to an end, I was promoted to operations officer. I considered myself fortunate in my role as an MIO, primarily due to the strong connections I had formed with the local community. This included officials from the local administration as well as generals and field commanders from different rebel groups. However, my close relationship with the imams and elders of the indigenous Muslim community stood out. Regular visits to the Tubmanburg Mosque and occasional donations to the poor helped foster this bond. Furthermore, while preparing a photo-based report on notorious and renowned commanders from the three armies involved in the recent civil war, I had the opportunity to host former generals and field commanders. As a result, I developed strong connections with these individuals. The report, which garnered significant interest and importance, was later received by the force headquarters, JEMEK, CIMIC, and the security section.

In the midst of the general election buzz in Liberia, I had managed to navigate through the morning rush and settle at the office. Just then, the UN landline phone rang, and Major Kwesi Aima of Ghana picked up the call and handed it to me. It was Robert Edebey, the regional administrative officer, also known as the civil boss of the area. Robert informed me that someone was looking for me and that he had sent her to my office. Intrigued, I asked who it was and why they were looking for me. Robert simply said, "I sent her to your office. You talk to her and see the rest." After exchanging greetings, we hung up the phone.

A few minutes later, a tall and slender young lady walked into the office with a smile on her face. She greeted me and confirmed if I was Jamil. I gestured for her to sit down, returned the greeting, and introduced myself. Curious about why she was looking for me, I asked her the reason. She explained that she had seen me before and heard my name frequently, but this was our first conversation. She then handed me a small folded paper, which I noticed was written in Arabic. I inquired about who gave her the message, and she replied, "An old man gave it to me and asked me to forward it to you."

I had met this young lady, Miss Danielle, at a previous conference but only got to know her better that day. She was the civil affairs team leader at the Bopolu site. As Miss Danielle left, I realized I needed the assistance of our houseboy Mohammad, who was my regular Arabic translator. Since coming to Liberia, I had learned that the local Muslims, who practiced Islam, were well-versed in Arabic and could speak, read, and write in the language. Before lunch, I located Mohammad and brought him to the office, making him feel special. I handed him the letter in Arabic, requesting his help in understanding its contents. Mohammad examined the old paper and exclaimed, "Ya Allah Major, there is horrific news here, interesting!" (After living with the UN peacekeepers for a long time, Mohammad had developed the habit of finding thrilling aspects even in the most horrific news and had learned to handle such news with ease.)

When I asked him about the news, he revealed, "There are many weapons stashed near Safulla town, and someone from that town has seen it." Intrigued, I wanted to know who sent this message, and Mohammad replied, "The imam of that town, Yusuf Sarlif. He wrote it."

"I know him; he is a good man," Mohammad added.

I asked Mohammad if we could locate him soon and where we could find him. We needed to talk to him. "And listen, keep this message under wraps," I instructed Mohammad.

Two days had passed since I received the letter when, on August 10, 2005, I found myself east of the Tubmanburg high-school field. There were three small and humble abodes with a courtyard in the middle. As our car smoothly slid on the muddy path towards the courtyard, Shavana, the local rebel group's field commander, approached me, extending his hands for an African-style handshake.

Equipped with paper, pen, an enlarged map of the Congba district, and a map marker, I stepped out of the car and approached Imam Sarlif. Imam Sarlif and I engaged in a conversation about the news. The courtyard was adorned with medium-sized mango trees, one of which provided shade from the sun. The sweet aroma of mango buds filled the air, creating a pleasant atmosphere. Under the tree's branches, a bench made of splintered wood hosted three men. They wore Tupis (caps Muslims use for prayer) on their heads and had beards on their faces. Their smiles were gentle, their eyes bright, and they exuded a sense of calm and wisdom. Though they hailed from a small town in the jungle, they spoke to Imam Sarlif with respect and admiration, as if he were the wisest and most esteemed individual in Tubmanburg, the shining capital of Bomi county. I, too, aimed to treat them with the same respect and reverence.

After a thorough conversation, I gathered the essential information and marked the map accordingly. As we exchanged parting salutations, I placed five dollars in Imam Sarlif's hand, acknowledging the distance he had traveled and the hardships he had endured. "Thank you very much. I am grateful," I expressed my appreciation. His smile softened and widened. Finally, I requested Imam Sarlif to stay for a few more days before I got back into the car, exchanging a few words with Shavana.

As soon as I turned the key in the Anmil-1163 Nishan patrol car to start the engine, a swarm of children waiting in the courtyard chased after the car, running and cheering loudly. I wished I had brought biscuits and fruit to share with them; their presence tugged at my heartstrings. I accelerated the car on the main road, and a whirlwind of plans filled my mind. Glancing at my watch, I pressed the gas pedal and, speaking to myself in a soliloquy, said, "Senior MILOB (Military Observer) needs you in the office right away. I want Senior MILOB in the office now."

It was half past one in the afternoon when I arrived at the shared parking lot of the office. I parked my car and hurriedly made my way to the senior MILOB's office, clutching the map with preliminary marks and the paper containing the latest information. As I entered his room, Colonel Sergei Spak boomed in his weighty voice, "What's the matter? Anything wrong, Jamil?" The senior MILOB had not been informed of the findings and subsequent events. Only my colleague, Major Quichi Aima from Ghana, had some idea.

The senior MILOB listened attentively to everything I had to say, giving a slight nod to indicate that he understood. To an outsider, it might have seemed that the senior MILOB was indifferent to my words, but I knew that this Russian officer had nerves of steel forged by three wars, including the Afghan War, and four UN missions. He wasn't one to easily show surprise or curiosity. Moreover, I had witnessed his true capabilities in several past incidents, where he even directly contacted Chief Military Observer Major General Tahir regarding extraordinary matters. This time was no different. However, what he said after hearing everything wasn't what I had expected. Following my briefing, his first words were, "It will be a MILOB-led operation, and you will positively lead it."

I took a mere two to three seconds to conjure up an image in my mind of how the entire operation would unfold. Without a moment's hesitation, I replied, "I will be honored to do it, sir. Your confidence in me also fills me with pride." Major General Tahir then instructed me to plan the reconnaissance quickly. Just as he began a conversation with someone on the phone, I saluted and exited his room.

The next day, around nine in the morning, my mobile phone rang. Lieutenant Colonel Khurram, the military assistant to the deputy force commander, was on the other end. He said, "Jamil, please talk to the general."

Without delay, I responded, "Right, sir!" and from the other end I could hear the voice of General Tahir.

We exchanged salutations, but at the end of our exchange I faced a barrage of questions. "I read your fire investigation report. You did well. But this is a different operation. I expect success. You will get all possible support," the General said. His words lasted only three minutes, but they felt like three hours.

I hurried into the senior MILOB's office, and he must have noticed the beads of sweat on my forehead. After listening to everything, he grinned and said, "Keep your sweat for the operation."

The next day, August 13th, at nine in the morning, our car sped along, leaving the Pakistan sector headquarters behind us on the left. Major Aima was behind the wheel, and I held a marked map, GPS, camera, and a Motorola set on my waist. These were the faithful companions of an observer. As we reached the helipad, I saw a Ukrainian MI-8 touching down. In the adjacent shed-3, six soldiers along with a lieutenant were seen. Major Aima gave me a curious look as he took in everything. I smiled and said, "I changed the schedule yesterday." Aima nodded and chuckled. For the readers' convenience, a typical air patrol would consist of two MI-24s soaring together in the sky, but the routine patrol schedule had been changed the day before. As the ops officer of the MILOB sector, I had the authority to make such changes.

Based on the coordinates provided by MOP, I shared the flight schedule and the intricate plan with the co-pilot, but I noticed a significant discrepancy between my reconnaissance plan and theirs. While my plan was to fly to Target Safulla Town via Juhi Town and potentially perch there, their air ops plan was simply to land in Juhi Town. I tried to calm the storm in my mind and called Squadron Leader Rashid, the flight schedule officer of Air Ops and my ISD one-course senior. We had established a cordial rapport before, and I urged him to reconsider the reconnaissance request and immediately email it to Lieutenant Colonel Tarek, the chief operations officer of the Pakistan sector. I then called Tarek and handed the phone to the pilot. Lieutenant Colonel Tarek explained the situation to the pilot at length, and I wondered how much the Ukrainian pilot grasped. He only muttered, "You have put me in trouble" and began conversing with someone, likely his superior. Meanwhile, I updated Senior MILOB on everything during this interval. A few moments later, the pilot declared, "You will get the third seat in the cockpit," to which I agreed. Additionally, Senior MILOB contacted me again and instructed me to fly with them. He had spoken to the Ukrainian colonel. It then dawned on me that a Russian senior colonel addressing a Ukrainian colonel was significant and influential considering their shared history and language.

Within the next 10 minutes, the rotor began to whirl at full speed. Everyone was on board, and I found myself seated in a small seat behind the cockpit with my marked map, somewhat stooped over. This position allowed me to conveniently observe the front row and both sides of the window. All the hassle was over, and the helicopter reconnaissance began. It was 12:30, roughly two hours behind schedule. Around 25 minutes later, we flew over Juhi Town, a place I had visited a few times before. We completed a full circle by flying left, keeping the old helipad and the market in the center, just as I had observed in the past. The field next to the market was quite large and filled with grass, making it an easy landing spot. The pilot flew directly over the field, indicating that he was inspecting his approach before landing.

After flying over the field and turning right, we still hadn't reached our target. I scanned my map again, feeling the frustration building up. The imam's description of Safulla village town was so vivid in my mind, but the helicopter's direction didn't align with my map markings, only adding to my confusion.

"The area looks barren, nothing but jungle," the pilot said, expressing his skepticism. Determined to find our target, I insisted on checking their bearing setting. The co-pilot handed me his chart, and I quickly examined it. "Please, let's go back to Juhi Town. From there, our target is seven kilometers away at a bearing of 220 degrees," I asserted.

The pilot seemed doubtful but acknowledged my expertise as a gunner officer. He said, "You have one shot, or we head back." I agreed, confident in my ability to locate the target.

Within fifteen minutes, we were back over Juhi Town. The pilot had my map before him, and I guided him as the navigator using my GPS and map. "Seven kilometers, 220 degrees bearing," I repeated. And then it happened. Fortune smiled upon me that day as I spotted five or six houses and a small triangular clearing. I informed the pilot, "Those houses ahead are our target, and that clearing over there is our helipad." Everything became clearer as we approached. The houses stood out, and people came out to watch us. But what surprised me the most was a white cloth fluttering on a long bamboo pole attached to a tree beside the triangular field. It was the same white cloth Imam Sarlif had mentioned, indicating the boys' playground. How did he know? How did he communicate and verify this information in such a short time? I marveled at his abilities.

I urged the pilot to attempt a landing on that field. The pilot, co-pilot, and crew looked amazed and impressed. However, as we got closer, we realized that the field was tangled with shrubs and jungle, making a clear approach impossible. The pilot declared that he couldn't land there. It was then up to me to find an alternative approach, a direction where the field was clear and accessible. The pilot pointed to a corner of the field where the hill sloped down gently.

Within half an hour, we returned to Tubmanberg. As we approached the helipad and descended low, we saw two vehicles from our MILOB sector waiting for us. The helicopter touched the ground, and I noticed Anmil-1163 beside the senior MILOB's car, Anmil-1175. I disembarked and witnessed Colonel Sergei Spak approaching with determined strides. The pilot saluted him and exchanged a few words in Russian, a brief conversation that lasted no more than five minutes. The pilot saluted again and returned to the helicopter. I expressed my gratitude to the security team leader and the exhausted members of the LMG groups who had accompanied us. They departed, and the senior MILOB joined us in the shade, away from the rotor's wind. Major Aima was also present, seated inside. When I saluted him, he came forward, greeted me with a handshake, and said, "I heard everything from the pilot. Well done! But now you need to go to your source with your camera images and confirm. Then come back and write a report."

After the successful mission, we returned to our headquarters where the Pakistanis had arranged some light snacks and juice for us. A major from Pakbat 3 was also present. Major Ayma and I enjoyed some nibbles and then asked for permission to leave. Taking the key, I slid into the driver's seat, and as usual, Ayma's banter and playful smirk set the tone for the journey. It was around four in the afternoon when I called Saban, the rebel field commander, to inquire if Imam was present. When I got out of the car, I turned on the camera, showed Imam the footage, and asked if it was his town and the designated playground. Imam's response was filled with awe and gratitude, affirming that it was indeed his town and that his people had hung the white cloth that morning.

Although I was curious about how Imam had managed to communicate and confirm the details within just two days, I refrained from asking and instead requested him to stay for a few more days. I also asked if he would be willing to join me on the helicopter in the upcoming days. Imam responded with a serene smile that spoke volumes.

After resolving some urgent matters, we headed to the office, and I longed for a refreshing shower to wash away the exhaustion of the day. The sound of the helicopter still echoed in my ears. The following days passed quickly, filled with meetings, planning conferences, and coordination efforts. The highlight was the final coordination meeting conducted by the senior MILOB at our headquarters attended by the leaders and officers of the teams selected for the operation. It was during this meeting that I personally confronted Captain Brian, the American captain of Team 4, and reprimanded him in front of his team leader.

The background behind this incident is that Team 4 was the team deployed closest to our headquarters, and the area of our operation fell within their assigned area of responsibility. I had proposed that one of their members join our team, and the senior MILOB had informed their team leader about it. Captain Brian, a marine officer with numerous rank badges on his uniform, always considered himself special. He introduced himself as "US Marine Captain Brian" and approached me with curiosity upon learning about our search for a weapons cache. The following day, during our final planning coordination meeting at the Pakistan sector headquarters, I informed the senior MILOB and the team leader about Captain Brian's inclusion in the operation. It was solely due to the suggestion and agreement of the senior MILOB and the team leader that Captain Brian became part of the operation.

During the coordination meeting, it was decided that we would land in Jui Town by helicopter and then trek for five or six hours through the jungle to locate and confirm the disposal of the cache on the same day. We would spend the night in Safuoula Town and prepare the field next to Safulla Town for helicopter landing by 10 am. The plan also included the provision that if the field was unsuitable for helicopter landing, we would trek back to Jui Town and take a helicopter from there.

The atmosphere turned tense as our US marine captain expressed his concerns and unwillingness to participate in the operation. As artillery officers, Colonel Tareq and I often engaged in playful banter and mock battles with the infantry, but his refusal to go caught us by surprise. He argued that the operation was insecure, too risky, and lacked cover and air support, stating that the US Army would not typically conduct such an operation.

Without hesitation, I quickly made up my mind and informed Colonel Tareq that I agreed with dropping the US marine captain from the mission. Colonel

Tareq instantly agreed and instructed me to inform the senior MILOB of our decision. We decided not to dwell on the matter any further, as dropping his story from our plan would be the best course of action. Let's return to the day before the operation.

After concluding our final coordination meeting at the MILOB headquarters that morning, I still felt tense, awaiting Imam's movement order permit (MOP) to fly on the UN helicopter. None of us had received the MOP yet, but having Imam by my side gave me a sense of confidence. I checked my Lotus Notes repeatedly, anxiously waiting for the mail. Finally, at 1:47 pm, the email arrived. I quickly scanned the heading for the MOP and found one. However, I realized that Imam's MOP was missing. Opening the email, I discovered a long list of names, and to my relief, the last name on the list was Imam's. They had included him in the same MOP. Additionally, two civil affairs officers, both Nigerians, were also listed to accompany us on the mission.

On the morning of the operation, which could be referred to as D-Day, Major Ayma assigned me the task of escorting him to the helipad. However, before heading there, I had to pick up Imam Sarfli as previously arranged. We arrived at the helipad by 8:15 am, accompanied by Imam Sarfli. Waiting for us at the shed next to the helipad were the Pakistani security team led by Captain Shahrukh, consisting of nine men as well as the four-member Pak-engineer's explosive ordnance disposal team under Captain Niyar. Majors Chima and Khayyam, the ops officers of the Pakistan sector and Pak-Bat 3, respectively, were also present, which was a rare sight as they seldom left their offices. We exchanged salutes as per military protocol, their presence serving to boost the morale of their teams.

We were about to embark on a mission that had never been attempted before: trekking through the dark African jungle on foot, at least not in that specific Liberia mission. This necessitated the special arrangements that had been made. Just as we were getting ready, another patrol led by the sector boss of civil affairs, along with his two associates, arrived. Finally, Dr. Major Kashif joined us to provide medical support, completing our group of 15.

At exactly 8:45 am, an MI-8 helicopter landed on the helipad. I hurried forward, but one of the crew members informed me that they wouldn't be turning off the engine as the captain wanted to take off as soon as possible. We were prepared as well, and within five minutes all of us boarded the helicopter. Dr. Major Kashif sat beside me, while the others were distributed among their respective teams. I managed to exchange a few words with the captain and co-pilot during the journey.

After a 25-minute flight, the crew announced that we had arrived. Within five minutes, I caught sight of the houses and markets of Jui Town. The pilot executed a smooth landing, making it feel like we had touched down on a pillow. As soon as the crew opened the door, I bid them farewell, saying, “See you tomorrow morning.” They smiled and nodded in response. Feeling confident, I chuckled as I stepped off the helicopter first, unaware of what awaited us.

It was the evening of August 16th when disaster struck. The ground trembled as the mountains erupted, throwing me several yards away onto the ground. I surveyed the scene and saw my fellow team members scattered like leaves. Thankfully, the hillside we had taken cover behind wasn’t too steep; otherwise, the outcome could have been disastrous. I wondered if anyone had ever witnessed such a devastating blast in this jungle before. As an artillery officer, I could swear that I had never heard anything like it. The simultaneous detonation of 11 rockets shook me to the core.

Despite the chaos, we quickly regrouped. Miraculously, my map, SATphone, wireless set, and camera remained intact. Our teams gathered together, including a new special team led by Imam Sarfli, with three members armed with machetes as part of his team.

Our journey was divided into three parts: from Jui to Tinada, from Tinada to the cascade leading to the mountain, and finally to the grassy Seyfula area nestled on the mountain’s flank. Our target destination was within 500 yards of Seyfula, according to the information shared by Imam. I estimated a total travel time of five hours, although Imam claimed it would be quicker. However, considering our diverse team and the fact that we were a fighting patrol, I knew we would likely lag behind. The multi-ethnic composition of our team meant that progress would be slower than a more homogenous group. Two men armed with machetes led the way to clear the jungle, followed by two scouts from the fighting patrol. Captain Shahrukh, Imam, and I walked side by side, with the rest of the team following behind. At the rear, a light machine gun group from the fighting patrol provided cover. Another group from Tinada led the way, always a step ahead of everyone else.

We began our journey at 9:55 am. As planned, we reached Tinada by noon, but the Pakistani team members needed a break. They were sprawled out in exhaustion, unaccustomed to the challenging terrain and dense foliage. Niyar and his corpulent comrades were particularly drained. In contrast, the civil affairs personnel were energetic and lively. We extended our break to 25 minutes instead of the originally intended 15 minutes, as it was necessary for the Pakistanis to recover.

The next two hours of our journey ended up taking almost three hours. However, when we came across a clear water stream flowing from a mountain waterfall, I couldn't resist. Four Liberians, two Nigerians, and I dove into the water, feeling the fatigue wash away instantly. I cupped my hands to drink the water, despite Niyar's warning that it might be dangerous. I responded, stating that I would rather risk drinking the natural water than the chlorinated water they had brought. If it were harmful, then all the Liberians would be affected as well.

We continued our trek, encountering small hillocks covered in dense and prickly shrubs. I was tempted to cut through them, but Imam cautioned that the Pakistani soldiers might struggle with such terrain. We deviated slightly from the planned route. In the afternoon, we ascended a muddy trail, and Imam announced that it would take another 30 minutes to reach Seyfula Town. Glancing at my watch, I expressed the desire to push past the target. With everyone's agreement, we decided to bypass Seyfula and climbed a hill to the right. The summit was relatively flat, with radish and bamboo bushes scattered around. The ground was moist from the previous day's rain. Imam pointed out a bush to me, and as I looked, I noticed the Pakistani soldiers securing positions nearby. They acted with precision, demonstrating their well-trained skills. Shahrukh, Niyar, and I approached the bush. Inside, we discovered large-caliber ammunition scattered everywhere—dozens of fresh rockets wrapped in thick blue polythene, along with AK-47s and thousands of rounds of ammunition of different calibers.

Captain Niyar, a professional officer, got to work right away. He extracted what was in the pit in 10 minutes. Shahrukh's two soldiers swiftly cleared one side, and two were assigned to tally all the ammunition. I counted the rockets: 11 fresh 40-mm rockets in their original packaging, greased. Three AK-47s were in a decayed state, along with one 7.62-mm heavy machine gun that had an extra barrel. We retrieved a total of 1300 rounds of ammunition of various calibers.

A prompt decision was made to seize the weapons, but due to restrictions, explosives or ammunition couldn't be carried on the helicopter. Niyar proposed disposing of the rockets and ammunition together in the pit, and we all assisted him. Imam's son had also joined us by then. I instructed Imam, his son, and two civil affairs staff to move to the town while we wrapped up the job and followed.

The next 15 minutes passed quickly. Niyar set up explosives, detonators, and detonation cables, completing all the necessary preparations. We took up positions on the slope next to the road, with an aerial distance of no more than 50 yards from the explosion site. Niyar mentioned that the sound would be dampened inside the pit, but there would be a lot of smoke. However, the noise was deafening. Despite covering our ears, the vibration and shock still threw us off balance. We realized that our estimation of the sound and smoke had been inaccurate.

Furthermore, even though we were only 50 yards away from the center of the explosion, our skin burned because Captain Niyar only had cables that could cover 60 yards. He used a non-electric system, a simple explosive charge with yellow thread (salt). He had set it up for about 30 yards, explaining that it would take approximately a minute and a half to burn that length. During this time, we could safely move away and distance ourselves. However, the smoke and shock resulting from detonating 11 rockets and such a large amount of ammunition in a deep pit affected us significantly.

I activated my and Captain Shahrukh's SATphone. I communicated with the senior MILOB while Shahrukh had likely conversed with the sector by then.

Finally, when we took the road to Seyfoula Town, we realized it was only about 300 yards away from our operation location. A small mosque stood in the town, where the Maghrib prayer was being called. There were six small and big houses, with approximately thirty people present. Everyone gathered in the middle of the town field, which was more like a big courtyard similar to those seen in large family estates in our country. They were all staring at us with fear and curiosity in their eyes and faces. I could discern extreme terror on their faces due to the loud noise and vibration. Men and boys approached us and clasped our hands.

I observed four people setting up two tents in the courtyard. I was astonished to see two brand-new US Aid tents being skillfully pitched. They quickly assembled long, thin tree branches as corner support. Then, they spread triple layers of new, thick blankets underneath.

Seeing all this, fatigue seemed to overwhelm our bodies and minds. I sat down next to the tent and took off my boots. Imam and his son brought us water. I realized they were summoning us for prayers. After performing ablutions, I entered the small mosque. The blind muezzin who had called for prayer led the Azan, and Imam Yusuf Sharif led the prayer.

After the prayer, I accompanied Imam to the nearby playground. There wasn't much light, but I could see that there were a lot of bushes. If they weren't cleared, it would be impossible to land a helicopter. I also thought about the trek back to Jui Town the next day. Everyone was so exhausted, and it's difficult to express just how tired we were. I made up my mind that the helicopter would come here. I asked Imam if it was feasible to clear those bushes by 9 am the next day. There was also a tree head that needed to be chopped off from one corner. Seeing Imam's concern, I said, "You organize the people, and I'll give them 20 dollars." Upon hearing this, Imam said no money was required and that the townspeople would take care of it.

When I entered the tent, I saw that Imam had brought rice, eggplant, and cassava leaf, African-style vegetables cooked in Bengali style. In one tent, we had three officers, and the soldiers stayed in another tent. Two Nigerian civil affairs staff spent the night in a small house. I joined Imam to have a meal, and we finalized the discussion for the following morning.

I don't know when I fell asleep, but when I woke up in the morning, there was activity and noise all around. I glanced at my watch and saw that it was 6:40. Everyone quickly got ready and ate some dry food we had brought. We made a makeshift area in a nearby bush with an old tent cloth for our morning work.

At exactly 7:25, everyone gathered at the nearby field. Boys and girls, young and old from the town, were all there, holding spades, machetes, and whatever tools they could find. I pointed to an area, and everyone started energetically clearing the bushes. We also helped to move the cut bushes to the right side of the slope. By 8 o'clock, the work had made good progress. Shahrukh himself turned on the SAT phone to contact the helicopter and informed the sector. The pilots had taken note of these coordinates during the reconnaissance.

Near the northwest corner of the field, a tree branch was being cut and lowered. A girl of around 10–12 years old was assisting in moving part of the branch. Suddenly, I heard a shriek. Two soldiers had spotted something. Shahrukh went ahead to investigate. He saw something and called me. I quickly ran over. I was astonished to see two rockets again, just like before. I called Niyar, and we quickly made a decision to disperse. We searched the entire field and bushes. After 10 minutes, there was another shriek. Two more rockets were found on the other side. I didn't know what to do – continue searching or finish the helipad work. One of the civil affairs staff members suggested that if we completed the helipad, the bushes would be clearer, making it easier to spot anything else. We all agreed with the suggestion.

Niyar delivered unfortunate news on this side. We had very limited explosives and a shortage of cables. Detonating the four rockets with these resources would be challenging. Nonetheless, at precisely 9:30, the field was cleared, and only loose ammunition was found in six different spots. No more rockets were discovered. However, Niyar and I were discussing the best way to destroy them. Meanwhile, Shahrukh brought good news that the helicopter would be arriving directly from Monrovia to the new helipad within five minutes.

The only remaining challenge was to destroy the four newly discovered rockets before the helicopter's arrival. We decided to dig a pit underneath the slope, ensuring our safety on top. Niyar's men wore full explosion safety gear, including Niyar himself. I supervised the work as we used the remaining surplus explosives. We attached the cable and hurried to the south side of the field, lauding it for burning.

Minutes passed while we waited anxiously. Initially, there was only a faint sound, and nothing happened. After another 10 minutes of waiting, Niyar and his men went to inspect the rockets. To my surprise, they were still lying there untouched. Niyar called me, and I joined them to explain how to activate or ignite the rocket fuse. Suddenly, I noticed hundreds of rounds of loose ammunition lying nearby. I inquired if they could be used, and one of Niyar's Havildars confirmed that they could. Time was running out on our side, and we had to complete the task within 10 minutes. Meanwhile, Shahrukh's soldiers had set up a windsock, which fluttered on a long branch tied to a tree, indicating the southeast direction.

Unfortunately, we ran out of explosive slabs, and only a few small cut cables remained. We used these cables to tie the four rockets, and some smaller cables were randomly arranged with the loose ammunition. The ends of each cable were taped together. Two cables, each about 10–12 yards long, were given to me on both sides. However, just as this was happening, I heard the sound of the helicopter's rotor. Yes! The helicopter was approaching from the right side and would arrive in three or four minutes. I shouted, "Niyar, hurry up!"

I saw that Niyar's Havildar had attached a detonator to the head of his cable. I yelled for everyone to move to the south of the field. Meanwhile, I noticed the helicopter circling and flying in from the north. The moment had arrived. Niyar's Havildar ignited the fuse. We all ran, holding our breath in anticipation. I gazed at the explosion site with wide eyes. "Ya, Allah, this is the helicopter's landing approach route." A loud noise reverberated through the entire field and jungle. The pit burst, and thick white smoke billowed upward. I saw the helicopter make a slight left turn, still about 500 yards away. It then descended straight and landed right where Shahrukh's men had marked an 'L'. We took cover from the helicopter's wind, but the freshly cut bushes, small branches, and leaves flew around, covering the area like a green blanket. As the wind subsided, I moved forward.

I exchanged greetings with the familiar crew, and they inquired about the earlier event. I chose not to reveal anything at that moment. If I were to mention the four rockets exploding in the helicopter's approach, it would surely scare them. I glanced at Niyar, who was accompanied by two men inspecting the explosion site. I moved ahead and greeted the pilots through the window. Niyar returned, giving me a thumbs-up sign, assuring me that everything was fine and that there were no more rockets.

As the preparations for departure were underway, I observed our men gathering on one side, resembling a flock of birds ready to take flight, while the villagers stood on the other side. Imam Sarlif and one of his men stood apart, watching me. I approached them and warmly shook their hands, expressing my deep gratitude. I thanked them profusely for their assistance. The blind muezzin clung to me tightly with both arms, pleading for me to return someday. Deep down, I knew that I might have to come back again. It had only been two months since my last visit, and yet here I was, contemplating another return. But that was a story for another time.

Before boarding the helicopter, I captured a photograph with Imam, the three individuals who had helped us clear the jungle, the muezzin, and a few others. It would serve as a cherished memento of our brief encounter. The clock indicated 10:55, and we all boarded the aircraft. The rotor blades spun faster and faster, creating a powerful gust of wind. The townspeople lined up on one side, waving at us with smiles and tears. I raised my hand and waved back, overwhelmed by a surge of emotions. The helicopter's door closed with a resounding thud that the townspeople would never hear. In a matter of three minutes, we ascended into the sky, making a graceful half-circle before heading towards Tabmanberg. The helipad and Sayfoulla Town gradually shrank and faded away beneath us like distant memories. I leaned my head against the seat, feeling a profound sense of relief and accomplishment. We had done it. We had successfully completed the operation. However, our task and mission were far from over. I had to write a preliminary report before the evening, outlining the surprising cache that remained a secret for everyone else.

SECURITY PRISON 21

MAJOR MD. ASADULLAH

(UNITED NATIONS TRANSITIONAL AUTHORITY IN CAMBODIA - UNTAC)

Our camp, nestled in the depths of Thala Baribhat on the western bank of the Mekong, was an enigmatic sanctuary. The ancient forest, cloaked in colossal trees and towering elephant grass, enveloped us, resembling a mystical reserve untouched by time.

A trail daubed with reddish mud emerged from the Mekong's ferry terminal, snaking its way through the vibrant tapestry of dense foliage. In the rainy season, this realm transformed into a labyrinth without paths, where canals of water meandered and vast reservoirs materialized in unexpected places. The road, punctuated with countless potholes, presented a perilous challenge even to the most seasoned driver. The turbid waters obstructed our view, making it difficult to discern what lay ahead, even within arm's reach. Yet we forged ahead, gripping the wheel tightly, seeking solace in the wind whispering through the encompassing trees.

It was a road that devoured vehicles, reducing them to skeletal remnants within a year, but for French Major Bertrand, it remained an enduring favorite. At irregular intervals, we encountered convoys of massive Russian trucks crammed with families squeezed into confined spaces, where even a feline would struggle to find room to maneuver. Children clung to these trucks alongside their families, carrying with them all the trappings of domestic life. Nearly everyone bore arms, yet their countenances wore an eerie serenity—a grotesque curiosity.

Upon inquiring with the female interpreter, we discovered they were all members of the Cambodian People's Armed Forces (CPAF) faction. War was imminent, and these families departed on their trucks to confront their adversaries. I marveled at how effortlessly an entire family committed themselves to the battlefield without a second thought. It was ingrained in their culture, for they knew not if they would ever return to their homes in this very place. The uncertainty of their future rendered any plans or preparations irrelevant. In that moment, my mind was eclipsed by recollections of the wilderness of Karbala. I saw a few tents amidst the boundless desert, and within my mind's eye, scenes unfolded—Duldul, the beloved horse of Imam Hussain, scarred and wounded, returning to the community to those closest to Imam Hussain bearing the burden of countless arrows. Shimar, bearing the severed head of Imam Hussain, entered the court of Ajid. And within the tent, young Zainal wept, tormented by an unquenchable thirst, while the Forat River flowed not too far away, just as the eternal flow of the Mekong River persisted, impervious to earthly pain.

Our camp, nestled within the controlled territory of CPAF, exuded an air of tranquility. The presence of Khmer Rouge forces led by NADK seemed absent, their movements concealed within the shadows. It was, in essence, a peaceful enclave. However, there were moments when the CPAF cantonment stood vacant, and it was during these occasions that our primary task was to oversee the surrender of arms.

When we inquired about the whereabouts of their army members, they would knowingly or unknowingly inform us that the soldiers had taken leave to tend to the crops in the nearby village. But we knew the truth—they had ventured forth to wage war against the Khmer Rouge.

Recently, a captain from CPAF paid a visit to our camp. Prior to my assignment here, I had purchased a pair of "Made in Thailand" long boots from Phnom Penh's bustling central market. Slipping into those boots was an exquisite sensation of comfort. Yet, when I gifted a pair of my Bangladeshi long boots to the CPAF captain, his face erupted with joy, radiating a radiant smile. In that moment, it reminded me of the soldier in Erich Maria Remarque's poignant novel, "All Quiet on the Western Front," who found solace in a fallen comrade's boots. In return, the captain offered me his bayonet—a token of our exchange. I couldn't help but wonder if that very "long boot" accompanied him into battle, serving as a source of security and courage. The future remained uncertain, and only time would reveal whether he would return wearing those same "long boots" or if they would become a poignant reminder of a past that had faded away, forever bound to that last visit to our camp.

As far as I knew, no agreement had been reached between the United Nations and the Khmer Rouge forces at that point. Nonetheless, the United Nations had already resolved to proceed with conducting elections in the country, pending negotiations and consultations with other factions. The future held its secrets, and the path to resolution remained veiled in uncertainty.

It was within that tumultuous backdrop, in the shifting period between September and October, that British Captain Mike and I, accompanied by the election registration team, disembarked from the helicopter in the distant Stung Treng area. We found ourselves standing in an empty crop field, a forsaken village veiled by forests. Yet, to our astonishment, we were swiftly encircled by a sizable contingent of Khmer Rouge militants armed with menacing automatic rifles.

The sight of this armed assembly forming a ring around our helicopter left me in a state of bewildered surprise. Our male interpreter, Nara, wept as he was beckoned to step down from the helicopter, joining me to prepare for the negotiation that awaited us. Initially hesitant to engage with them, Nara succumbed to tears, as if his sobbing served as an omen of imminent hardships.

Through his tearful lamentations, Nara imparted crucial information: those who wore crimson bandanas around their necks were not affiliated with CPAF; they were the formidable and impregnable Khmer Rouge—the guerrilla force associated with Pol Pot. They were the force that had either claimed the lives of his wife and only daughter or subjected them to the anguish of forced disappearance. Nara himself had managed to escape, seeking refuge in a camp across the border in Thailand.

Since his return in 1979, he had tirelessly searched for his lost loved ones, yet all efforts proved futile. Thus, summoning the courage to approach and negotiate with the Khmer Rouge was an arduous task for him. For the first time, a sense of trepidation, unease, and confusion washed over me. Still, attempting to flee in the helicopter at that very moment would undoubtedly invite the terrorists to unleash a missile from their ready rocket launcher, sealing our fate and bringing an abrupt end to my existence. My solitary struggle, having only just begun, would be prematurely extinguished. Moreover, this would transpire in a foreign land where I had arrived with the noble intention of raising the banner of peace and ending their internal strife and conflict.

As I stepped off the helicopter, I could feel the weight of the situation bearing down on me. The Khmer Rouge stood before us, armed and ready, their presence suffocating and menacing. I was the only member of our team who could speak their language, so it was up to me to communicate with them.

With a limited vocabulary, I struggled to convey our message of love and solidarity. But I knew I had to try. As I greeted their leader with a gentle smile and asked "Sok Sabay?", the air was filled with a momentary hush. It was a silence I understood well—one that could easily turn to violence.

To my surprise, they responded positively, and I could see the elation on their faces. But it was a delicate situation, and one wrong move could lead to disaster. So, I made a mistake when I uttered the words "Khmer Rouge," not realizing the weight those words carried.

Nara warned me of my mistake, telling me that even a single mention of those words could mean the end of our lives. I felt a pang of fear and regret, realizing the gravity of the situation. But I knew we had to continue with our dialogue, even if it meant acquiescing to their every desire and dictate and potentially being kidnapped and disappearing forever.

As I looked around at my team, I knew the responsibility was on my shoulders, and I could not let them down. I had to be strong and careful and do everything in my power to ensure our safe passage out of this foreign land.

On my journey back, I turned to Nara, seeking clarification on the significance of the term "Khmer Rouge." Nara explained, "The word 'Rouge' is French for 'Red.' The term 'Khmer Rouge' refers to the 'Red Khmers.' However, they dislike this name because it was given to them by Prince Norodom Sihanouk, the first ruler of independent Cambodia, who once oppressed them severely. The Khmer Rouge pride themselves on being identified as communist guerrillas."

Behind the name Pol Pot there was a man called Saloth Sar. Born into privilege in 1925 in Cambodia, Saloth Sar was influenced by the teachings of Gautama Buddha, particularly the principle of non-violence. Unfortunately, his chosen pseudonym could not shield him from becoming a figure of fear and dread among his opponents.

Under Pol Pot's leadership, over a million Buddhist monks were brutally murdered, a tragic betrayal of the non-violent principles Saloth Sar had once embraced. It is disheartening to see echoes of this disregard for human life in present-day Myanmar, with the dictatorial regime led by General Min Aung Hlaing. Looking back further in history, we encounter General Yahya of Pakistan in 1971, who committed similar atrocities, adding his name to the roster of history's most ruthless killers. Dictators throughout time have transformed into ruthless murderers, as if their brutality escalates with each passing era, regardless of the ideologies they claim to uphold.

The concept of "Room 101" in George Orwell's novel 1984 comes to mind, symbolizing the ultimate horror and nightmare. Faced with the prospect of being taken to that room, the protagonist cries out, "Not me, send Julia!" Even though Julia was his lover, Winston not only sacrifices her but also consents to her suffering in the depths of that chamber, all in an attempt to save himself.

Such sacrifices, such choices, test the limits of our humanity and confront us with the complexities of love and self-preservation in the face of ultimate terror.

In delving into the true horrors of "Security Prison 21" once created by Pol Pot, I must clarify that it surpasses even the nightmarish Room No. 101 of your imagination. However, I believe that before delving into the darkness, providing some background knowledge will help set the stage for my reader.

Let us travel back to the year 1948.

A young man named Saloth Sar ventured to France to study radio electronics. During that period, Europe had been engulfed by the allure of Marxism. The fervor for socialism swept across the continent, captivating even intellectuals like Jean-Paul Sartre, who saw Marxism as the ultimate remedy for all social ills. Sar, too, fell under its spell and wholeheartedly embraced socialism. After immersing himself in its ideology, Sar, now known as Pol Pot, returned to Cambodia in 1953.

Upon his return, Pol Pot sought to establish connections with the Viet Cong of Vietnam, finding like-minded individuals who shared his fervent dedication to socialism. The same year, fueled by a burning desire for independence, the Cambodian people broke free from the grip of French colonialism, snatching their autonomy from the hands of their oppressors.

Norodom Sihanouk ascended the throne as king, but being a traditionalist politician, he struggled to grasp the essence of Marxism. Unaware that the Cambodian independence movement had been directly influenced by Ho Chi Minh, the charismatic leader of the Vietnamese working class, Sihanouk failed to comprehend the underlying currents at play.

Sihanouk's ignorance was compounded by his brutal suppression of the communists, inadvertently fueling the flames of the Communist movement. Under Pol Pot's leadership, the 1960s witnessed the exponential growth of the Communist movement, with guidance provided by the wise counsel of the Viet Cong.

This set the stage for the dark days that were yet to unfold, where Security Prison 21 would become a harrowing testament to the depths of human suffering and the brutal realities of Pol Pot's regime.

In the year 1967, under Pol Pot's leadership, communist activists, aided by the Viet Cong, persisted in their campaign to destabilize and weaken the government forces led by Prince Norodom Sihanouk in Cambodia. Eventually, they succeeded in overthrowing the government, forcing Prince Sihanouk to seek political asylum in China.

Then, on the fateful day of April 17, 1975, after eight years of guerrilla warfare, hordes of Khmer Rouge rebels, accompanied by a fleet of tanks, marched triumphantly into the city of Phnom Penh. Cambodia was reborn under a new name, "The Democratic Kampuchea," as decreed by the Pol Pot regime.

During their reign from 1975 to 1979, the Khmer Rouge government implemented various social engineering projects and reforms, including the infamous "Sanctification and bringing correctness in real" campaign. These initiatives, however, concealed a nightmarish killing spree that claimed the lives of nearly three million Cambodians. One of the primary motivations behind this genocidal campaign was the Khmer Rouge's deep-seated anger and resentment towards the inhabitants of Phnom Penh, whom they regarded with suspicion and contempt, labeling them as nothing more than bourgeois parasites.

To the Khmer Rouge, the capital's residents symbolized decadence, disloyalty, and a shallow pursuit of material pleasures. They were seen as intoxicated with Western culture, indulging in drinks, nightclub dances, and a taste for American luxuries like Coca-Cola and hamburgers. In the eyes of the Khmer Rouge, these urbanites lacked the moral fiber and revolutionary spirit required to support the communist cause, as they seemed more concerned with serving American interests than the welfare of their own nation. They neglected the plight of their impoverished rural compatriots, fixating instead on securing jobs after graduation and the allure of migrating to America.

Thus, the Khmer Rouge guerrillas scornfully referred to the residents of Phnom Penh as the "whores of the Mekong River," encapsulating their disdain for those tainted by American influence. Rather than embracing American culture, the Khmer Rouge saw it as a corrupting force, hindering their pursuit of a society built upon their revolutionary ideals.

Pol Pot, with his poetic yet profound slogan, "The spade is your pen, the rice field your paper!" unveiled his political manifesto, determined to reshape society. However, for the America-addicted bourgeois of Phnom Penh who found themselves at odds with Pol Pot's ideology, the consequences were dire. Faced with the menacing Khmer Rouge guerrillas brandishing their rifles high in the air, the pro-American bourgeois had no choice but to comply and seek refuge in the countryside, abandoning their city lives. Defiance would only invite the wrath of the Khmer Rouge, with bullets as their response.

Along the road, an ominous black truck would periodically wait, foreshadowing the impending tragedy. In brazen acts of audacity, the Khmer Rouge would select hapless locals and load them onto the truck, destined for the dreaded detention center known as Toul Sleng, or "Security Prison 21."

'Security Prison 21,' originally the covert headquarters of Pol Pot's secret police, had been established within the walls of an old primary school. It served as an interrogation center, but above all it became a site of unutterable horror, where countless individuals were subjected to interrogation, torture, and merciless death.

Unlike the characters in Orwell's Room 101, who occasionally emerged with their spirits shattered yet physically intact, the fate of those who entered Pol Pot's 'Security Prison 21' remains shrouded in mystery. The exact number of survivors, if any, remains an enigma—an everlasting testament to the brutal and oppressive regime that engulfed the people of Cambodia under Pol Pot's reign. Many never returned to their loved ones and are forever lost.

The prisoners at Toul Sleng endured the agony of interrogation. Their torment only ceased when their answers satisfied the interrogators. The pursuit of truth held no significance within those walls. The paramount importance was solely the satisfaction of the guerrillas with the responses provided by their captives.

Subsequently, the near-lifeless prisoners were once again loaded onto trucks. These vehicles careened with deadly purpose towards ‘Choeung Ek,’ one of many killing fields utilized by the Khmer Rouge. It was there that nearly 20,000 Cambodians met their tragic demise.

Prisoners were coerced into digging deep pits in the unforgiving earth. Once the pits were painstakingly prepared, the heartless Khmer Rouge commanders forced the prisoners to stand beside these grim graves. In their relentless pursuit of brutality, the Pol Pot regime had no intention of wasting precious bullets. They conserved their ammunition, choosing instead to bludgeon their helpless victims to death. The sounds of rods, shovels, hammers, and other tools striking human flesh echoed in the air, mirroring the horrific efficiency of Hitler's gas chambers in their quest to spare bullets.

Through this campaign of mass murder, the Khmer Rouge extinguished the lives of approximately three million people, effectively halving Cambodia's population. Pol Pot's deluded belief was that his revolution could succeed with a much smaller population.

As time passed, what had once been a friend—Vietnam—transformed into Pol Pot’s most formidable adversary. The relationship between Democratic Kampuchea and Vietnam deteriorated, and when tensions reached their breaking point, Vietnam intervened. The Vietnamese army infiltrated Cambodian territory, permeating the land in 1978.

In 1979, as the irresistible tide of Vietnamese troops rapidly advanced towards Phnom Penh, Pol Pot and the Khmer Rouge—inevitably inseparable from him—fled towards the Thai border, seeking refuge from the encroaching danger.

In the autumn of late October 1992, during a brief respite lasting thirteen days, I made my way to Phnom Penh’s haunting ‘Tool Slang’ or ‘Security Prison 21.’ Stepping into that aged primary school with my entrance card in hand, I sensed an eerie presence enveloping me. It was as if not hundreds, but thousands of spectral figures, shackled prisoners, trailed behind my every step. Among them, I couldn’t help but wonder if Nara’s wife and daughter found themselves entangled in this ethereal procession. The sultry yellow afternoon added an air of surrealism to the experience, heightening the weight of history that filled every corner of that solemn place.

Another Heart Bleeding White

Major Shaikh Khalid Raihan
(United Nations Mission in Ivory Coast - UNOCI)

I often wonder why people resort to splashing cold water in their eyes when they experience a burning sensation. In my experience, it only intensifies the discomfort and results in a quick reddening of the eyes. This, of course, invites a barrage of questions from concerned onlookers. What happened to your eyes? How long have they been red? Do you require medical attention? The inquiries flood in relentlessly. Seeking respite from the inquiries, I sought solace by pulling a pillow close and declaring it a night well-spent. However, sleep eluded me that particular evening. Restless, I tossed and turned, finding no relief. I rose from bed to quench my thirst, indulged in a cigarette, and made another futile attempt at sleep. Alas, my efforts were in vain. As I stared into the darkness beyond the window, the scenes of my life began to unfold before me, akin to a captivating film.

On the fateful day of December 30, 2005 in Cheravine, Ivory Coast, I found myself standing atop the observation tower of my checkpoint, a vital marker within the zone of confidence. Through my binoculars, I meticulously surveyed the expanse that lay before me. The landscape revealed nothing more than the majestic Fromager trees swaying gracefully in the winter breeze and the radiant heat emanating from the shimmering asphalt road as it disappeared into the distant thickets. Earlier that morning, my company commander had embarked on a crucial patrol, accompanied by the UN Military Observers (MILOB), venturing into the border regions of Ivory Coast and Liberia. As the second-in-command, it was my duty to hold steadfast at the Cheravine checkpoint while my fellow coursemate, Captain Hassan, assumed responsibility for overseeing the company headquarters in Danane. In recent days, our intelligence had alerted us to potential unrest brewing among the rebel forces, driven by their aim to destabilize the UN presence in Ivory Coast.

The morning hours seemed uneventful, devoid of any noteworthy incidents. However, an eerie stillness hung over the small town of Danane, as though an impending storm lurked beneath the surface of the silence. Within Danane, the local headquarters of the rebel forces, known as the Forces Nouvelles (FN), was commanded by Captain Eddie Meddy, who oversaw a formidable battalion-sized troop. Intelligence reports had led us to anticipate FN activities in the vicinity of my checkpoint, with the objective of disrupting the movement of unarmed personnel. Yet, until that moment, no signs of such activity had materialized. Consequently, we allowed ourselves a momentary respite, easing our vigilance and temporarily relaxing our heightened state of alert at the checkpoint. Little did we know that a sudden and unexpected confrontation was about to erupt at our company headquarters in Danane.

Captain Hassan was stationed at that very headquarters, situated approximately 5 kilometres away from my location. He commanded a well-equipped force consisting of an armored personnel carrier (APC) and a contingent of 40 soldiers armed with a potent 12.7-mm coaxial machine gun mounted on the APC complemented by two rocket launchers and an arsenal of anti-personnel grenades. Given our troops' fortified position within the central post office building of Danane, shielded by concrete bunkers and encircled by a labyrinth of barbed wire, it appeared we possessed ample means to withstand the threat posed by a few hundred FN gunmen. However, the reality of the situation would soon reveal an unforeseen twist in the unfolding events. A sense of unease permeated Captain Hassan's voice as he urgently contacted me over the crackling radio waves.

"Charlie Two, do you copy?"

"Charlie One, loud and clear. What's happening?"

"Khalid, they're all gathering around the HQ building."

"Who are they?"

"The FN gunmen. They're armed with submachine guns, and more are on their way."

"How many of them?"

"About thirty to forty. They appear highly agitated and are causing chaos outside our camp."

"Can you guess what they want?"

"I don't know. They're asking for the BANBAT (Bangladesh Battalion) company commander to come out and talk to them."

"Inform them that he's out on patrol today."

"For God's sake, I've already told them that. Now they're demanding that I come out and face them."

I was well acquainted with these trigger-happy FN gunmen by then. I knew that even a casual conversation could escalate quickly if things went slightly wrong. They wouldn't hesitate for a moment to put a bullet through my head or slit my throat with a rusty knife. What made matters worse was that in such tense situations, they were often heavily under the influence of alcohol and hashish, which only fueled their aggressiveness.

"Stay calm. Lock the iron gates. Get everyone geared up and to their posts. Do not step outside the camp. Observe the situation, and I'll call you back in ten minutes."

"I've already locked the gates. Mate, things are not looking good here."

The high-frequency radio net for communication with Battalion HQ was jammed during the daytime. The telephones were kept in the military observer's office, locked and inaccessible, and all of them were out on patrol. I decided to instruct Hassan to call Battalion HQ first. I radioed him again,

"This is Charlie One, do you copy?"

"Yes, loud and clear. Go ahead."

"Break into the military observer's telephone room and call Battalion HQ to report the situation."

"Oh... I forgot to tell you. I already did that."

"What did they say?"

"They said the same thing is happening at their location. It's a coordinated movement across the entire FN-held territory to unsettle the UN forces."

"And...?"

I think I heard Hassan sigh.

"And we should handle this on our own since we were warned of such a possible development. The operations officer wanted to know why the company commander was out on patrol on a day like this."

"Yeah! As if he had been secretly given the FN unrest program in advance by the FN!"

I muttered to myself and ended the conversation to start contemplating other options. In the next ten minutes, the tension escalated, and the possibility of a direct confrontation at our company HQ by the FN became increasingly apparent. Given the circumstances, I couldn't dismiss the suspicion of a similar scenario unfolding at my location simultaneously. I could sense Hassan's mental state. Then, Hassan called again and got straight to the point.

"Khalid, there are nearly a hundred of them now, and they have completely surrounded us. They want to breach our gate and force the camp commander to face them. My troops are on edge. Anything can happen at any moment. I wouldn't be surprised if a gunfight breaks out between them and us. Either you send reinforcements to make them retreat or ask Eddie (the rebel commander) to restrain his men."

"Stay where you are, mate. Help is on the way. Just don't open up until it arrives. Can you do that?"

Hassan's voice cracked as he responded to me.

"They just set fire to the boundary fences... I'll talk to you later."

The gravity of the situation left me with limited options. I resorted to my last option: calling for the French Quick Reaction Force (QRF). Captain Busch, the French company commander, promptly responded to my call and set off from his camp with a patrol consisting of four fully armed VAB APCs.

These bulletproof French APCs were highly mobile and had enough firepower to turn a few hundred gunmen into ashes within seconds. He arrived at my camp precisely 10 minutes later. In the meantime, I assembled a C-type patrol of 9–10 personnel in a pickup truck, instructed my senior JCO at the checkpoint to assume command, and prepared myself to head to the company HQ in Danane.

However, my senior JCO arrived with even graver news. The FN gunmen had completely blocked the entrance to our camp, set fire to the surrounding areas, and their numbers had swelled to almost two hundred. Swiftly, I briefed Captain Busch on the situation and asked if he could establish a roadblock with his patrol between Danane and Cheravine.

The plan was straightforward. According to the memorandum of understanding between the UN and the French government, the QRF couldn't intervene or employ lethal force unless specifically requested by the UN forces or in a life-threatening situation. Thus, I would venture in to negotiate with the FN members while Busch would wait, setting up a roadblock between Danane and Cheravine.

I still harboured doubts about a similar clash simultaneously occurring in Cheravine. Hence, I positioned Busch's patrol in the middle, and I took nine troops with me from the checkpoint. If the negotiation failed, I would request Busch to breach the camp with Hassan's support from within the concrete bunkers, marking the end for either them or us.

Even if we survived, I could face a court-martial for calling in the French QRF to deal with the situation. As the acting company commander, I might be questioned for failing to handle the situation tactfully.

Before I got in the front seat of the pickup truck, I glanced at my troops. Tense and already filled with anticipation, all of their eyes were fixed on me. They knew that this young captain was leading them, but they were unaware of what lay ahead. I felt the weight of responsibility for my troops solely on my shoulders—an immense responsibility I had been trained to shoulder for a long time but had never truly encountered.

As the wheels started rolling out of Cheravine, I imparted my final instruction to my senior JCO, "Maintain control of the checkpoint until my return. Everything will be all right." He nodded silently.

Unlike the French QRF, my patrol was armed with Chinese carbines, with one SMG serving as my weapon and a light machine gun cynically secured to the iron frame of the Toyota pickup truck's rear roof with a rope. We were equipped with bulletproof jackets and fibreglass helmets, which offered limited protection. However, the pickup truck itself was vulnerable to small arms fire as it was essentially a soft skin vehicle. Aware of our weaknesses and bound by the rules of engagement under the UN peacekeeping forces protocol, I meticulously planned my actions to avoid being drawn into a gunfight at all costs.

During our journey, Busch halted and signalled that he had reached his designated roadblock position while I proceeded towards Danane. A raging inferno accompanied by billowing black smoke loomed in the distance. I remained in constant communication with Hassan, who was occupied with organizing our troops. I had already instructed him that, unless lives were in imminent danger, we were not to initiate violence with our weapons. Our purpose was to uphold peace as part of the peacekeeping forces. I also advised Busch to patiently maintain his roadblock position until I signalled him for assistance.

As I approached the HQ camp in Danane, our pickup truck came under attack from burning torches and stones. I couldn't fathom their true intentions! The rebel gunmen were shouting, demanding that the BANBAT troops open the main gate. Over a hundred of them encircled the entire camp, blocking all entrances and exits. With bloodshot eyes and sinister, sweat-drenched faces, they seemed possessed by some malevolent spirit that had haunted their bloodlines for centuries. They were frenzied, on the brink of unleashing a violent massacre.

At that moment, I was at a loss for what to do, let alone engage in negotiations. It was difficult to comprehend that these were the same individuals I had shared meals, sang songs, laughed, and embraced with. Somehow, our driver managed to manoeuvre the battered pickup truck to the front of the main gate amidst the chaos and deafening noise. My men and I leapt off the truck, forming a half-circle, with our backs against the main gate and facing the FN gunmen. We resembled a group of prey encircled by ravenous hyenas.

In Bengali, I commanded my men to keep their weapons lowered while I directed the pickup truck inside the main gate. By then, it bore dents on the bonnet cover, both side mirrors had been torn off (surprisingly, the windshield remained intact despite the blows from shards of stones and SMG buttstocks, a testament to the Toyota's durability!), and the tires were punctured in multiple places. Desperately searching for familiar faces, I spotted Ibrahim, the Imam of the local mosque; Jacques Delly, the Mayor of Danane; Bamba, the intelligence Chief of FN; and, at a distance, Captain Eddie, the FN Commander, with his hands folded against his chest.

I glanced back to see that Hassan had arrived at the main gate, and I instructed him to stay inside. Our group's arrival and positioning in front of the main gate caught the attention of the FN gunmen, further fueling the chaos with newfound enthusiasm. They emitted bone-chilling, eerie howls and made grotesque throat-cutting gestures. On the other side of the gate, the BANBAT soldiers also observed us skeptically. I was yet to devise a strategy for how to initiate negotiations, but in the midst of the uncontrolled pandemonium and shoving, our prospects seemed grim.

I called for our UN interpreter, Quoa Jerome, but he sensibly gestured with bulging eyes, pointing towards the FN gunmen, and declined to come out by my side. I decided to proceed alone. Taking the first step, I beckoned Captain Eddie to approach me. With a self-satisfied expression, he strolled toward me, relishing the scenario he and his men had crafted, with him playing the role of the master orchestrator. Locking eyes, he motioned for silence, and the frenzied crowd appeared to instantly heed his command. In that moment, Eddie and I, facing each other, became the singular focal point.

The captain of Liberian origin spoke fluent English with an African accent.

“What are you doing, Eddie? What is this?” I asked, puzzled by the situation unfolding before me.

“What does it look like? You’ve come here to tell us lies. We want an end to it,” Eddie retorted with frustration.

“What do you mean?” I inquired, hoping to understand his perspective.

“The UN forces have been here for two years now, and Laurent Gbagbo is still the president of Occupied Ivory Coast,” Eddie exclaimed, raising his voice for others to hear.

“And our great leader, Alasan Quattara, is in exile in France,” he continued, his voice growing louder, resembling a populist leader addressing a crowd hungry for answers.

“And you’re fooling us to gain time so that Gbagbo can come and loot our properties, rape our women, and kill our children?” Eddie’s words became accusatory, and the gunmen behind him showed their support by raising their weapons. However, I doubted how many of them truly understood the English conversation between Eddie and me. Some of them kicked the air and pounded their fists on their thighs, displaying their frustration. Despite the escalating tension, I maintained my composure.

“You know that’s not true, and you know how hard we’re trying to address the issues...” I tried to interject, but Eddie interrupted me abruptly.

"You're not trying anything. You're just keeping us deprived of what is rightfully ours. If you guys weren't here, we would reach Abidjan and bathe in Gbagbo's and his soldiers' blood," Eddie declared, his words dripping with frustration and anger. It was disheartening to hear such statements coming from a responsible commander like Eddie. Nevertheless, I had to remain patient in the face of his intimidating words.

"I hear you, Eddie, but this is not the place, and this is not how we discuss things," I calmly responded, attempting to redirect the conversation.

"This is my territory, and you're in my country. You don't have the right to tell me how and what I discuss," Eddie retorted, giving me a piercing glare.

In my mind, I searched for words that could effectively counter Eddie's remarks. Unfortunately, I often struggled to find appropriate words or logic to establish my claim, whether it was with my wife, with my boss, or during an argument. Observing Ibrahim and Jacques making their way towards us through the crowd, I took a deep breath and tried again.

"I will listen to everything you have to say. But not here," I proposed, hoping to find a more suitable setting for a constructive dialogue.

Eddie laughed maniacally, glanced back at his gunmen, and shouted in French.

"We've wasted enough time. The UN has deceived us. No more of this dirty game. You promise us a deadline to evict the devil Gbagbo from his chair, or your men will see hell today," Eddie proclaimed, prompting the crowd to respond with approving roars. I clenched my jaw in a sense of hopelessness, realizing the magnitude of the task ahead.

Eddie was well aware of the irrationality of his demands. He expected a captain from the Bangladesh Army stationed on the remote border of Ivory Coast and Liberia to declare an eviction date for the ruling president of Ivory Coast. Yet, his delusional gunmen seemed to nod their heads foolishly in agreement.

As Eddie and I continued our conversation in English, it became apparent that my soldiers were struggling to understand the exchange. However, when the FN soldiers raised their guns and Eddie's words grew louder and more frenzied, my worst fears were realized. One of my soldiers, gripping a carbine, leaped onto the APC positioned inside the main gate. Witnessing this, Lance Corporal Mizan, who stood beside me, instinctively raised his own carbine and loaded it. The control I had been desperately trying to maintain between our soldiers and the FN soldiers slipped from my grasp entirely. The FN soldiers reacted furiously to this turn of events. The mechanical sound of countless Kalashnikov rifles being readied filled the air, as a dozen of them, with faces twisted in rage, aimed their weapons at Lance Corporal Mizan. Their fingers hovered over the triggers as they closed in, ready to open fire.

I had to take action, and I had to do it quickly and decisively. There was no time for contemplation; I relied solely on my instincts and stepped outside of the conventional approach. Without hesitation, I positioned myself in front of Lance Corporal Mizan, tossing my weapon to one of my soldiers. Swiftly, I turned around, tore open my bulletproof jacket and helmet, discarded them on the ground, and stood there, ready to face the advancing FN soldiers. I believe that neither side expected such an unconventional reaction, and for a brief moment, confusion clouded their minds as I exposed myself completely, without any protection, to their pointed guns.

In that fleeting moment, perhaps, though I cannot be certain, the tension between both sides and the desire to display unnecessary strength subsided, and thoughts of the potential consequences crept into their minds. After all, killing me would only unleash the bloodshed that no one desired. Additionally, the French QRF stood prepared to respond within minutes with a notice to move. By risking my own life as a trump card in this death-defying gamble, I stood in absolute silence, projecting an unwavering resolve, making it clear to all that whatever ensued would have to pass through me. In those moments, I didn't think of my country, glory, family, or friends. It was solely me and my men. I bore the responsibility for their well-being. They had followed me with unwavering loyalty, and it was my duty, or whatever you want to call it, to ensure their safety as long as I could keep a clear head.

During those critical moments, I sensed that my fate hung on a razor's edge. A slight increase in pressure on the trigger from any of those intoxicated FN gunmen, with their loaded Kalashnikovs at point-blank range, would not be accidental. If that were to happen, I would be reconciling my karma in the afterlife instead of recounting this story to anyone.

After the shockwave passed through my body, the first sound that reached my ears was the loud scream of my captain, Hassan, from the other side of the main gate.

"K-H-A-L-I-D!"

His voice resonated like a scene from a Bollywood movie, where separated lovers reunite and run towards each other, shouting their names.

Then, I felt someone grabbing my arms from the side and standing beside me. It was the Imam of Danane Mosque, Ibrahim, stepping up to confront Eddie's men on my behalf. Within seconds, I also noticed Jacques Delly, the Mayor of Danane, standing by my other side. The presence of these two influential figures seemed to slightly weaken the rebel gunmen's grip on their weapons. Eddie appeared somewhat unsettled by this response, raising his hands and furrowing his brow in a gesture that seemed to say, "Now what the hell are you two doing here?"

The next words came from an unexpected source. Amidst the chaos, I hadn't noticed that women, children, and elderly men had emerged from the crowd. One young mother, with a child strapped to her back, stepped forward and spoke in French,

"Il est le capitaine Qualid" ("He is Capt Khalid," unfortunately, no one in Ivory Coast could ever pronounce my name, Khalid, properly). Then she looked around and asked the others, "What is happening here?" (At that moment, I couldn't grasp the full extent of her words, but later, my interpreter, who stood behind the gate witnessing everything, described what she had said to me.) She turned to the onlookers and continued,

"He is the BANBAT captain Qalid. He provided us with food when we were hungry at night. He helped rebuild our houses when they were damaged. And just yesterday, I could feed milk to my baby because the BANBAT captain provided us with milk for the children. What is wrong here?"

Then, the elderly men and the local Dojo chief, who nurtured the tribe, stepped forward and said,

"Yes, she is right! BANBAT is here to help us. They have assisted us in every way." He looked at Eddie and added,

"Last month, I asked for your help in clearing some bushes where evil spirits were hiding. Neither you nor your drunken soldiers ever came to our aid. It was the BANBAT captain who did it all for us. Now we are free from those malevolent spirits!"

This time, the Mayor of Danane spoke up,

"Eddie, what is wrong with you? Please instruct your men to lower their guns!"

Eddie appeared even more disturbed and took a step back. Regaining my composure, I turned to Ibrahim and said,

"That's enough now. May I speak, please?"

Ibrahim responded with his deep voice,

"Go ahead, Captain. I will interpret for you."

I directed my gaze toward the crowd in front of me. By that point, we had more than just gunmen present. It was a mixed crowd of elderly men, women, children, and armed individuals—people who had emerged from their homes and barracks, either in opposition to or in support of the UN, to witness this year-end spectacle. I looked at them and said,

"Bonne soirée, dear friends. Do you all know me?"

They nodded in response.

"I am your Muslim brother, Captain Khalid, standing in front of you and speaking in the name of Allah," I said. Ibrahim interpreted my words, and the crowd nodded in agreement. I continued,

"For the sake of Muslim brotherhood across the globe, please believe me when I say that we are here to live together in harmony and assist you. We have left our homes and families to bring peace to your country."

"Your father is my father, and your mother is my mother. At the end of the day, we are all brothers and sisters from the same family."

In any African crowd, referencing their tribe, religion, or ancestors has a significant influence. If you can establish a connection with any of these elements, you are considered a winner. They will embrace you in an instant.

"If you believe, even for a moment, that the members of BANBAT here have helped you, supported you, and cared for you, then raise your voice and let us know."

They immediately responded with a resounding "Oui Oui" (Yes, Yes) much louder than I anticipated. Meanwhile, Eddie had retreated to join his gunmen. As I continued to speak and Ibrahim interpreted my words, he realized that his momentary advantage was slipping away. He looked foolish, as if Imam Ibrahim and I were stealing his jackpot!

However, I said, “I want us to forget whatever has happened this afternoon. I want to ask all of you, brothers and sisters, to listen to me for a while. Will you?”

They responded with a collective “YES,” but Eddie once again interjected, claiming, “We have things to say, and you guys have to listen to us, too.”

I didn’t even acknowledge him this time as I continued, “Then let’s move away from this place because my soldiers are tense here, and the French soldiers on the road are not comfortable waiting.”

The rest of the task seemed relatively easy. The crowd agreed to move in front of the Danane central community centre, where we had a large space to address people with a sound system. However, they wanted me to be with them. I threw my SMG inside the gate, and a few moments later, accompanied by my two unarmed men, I led the bustling local crowd towards the community centre. The rebel gunmen walked beside us. Seeing all of us marching together, no one would ever imagine what had happened just 15 minutes ago.

I suggested that they sign a memorandum expressing their demands, which I promised to send to the UN Forces Headquarters in Abidjan. Three hundred people in total signed it. We exchanged views over the next couple of hours, with local elites from Danane delivering speeches, followed by my second round of uplifting brotherhood-themed speech, and finally Eddie had his turn. Lastly, a ceremony was held to hand over the memorandum, and the event concluded with cheers, dancing, and expressions of appreciation.

Later, we learned about incidents in other UN camps within the rebel-held territory. There had been firefights and injuries in UN camps led by African nations, looting and riots in some BANBAT camps, and the Level 1 hospital at our Battalion HQ had been ransacked before being brought under control by force. Fortunately, nothing of that sort happened in Danane. Some Ivorian TV channels broadcasted glimpses of me and my men peacefully interacting with locals and handing over the memorandum, serving as evidence of the peaceful activities by the rebel forces in establishing their legitimate claims.

Still standing inside the gate, when I returned to the company HQ, Hassan embraced me tightly until I was almost out of breath. Captain Busch came and offered me a special dinner at his camp. My company commander, Major Momin, returned in the late evening and said he had heard about the incident from the MILOBs because it had been proudly transmitted over the radio network. He chuckled and patted my shoulder.

“I heard that you have won the hearts and minds of the locals?”

I smiled a little bit.

Most importantly, my soldiers tossed me up into the open sky for a jolly-good-fellow celebration (this is a cheering event for soldiers where they lift a senior officer and throw him up in the air, catching him with loud cheers). Lance Corporal Mizan came to my room that night to express his eternal gratitude to me for guarding him.

Later, Bangladesh Road was inaugurated in front of our camp in appreciation of BANBAT’s contributions to the local people. A village was named Dhaka in honour of the capital city of Bangladesh. This made headlines in national and international newspapers. On May 18, 2006, I was urgently repatriated to Bangladesh on medical grounds due to the neuro-psychiatric side effects of Mefloquine, and the doctors suspected I would go blind due to my retinal injury. Luckily, that hadn’t happened by then, and returning soldiers from BANBAT reported that some locals in Danane had named their children “Khalid.” Trust me, I wasn’t at all biologically responsible for that!

Almost 6 Years Later

As I rested a bit after my 2.5-hour car journey from the office to home, I received a phone call. An unknown voice asked if I was the retired Major Khalid. After I confirmed, he identified himself as Sergeant Mizan, the former Lance Corporal Mizan from BANBAT, the soldier whose life I offered to save first. We exchanged pleasantries and spoke to each other for a long time. I asked about his family, military career, and the whereabouts of a few other BANBAT soldiers. Eventually, the conversation reached its conclusion. I was lying on the couch, speaking on my cellphone, when Mizan asked,

“Sir, one last question before I hang up.”

“Yes, go ahead!”

“We, the BANBAT soldiers, are proud of you. We still talk about that incident in Ivory Coast. You have no idea how much we loved you. Do you miss the army? Do you still have the same courage to stand up for your soldiers?”

I sat up on my bed. I don’t know what happened to me. It felt as though Mizan had struck a wounded place that had been awaiting massive bleeding for a long time. In an instant, memories of my days as a soldier in uniform, handling guns and bullets, flooded back, knowing they would never return to my life. Between 2005 and Mizan asking me this question, I had lost so much and changed irreversibly, but for the better. I had thrown myself in front of loaded guns to save a soldier, but I had lost 58 of my fellow officers and could do nothing to save their lives. The camaraderie and pride I had once felt as a soldier had been extinguished. The strength of the oath I took on the Holy Quran during my graduation as an army officer was fading away. I was no longer the same person. I was just another wounded heart, bleeding and yearning to settle unfinished matters in life. I wanted to tell him all of this, but instead, I said,

“No, Mizan. I have no courage left in me. I am not a soldier anymore. Please give my regards to everyone, and take care, all of you.”

The organic wire connecting my heart to my brain began transmitting signals. I felt a dampness around the corners of my eyes as my vision grew hazy. It was time for a cold splash of water on my face and to prepare for another day.

LET MY PARTY BEGIN

LIEUTENANT COLONEL MD. FERDOUS IQBAL

UNITED NATIONS MISSION IN LIBERIA (UNMIL)

Peacekeeping, a noble endeavor, even when contributing only a fraction towards global harmony, had always been a cherished aspiration of mine. The mere thought of making a meaningful difference in the pursuit of peace filled me with pride. In 2007, during my tenure as a captain, my name graced the list of candidates for a peacekeeping mission, sending a surge of exhilaration through my veins. Yet, as fate would have it, destiny had other plans in store. An unfortunate accident resulting in a leg fracture dashed my hopes of joining the mission, leaving me sidelined and yearning for another opportunity to fulfill my calling.

Fast forward to 2010, a year marked by my complete recovery and a renewed sense of purpose. Once again, fortune smiled upon me as I received the invitation to partake in a peacekeeping mission. By then, my rank had elevated to major, which was accompanied by greater responsibilities as an operations officer for Battalion 18, UNMIL (Liberia). Despite the anxiousness that coursed through me, akin to the anticipation preceding a monumental moment, I remained steadfast and dedicated myself to meticulous preparations.

Amidst the excitement and anticipation, I recognized the importance of tending to my family's needs. Ensuring the well-being of my wife during her exam period and providing the necessary care for my six-year-old daughter became paramount alongside the official preparations. While I shall not delve into further details, suffice it to say that I stood ready to discharge my duties with unwavering commitment.

As BANBAT 18 assumed the mantle, the peacekeeping mission in Liberia was gradually drawing to a close. After an arduous 19-hour journey, we arrived at the Monrovia airport, wearied and greeted by a gentle drizzle. Following additional hours of formalities, we embarked on the next leg of our voyage to the destination. Boarding the coaster, fatigue gripped us, and we succumbed to slumber within minutes.

Suddenly, I awoke to the coaster jolting violently, reminiscent of a boat battling tumultuous waves amidst a raging storm. Initially disoriented, I struggled to comprehend the unfolding situation. Glancing outside, I realized that we had veered off the paved road onto an unpaved path slickened by earlier rain. The coaster careened left and right, jostling us with each lurch and undulation. On one side of the road yawned a menacing abyss, heightening my unease. Determined to remain vigilant, I resisted any remnants of drowsiness that threatened to overtake me for the remainder of our journey. Curiously, my fellow officers appeared unfazed by the rugged terrain, seemingly undisturbed by the bumpy ride.

The first day with our unit unfolded dramatically. It commenced with a ceremony of consignment and orientation, where enthusiasm permeated the air. Later that evening, a grand celebration was held to bid farewell to the outgoing BANBAT chief and welcome the new appointee, an occasion brimming with camaraderie and guidance from the incoming leader.

April 30, 2011, the clock approached 7:30 in the evening, casting an aura of anticipation. A mess attendant approached me, his voice hushed as he whispered, “Sir, the duty clerk from our office urgently requests to meet with you. He claims it is an emergency.”

Without a moment’s hesitation, I abandoned my half-eaten dinner and hastened to the office, where a United Nations police officer awaited me, a sense of urgency etched across his face.

In a curious twist of fate, just a stone’s throw away from our encampment in Liberia, a tragic accident had unfolded. Approximately one and a half kilometers away, a collision between a civil truck and a motorcycle had claimed the lives of two motorcyclists. The gravity of the incident incited a hostile local mob, driven by a thirst for brutal revenge, to surround the truck and its occupants. In Liberia, such incidents often ignited violent clashes among different factions, and the situation had the potential to escalate rapidly. Despite the efforts of the Liberian police to defuse the tension, the ANPOL representative emphasized the urgent need for BANBAT intervention to quell the turmoil.

Without hesitation, I swiftly informed the camp commander, urging him to have the Quick Reaction Force (QRF) prepared to mobilize within 10 minutes. Next, I made my way to the field mess, where I briefed the new chief about the unfolding situation. However, a minor setback arose concerning the deployment of the QRF officers. With most senior officers occupied with formal ceremony obligations, a junior officer was overseeing the operation. Yet, the chief expressed his preference for a senior officer, preferably of major rank, to be involved. In anticipation of the chief's directive, I patiently awaited further instructions. As the party continued and attendees indulged in food and drink, all eyes turned to one another as the chief made his intentions clear. Without hesitation, I assured him, "Sir, let me take charge of this. I will handle it." With that, I promptly retreated to my room.

Within me, a resounding soliloquy echoed, "Let their party continue, for mine is about to commence."

I seized the opportunity to be part of the operation on its very first day. Donning my uniform, bulletproof jacket, and helmet, I made my way towards the camp's main gate, where our QRF team awaited. A mixture of fear, excitement, and uncertainty was palpable among the team members. It was indeed the first day, and none of us knew the nature or objectives of the operation that lay ahead. However, I was determined to conceal any apprehension from the officers under my command. It was crucial for them to believe in me and my leadership, and displaying fear would only exacerbate their own anxieties.

The sight of their operations officer assuming command of the mission filled the platoon with a sense of awe and curiosity. Yes, in that moment, I resolved to personally oversee the operation. Almost instantly, I noticed a remarkable transformation occurring within the team—a surge of intrinsic motivation and newfound confidence radiating from their demeanor.

Bringing everyone together, I gathered them closely and delivered a concise briefing on the mission's objective—to safely retrieve the truck driver and his assistant—and outlined the process and rules of engagement. With the briefing concluded, I took the lead as we set off for the scene. Our convoy consisted of three armored personnel carriers (APCs), with me personally commanding the leading APC, while a JCO and an NCO led the remaining two vehicles.

As we arrived at the scene, chaos engulfed the area. A scattered cluster of Liberian police officers surrounded the truck, while an agitated crowd wielding an array of sharp weapons, sticks, and cudgels gathered around them. It was impossible to dismiss the notion that some among them might be concealing hidden pistols or revolvers.

Without hesitation, I made my way toward a UNPOL officer present at the site. Through his intervention, I arranged a discussion with a representative from the motorcycle association and a Liberian police officer. It was worth noting that many of the motorcycle association representatives in Liberia had once been affiliated with various rebel groups.

The representative of the motorcycle association adamantly demanded the surrender of the truck's driver and assistant to their custody. I endeavored to explain and reason with him, hoping to find a peaceful resolution. Despite my best efforts to clarify the situation, he remained resolute in his stance.

Assuring him that the police would ensure legal consequences for the driver and assistant if taken into custody failed to assuage his concerns. In light of this, I resorted to using the megaphone on the APC to broadcast an alert message. I sternly warned that non-compliance with the police in carrying out their duties would necessitate the use of force.

When our attempts to dissuade them from resorting to violence proved fruitless and we were left with no alternative, discussions came to an end, and action ensued. I had no choice but to issue a clear warning that non-cooperation could result in a forceful response. As a display of force, I assigned individuals from each APC to strategic positions.

To our relief, this maneuver seemed to have a positive effect. The enraged mob slowly began to disperse, one by one. Seizing the opportunity, I positioned an APC next to the truck and commanded a slow movement forward. The APC inched forward cautiously, while the agitated individuals seemed less inclined to engage in a violent clash. Gradually, they faded away, with only a few remaining. In that moment, with the assistance of UNPOL, I invited the representative for another conversation. I reassured him that a thorough and fair investigation, trial, and prosecution of the incident would be conducted by the Liberian police in cooperation with UNPOL.

Perhaps due to the overwhelming circumstances or our concerted actions, the representative of the motorcycle association appeared somewhat pacified. With the situation now more controlled and calm, we arranged for the truck to be escorted to the local police station.

Upon returning to the camp, to my disappointment, the party had concluded. Regrettably, I had missed the opportunity to partake in the festivities and satisfy my appetite. However, the kind-hearted mess waiter awaited my arrival with a meal, and his thoughtful gesture provided sustenance for me.

Before retiring to my quarters, I briefly met with the chief, offering him a succinct summary of the day's events. Eager to alleviate my considerable hunger, I promptly consumed my meal and sought rest.

Unfortunately, the events of that day left me with no opportunity to contact my loved ones miles away. I longed to call my mother and wife, but exhaustion overcame me, and I surrendered to slumber as soon as my back touched the bed.

Over the course of the following year, a series of events unfolded that significantly impacted the schedule of our mission. Our responsibilities included providing assistance in preparation for the national elections in Liberia, conducting ceaseless long-range patrols in border areas during the crisis in Ivory Coast, and establishing temporary operating bases in remote regions. As I previously mentioned, Battalion 18 inherited the peacekeeping mission in Liberia during a contraction phase. Therefore, a major focus of our mission was to build the capacity of Liberian individuals and institutions.

Prior to Battalion 18, a total of 17 Bangladeshi battalions had successfully fulfilled their assigned tasks at different times. Considering the continuity of past battalions, our commanding chief at the time, Colonel Mizanur Rahman Khan (now major general), led various peacekeeping military operations in Liberia. Alongside UNMIL and other UN agencies, he worked tirelessly to implement initiatives that improved the lives of the local population and spearheaded socio-economic development throughout Liberia, leaving a lasting impact on thousands of lives in local communities.

Among the numerous commendable efforts undertaken by the peacekeepers, the establishment of four vocational training centers, an ideal city project, a children's entertainment center, and eight agricultural development projects stood out. These initiatives received extensive coverage in various national daily and weekly newspapers in Bangladesh, highlighting the significance of the peacekeepers' endeavors in fostering progress and growth in Liberia. Allow me to delve into a few of these noble pursuits. Here are brief descriptions of some of these projects.

The Daily Observer (October 4 & 16, 2010):

> On July 24, 2010, under the overall management of BANBAT 18, the "Bangla Nimba Capacity Building Center" (BNCC) was inaugurated by the former honorable president of Liberia in the city of Tappita. In his speech, the president lauded the BNCC for its role in creating welfare-oriented initiatives that served as a testament to the capabilities of Bangladesh's

peacekeeping forces. The Daily Observer, one of Liberia's prominent dailies, published a comprehensive report on the endeavors of Bangladesh's peacekeeping forces, emphasizing their significance.

The New Republic (October 15, 2010):

Addressing the issue of unemployment in post-conflict Liberia, the recently established "Bangla Nipa Capacity Building Center" was officially launched on July 27, 2010, in Sanniquellie, a war-torn city of Liberia.

Insight (December 14 & 29, 2010):

One of the primary concerns in war-torn Liberia was the shortage of food. Despite the abundance of agricultural land, a lack of proper planning had rendered much of this land unused and unproductive. To encourage local communities in the region to engage in farming and reignite their passion for it, the Bangladeshi Contingent 18, along with previous Bangladeshi peacekeeping forces, undertook several agricultural projects. In 2010, these projects yielded impressive results, with bumper crops being produced. 'The Insight' featured a meticulously detailed comprehensive report on these agricultural initiatives in its special edition on December 14 and 29, 2010.

RUPOSHI[1] BANGLA

MAJOR MASUDUR RAHMAN
(UNITED NATIONS MISSION IN IVORY COAST - UNOCI)

"Assalamu Alaikum, sir," I greeted respectfully.

"Walaikum Assalam. Listen, I have an assignment for you. You are to go on duty at a prestigious five-star hotel."

"Understood, sir. When and which hotel is it, sir?"

"Hotel Kot Divoire."

"Where can I find it, sir? How far is it?"

"Check the map. Time is of the essence. Hurry up and make all the necessary preparations. You only have one week. A letter will follow shortly. Stay safe and take care. Allah Hafez."

"Message received, sir."

The Bangladesh Army thrives on achieving what others consider impossible. And so my exhilarating race began. With just one week at my disposal and only two days off in between, I was determined to wrap everything up swiftly, bid farewell to my loved ones, and board the plane destined for Africa. The training I had received on changing attire swiftly and navigating five floors in five minutes during regimentation proved to be incredibly useful on that eventful day.

Since my childhood, I had held a romanticized view of foreign countries, perceiving them as beautiful, superior, and highly advanced in every conceivable way. However, it wasn't until I embarked on my journey to Africa that I realized the immense fortune of living in a modern and progressive nation. It was an eye-opening experience that made me truly appreciate the level of development we enjoyed back home. In comparison

[1] Ruposhi is a Bangla word which roughly means 'beautiful'.

to the country I was in, we were miles ahead—something for which we should be immensely grateful. After witnessing the devastating consequences of a prolonged 13-year civil war, I couldn't help but contemplate the fate of our own land had we endured such a tragedy for even a fraction of that time. The toll such conflicts take on a nation's foundation is unimaginable, capable of shattering its very backbone.

Not everyone may be familiar with the geographical landscape of Africa. Allow me to introduce you to the enchanting Republic of Cote d'Ivoire—a land nestled between the nations of Ghana and Liberia on the coast of the North Atlantic Ocean. Spanning a vast expanse of approximately 322,460 square kilometers, this gem of a country is several times larger than Bangladesh.

Despite its considerable size, the population of this nation amounted to a modest 24 million, 13 thousand, and 104 souls at the time of my writing. Among them, 40 percent identified as Muslim and 35 percent as Christian, while the remaining quarter followed traditional beliefs or their own faiths. French served as the lingua franca, a legacy of its colonial past under French dominion.

In the annals of history, the Republic of Cote d'Ivoire held distinction as the world's third-largest producer of coffee. It was in 1842 that the country came under the rule of French colonialism, which persisted until 1882. On the dawn of January 1, 1986, the nation adopted its current name, Cote d'Ivoire. With a ranking of 68th largest country in the world, the Republic of Cote d'Ivoire possessed a rich and captivating history.

Following a prolonged period of civil unrest, the nation had been denied the fundamental right to hold democratic elections for a significant period. It wasn't until 2010 that the Ivory Coast finally had the opportunity to exercise this essential element of a functioning democracy. Alassane Ouattara emerged as the clear victor in the initial stages of the election. However, the incumbent president, Laurent Gbagbo, refused to accept the results. In a parallel move, Gbagbo's loyalists in the Constitutional Council declared him the rightful winner, while Ouattara took the oath of presidency.

This unique predicament resulted in both Ouattara and Gbagbo being sworn in as presidents, effectively dividing the nation into two factions, each with their own armed forces. It was a situation akin to one country with two presidents and two armies. Gbagbo garnered support from the Young Patriots, while Ouattara held command over the northern region of the country. The southern region remained under Gbagbo's control, and as the two presidents vied for power, the situation quickly escalated into a full-blown civil war.

The consequences of this power struggle were catastrophic, with violence spreading throughout the land. Villages were attacked, and the streets were plagued by killings, looting, arson, torture, and oppression. The situation was nothing short of horrific, and the people of Ivory Coast were in desperate need of assistance. In response to this critical situation, the United Nations launched one of its most significant and extensive peacekeeping missions, aimed at restoring stability and governance to this West African nation—the mission in Cote d'Ivoire. Initiated in 2003, this mission continues to this day.

The African night whispered a haunting tale, a story of a land torn asunder, ravaged by the ruthless grasp of foreign powers. The English and the French had claimed the lion's share, while the Spanish, German, Italian, and Portuguese had taken their own slices. Africa had been divided as spoils of war. The wealth that flowed from this land had built empires and enriched many of the Western nations. Europe's splendor is owed much to Africa.

The scent of the African soil, infused with the sweat and toil of millions, still lingers in my nostrils. The resilient African wind tousled my hair, a testament to the indomitable spirit of those who once called this place home. And under the enchanting moonlight, so luminous and captivating, Africa cast its spell upon me. In those moments, I found solace, momentarily forgetting the pain of a world marred by war and unrest. The sparkling tapestry of countless stars in the African night sky bestowed upon me a sense of tranquility.

Yet, beneath this beauty, a simmering rage and sorrow coursed through me. The black people of Africa had been forced to endure arduous labor, giving until there was nothing left to give. In return, they received hollow promises of "respect," no more than the chains of slavery. The scars of this history run deep and are etched into the very fabric of the land.

Amidst ongoing peacekeeping efforts in Ivory Coast, I joined the mission. My first assignment led me to 'Ma'an,' a town nestled among the mountains in the west. I learned from the locals that it was home to the Yakuba people, one of the 66 ethnic groups in the country. The name 'Yakuba' meant 'offspring of the mountain' in their language, a testament to their heritage. Legend had it that the chief of Kandopo village sacrificed his daughter

Ma'an to their god following a dream that prophesied prosperity for the region. The city took its name from that ebony beauty. It was a place of extraordinary magnificence, embraced by the splendor of nature. At dawn, as I stepped outside, a fluffy cloud rested upon a colossal mountain, a celestial sight. Occasionally, the cloud would stray and grace our golden flower garden, casting a gentle shadow over our camp. A deer would visit, grazing on fresh grass, seemingly as at ease with the cloud as it was with us.

The enchanting city of Ma'an boasted boundless beauty, guarded by the Tunkui and Ture mountains. On the way, the renowned 'Teeth of Ma'an' mountain grinned, while a captivating cascade murmured throughout the day. When I traveled to the distant city of Daloa from Ma'an, serene lakes awaited, offering respite to my weary eyes. Tiny boats dotted the surface, skillful young men netted fish, and the air carried the scent of freshly harvested crops and cashew nut gardens. Exploring the northern city of Tuba revealed an African spectacle resembling a masterpiece painting. Round mud houses with humble doors held tales of simple joys and sorrows and palm wine jars, and the aroma of sweat mingled with the essence of the land.

I had known no bounds of joy and wonder since I set foot upon the African soil. The special handshake with groups of farmers on the pebbled path became a familiar custom. Each person I met felt like a kindred spirit, as if we had known each other for ages. The Bengalis, with whom I shared a deep connection, forged a unique bond with us. Their faces would light up upon seeing a Bangladeshi, and in moments of homesickness, I would turn my gaze to the mountain. The golden foliage and breathtaking scenery made me forget that I stood in a foreign land on a different continent.

As the distant sound of a tambourine carried by the winds reached my ears, or when the thunderous blast of a hunter's rifle awakened me from slumber, I would venture to the veranda. There, I witnessed the moonlight cascading gently upon the glistening petals of golden flowers, a celestial beacon of hope.

The genuine kindness of the locals forged an unbreakable bond between us. The warmth of their hospitality enveloped me, making me feel as though I had discovered a second home. As the saying goes, "Hospitality is the greatest gift a guest can receive." Just as I began to feel a deep connection with the people of Ma'an, fate called me to a place called Danan, where I had the privilege of meeting Major Mostofa. He was a wise sage eager to impart his pearls of wisdom to a curious young pupil like me.

The scorching sun that burned relentlessly during the day in the camp, the moonlit bathing at night, and the melodious voices of Captain Mirza, Captain Raihan, and the two other officers. Ah! What more could I have asked for in the land of Africa? Their voices were akin to the celestial harps of angels, soothing our weary souls and transporting us to another realm. I marveled at the beauty of this country, where the people, though dark-skinned, possessed hearts of kindness. I wished that some of them could share their rich culture with our homeland. However, a harsh reality abruptly shattered my words and the serenity of the moonlight. I gasped as I witnessed their plight. One of them spoke with a voice heavy with emotions, saying, "The sound of a bullet whizzing past your ear ignites a frenzy within your mind, evoking a mixture of fear and relief that you are still alive. It is a melody of scars that one cannot easily forget. The sight of lifeless bodies strewn across the roads, their anger and despair—how can we overlook such experiences? Are these the aspects of culture you speak of? The world truly is a stage where good and evil play their parts, coexisting within our very habitat."

It was a day in late August 2012, the 20th to be precise. A battalion of Bangladesh's finest soldiers had been entrusted with the mission of restoring peace and order in several villages near Ma'an Camp. Their objective was to halt the looting and arson and rescue the UN civil members who had been taken hostage. Under the leadership of Major Arif and Captain Tanvir, a team of 30 courageous men embarked on their mission equipped with two armored personnel carriers (APCs) and three other vehicles. The streets teemed with mobs of rebel supporters buzzing like angry bees. To evade their wrath, the team set out at the break of dawn. As the communications officer, my responsibility was to ensure the communication channels remained open via the HF set. Upon reaching our destination, we were confronted with an unexpectedly high number of checkpoints. The area had been transformed into a fortress guarded by the Young Patriots, who had erected barricades using wooden sticks, stones, and debris. We attempted to reason with them, but when all else failed we adhered to military protocol to clear our path. We successfully navigated through eight to ten such barricades before reaching our intended destination.

On our way back to the APC, after rescuing the captive team it was almost six o'clock in the evening. In that moment, I caught a glimpse of the dark side of Africa. I bore witness to the depths of their cruelty. The first barricade we encountered was swarmed by a sea of people armed with weapons and concealed behind masks. We endeavored to reason with them, but their obstinacy was unyielding. We adhered to our customary approach:

a request, followed by persuasion and negotiation. Yet, when all peaceful avenues failed, the harsh command was given.

Commander Arif issued the order to break through the barricade. The APC plowed ahead with the force of a bulldozer. The Bagbo army unleashed a barrage of bullets from both sides. We responded in self-defense. In the face of this onslaught, we managed to cover two kilometers of road, navigating through the storm of gunfire with two APCs and three cars. Our first priority was to ensure the safe delivery of the UN civil members before returning to the camp. The clock struck 6:45, and it was then that we realized Foylul Karim, our dedicated interpreter, had suffered a minor injury. This battalion had accomplished numerous remarkable feats before, and their success in this operation earned them the commendation of the force commander in the form of a recommendation letter.

The village near Ma'an Camp, now known as Ruposhi Bangla, was once called Kasiyapu. It had been several years since our predecessors, the peacekeepers who were previously stationed there, had built a mosque for the villagers. The majority of the villagers were Muslims, and they expressed their gratitude by naming their village after our homeland, 'Ruposhi Bangla,' meaning 'Beautiful Bengal.' Upon stepping into the village, we were welcomed by the festive beats of drums and the vibrant sight of women dancing and singing. The village consisted mostly of small houses constructed of clay, some with tin roofs, and others thatched with straw. It resembled many Bangladeshi villages, with mango trees bearing abundant fruit. If a mango fell to the ground, it remained untouched, as the saying goes, "One man's food is another man's poison." It was reminiscent of the sight of a crow refraining from pecking at a fallen coconut. Nevertheless, the villagers seemed content in their own way.

I observed a village woman drawing water from a deep well, carrying her little child on her back, wrapped in a cloth. The child remained quiet as a mouse. In a spacious courtyard, most of the villagers had gathered, while their village head sat on the porch. Our contingent commander, Colonel Ambia, joined them on a makeshift stage, assuming a relaxed posture as if he intended to enjoy the spectacle. The older women began the entertainment with their graceful dance and melodious songs. They were followed by the younger women, adorned in different attire. Finally, the teenage girls, some carrying babies on their backs, took the stage. The village head expressed heartfelt gratitude to the Bangladesh peacekeeping force on behalf of the villagers. He then bestowed upon Colonel Sir the Alkhalla, a traditional African honor, and bestowed upon our battalion the title of 'The Guardians of Ruposhi Bangla Village.' With their love and blessings, we bid farewell to the village and returned to Ma'an Camp, cherishing the feeling of a home away from home.

The country had endured the ravages of a prolonged and bloody civil war that had left its economic infrastructure shattered. Hunger and disease loomed over the land like ominous specters of death. Whenever our vehicles passed by, children and elders greeted us with radiant smiles and outstretched hands, chanting 'Bangla good' and 'Monami bisque.' They yearned for food like parched land craves rain. Perhaps they felt a special connection with the Bangladeshi peacekeepers, who, unlike others, responded to their cries for help. Though contingents from Pakistan, Niger, Chad, and elsewhere were also stationed across Ivory Coast, our efforts may have outshined them. The country suffered from a severe lack of hospitals and clinics, resulting in countless lives being lost to various ailments every day. The death rate was alarmingly high, and AIDS ravaged the population, cutting short many lives. Our BANMED not only distributed free medicine but also provided regular services and treatments to Ivorian patients. In addition to our peacekeeping duties, our battalions went above and beyond, leaving an indelible mark on the hearts and minds of the Ivorian people. Undoubtedly, our exceptional dedication and commitment set us apart from others in the eyes of the people of Ivory Coast.

As a signal officer, I ventured beyond the camp only infrequently. However, fate smiled upon me on one occasion when I was tasked with securing a location 8 kilometers away from Ma'an Camp for Eid prayers. Early in the morning, I departed with one APC and three other vehicles for this assignment. There, I encountered an elderly man who engaged me in a lengthy conversation on various topics. Though he had a limited command of English, we were able to communicate through the assistance of an interpreter. He held great admiration for the Bangladeshi peacekeepers, describing us as a godsend for his country. He added that we were truly blessings from above, but what delighted me even more was when he assured us that we Bangladeshis cared for and dreamed about their country in a way that they themselves were not prone to. He commended our sincerity and dedication, asserting that no other peacekeepers in the world matched our level of commitment. He praised us for transforming our population into strengths and resources rather than weaknesses. During our conversation, we briefly discussed topics such as food crops, mineral resources, imports, exports, the education system, the garment industry, and other projects in both countries.

As I concluded my conversation with the gentleman, he approached me with heartfelt gratitude for the security provided by BANBAT. He had been captivated by the Bangladeshis he had encountered over the past nine years. He even extended an invitation to his home, expressing his fondness for a distant land he had never seen. He vowed to visit Bangladesh if the opportunity to travel to Asia ever presented itself. His words filled me with immense pride, and I sincerely thanked him, encouraging him to explore our country someday. With a sense of elation, I bid him farewell and rejoined my team, carrying the knowledge that the name of Bangladesh would forever be etched in the hearts of not only this man but countless Ivorians. Truly, a good name is better than riches.

On September 1, 2012, we had already covered more than half of our mission area and proceeded to inspect various camps along the way. Our convoy's next destination was the Juhanwe camp situated adjacent to a gold mine. The mine did not belong to Ivory Coast itself, however, as the precious metal was being diverted to foreign lands. The contrast between the wealth and poverty of Ivory Coast reminded me of the words of our poets, who emphasize that wealth does not reside in a poor man's house. The simplicity, innocence, and kindness of the Ivorian people resonated deeply with me. They had embraced our Bangladeshi brothers as their own, leaving me with a sense of awe. We were fortunate to have the opportunity to witness the gold mine thanks to the initiative of our late Major Mahfuz. Although he was no longer with us, his efforts would always be remembered. People may pass away, but their legacies and memories endure.

Stepping into the gold mine was an awe-inspiring experience. It felt as if I had entered another world, a dream I had never imagined. As a science student, I had learned about the extraction and refinement of gold through books, but witnessing it firsthand filled me with wonder. Gold beneath my feet! The reality was difficult to grasp, a testament to the marvels of nature's abundance. After the tour, we returned to our camp, but the memory of that experience lingers in my mind to this day. I completed my mission and returned in March, but the love of those black people and the natural splendor of Ruposhi Bangla, the Beautiful Bangla, an Ivorian village, remain in my heart. The air was no longer tainted by death, and there were no more corpses on the road. Ruposhi Bangla stood as a testament to the genuine affection for Bangladesh, a love that resonated deeply within our hearts. Their culture reflected their selfless friendship, all made possible by the Bangladeshi peacekeepers. We Bengalis have the potential to achieve great things if we set our minds to it.

Let the glory of Bengal shine across the globe. May it inspire everyone with pride. Each member of the Bangladesh contingent, the peacekeepers, is a vigilant protector of world peace. Let this message of peace spread throughout the world, resonating with nations far and wide. My army is my pride, my inspiration.

Despite all the challenges it faced, Africa endured. It remained a place of beauty and wonder, a testament to the strength and resilience of the human spirit. As I stood there, I knew I was a witness to something truly remarkable. It was a story shaped by the hands of many, yet it endured because of the indomitable spirit of the few who refused to be broken.

THE NOR'WESTER TEMPEST

COLONEL MD. SHAHJAHAN MOLLAH

(UNITED NATIONS TRANSITIONAL AUTHORITY IN CAMBODIA - UNTAC)

In the depths of my being, an overwhelming void stifled any desire to express myself through the written word. Everything appeared disarrayed, unworthy of my attention in that particular moment. It seemed as if the world had lost its luster, leaving me adrift without purpose. A sense of lethargy enveloped me, as if the will to create had vanished entirely. Was this the grip of depression? I struggled to articulate my thoughts, for even a single word evaded my pen's grasp.

While many writers managed to carve out time amidst the ceaseless demands of their lives, I found myself grappling with the challenge of maintaining a consistent writing practice. Yet, amidst my own struggles, one individual stood out in my memory—the ambidextrous scribe at Dulal Bhai's esteemed Bangkok office. Observing him effortlessly juggle his tasks, I marveled at his ability to seize every moment of respite to put pen to paper. If granted an extended break, his well-worn notebook would bear witness to his prolific musings. He was truly an astonishing and captivating character.

During each visit to Bangkok, a pilgrimage to Dulal Bhai's office became an integral part of my itinerary, an almost natural phenomenon that unfolded with each sojourn. The genuine warmth and earnestness emanating from Dulal Bhai compelled me to accept his invitation to visit his home—an invitation he extended without fail. Located in close proximity to the bustling Pahurat ATM market, I received his address from a relative, planting the seeds of our destined encounter during my inaugural visit to this vibrant city.

The first day of my Bangkok adventure passed within the confines of a hotel room, teeming with anticipation for the upcoming explorations. However, an unexpected turn of events awaited me as the hotel reception alerted me to a guest eagerly awaiting my presence in the lobby. With a mixture of curiosity and intrigue, I descended the staircase and was greeted by the unassuming figure of Dulal Bhai—a man of humble ingenuity and pure authenticity. Although possessing his contact information, I hesitated to make the call, rendering our initial meeting somewhat awkward. I couldn't help but wonder how he had acquired knowledge of my hotel address. After the exchange of formal pleasantries, I learned that he had traced me all the way from Dhaka.

Dulal Bhai's mild disappointment upon learning of my hotel check-in was palpable, a reflection of his unwavering commitment to hospitality. Paradoxically, his gentle reproach brought us closer together, bridging any remaining gap. However, what unfolded next caught me completely off guard. With assertiveness, he persuaded me to request an early check-out from the hotel, insisting I follow him to his office after arranging for a taxi.

His office, aptly named Bounty Travel Limited in honor of his beloved son, stood proudly in Pahurat. "You're here in Bangkok to enjoy its wonders, aren't you? Leave your luggage in our care," Dulal Bhai proposed, his smile exuding warmth. "After a day of exploration, you can make your way to my home." This arrangement promised a more efficient use of time, sparing me the need to shuttle between his house and the hotel—a proposition that resonated with my spirit of adventure. As I embarked on my day's journey, Bounty Travel became the temporary guardian of my belongings.

Amidst my wanderings, I found myself seated at a Bengali restaurant, indulging in a flavorful lunch of rice and Thai climbing fish—an encounter that brought to life the vibrant tales I had heard. Its imposing presence, akin to an impregnable fortress, made it a formidable companion to the accompanying vegetables, like ants seeking refuge behind towering concrete walls.

Yet, with each subsequent visit to Bangkok, Dulal Bhai's orbit became my sole focus and his home my preferred sanctuary. There was a time when I arrived in Thailand only to discover Dulal Bhai's absence. Upon contacting his office, I learned that he had momentarily left Bangkok. Two days later, my hotel received a call requesting my availability for a conversation. It was none other than Dulal Bhai himself, returning to the city. I met him at the earliest opportunity and asked, "Bhai, where have you been? Is everything alright?"

With a hint of jest and frustration, he replied, "Oh, my dear friend! You people travel abroad and become entangled in love affairs. And it's always me who must sort out the mess."

"Bhai, believe me! I am married, and my wife is sitting right here with me. I have no idea what relationship you're referring to," I responded, perplexed.

"No, no! It's not you. It was those military folks who caused the trouble. Most likely he was a police officer. You know how it is... a military observer, perhaps, associated with the Cambodia Mission," Dulal Bhai clarified, assuring me that I was not the subject of his exasperation.

"Ah, I see! That explains it," I replied, regaining my composure.

"The story is quite intriguing. If you're a writer or inclined toward writing, it could make a great novel, I tell you," Dulal Bhai remarked, his frustration dissipating through a wry smile.

"But, Bhai, you know the story. Why don't you write it yourself? Besides, I'm not confident in my writing skills, and even if I were, no one would read it. By the way, what's your story?" I inquired, genuinely curious.

"All right, listen closely," he began, inviting me into the realm of his narrative.

To fulfill the mission of establishing peace in Cambodia, the Bangladesh Police Force, alongside the Bangladesh Army, has been actively contributing to maintaining peace as part of the multinational force of the UN peacekeeping mission. Rahat, a smart and sincere police officer, had been deployed for a three-month mission in a remote village near the Thailand border.

Dedicated to his duties, Rahat found solace in appreciating the beauty of nature, pursuing his hobby of photography, and immersing himself in the foreign land. However, he never allowed his duties to be neglected. As he explored the hilly regions and countryside, Rahat became enchanted by the charming elegance of Cambodia, developing a deep affection for the country.

"You know, in our country, there are no remarkable mountains or hilly regions except for the Chittagong hill tracts. It's all flat plains. If one wishes to develop an affinity for hilly landscapes, the only option is to travel to the

Chittagong hill tracts. Perhaps Rahat will have a posting there one day due to the nature of his profession," Dulal Bhai explained, shedding light on Rahat's longing for such scenery.

During his college years, Rahat went on a picnic to Rangamati and was captivated by its breathtaking beauty. However, the responsibilities of his occupation and the picnic experience remained distinct from each other. Being in the army meant venturing into deeper, more challenging areas where danger lurked among the trees.

One day, while strolling on a road in Cambodia, Rahat felt as if he were walking along a remote road in a Bangladeshi village. The fading light on the horizon signaled twilight's approach, and as darkness fell, lamps illuminated the temples and homes of the villagers.

Nearby, Rahat noticed children learning the Bengali alphabet (o, a, ko, kho—the phonetic sounds of the Bengali alphabet). He couldn't believe his ears—the Bengali alphabet being taught in a remote region of Cambodia, in tune and rhythm! It seemed surreal, but it was indeed the sound of the Bengali alphabet resonating in the air. Dulal Bhai humorously remarked that leaving that place would surely remind Rahat of the Bengali Doppler effect.

Curious, Rahat followed the sounds and approached the abode where two students were receiving Bengali alphabet lessons from a female teacher. And then, in that moment, their eyes met—the teacher's transcendently stunning beauty instantly captivating Rahat, like experiencing a crush at first sight.

She delicately caressed her hair and glanced at herself in a mirror from her vanity bag. Rahat couldn't help but think, "If she asked the mirror, 'Mirror, mirror, who is the most beautiful girl in the world?' the mirror would surely respond, 'You are,' and even if it didn't, I would."

It was evident that she didn't realize the extent of her beauty. Rahat couldn't resist the urge, so he knocked on the door of the abode and asked, "May I come in?"

The six curious eyes inside stared at him until he put them at ease with a friendly "Good evening." Then, he gently tapped on the door and, as they grew more comfortable with his presence, he inquired if they would be willing to share what they were reading. They showed him their books, which turned out to be Cambodian alphabet books, not Bengali. It dawned on Rahat that both Cambodian and Bengali languages shared the same origin, Sanskrit.

Driven solely by curiosity, Rahat showed up at that moment, making every effort to convey to them that his presence was purely out of profound curiosity. However, they appeared unperturbed and stoic, their puffy jowls and trifling eyes revealing nothing about their understanding of his words.

As soon as Rahat left the room, a girl with a lustrous, glossy voice exclaimed, "Thank you so much!" And with Rahat's departure, laughter filled the air, flowing like a fountain. After that, visiting the area became a secondary profession for Rahat, whether for official or personal reasons, as he hoped to encounter that enchanting beauty from heaven once again. Perhaps she, too, awaited the return of the foreign admirer.

However, opportunities to catch another glimpse of her became increasingly rare. Rahat wondered when that sacred moment would arrive, the moment of their rendezvous. Days passed as Rahat ambled around the house, but his efforts were in vain. Sometimes he would find the two students studying without their teacher, and other times the abode remained dark when he passed by.

Deep inside, Rahat felt like a foolish and clueless loggerhead, regretting that he hadn't asked the girl's name that day. "She must have known English since she thanked me. Why can't we think more clearly? Damn! Why does our brain fail us when we need it the most?" Rahat muttered aloud, engaging in a perfect soliloquy while pondering and berating himself.

The war-weary land of Cambodia was plagued with landmines and unexploded ammunition, and the UN peacekeepers were assigned the crucial task of mine clearance. However, the scale of the problem required not only a significant amount of time but also substantial financial and human resources. To address this challenge, the UN decided to train local people to assist in demining operations. With the selected participants, the demining efforts were progressing well. Simultaneously, awareness lectures and training sessions were conducted across Cambodia, regularly reminding citizens to be vigilant of the mines.

At the camp where Rahat was deployed, a session on mine awareness was taking place, with an emphasis on the need for caution in daily activities due to the presence of mines until the demining team could completely clear the area. This session had been ongoing for nearly a week. During the orientation, amidst the sea of faces, Rahat spotted the person he had a crush on—the teacher he had encountered.

Rahat's mind became consumed with thoughts of that girl, completely forgetting everything else except for the moments when he caught her eyes and heard her tender voice saying, "Thank you so much." The lecture continued, but Rahat was in a different world, his thoughts entirely focused on her. Questions raced through his mind—how would he approach her, when would they meet? Would it be during the session break or at the end of the day?

He yearned for the moment to come, perhaps after the training ended during the weekend, yet that moment felt distant. Rahat couldn't afford to be patient or waste time waiting. Thus, he decided to approach her at the end of the daily session, as engaging in a meaningful conversation during the break seemed like an elusive luxury. Unfortunately, fate had other plans. Just as the last lecture of the day concluded, she left with the group she had arrived with.

Rahat made up his mind that he would approach her the next day, meticulously planning his move behind the scenes. It's worth noting that UN peacekeepers are expected to foster strong relationships with the local population, cultivating a sincere and cordial bond. While personal connections and private relationships were permitted and possible, those who faced complaints jeopardized their safety. With the determination to meet her, Rahat had a sleepless night, eagerly anticipating the moment when their eyes would meet again. And the next morning, their longing gazes finally converged.

Rahat found himself strolling near the camp gate, where his encounter with Soma began. It started with a simple "Good Morning" and blossomed into a light conversation before they both went about their respective duties. Eventually, her name was revealed—Soma, named after the first queen of ancient Cambodia. Soma's English proficiency was not extraordinary, but it was sufficient to facilitate communication. She worked as a schoolteacher and was attending a session for educators when their paths crossed again. The remaining six days of the training seemed to pass in the blink of an eye, and Rahat's presence in Soma's life grew exponentially. Overwhelmed by Rahat's affection, Soma finally succumbed to the irresistible allure of love.

As Soma shed her bashfulness and hesitation, she drew closer to Rahat. To Soma, the vibrant and sun-kissed Rahat, with his dark sunglasses, became her favorite. His complexion may not have been fair, but his lively personality radiated a charming aura. In their minds, they both played the protagonists of a romantic movie, where a village girl and a foreigner embroiled themselves in animated affairs, traversing forests, rivers, and enchanting natural landscapes. Their metaphysical connection was undeniable. Rahat showered Soma with gifts, and though she appreciated them, her true joy resided in his mere presence. Rahat's heart overflowed with love.

However, their time together came to an end as the mission concluded, and Rahat was compelled to return to his home country. This marked a dramatic turn in their story, as the circumstances grew more complex. Rahat proposed the idea of marriage to Soma, seeing it as the only viable way for them to be together in his homeland, Bangladesh. He shared his plans, attempting to convince Soma that this was the most practical solution.

Yet, the societal and cultural barriers that Soma faced, rooted in her country, religion, language, and family, presented formidable obstacles. She would have to relinquish everything she knew to join Rahat in a foreign land. To express the sensitivity of the situation, Rahat had never been more delicate. He was prepared to sacrifice everything—his job, his country, and more.

However, reality proved to be different. Would Soma's parents even allow him into their lives? Love may be heavenly, but practical matters such as income were also crucial for sustaining a family and ensuring everyone's happiness. Moreover, Cambodia was still a war-torn country at that time, making such prospects even more unlikely. Additionally, how would Rahat's parents react to Soma? Furthermore, as a government officer, he would need official permission to marry a foreigner, a process that couldn't be expedited. Rahat's mission formalities to return home commenced as it neared completion. One day, Rahat and Soma made an oath to stay united, with the sun and moon serving as witnesses to their love.

Rahat returned to his country, promising Soma that he would soon return to Cambodia after convincing his parents and obtaining the necessary approvals to marry her and bring her home. They vowed to marry on Pohela Boishakh, the first day of the Bengali year.

Filled with both longing and uncertainty, Rahat embarked on the journey of completing the remaining marriage formalities with Soma's family. Unsure of how to proceed, he felt unnerved. Consequently, he flew back to Bangkok and sought the assistance of his trusted confidant, Dulal Bhai, in navigating this intricate process. Dulal Bhai assured Rahat of his support.

Dulal Bhai understood the challenges of convincing a Thai-speaking Cambodian family. Therefore, he arranged for two other trustworthy acquaintances to accompany him in meeting Soma's parents. Fully aware of the intricacies involved, they proceeded with utmost caution, knowing that hidden pitfalls often accompanied such endeavors.

Against all odds, Dulal Bhai successfully met with Soma's parents, encountering the anticipated adversities and reservations. However, the foreboding they had imagined soon began to dissipate. After all, what could stand in the way when the betrothed couple themselves were convinced and eager to start a life together? Soma's parents, the local administration, and the society to which she belonged had all been won over and rejoiced in their shared decision.

Ecstatic jubilation filled the air as family and friends gathered in Bangkok to celebrate the momentous occasion of their daughter's new chapter in life. Dulal Bhai, with a gleam of triumph and delight on his face, recounted the story to me as if he had just returned from a glorious victory in battle. The love story of Rahat and Soma, as narrated by Dulal Bhai, felt more captivating than any work of fiction.

Anticipation swirled in the air as to how Soma's life would unfold in her new surroundings. The long-awaited moment of Pahela Baishakh, the Bengali New Year, held a special significance for them. Dulal Bhai, the ever-witty matchmaker, joked about his newfound international matchmaking title, making us all burst into laughter.

Little did I know that the story was far from over. A few days later, upon my return to Bangkok, I discovered Dulal Bhai's absence from his office. After numerous calls, he finally answered, his voice heavy with sorrow, and he invited me to his home. Crossing the river, I reached Dulal Bhai's house, where his beloved dog warmly greeted me with a wagging tail.

"Hey, brother, the matchmaker! How have you been? Have you come up with any new tricks?" he greeted me, his face paler than ever before, as if carrying the weight of the world's sorrows.

"I failed in the first attempt. What can you expect from the second?" he replied, his voice laden with disappointment.

Shocked by his words, I asked, "What happened? Did their marriage crumble and succumb to futility?"

"Yes," he replied, leaving me utterly surprised.

"Why?" I inquired, wondering if it was a betrayal or interference from their families.

"His voice tinged with sadness, he said, "I feel hesitant to share the details, but since you insist, I will tell you."

He began recounting, "Rahat returned to Bangladesh with Soma, and his family embraced her with open arms, showering her with love and adoration. The wedding date was set for Pohela Boishakh, and Rahat took Soma to explore the enchanting beauty of Bangladesh, visiting its renowned tourist destinations."

Dulal Bhai described how Soma was captivated by the alluring charm of Rangamati, where lush green hills peeked through vibrant waters, and how she marveled at the hospitality of the people, who were soon to be her new family. Rahat left no stone unturned in acquainting her with the customs and people of his country.

As the wedding preparations were in full swing, Rahat, Soma, and their relatives returned from their shopping expedition at New Market in the capital. However, their journey took an unexpected turn when their car halted at Sobahanbag. The atmosphere grew dark, and a ferocious nor'wester storm unleashed its wrath even before the dawn of the Bengali New Year.

In the midst of the chaos, Rahat stepped out to buy a few specific items, and suddenly a swirling cloud of dust enveloped the surroundings. The fate of Rahat's attempt to cross the road remained unclear. After the storm subsided and the lights flickered back to life, it became evident that tragedy had struck. Soma anxiously awaited Rahat's return, but as the minutes agonizingly went by, she grew increasingly worried.

Sabahat, Rahat's younger brother, ventured to investigate the commotion nearby. His anguished cry shattered the silence as all rushed to the spot. Rahat's lifeless body lay on the road, drenched in rain and blood—a victim of a speeding bus that had careened through the storm.

Soma fell silent, her world shattered in an instant. She knelt beside Rahat's lifeless form, clutching the hand of the man for whom she had forsaken her homeland and everything she held dear. Like the final day of spring that bids farewell with grace and blessings, all hopes, dreams, and yearnings vanished for Soma. The next day was Baishakh, and the wedding flower would never bloom.

Rahat's parents, desperate to salvage the situation, implored Soma to consider marrying Sabahat, Rahat's younger brother. However, Soma remained unconvinced, offering no positive response. One day, she managed to pass through immigration at Dhaka airport and disappeared forever.

Dulal Bhai's narrative came to a momentary halt, his eyes welled up with tears, which streamed down his cheeks. Silence engulfed the room, filled with the weight of grief and loss.

THE POETICAL AMBROSIA: GIVE ME SOME TEARS

COLONEL G.R. MD. ASHRAF UDDIN
(UNITED NATIONS MISSION IN SIERRA LEONE - UNAMSIL)

What unfathomable consequences can internal conflict wreak upon a prosperous nation, shattering its thriving culture and disrupting a once-vibrant way of life? This question found its answer when I joined the UN peacekeeping mission in Sierra Leone, a country known for its wealth in West Africa. Freetown, the capital, nestled on the gentle and tranquil shores of the Atlantic Ocean, basks in perfect year-round weather.

Despite the breathtaking beauty of Sierra Leone's natural resources and landscapes, they proved insufficient to shield the nation from the unfortunate reality of civil war, which tore apart its unity and left a trail of destruction and devastation in its wake. Democracy withered, replaced by a cruel dictatorship that held the country in its grip. Fueled by neighboring Liberia, young rebels turned against their fellow citizens, unleashing a brutal and bloody war that claimed the lives of over 50,000 individuals and left countless others impoverished and homeless.

Sierra Leone's fertile and highly productive land, spanning a vast territory, was enriched with valuable mineral resources such as iron, rutile, bauxite, gold, diamonds, and more. Palm trees adorned its landscape, yet despite its abundance, a population of only five million called it home. The majority of its inhabitants, once colonized and oppressed, were of African descent, particularly Black Africans.

Tragically, even after achieving independence, the nation struggled to maintain unity, allowing foreign powers to exploit its resources through cunning tactics. Envy, resentment, violence, hatred, aggression, conflict, and resistance took hold of Sierra Leone's collective consciousness, culminating in a barbaric war to settle their differences. In this tale of Sierra Leone, the abundance of resources brought forth successive suffering and unrest, where beautiful beginnings did not always lead to happy lives, and the happily-ever-after depicted in fairy tales remained elusive.

Democracy was buried in the graveyard of the nation, as a single leader's dictatorship continued to rule with an iron fist. In response, a ferocious gang of young rebels launched a brutal insurrection against their own innocent people, fueled by direct support and instigation from neighboring Liberia. What ensued was a fratricidal war characterized by rampant violence, bloodshed, destruction, mass shootings, killings, and arson—a devastating annihilation. As time went on, the number of opposing factions grew, escalating the conflict. Such is the nature of any civil war.

Innocent civilians, resigned to their fate, were not spared from the ferocity of the vengeful rebels. The uncorrupted and unsuspecting population always bears the brunt of the brutality in any civil war.

The ravages of civil war left Sierra Leone in a state of lawlessness, economic collapse, and widespread hunger. In this chaotic environment, arms and ammunition flooded the territory unabated, fueling an unending cycle of violence. The civil war, which lasted from 1991 to 2001, claimed the lives of over 50,000 people and forced millions to flee their homes, leaving behind a nation plagued by poverty and homelessness.

It was amidst this dire situation that we took action, arriving in Sierra Leone in March 2001 with guns, ammunition, and grenades. The people of Sierra Leone had come to understand the gravity of the situation and were weary of the prolonged conflict. Though the war had officially ended, weapons were still in abundance, and the process of disarmament faced resistance and protests. Despite operating in a war-torn landscape, our only choice was to persist in our mission to establish peace. Through delicate negotiations, we succeeded in persuading rebel groups to lay down their arms.

Even in the midst of the battlefield, our unwavering commitment as peacekeepers remained steadfast. We established fortified camps, which served as beacons of security and symbols of our dedication to the principles of peacekeeping, human dignity, social progress, and service. Our objective was to win over the hearts and minds of both war-weary soldiers and vindictive rebels as well as the vulnerable and helpless civilians caught in the crossfire. We aimed to create a safe space where they could freely express their thoughts and concerns, understanding that trust was a mutual exchange that needed to be nurtured.

It is with great pride that we can claim to be the first to have earned the complete trust of all rebel groups. The surrender of the rebellious forces stationed in Lunser, who had initially refused to relinquish their weapons to our peacekeeping force (BANARTI-1), marked a crucial turning point towards peace. This achievement, spearheaded by Bangladeshi peacekeepers, was a significant milestone for the UN peacekeeping force. The international community celebrated this resounding victory for humanity, achieved through discussion, negotiation, and the dedicated efforts of the Bangladeshis.

Lunser, located approximately 100 kilometers from the capital, Freetown, housed the notorious and formidable 5th Brigade headquarters of the Revolutionary United Front (RUF). Previous peace agreements with the RUF had been marred by accusations of betrayal and broken promises. However, armed with newfound faith and determination, we refused to be swayed by past disappointments. Gradually, we replaced their disbelief with trust and faith. We worked tirelessly to restore dignity, honor, and the social fabric of the nation, instilling a renewed sense of hope and security where despair and darkness once prevailed.

We embraced our role with utmost seriousness, viewing it as an opportunity to serve and make a meaningful difference. Through the power of unity and peaceful negotiations, we succeeded in rekindling hope and trust in humanity, fostering social development and security. Together, we revived faith in the inherent goodness of mankind and the promise of a better world, heralding a new era of optimism for our collective future.

In a city devastated by war, a flicker of life began to emerge as residents who had fled elsewhere slowly returned. Overcoming their fear, they reintegrated into the community, resuming their pre-war way of life. The young people, in particular, had an insatiable love for football. To infuse some excitement and energy back into their lives, we organized regular football tournaments at the Lunaar Stadium, distributing footballs to neighboring villages and neighborhoods.

Music and dance held a special place in the hearts of the people of Sierra Leone, regardless of age or gender. Watching them move to the rhythms, especially the unique African dances, was a joyous sight. It was evident that music bands had gained widespread popularity in the country.

During our time in Lunaar, we had the opportunity to organize a band concert in the town. I vividly remember an evening when the music of the band broke through the silence, engulfing the entire city with its enchanting melodies. People from all walks of life came together to be part of the concert, and it felt as though the music breathed life back into the war-torn city, offering a glimmer of hope amidst the ongoing conflict. Slowly but surely, Lunaar began to show signs of life once again.

As a year passed in Lunaar, we witnessed a quiet revival taking place. Despite the city's stillness, it was reawakening, with residents returning and the streets coming alive with activity. Each day, we noticed small but significant changes that indicated the city's gradual rejuvenation. It was a testament to the resilience of the people and their determination to rebuild their lives. With every sign of life returning to Lunaar, our hope for a brighter future for its people grew stronger, motivating us to continue our mission with renewed vigor.

Before our departure, the Bangladeshi peacekeepers stationed in Lunasar organized a cultural festival in collaboration with the local community. This celebration served as a tribute to the revived spirit of Sierra Leone and showcased the diverse cultural heritage of both Sierra Leone and Bangladesh. The festival aimed to bring people together, transcending their differences and celebrating the unity that emerged from the city's resurgence.

Sierra Leone was home to several clans and linguistic-cultural groups, each with distinct languages and cultural practices. These differences were profound and diverse, leading to intense conflicts during the civil war. The Timne and the Mende, the largest clans in the Lunasar area, were particularly significant.

In an effort to foster cultural exchange and understanding, we organized joint cultural programs. Our soldiers learned Timne language songs, jokes, and dramas, while local Sierra Leonean youths were trained to perform Bengali country songs, comedy acts, recitations, and more. Through these collaborative efforts, we aimed to bridge cultural gaps and promote unity among diverse communities.

Above all, our officers and soldiers orchestrated a program filled with entertainment and comedic performances, showcasing the experiences they had acquired throughout the year. The program included marching, Zari gaan, Shari gaan (traditional Bengali songs), country songs, and lighthearted acts that brought laughter and joy to the audience.

The enchanting scene of dark-skinned Africans singing Bengali songs and delivering jests left the audience in awe. The locals were ecstatic to witness the Bangladeshis, who had traveled from distant lands, captivating them with heartfelt performances in a language foreign to their ears. Visitors from the United Nations headquarters and contingents from all corners of the globe were spellbound in witnessing the remarkable achievements of the Bangladeshis in the face of adversity and hostility.

The program unfolded as a delightful medley of song, dance, drama, and humor, with the poetry recitation standing alone in its pure form at the end of the cultural ceremony. As the curtains drew closer, I took the stage to recite two poems of my own creation: one in Bengali and the other in English. The Bengali poem, infused with cultural nuances only a native could fully comprehend, was an exclusive tribute to the Bangladeshis.

Then came the moment to recite the English poem. The words resonated deeply with both the local attendees and the foreigners gathered in the grand hall. A serene silence enveloped the room during my delivery. Tears streamed down the faces of elders, youth, and even women, who were no strangers to the perils of armed conflict, including those visiting from other countries. Alammi, a young leader, approached me and expressed how deeply my poem had moved them, as if it had lifted a ten-year burden from their shoulders. He said, “Your poem has touched us so deeply. It’s as if you’ve liberated us from the trials and tribulations of the past, bringing us together in this very moment. Tears were the only language left for us to resonate and respond to your poetry.”

Fatimatou, a middle-aged woman and NGO leader, requested a copy of the poem so that she could share it with abused girls, hoping to provide solace and inspiration. The local church’s father also sought a copy to use in Sunday school.

As the author of the poem, that was the pinnacle of achievement for me. I felt a profound sense of fulfillment and gratitude. What more could an author ask for? All credit goes to the Almighty!

When I delivered the poem, even I couldn’t resist shedding tears and expressing the depth of my emotions. Later, the poem was published in “Tohoshangha Odhikar,” a publication by the United Nations Secretariat in Freetown. It was also included in a book of short stories titled “Shada Kak o Kalo Manusher Golpo - The Stories of a White Crow and the Black People,” a compilation inspired by memories of Sierra Leone, which was published upon my return to the country.

That poem is appended below:

GIVE ME SOME TEARS

Give me some tears, O world! Grant me a single tear,
Regardless of your color, status, or near or far;
O people on this Earth, can you hear?
Give me some tears, O world! Bestow upon me a single tear.
Within me, all is dry and void,
That is why I cannot weep, nor avoid!
No matter the pain you make me endure,
You squeeze my heart and my body impure!
My eyes can only stare, devoid of emotion,
Unmoved, dry, showing no devotion!
Oh my God! A desert within me lies!
I cannot shed a tear, no matter how hard I try!
Do not fear when you see me so,
I am not a ghost, simply lacking tears, you know;
For in Sierra Leone, over the past decade and more,
I have shed and shed and shed scalding tears galore!
I wept when my toy was broken, shattered,
I wept when flames consumed my home, it mattered;
I wept when forced into the bush to dwell,
Starving, sleeping amongst creatures, I can tell.
I wept when they violated my mother's grace,
Silenced me and took my sister from this place;
I wept when they dragged my father away,
We all wept when fear held its sway!
We returned, only to be displaced by war's strife,
Helpless, losing all, as chaos raged in life!
Then, we forgot how to cry when they killed each other,
Life turned to emptiness, consumed by war's smother.
Now, I cannot cry, I have no more tears to shed,
The civil war has scorched and ravaged, left us shred!
Why all this occurred remains unclear,
Lost in the fog, it's not quite here;
In the end, we are left wounded, battered,
Bloody and weary, like a beast tattered.
Their selfish power struggles brought hell upon us,
O world, witness today, the shattered peace we discuss!
O brothers, fathers, sisters, mothers! Offer me a tear,
So I may weep again, laugh, and feel life's cheer;
The burden of pain and loss I cannot bear alone,
Your helping hand has given me hope, dear world, known!
In times of need, O brethren of the world, you did not scorn,
Grant me tears to face life, I shall return, reborn!

Lunsar, Sierra Leone, 04 Feb 2002.

To the Ears of Bangladesh

Colonel Nazma Begum

United Nations Mission in Ivory Coast (UNOCI)

The tumultuous alliance of the howling wind and torrential rain sought to tear the dark night apart. Yet, amidst this chaos, I remained vigilant. Clad in a bulletproof helmet and jacket, I stood at the apex of vigilance. The darkness enveloped us, but my keen eyesight pierced through, as if capable of perceiving the unseen beyond the veil of obscurity. Perhaps it was a desperate attempt to reassure myself that I was fulfilling my duty as a contingent commander to the best of my ability. Confronted with daunting adversity, I steadfastly embraced my role, finding solace in the knowledge that I exerted every effort to navigate tumultuous circumstances. The impending night loomed menacingly, harboring unnameable perils. In the early hours of darkness, we received an ominous intelligence report that made us privy to the possibility of a suicide attack upon our camp that night.

Within the Bangladesh Level Two Hospital medical unit, the unit under my command, there existed a small but dedicated team of only 56 individuals. Lacking formidable weaponry, we relied solely on a group of skilled physicians, abundant medical resources, ample supplies of medicine, and cutting-edge medical equipment. I am acutely aware of the unparalleled recklessness exhibited by suicide attackers in this nation. Their determination knows no bounds, rendering even the most fortified security measures futile. Multiple layers of security would not withstand their assauLieutenant They were unruly forces, and we were not exempt from their reach. There was no reason for them to consider us differently, and thus we found ourselves equally vulnerable.

As I scrutinized the enigmatic interplay of twilight shadows that had settled half an hour earlier encroaching upon the realm of obscurity outside my window, my ruminations were abruptly interrupted by the ear-piercing trill of the landline telephone. Deputy Contingent Commander Lieutenant Colonel Solayman urgently conveyed the gravity of the situation: "Sir, your presence is immediately required. A Moroccan captain is here with a message of utmost urgency."

I quickly changed my attire, ensuring that the discarded clothes were neatly returned to the wardrobe, and swiftly left my room. The Moroccan captain, Jisan, appeared slightly nervous, his face exuding palpable trepidation. Upon deeper reflection, I too became consumed by an overwhelming sense of anxiety. We had arrived in this place in February, a month dedicated to celebrating linguistic uniqueness and the power of language. It was a month recognized on the global stage as International Mother Language Day, a day to commemorate linguistic diversity and honor the language martyrs of Bangladesh who sacrificed their lives for the freedom of language. It was a source of pride for all Bangladeshis.

In that distinguished month, I had the privilege of unfurling the majestic flag of Bangladesh in a momentous ceremony held in Dalwa, Ivorikost (commonly known as Kuddibhuya). With great pride, conviction, and a multitude of commitments, I assumed the role of the inaugural female Army contingent commander from Bangladesh. Our mission in this conflict-ridden land was to foster harmony, peace, and a sense of truce. Amidst the tempestuous environment, our aim was to imbue the essence of serenity and spread the fragrance of tranquility, striving to bring an end to the war.

In that fleeting moment, all unease was dispelled, akin to emerging from a slumber. A silent soliloquy echoed in my mind, "I am the first female contingent commander here. In my footsteps, a new generation of young female commanders will follow. My success will inspire them. I must leave no room for frailty or lapses in my actions. In this international arena of multilateral peacekeeping, I must uphold the esteem and dignity of our homeland."

Inspiring Captain Jisan, I assured him, "Do not falter; we stand united with you. We are present and ready. Rest assured, your formidable armored division commands respect, and your vast fleet of tanks makes any challenger think twice. Even if someone dares to confront you, their consequences will be dire and their defeat imminent."

Expressing gratitude for his timely disclosure of the urgent news, I instructed Deputy Contingent Commander Lieutenant Colonel Solayman to convene an immediate meeting. Demonstrating exceptional insight and foresight, the capable and visionary deputy commander swiftly organized the emergency gathering and informed me of the proceedings. Adhering to martial protocol, I assumed my position before my fellow BANMED members and provided a comprehensive briefing. With clarity and precision, I elucidated the current circumstances and our strategic approach for the day, ensuring transparency and understanding. Pointing to the elongated axis depicted on the wall-mounted map, I highlighted the potential area of assauLieutenant

We all donned sturdy military attire, with bulletproof jackets and helmets distributed to everyone. Although armed standoff drills had been conducted multiple times before, each individual remained resolute, prepared for the imminent armed confrontation and positioned themselves according to their training. With unwavering conviction, I addressed the assembly, “Fear not, for we are a valiant and illustrious race, possessing the strength to reduce any adversary to mere dust. None shall prevail over us.”

Upon hearing the commander’s words, the eyes of those in the line shimmered with renewed confidence, their trepidation dispelled. Having instilled a sense of trust within them, I confronted my own mounting fears, seeking solace in the protective gear as I adorned the helmet and the weighty bulletproof armor. Yet, the burden of my increasing apprehension made it nearly impossible to lift the armor alone. I sought the assistance of a fellow female officer, relying on her to help me with the cumbersome vestment, suppressing my inner concerns that remained concealed from the rest.

As midnight marked the commencement of our nocturnal vigil, an eerie silence settled, accompanied by the first hours of the night. The heavens were shrouded by thick clouds, while armed officers maintained their vigilant positions. Adjacent to me were the female quarters, where I inspected the positions of each woman, bolstering their morale along the way. Once satisfied with the thorough inspection of the hospital courtyard, military lines, and duty posts, I returned to my own position.

However, faint voices whispered in the distance, obscure and unsettling, compelling me to temporarily leave my post once more. In the absence of electricity, the surroundings were engulfed in oppressive darkness, intermittently illuminated by vibrant bolts of lightning. These sporadic flashes revealed the determined faces of Bangladeshi peacekeepers, their hands firmly gripping their weapons. Even their firearms seemed to embody their focus, determination, and unwavering commitment to their duty and purpose.

Undeterred, I held steadfast and returned to my post, drawing strength from the wellspring of bravery within me. As if in harmony, the sky above ruptured, releasing a torrential downpour that merged with the frenzied gusts of wind. Prior to this, I had diligently activated and placed all available communication devices within easy reach, ensuring that the single press of a button could transmit an emergency message to the highest echelons of the United Nations or enable direct contact with the headquarters in New York. Every device, including each mobile phone, was operational and ready. However, in a twist of irony, internet connectivity was painfully absent, rendering all communication efforts futile.

Surrounded by stygian darkness and engulfed by the unrelenting storm, conventional means of information exchange were extinguished. It felt as though we were isolated from the rest of the world. At this crucial juncture, primal fear gripped me. I was acutely aware that the insurgents inhabiting this land showed a flagrant disregard for human life. Their lack of attachment to worldly possessions made them prone to carrying out reckless suicidal attacks, targeting friend and foe alike without hesitation.

This hostile terrain was their domain, and they possessed an intimate familiarity with every intricate alley, intersection, and crevice. The tempestuous, pitch-dark night became their ally, rendering our erstwhile security measures ineffective. We were not a combat force but rather a support unit, lacking heavy armor and potent weaponry. In the midst of this precarious situation, the resonating sound of a powerful motorcycle engine reached my ears, sending shivers down my spine. My heart pounded relentlessly, reverberating like a hammer inside me.

Two hundred yards ahead of my room, on the road that passed the sector headquarters, two cars drove side by side, seemingly targeting us with their slow and deliberate approach, undoubtedly harboring malicious intent. I swiftly dialed the deputy commander's mobile phone, wasting no time in establishing urgent contact.

After a brief pause, he called back, his voice trembling with trepidation. "Sir, a pair of large vehicles loaded with substantial amounts of timber or similar material have passed by due south of the avenue, near the Sector Headquarters."

"Sir, although they have crossed the intersection, we are not out of danger. We should not let our guard down. Despite this intelligence, I struggle to dispel my lingering anxieties. I can't help but wonder why I have never heard such loud vehicular noises during the night before. Why is it that today, on this disastrous and turbulent midnight, I am hearing these sounds for the first time?"

All the soldiers were armed and ready, positioned for action. I proceeded to provide directives to the officers, emphasizing heightened vigilance and reiterating their responsibilities in the event of mutiny or insurgent aggression. I briefed them once more, ensuring that they understood their roles and duties.

Suddenly, the tranquility was shattered by the sound of gunshots echoing through the air—a fleeting volley of gunfire. Before my heartbeat could even subside, I realized it was a false alarm. The commotion was not caused by firearms but rather by an explosion in close proximity. It was more likely that lightning had struck nearby.

Following that, sporadic gunfire could be heard in the distance at intermittent intervals. Amidst the profound dread, I found some solace in knowing that our location was not the primary target of those weapon discharges. Another party bore the brunt of the violence. From then on, I frequently mistook the roll of thunder for the sound of firearms, and vice versa. That tumultuous night felt interminable and profoundly terrifying, resembling the sentiment expressed by football club fans who claim that 90 minutes is long at the Bernabeu. It was the longest, most horrifying night of my life.

Uncertainty plagued my mind, as I questioned whether the obsidian shroud of darkness would ever yield to the nascent radiance of daybreak, if the night would ever give way to a new morning. Time seemed to have halted its inexorable march, and the deep darkness lingered. Anxiously and restlessly, I endured the night, anticipating the end of darkness, waiting for the break of dawn.

The transition from midnight to the early hours of morning barely registered with me. I pulled back the curtains and observed the nocturnal veil receding. The rain had also ceased. The night vanished, taking with it all the malevolent energy it carried.

The precipitation had subsided, and despite the harrowing ordeal, time continued to march forward—a reaffirmed truth. Eventually, the revered hues of the rising sun appeared at the break of dawn, seemingly granting a renewed lease on life. While aware that insurgents could launch a suicidal assault at any moment, disregarding the distinction between day and night, such incursions that occurred here took place mostly in daylight.

Nevertheless, I still couldn't fully comprehend why the daylight brought me a sense of relief and eased my apprehension. It was as if the illumination of day had a calming effect on me, alleviating my dreadful fears. Not simultaneously but gradually, my medical compound regained its liveliness and vigor. Our premises became bustling once again, and patients flowed in as they usually did.

We immersed ourselves in the provision of healthcare services, wearing congenial smiles on our faces. The outdoor patient area thrived, witnessing a fourfold increase in the number of patients seeking assistance.

Our achievement lay not just in providing medical care but in earning the trust and confidence of the people we served. Through our respect, empathy, and accurate diagnoses, we formed deep connections with them. In a short span of time, we became more than just healthcare providers—we became their friends and even surrogate relatives.

The patient demographic was predominantly composed of women and children, mirroring the disease patterns we were familiar with. AIDS and malaria cases were prevalent, and there was a significant number of single mothers and children. Their social system allowed for cohabitation without marital relations, with men having multiple wives under the same roof. This patriarchal society often resulted in vulnerable young girls and women bearing the burden of fatherless and insecure children at a tender age.

During a momentary pause, I noticed a teenage girl, likely around 16 or 17 years old, standing close by. She carried three children with her, one tied to her back and another on her chest. Cradled in her right hand was her precious two-year-old eldest child, basking in the love and warmth of her mother's embrace. Her shy smile and clear voice revealed that she was a single mother.

The father of her eldest child was a childhood friend, while the father of the twins was a wealthy married man. Both men had promised her marriage and professed their love but ultimately left her alone to navigate the stormy waters of love. They had returned to their own lives, leaving behind the shattered dreams and lost innocence of this young girl.

With a pale smile and a sense of gratitude, she shared that her eldest child had fully recovered under our medical care. However, on this turn, it was the twins who were afflicted and suffering from fever. This teenage girl shouldered the responsibility of raising three children alone, like a lone captain in a tumultuous ocean, with an unmarried and uncertain life. My heart swelled with indignation towards the inhumane social system that allowed such circumstances to prevail.

That particular day began with a sunny morning, but the sky soon darkened as black clouds rolled in. In an instant, the clouds unleashed heavy rainfall, with African raindrops falling like chuckles from the sky. It resembled the heavy monsoon rains of my beloved motherland, where the clouds first gather before pouring over the gray dust.

The morning sky had been a golden blue of autumn, but within moments it transformed into the rainy season. It felt as if luck had brought us to a country where we experienced a blend of Bangladesh's six seasons every day, from morning till evening. Summer, monsoon, autumn, late autumn, winter, and the fullness of spring—there was no obsession with seasons as we have in our country. They did not have to wait for specific weather changes; their climate remained consistently hot and dry, wet and dry, or warm and dry throughout the year. It seemed the pleasant weather had prevailed there, at least up until that day.

I had dubbed that country "The Land of Eternal Spring". Its inhabitants were generally kind, gullible, cheerful, and amiable, though I had also witnessed extreme ferocity among them. One of their preferred foods was a rodent-like animal known as the agouti, adding a unique aspect to their culinary preferences.

In that country, a stark contrast existed between the elite living in opulence within the capital and the increasing destitution of the poor. The rich continued to amass wealth while the poor grew poorer. Despite this darkness, we planted our shining flag there, a symbol of hope amidst the despair.

The land itself was magnificent, adorned with emerald cocoa trees and rich natural resources such as gold, diamonds, coal, and manganese. It was renowned as a major producer of coffee, cocoa, and palm oil. The majority of the population, around 68%, was engaged in agriculture and related fields. The country covered a significant area and had a diverse population that was predominantly Muslim (35–40%) and Christian (20–30%), with other religions also followed. Although French was the official language, various regional languages were spoken by different communities.

The statistical data revealed a concerning picture of the region's literacy rate. Only 54.34% of males and 45.12% of females were able to read and write. Additionally, the average age of the population was relatively low, with males averaging 44.86 years and females averaging 46 years. This was primarily due to the prevalence of AIDS and malaria as well as a high infant mortality rate.

While cassava was their main food staple, their daily menu also included rice, bread, boiled corn, vegetables, fish, meat, and fruits. Village girls wore tops on their upper bodies and lungi-like cloth on the lower part, similar to girls in our hill tracts. Interestingly, natural hair growth was rare among them, and girls resorted to artificial hair made of fiber or synthetic materials to compensate.

For many years, Ivory Coast was a French colony, which resulted in the suppression of their own language and culture under the dominance of French culture. Political instability, conflict, and violence had been recurring events in the country's history. The days were marked by upheaval and inter-ethnic conflicts that shook the world. To establish peace and stability, several thousand troops from various nations were deployed for 12 years, including countries from Asia, Africa, Europe, and America, including Bangladesh, India, and Pakistan.

Finally, the people of Ivory Coast had their voices heard through a much-awaited election, and peace prevailed despite some distractions. Most of the peacekeeping forces returned to their respective countries, while a small number remained to ensure the continuity of peace. As a medical unit, we also stayed behind to address the healthcare needs of the people. We remained until the end of the designated mission days, knowing that our time there was limited.

Soon, we would bid farewell to that captivating land of timeless beauty, our loved ones and the fertile soil of our motherland. In a fleeting moment, we would leave behind the evergreen landscapes, the fragrant pathways, and the joyous children playing in the rain. We would no longer have the privilege to cherish the earth soaked in the ecstasy of torrential downpours.

The day of our medal parade was rapidly approaching, as was our departure back to our homeland. Amidst the preparations, we received alarming news of isolated rebel assaults. A sense of dread enveloped the unit, casting a dark shadow of terror over us. With our weapons and ammunition packed, we realized our limited means to defend ourselves as military personnel. The fear in our hearts remained concealed as I rallied our team, reminding them that we had faced similar situations before and should not succumb to fear. We worked tirelessly throughout the night, transforming it into a double day of packing and organizing.

The rows of seats were filled with spectators oblivious to the fact that only a few peacekeepers remained in the country. Members of the Force Headquarters had arrived from the capital city of Abidjan accompanied by the SRSG. The presence of the force commander, deputy force commander, and deputy SRSG added significance to the event. Local dignitaries, including the mayor, deputy mayor, Ivorian army, police, media workers, and other officials, attended with great interest and enthusiasm. It was the day of our medal parade, and the members of BANMED 12 took their positions at the center of the field. Leading the parade was Lieutenant Colonel Solayman, the deputy contingent commander.

The parade concluded splendidly, followed by a captivating cultural program and a photo session. Each segment received thunderous applause from the spectators, who were captivated by the performances. Every member of our unit was honored by the special representative of the secretary-general, with acknowledgement of their unwavering commitment to serving humanity thousands of kilometers away from their loved ones and homeland. It was a joyous occasion that elevated the beloved motherland and the national flag to the highest position of recognition.

To my astonishment, my name resonated through the microphone. The special assistant to the SRSG announced, with a jubilant voice, that I had earned the Force Commander's Commendation. Before presenting me with the medal and certificate of commendation, he eloquently read out the entire certificate in impeccable English. I was overwhelmed by the magnitude of this unexpected achievement and felt like I was soaring on cloud nine. The SRSG congratulated me in a convincing voice, recognizing me as the first female army contingent commander in the history of the United Nations. He praised not only my dedication as a medical commander but also my competence and dignity in fulfilling the responsibilities of a senior representative of the country when it was most needed. This milestone established an exemplary role for women worldwide in this golden age of empowerment.

The SRSG expressed that the United Nations would always cherish the Bangladesh Medical (BANMED) Contingent for their expertise and sincerity and the guidance provided by Colonel Nazma. Applause filled the air once again, accompanied by congratulatory messages. I found myself participating in interviews with numerous newspapers and TV channels. These interviews were later disseminated to various media outlets in Bangladesh, emphasizing the achievements of our mission.

The countdown to our departure had begun, with only two days remaining. The BANMED makeshift hospital, a close friend to the people of Dalowa, was neatly packed and ready to be loaded onto covered vans. On the scheduled day, a convoy of large trucks entered the BANMED compound one after another. My heart shattered into pieces, as I realized that BANMED would no longer grace that place. We were truly leaving. The vans carrying the Bangladesh Hospital, which had been nestled in the heart of Dalowa for an era, would soon hit the road to Abidjan, then sail across the Atlantic for a long journey back to Bangladesh. The name of Bangladesh Hospital would be erased from the map of Ivory Coast, but the memory of Bangladesh's medical service would remain etched in the minds of millions.

As I inhaled the familiar scent of the Atlantic, a bird's chirping reached my ears, carried by the gentle morning light. That fragrance had become a permanent part of my emotions, akin to the raw rain-soaked earth of Bangladesh. I could easily identify it in the calm or breezy wind. Even if I never returned to that place again, I knew the fragrant wind of the Atlantic would forever reside in my heart.

Glancing at the road, I watched as the vans neatly moved in an organized line. The heavy wheels churned up dust, obscuring the road in front of the sector headquarters. As the dust settled, I caught glimpses of the covered vans carrying BANMED signs disappearing from view. Saying goodbye to the hospital, I felt an overwhelming sense of emptiness, like a wanderer crossing mirages in a desert. Suddenly, I felt like a stranger in a distant land, standing there alone and invisible. But then I realized, I was not uninvited or unwelcome in Côte d'Ivoire; in fact, I might have been dear to her. I had always been loved there. Côte d'Ivoire, whose nature had embraced me like her own offspring whenever I announced my departure and made me weep—how could I be anything but close to her? Was I not her true relative?

I knew that no matter how far I went or where I ended up, Côte d'Ivoire would always hold a place for me in her heart. When the lush green leaves danced in the rain, when the cuckoos sang in the distant forest at noon, and when the young cowherd returned home with his cows bathed in twilight's gentle sunshine, Côte d'Ivoire would be eagerly looking for me. Wouldn't she? Would she really search for me with care? These thoughts lingered as I inhaled the everlasting scent of the Atlantic, a fragrance that would forever remain in my heart.

VILLAGE CHIEF MAGURO

MAJOR MD. HASAN SHAHRIAR

(UNITED NATIONS MISSION IN THE DEMOCRATIC REPUBLIC OF THE CONGO - MONUSCO)

Maguro had been pinioned. He was tightened with a stake, hooked and twisted. He was powerless to wriggle free. Every movement only stiffened the bonds around his hands. For 14 hours, he had been in such a condition. His hands had turned a livid hue of blue, and his body color was so dark that it blended with the shadows. They had deprived him of food and water since they captured him for reasons unknown to him. Not even water! He had no idea what fate awaited him at their hands. They didn't know what they would do with him either. A searing pain tore through his chest. All he needed was water then. His head spun like a top. As usual, darkness overshadowed his eyes like a curtain falling on his life.

Maguro had been shackled in a secluded spot somewhere amidst the jungle. How dark, gloomy, and eerie the place was, even in daylight. The temperature plummeted as nightfall approached, contrasting with the sweltering and muggy daytime. Maguro was aware that the cold would bite harder as the light waned. He had endured a terrible ordeal the previous night in the frigid weather. A sudden thunderstorm had lashed out at him to add insult to injury. He had survived the night in a deplorable condition. Only God knew what was in store for him that day! That morning, he had overheard the militia group talking about him. The group members had uttered something about Purnima (the full moon night) and cast him malicious glances. Their cruel look only incited more fear in him. He had sensed that the group had a sinister plan for him. A young man had approached him shortly before and murmured in his ear, 'You will meet your doom on Purnima (the full moon) night.' The sky had almost had a full moon the previous night, so that day would be Purnima. That day would seal his destiny, then!

He had lived for four decades never knowing the taste of fear. Yes, he had passed 40 years of his life, and he could not remember ever being afraid. But on that day, as dusk crept over the horizon, the twilight was slowly fading and giving way to the night, gradually hand in hand with the darkness; his dread dug deeper, and the fear seemed to have reached his veins.

Bizarre apparitions plagued his vision, and he suffered from occasional delusions. He beheld a scene of innocent joy: a group of children frolicking together. They were laughing and smiling. Their laughter was like music to his ears. He admired their beauty and creativity as they devised new games every minute. One uttered a command, and another performed a gyrating dance. His little girl was also among them, a flower among the weeds. But no, they were not weeds; at the moment, he seemed to have corrected himself—all the children were more beautiful than Mother Earth. He wondered if it was all a dream or a cruel trick of fate.

It seemed like they had quarreled over something very trivial. Suddenly, one of them brandished a shiny knife and strode menacingly toward his daughter. "Hey! What are you, what are you doing, boy? What is this madness?" Maguro shouted. He was screaming desperately, howling, "Mairo, get out of the way, get out of the way!" The girl didn't hear him, as she was oblivious to his cries; she just raised her hand as if to ask, 'What happened? Why are you calling?' The knife-wielding youth drew nearer, the sharp point of the blade poised to pierce her flesh at any moment. Suddenly, a resounding slap rang out, striking his cheek. He was jolted back to reality, toppling over without the ability to maintain his balance. He fell down without being able to balance himself. He still had a smile on his face. He thanked God in his mind. Luckily, it was just a dream, not real. Well, what would happen to his family if he died? He contemplated the potential ramifications of his demise on his loved ones: his seven children, wife, and old mother. What would happen to them all? His wife might start a new life with someone else. Would their new father accept the children? Would he love them like he did? He didn't want to die; he didn't want to die at all. Although he had very little in life, the fear of death had firmly taken hold of him on that particular day. He didn't have anything in life, he had nothing to lose, had never feared anything, even death! But that day...........

Maguro had been returning home from Kagaba market with a week's worth of groceries the day before yesterday evening. The evening was thickening. As the evening deepened, he knew he had to hurry back home. On market days, the road was often occupied by the militia, who would plunder and leave people destitute without a second thought. They took away whatever the returning people had with them and left them with nothing. Their brutal tactics would leave no room for dissent, and even a slight protest would result in a fatal shot. They would have shot down anybody in their way without hesitation. Maguro hired a motorcycle to go home quickly and had driven from Kagaba to Nomba and stopped there. The road ahead turned right off the hill and was notorious for the two robberies that had occurred there last month. A militia group had killed a village chief, leaving his severed head on the road as a warning for all who dared to cross their path. The boy driver was visibly afraid, and Maguro had no issue understanding why that young lad was nervous. The boy was trying to understand the situation. It would have helped if they had been accompanied by two or three more motorcycles that day. But there was no one around. Maguro was a brave man. He told the boy driver in a reassuring tone, "Don't be afraid, I am the village chief of half a village. I am with you; you have nothing to fear. We are alone here, so what? Nothing to be afraid of, young lad... move ahead!"

The road to the right led to a coal mine on the mountain. Maguro had never committed such a grave error in his life. He had taken the road that curved to the right, where the mountain's coal mine loomed. The muddy road was slippery because there had been rain the previous day. Although his driver was young, he was skilled, an expert in driving. He had navigated the tricky spots of that area with caution. But as he moved past it, a sudden blaze erupted in the jungle beside the road. A flash of light was followed by a sound. Yes, they were being shot at. Someone had fired at them. Startled by the gunshot, he had lost his balance and toppled down with the motorcycle into the mud. Several more rounds of fire followed. By a stroke of luck, none of the bullets hit him. A group of seven or eight men came out of the jungle. He and his driver turned to stone at the sight of them. Maguro could not escape because his leg was twisted when he tumbled from the motorcycle. But as soon as the driver got back to his feet, he seized the opportunity and made his way to the jungle.

As he looked up, he felt a sharp blow on his jaw from a weapon. Maguro had blacked out right at that very moment. When he regained consciousness, he found himself tied to a stake. Darkness was all around. Only a faint light dimmed above his head. After a while, when the clouds left space, the sky sparkled in the moonlight. He gasped when he saw what lay on all four sides.

The news of the village headman's abduction must have had spread like wildfire by then. The brave young men of the village youth defense team had taken up arms (sticks and bows and arrows) and ventured into the dark jungle to search for him. But they were clueless about his whereabouts and could not contact him. How would they? They did not know where he was then or where to find him. The jungle was a vast and treacherous labyrinth full of dangers and mysteries. Still, they had a glimmer of hope that the Congo army (FARDC) would come to his rescue, accompanied by the ever-ready, vigilant, and fearless BANBAT of the United Nations. Perhaps they could free him from his captors. But how would they locate him? They vowed to scour every inch of the jungle, comb every corner if necessary. But they were running out of time. The night had fallen, and the full moon had risen like a silver disc in the sky. Those militants would have had slaughtered him as an offering by then. The moonlight reflected on the sharp blades that had been used to torture him earlier that evening. And those knives sparkled when sharpened. He had endured two rounds of agony, and his life hung by a thread. The peacekeepers would have to bring back his corpse if they ever found him. But maybe they would never find him; maybe he would be devoured by the beasts of the forest. And so, the name of Maguro, the unfortunate father, husband, and village leader, would disappear from the face of the earth, unnoticed, forgotten, ignored, and unremembered.

Major Ashraf, the camp commander of Bogoro, was beside himself with anxiety. He had received the news the previous night that the village chief of Ada had been kidnapped by the militia. The messenger was a young motorbike driver named Songalo, who had arrived at the gate of Monusque BANBAT camp at around half past nine, drenched in blood and mud. He had walked the 14-km road on foot. Post No. 1 had informed that a Congolese who had something to say had come, but his condition was extremely critical. At that moment, the camp officers were just about to sit down for dinner, but they all rushed to see him. The boy could not speak normally or coherently; he was giggling, fidgeting. Language Assistant (LA) Emmanuel was called for. All decided not to have dinner until they discovered the actual event.

That was the third time such an incident had happened in the last two months. The previous two times, those who were abducted had met a gruesome fate and were found as half-rotten corpses a week later; some devoured by the beasts of the forest were dragged by foxes and dogs to the locality. Some corpses were seen floating in the murky waters of the drain. No one could be rescued alive. Sigh! None had survived. But this time, those militants earned Major Ashraf's obstinacy. He vowed to rescue the village chief at any cost. He was ready to turn the whole region upside down, ransack the territory by scouring every inch of land until he found him, and get him released.

The investigation began. The intelligence team was already on the run, trying to find any way to grab any trace of Maguro. With the help of the locals, the search was on. Major Ashraf offered a generous reward for any information. He mobilized his men for a full-scale rescue operation. But there was a challenge: there were no witnesses to the incident, no traces of the culprits, no clues to their whereabouts. There were no houses around either. The militia group had vanished into thin air with their prey, the village chief Maguro.

Major Ashraf was not a man to give up easily; he only became more relentless after these incidents. He had sworn to find and rescue him, no matter the cost. He had launched a helicopter recce and scoured the skies for two hours that morning, spotting some dubious movements following the direction of Souke Mountain. He had dispatched a drone to probe further but was yet to receive the images it captured. The computer scanned those images, and then he analyzed the images. If his suspicions were confirmed, he would venture out himself that day and ride an APC as far as he could. Then, he would switch to a softer vehicle like a pickup or a jeep. But beyond that lay a thick jungle. A softer vehicle would not be enough to penetrate deep into that jungle. His last resort was a foot patrol. To walk through such a dense jungle was like walking on a razor's edge. The area was a haven for the militia. Moreover, news of the large group of searchers would not escape their notice. But he was determined to save the village chief, whatever it took. He had staked his own honor, his own life, on it. If only he knew the location, he would lead a special force to storm it and free him. But he had searched all around, and the chances of getting any news from there were slim but not impossible. How could they snatch a man and leave no trace for anyone to see? How could this be possible?! Nobody in this world had noticed any part of it!

The youth defense team of the Ada village was not alone in the rescue operation. A group of brave soldiers of the Congo military joined them under the command of Colonel Bulabula. He had sent Captain Ilunga Pol with a team of 20 men who were armed to the teeth. They carried light machine guns and rocket launchers, and one of them even wielded a machine gun single-handedly. Link belts of ammunition hung around their necks like mufflers and wrapped around their waists like belts. They were ready to face any danger with the BANBAT team. All of them were united under the BANBAT command and did not set out to fail or come back empty handed.

Clerk Nachir was seen leaving the office, speeding like a bullet. It was urgent to report the news to the camp commander. An e-mail was received from the Bunya Northern Sector headquarters with an operational order and some attached files. They had hit the jackpot! They were not completely sure, but they had a strong hunch that Village Chief Maguro was being held captive in that area. But it seemed like overkill to send a helicopter and a drone to rescue a village chief. It didn't add up! Only a few days ago, some European journalists and UNDP workers had fallen prey to militias from the Beni vicinity. They also had some Congo army members on their security team. But they were caught off guard by the surprise attack, and two FARDC soldiers lost their lives. The only clue they got was from a wounded soldier who was rescued and brought back. The wounded soldier had witnessed the captives being dragged to the Ilumu Hill area, and he was spared only because the militias thought him nothing more than a corpse.

Was the drone and helicopter reconnaissance meant for them? Had they really tracked down the abducted Europeans, or was there another motive? Nasir had no clue. Who knew? He was a military clerk, a soldier. A soldier's duty was not to question or think too much but to report. The camp commander would make sense of the rest.

Major Ashraf's eyes lit up as he looked at the mail and the files in his hand. He had struck gold! He had found it! He had a good inkling of where Maguro had been whisked away to. He was then more or less sure about the whereabouts of those abducted. The drone images revealed that it was a militia stronghold. There were guards with guns everywhere and scattered newly built makeshift huts dotting the landscape. A few African militia members in flashy European outfits were clutching FORTY SEVENS. He recognized those outfits. The clothes they were wearing looked very familiar. They had been worn by the journalists who had been snatched from Beni. They had received an email with their photos. Could they have been brought there, too? Was that the militia's den of captives? It added up. Yes, this time it added up! It made sense! He had not only employed the helicopter and drone reconnaissance facility to locate a village chief but also to get a hint of the abducted Europeans and find a whole camouflaged militia camp. Whatever it was, if he could kill two birds with one stone, why not! He gave the order to get ready for the operation and went up to pray. He always prayed two rakats of nafl (optional salat Namaz) before any critical mission. It was his old habit to offer that short prayer session.

Maguro lay prostrate on the ground, writhing in agony, clutching his stomach, as he endured the unbearable torment inflicted upon him. Each fiber of his being throbbed with excruciating pain, and he shuddered uncontrollably. Every inch of his body ached. His left eye was swollen shut, blurring his vision, while his left foot seemed shattered beyond repair. Helpless, he could not even make a feeble attempt to break free from his bonds. His hands were too weak to resist.

As dusk descended, a militia member came, and with a mere flick of his wrist liberated Maguro from his chains. He was finally released. But then a different sort of torment started upon Maguro as the militiamen ordered him to divest himself of his garments, and, despite fervent appeals for clemency, Maguro was subjected to a ruthless pummeling, ending his pleading. Ultimately, he was stripped bare, and a copious quantity of honey was maliciously poured onto his privates while he might have felt like jolts of electricity delivered to his cranium. In the throes of agony, shame, and debasement, Maguro beseeched a higher power, Maguro screamed out, "Oh Lord! Art thou truly present? If so, I implore thee, save me, save me!" (Do you really exist? If you do….! Please, God! Save me! Please, do save me!)

The evening was approaching fast. There was not much time left. Maguro's moments were fleeting. This time, Maguro knew what he was about to face that night and what the militia members were up to. Everything was clear by then. But what more he could do? According to the ancient customs of the Congo, to appease nature for a great undertaking, one must first satisfy its powerful creations – the ants.

The one to be offered was given a bath, and then honey was poured on his/her genitals. Then, they put honey into the ears of the sacrifice they planned. The nectarous scent enticed the demonic ants in a moment's time. Countless ants crept slowly toward the source. They ascended over the body and reached the private parts. They clung to it with force as if those ants competed. When the honey was depleted, they bit the tender spot, the most sensitive part of the human body; they gnawed the delicate membrane. The captive howled and thrashed in agony in this phase. As a result, the ants in the shoals climbed further up, glided into the ears, and bit. It was more than eating someone alive. When the honey in the ears was depleted, those f**** hungry ants ate the eardrums and became madder for sweetness. The captive, too, kept leaping off like a madman in severe pain. Then, the horde of ants marched through the ear to the brain, seeking more sweetness. Meanwhile, the captive writhed in agony, unable to withstand the excruciating pain inflicted by the incessant bites. The captive's writhing only provoked more bites.

The more he struggled, the more viciously the ants attacked. At that phase, the offerings, the ears of the human sacrifice, were plugged with wax, which only served to agitate the ants, causing them to seek a way out of the darkness in search of light. Some ants emerged through the nostrils, while others consumed the eyes. Eventually, the eye membrane ruptured, leading to the captive's demise. Meanwhile, the ants reveled in their success, feeling content as deities of nature. With the arrival of the full moon, the remaining parts of the body were grilled and eaten, as it was believed in certain areas of the Congo that consuming the flesh of a slain enemy imbued one with immense power. Perhaps, in some parts of Congo, in its hidden corners, this faith is still as alive as it once was.

He felt a sharp pain in his foot like something had bitten him. He opened one eye and saw a long trail of ants marching toward him. His legs had swollen up like balloons. It seemed as if a spell had been cast on them. Suddenly, it felt as if a thorn pierced his most private part. He shrieked, "Oh, mother," and jumped up with a cramp. He started to move over and hover helplessly with his hands tied.

The militiamen were standing, gathered around their spiritual leader. They were preparing for the main ceremony. They brought five or six foreigners clad in white tones and forced them to stand. Everyone ignored Maguro and got busy with them. This was his chance; he had to act then. He attempted to get up on his feet. He failed. Suddenly, he visualized his wife and children. He had to live. Yes, he had to live. He had to find a way to escape from this place. If he remained like this, a ghoulish death awaited him, and he did not want to die in such a manner. How could he let that happen?

Maguro, the village chief, deserved a better end. His body was a furnace of agony. The extent of pain in his body knew no bounds. He somehow peeled his eyes open and looked around. In front of his head, a thick forest lay in that direction. He would be safe if he could only make it to that jungle. That's it! He saved the last vestige of his strength and sprinted towards the green sanctuary. Did he survive then? Did he make it out alive? But if they caught him, he would undoubtedly be chopped to bits. He shuddered at the thought of such a gruesome death.

Twilight dissolved and the evening began. Darkness had quickly fallen. The red glow on the distant horizon had been like a lover's promise of meeting the next day, and he found hope. The birds were preparing to fly back to their nests. And just then, Maguro dashed like a madman through the dense forest. He had pushed forward with excruciating pain in his body and tremendous mental pressure. He had to escape that area as fast as he could. But his body had betrayed him. He had tripped twice while running like a maniac. He had spotted a footpath ahead. The road had ascended by brushing the side of the hill. If he could cross the bend of the mountain and reach there following the way left to the mountain, he would have found a way out and been able to arrange something for himself.

Maguro was running for his life. His body had given up, but his mind did not want to surrender. He had made it through the forest to the bend of the hill with great difficulty. He was breathless, gasping, clutching his naked hip with his hand, and leaning forward to catch his breath. Honey still oozed from his ears. As the honey dripped, a black silhouette appeared in front of him. Wheezing, he lifted his gaze at him, and at the same time a terrible blow plunged his whole world into darkness.

With utmost caution, Major Ashraf and his special forces had advanced through the dense jungle, a green labyrinth of danger and mystery. He had deduced from the pictures that had arrived in the mail, silent witnesses of a crime zone, the location where Maguro had been taken. And there had been a glimmer of hope, a faint ray of light in the darkness, of rescuing the abducted European journalists and UNDP members. He had to tread carefully on every step as if walking on a minefield. If their presence had been exposed, the first thing they (the militia) would have done was kill all the hostages in a bloodbath of innocent lives. Several plans had been devised to raid the militia camp, but somehow they had anticipated their plan as if reading their minds. Previously when he raided, the militia camp had been left deserted, devoid of human life, a ghost town of abandoned tents and worthless weapons. So Major Ashraf had kept his operational plan to himself this time, a secret locked in his chest. Only he had possessed the knowledge of the final plan.

Major Ashraf peered through the night-vision binoculars as he caught a whiff of smoke in the jungle. He grinned to himself as he spotted the smoke. He had guessed that spot earlier, and this had confirmed it. Major Ashraf had signaled everyone to halt using the army's field signal. He had to brief the operation plan to the selected ones in charge. Along with him was Congolese Army Commander Ilunga Pol. This squad of FARDC was their country's special forces commandos. That fearless squad had taken part in more than 30 operations. Then, Major Ashraf shared his plan with everyone. The suspected militia camp would be raided. First, he and Captain Ilunga Pol would return after ground recce (reconnaissance). Then, they would await Major Ashraf's signal after the team selected for the raid positioned themselves in different strategic positions.

Major Ashraf and Captain Ilunga had proceeded with utmost caution. It had been about a kilometer away but felt like 'walking the path for a thousand years.' They had reached near the militia camp. A little depression amid the dense jungle resembled a dried-up pond. That had been where the camp was. Then, they had seen what had made their eyes sparkle: a small open flat area. There had been five poles there. Each one had been tied up backward with abducted foreign hostages. And next to it had been something like a lump of flesh with black stains all over it. Ilunga tried to communicate something by gesturing toward that lump of flesh, but he (Major Ashraf) had not understood. But seeing his (captives) restlessness, Major Ashraf had quickly spotted and determined the raid's location and crawled away. Anything could happen that would have led the whole plan to be exposed to the militia, so Major Ashraf did not take the risk, and, finding no other choice, he crawled away.

After reaching a safe distance, what he had heard from Captain Ilunga made his eyes widen in horror. The thing that had looked like a lump with numerous black stains were millions of feverish and demoniacal ants and the Ada village chief Maguro lying like a lump of flesh! And the full moon would appear in the sky in a few moments. What would happen to them then? Hearing that, Major Ashraf had felt a cold stream running through his body. Hence, there was no time to waste. The lives of six innocents were in their hands.

Major Ashraf had instructed his Bangladeshi Special Force comrades to flank the enemy from the one side, while Captain Ilunga ordered his men to position themselves on the other. Then, both Major Ashraf and Captain Ilunga advanced toward the target area. They swiftly communicated their plan of action to each other. Major Ashraf and his BANBAT team would make up the main body of the attack, while Captain Ilunga and FARDC commando team were supposed to ensure the covering fire from a distance. Though Captain Ilunga yearned to be in the thick of the battle, where the bullets flew like angry bees, Major Ashraf positioned Sergeant Vijay to stay behind with four others in the holding party and rejoined the position with FARDC in the cut-up party. Then he returned to the main body for the final touch approach.

The evening clock had struck seven twenty. The operation was set to commence at the half-hour mark with Major Ashraf's signal fire. Everyone was on edge, up and ready. The unit had practiced raiding scenarios countless times in night and summer training, but this was their maiden real operation. Moreover, this was not a mere raid with hurdles but a full-fledged rescue mission. On the flip side, this was a walk in the park for the FARDC members. Two minutes before the execution, Major Ashraf glanced to his right and noticed Captain Ilunga and 8–10 more others had arrived and were positioned beside him. With Ilunga's broken English and gestures, he tried to signal to the major that they, too, would have loved to join the main body. Major Ashraf felt happy admiring Ilungu and the team for their bravery. Besides, they were seasoned veterans of jungle warfare. The reason they were excluded from the action party was the language barrier. If they misinterpreted his commands, the whole plan could fall apart, he had reasoned, and that is why they were not part of the initial plan to be in the main body of the attack. Not wasting any time, the major flashed a thumbs-up and waited. The full moon was glimmering luminously in the sky at that very moment.

The silence of the night was broken by the first bullet that flew from Major Ashraf's SMG at exactly half past seven. A wave of terror swept, and panic fell on the whole area with the sound of bullets. For three minutes, the bullets rained down like a storm. The militia members were caught off guard by the sudden attack. They had no time to resist or fight back with a counterattack. Some militia members fell right on the spot where they were shot down, but most of them ran away. They fled in all directions, leaving behind their fallen comrades. The smoke obscured everything from sight. The only thing that lingered was the smell of gunpowder.

That night, all six European journalists and UNDP workers were rescued from the clutches of their captors. They had also found Maguro, the village chief; he was saved in a highly critical condition, barely clinging to life. They rushed him to the UN level-one hospital with utmost care after they managed a special consideration letter for him. But his condition had deteriorated rapidly. He had been transferred from level one to level two, but the possibility of staying on this earth was getting narrower; the doctors could not keep him there, even in the level-two hospital, and referred him to level three in Kinshasa for advanced treatment. The doctors pronounced that he would make it. Yes, he might survive his wounds, the doctor reassured. I also had hoped that he would make it. I also felt that this brave village chief, who grew up fighting adversity, endured unimaginable pain, possessed a strong character, and was infinitely courageous, could not end his life story in such a tragic manner. He did not deserve such a tragic end.

What is a Country?

Major S.M. Saidul Islam
(United Nations Mission in Sierra Leone - UNAMSIL)

After almost thirty-six hours, Abdul Rahman finally lifted the tarpaulin and glimpsed the sky. He had not had a glimpse of the sky since he had left Baydoya at dawn the day before. The sun was shining bright outside. A gust of sea breeze blew through the chinks of the tarpaulin that was getting thrashed on the outer portion of the covered truck. Due to the extreme heat of summer, he smelled the sticky odor of banana and tarpaulin in his nostrils. He was not breathing properly. He breathed evenly, taking long pauses, and he did not dare remove the lid and expose himself that way entirely.

Mohammed Osman had cautioned him multiple times about not exposing himself prior to his departure from Baydoya, "Don't even think about showing your head when you get to the port. There's tight security, strict checking, with Bangladeshi peacekeepers guarding the gate. No one gets in without a card, not even a bluebottle or mosquito."

He didn't need to lift his head anyway. He would just sleep through, Abdul Rahman thought. He had been planning this for three months. He had to pay Mohammed Osman 25 dollars in installments and was not going to give up for such a trivial reason.

He had slept through most of the 400 kilometers of road. He had chewed plenty of Khat before boarding the truck to stave off hunger. Khat was a kind of tree branch, similar to grass. Somalia was the place of Khat, as it used to be widely produced and exported to many African countries. It induced a pleasant euphoria, especially when accompanied by liquor tea without milk and sugar. Khat had quelled his hunger well enough, but it took his sleep away from his eyes as well. He later dozed off on the truck's swing and his slumber was interrupted now and then when the truck traversed along the hilly trails. The truck had halted its journey twice on the way, as he had sensed it half asleep. But he did not feel like lifting his eyes to open.

His neck was placed in the crevice of the banana box, and he dreamed occasionally. In his dream, he had a vision of walking hand in hand with 'Ifrar' (his lover) on Baydoya's banana garden.

Another time he was jolted awake by a dreadful terror, a nightmare in which soldiers of SIAD (Syed) Bari had stormed in with rifles. He had felt a wave of upsurging emotion in his heart as his old mother's face popped up, the scenes of his brothers' death and his sister's watery eyes. But nothing could turn him back. No image could take a permanent place in his mind. Nothing! "This abattoir is not my country," he convinced himself whenever he thought of his country, his eyes shut with hatred.

> To him, a country meant:
> Up thrusted-upended bayonet of the Seven Clans,
> Barbaric rule of Syed Bari;
> The vein vaunting of Ali Mahdi,
> The havoc of Farah Aideed's bandit force - the thug army
> And the fleshless frame of the famished folks.'

Living in that country meant no less than facing death daily, at least for him. He thought it was better to run away, and he had chosen to flee, regardless of those who called him a coward. Suppose anyone considered him a coward, there were no issues. He felt fine about letting them do so. Nothing mattered to him anymore, let alone their opinions of him.

He had a connection, as he occasionally moved from Baydoya to Bangla Camp and back and ran errands for the officers. He would also bring supplies and intelligence about the guerrillas. He would earn a tip of one or two dollars for his services. That was a fortune in those days. One dollar is a lot of money now. Before the war, a dollar was equivalent to four shillings. Now (at the time of writing) a dollar could be exchanged for four thousand shillings. Thousand-shilling notes were common, but one- or two-shilling coins were so rare that foreign collectors or enthusiasts would not hesitate to pay 500 shillings to buy them.

In the course of discussing the intricacies of selling and buying coins, Abdul Rahman had the pleasure of meeting a camp officer of the Bangla camp, who became his acquaintance. When Abdul Rahman shared his ardent desire to leave the country, Captain Masud had tried to dissuade him from his wish to emigrate. He expounded on the significance of not departing one's homeland, He tried to convince Abdul, confessing that the act itself could be construed as a betrayal.

He articulated how deeply he missed his country and yearned to return to its embrace.

"How can you say that? Who the hell abandons their country? Look at me, I'm stranded so far away! Do you have any idea how I feel when I remember my homeland? It feels like it would have been better if I could have responded to my urge to escape," added Captain Masud.

In response, Abdul Rahman echoed his sentiment, stating that he, too, longed to embark on a new journey. "I have the same urge," replied Abdul.

A smile adorned Masud's countenance as he proffered an analogy that posited his departure and Abdul Rahman's as equal. Masud chuckled and said, "Your situation and mine are not comparable. I long to return to my country. I long to reunite with my parents. In contrast, you want to escape from them."

Rahman was disheartened by Masud's words, 'You, sir, do not understand my boblem (problem). This Somalia is not my country. I pom Yemen. You see my skin, it is more fair like you, Bangla." Abdul Rahman was well accustomed to that kind of English. He had parted ways with Abdul Rahman after a lecture of a few words on patriotism.

Abdul Rahman had served in the army for two years during Siad Barre's time. He had also fought in the war in Ethiopia. A war frenzy then gripped the entire country. Even the girls had donated money to the army fund, the very money they earned working so hard in the cornfield. Siad Barre had tried to attract or convince the Somalian people by calling upon them, i.e. promising them to have faith in him in regard to regaining their ancestral pride. The Somalis had no respect for a fallen leader. They did not take a defeated hero. He had been ousted from power after losing the war. That's when Farah Aidid took over. She had declared, "The vague promise of regaining the ancestral pride was nothing but Siad Barre's last resort to hold on to power."

Abdul Rahman was retrenched from the military after the war. The leader of his clan's militia wanted to take Abdul Rahman into his group. But he, who had lost three of his five brothers in the war, would naturally hate war. He had escaped the horrors of Mogadishu and sought refuge in Baydowa. After he fled away, the militia took his father, Abdul Mottakin, away. He had been living in exile ever since.

One day, as he was working on the banana farmer's banana garden farm, he had a flash of inspiration as an idea struck his mind. The bananas were transported from Baydowa to the port of Mogadishu, where they were loaded onto ships bound for export in foreign lands. If he could somehow sneak into a banana crate and board a ship, he could leave behind the traumas of Somalia. However, this idea did not captivate his mind for long. One day, he had shared that idea with Ifrar, but she had shown no interest and would not even try to understand why he would think that way! She clearly told him, "If you have such greed in your mind, you can forget about me right now."

Abdul Rahman had tried to make her see reason. He endeavored to explain with more effort, "We will die here sooner or later anyway. But I'm not afraid of that. My country is Yemen. My forefathers were kin to the Prophet."

Ifrar showed her fury that time with a scowl on her face, "This is your country of birth; you were born in this country, but your kinship to the Prophet will not avail you now. Do you heed any of the Prophet's teachings? All I see you do is marry four times, usurp your wives' wealth, and engage in strife."

He was familiar with Ifrar's scowl but not with those words. The girls had changed after Sayyad Bari had amended the inheritance law and proclaimed equal rights for daughters in their father's estate.

He had no headache about the distribution of property. They had become fragile anyway due to war and disaster. He had lost four of his five brothers in war and violence; only his sister remained among the survivors. He had intended to give her half the estate even if there was no law. And if he could leave this country, he would take his mother and Ifrar with him someday. Then his sister would inherit everything. These thoughts had resurfaced in his mind after speaking to Ifrar, but they were overshadowed by Ifrar's tight-clenched throat and the tears she carefully concealed in the corner of her eye. He had put aside the notion of leaving this country and them behind for a while.

Life in Somalia had turned tyrannical after the previous month's war at the candy factory. The Americans had turned into frenzied beasts. They had gone mad like rabid dogs. They had begun rounding up people in Mogadishu earlier. Now that wave had reached Baydoya as well. How could one survive in one's own land by evading foreigners' evil eyes? A voice kept nagging at Abdul Rahman's mind, "Why do you grieve so much? What is there to grieve? Your forefathers came here from Yemen. You didn't have to pay a lower price for that. Did you pay any less? Is this country yours, Abdul?" One day, after leaving his work at the banana plantation, he found Mohamed Osman alone on the shore.

Mohamed Osman had a good rapport with the banana traders. The truck drivers knew him, too. Osman had not agreed at first. Later, he had pocketed Abdul Rahman's wad of cash. The haggling had begun with 10 dollars and settled at 25 dollars. It was almost one and a half months' salary for Abdul Rahman. Osman had calculated that he could get a Kalashnikov (rifle) with that money. That was much better than a camel or a cow. He had to take care of the animals, feed them, and water them. And a gun, just lease it out and count the money every month. Osman's job was only to load him on the truck, and the deal was that if Abdul Rahman got caught at the port, Osman would not acknowledge that he knew him.

Abdul Rahman's stomach was growling with hunger because the Khat had ended its course. He could have grabbed a banana from the crate if he wished, but he was annoyed by the odor of bananas that had dried up after being exposed for an hour and a half. The port was relatively quiet. Abdul Rahman mustered courage and poked his head out of the tarpaulin. He tried to locate where his truck had halted. He was familiar with the container terminal of Mogadishu port fairly well. He had worked as a security guard there before enlisting in the army. There were crowds of people when cargo was being loaded on the ship and occasional throngs in between. He could slip away in one of those gaps. He had learned that there were many hiding spots on the ship. He would be safe if he was not caught before the ship departed from the port. He also recalled the small shops next to the container terminal. There was not so much hassle before. He used to eat at those shops during his work breaks. Rice, spicy fish, tea, biscuits—everything was available there.

His hunger intensified as he recalled all this. Abdul Rahman scanned the surroundings cautiously; the trucks were arranged in rows, and there was no trace of anyone. Suddenly he caught the sound of a big drum in the breeze. It was not distinct, but he knew the sound. He had learned it during his stint in the army. The band that plays during the march, that big drum. That indicated to Abdul that the guard was not very strong that day. There would not be much hassle once he got into the port. There were so many trucks and rows of containers, so hiding there did not seem hard. He looked around a little more and leaped off the truck with a thump.

The port was almost empty. Their truck had parked at the south end of the port. As far as he could remember, the food shops were on the north end. That meant he had to go north through almost the entire container terminal in search of food. There was a risk of getting caught, but he had to take a chance. He slowly started moving north, tiptoeing. As soon as he got a little further, he spotted an Irish soldier with a rifle slung over his back sitting on a campus stool with a book on his lap. He could not see anyone else around. Abdul Rahman swiftly moved to the side of a container. The soldier glanced up for a second and then immersed himself in the book again. Abdul Rahman waited for a few more seconds and then advanced again.

After a while and advancing a little more distance, Abdul got caught in the clutches of a Somalian police officer. He was not afraid of the police. He knew that Somali police would make a fuss, but money could make them do anything. Even the police could arrange for him to get a pass into the port if feasible.

He frisked Abdul Rahman and found nothing in his pocket except some banana peels. Abdul Rahman chuckled to himself. The war had made him shrewd and smart, clever and cunning. Nobody would discover anything by searching his body as he had stashed the money in a secret pocket. That very police officer revealed the reason for the lax security at the port.

The Bangla camp was having the final rehearsal of the model parade. Many of the local Somalis were invited to the camp; those who were not went as spectators.

Bangladeshis were not like others; if one could get into the camp once, regardless of anything, they at least got food for nothing. Acknowledging that Abdul Rahman was a vagabond, the policeman almost let him go, "Tell me, if you had any motive for stealing, how did you enter here?" Abdul Rahman had begun his sorrowful tale, saying that he had come to see his mother in a Banana truck from Baydoya; the policeman's mind might have been slightly moved by pity. The police officer's voice had softened a bit. Making the best out of that chance, Abdul managed to reach a settlement with the Somali policemen in regard to the port pass.

The policemen also gifted Abdul with the information he needed most—two ships were supposed to depart from the port that day. One of those was a Kenyan ship heading for Mombasa, and the other was bound for Dar es Salaam in Tanzania. He had no interest in boarding the Kenyan ship. It was a neighboring country, but the situation there was not ideal, as that country was struggling too.

On the contrary, Tanzania was a poor country, but he thought once there he might be able to avail himself a chance to move to another country using the Dar es Salaam port as a layover. In fact, it would have been best if he had found a Bangladeshi ship, he thought. He had had the opportunity to mingle with people from many countries since the United Nations had arrived in Somalia. No one was as friendly as Bangladeshis. The police officer had assured him that he could arrange a pass for Abdul Rahman for 10 dollars and he could get food at the Bangladeshi camp on request, but he would have to share a portion of that food.

Abdul Rahman delayed no longer; soon after, he hastened towards the Bangladeshi camp. A short distance from the gate of the container terminal was the camp. The sound of the big drum was gradually becoming distinct. Along with that, a familiar melody of a song was reaching his ear but was entirely clear. It was tapping on his eardrum once in a while, Sumalie to su' (Ja go Somalia) – 'Somalia is good', then it was drifting away in the wind again. He had learned a bit of that song from his mother before he enrolled in school, 'Sumalie to su, to su squdir sadaye.' When in his mother's voice the song elongated, and his mother sang ' Sumalie to su - Somalia is good', it was as if all the blood would gather on his face, as he would brighten up. Abdul Rahman felt as if everyone in Somalia was waking up to his mother's voice. His mother would say "this is our authentic song." Rahman did not comprehend much then, but his heart would leap when he heard that song. He would start dancing by shaking his hips. Not only at the school assembly but also at Eid-e- Miladunnabi (SA) and the Somali National Day parade, everyone would sing this song. In short, this song meant a day of celebration. Only on special occasions was that song played or sung by all; that's all he knew.

As hunger gnawed at him, he found solace in the song. The more he got weak or troubled for food, the more refreshed he felt listening to that song. His mother's face flashed in his mind. He longed to sing the song once. But he chided himself, grumbling, "This is not fair. Why do you long for this song that much, Abdul? Is Somalia your homeland?" But the melody lingered in his head. He imagined going to the Bangla camp and joining everyone in singing this song once, "Today or tomorrow, my departure is written, my relationship with this country will end. What harm will it do if I sing this song once more for one last time?" Of course, there was a risk if someone he knew saw him! If he ran into his mother, she would not let him leave the country. He dismissed the thought on the spur of the moment; how could his mother be there? The urge to sing grew stronger. He quickened his pace.

The big drum's sound grew louder and clearer. He had learned to march in sync with the sound of drums during his army days. That rhythmic patterned sound used to stir his limbs into motion and make him move with the rhythm.

But Abdul Rahman subconsciously started running towards the camp without even realizing what was happening to him and what was awaiting him. The road to the camp was adorned with flags of Somalia and Bangladesh. A crowd of people gathered a short distance from the camp's gate. But the main event was in the open space right in front of the camp. A gallery of wooden planks was set up on the mountain slope behind. There, too, people thronged. Hundreds of people. "Had the whole city of Mogadishu come here?" Abdul Rahman questioned himself. Abdul Rahman thought that if he could only reach the Bangladeshis, he would get some food. If he could get a bit more, he could take it aboard the ship. There was no hope of finding any food on the ship before departing the harbor. If he was caught, his days of planning would go to waste. He quickened his pace unwittingly. He had barely reached the gate of the Bangla camp when the darkness of his surroundings enveloped him. The ground beneath his feet suddenly gave way.

Abdul Rahman had no recollection of how long he had slept. He was even struggling to fathom whether he was in sleep, somewhere else, or in a dream. He couldn't be sure if it was sleep or something else he could not resist. The room he found himself in was unfamiliar to him. When he opened his eyes, he saw the face that had bent over him; it was his mother's. But how had his mother come there? He heard another voice. "He's awake, in a sense," the voice reached him from the direction of his feet.

His mother's face came closer, "My beta (son) has got his sense back! Alhamdulillah," she whispered as she lowered her face. A teardrop rolled down from her eye and dropped on Abdul Rahman's cheek. Abdul Rahman tried to get up to sit.

A Bangladeshi soldier nearby ran over. He said, "Stay as you are. Do not try to move. You have nothing in your body. Just nothing! Tell me, how long have you made it without food? How many days have you not eaten anything?"

He heard his mother's voice again, "Where have you been, my precious one." Abdul Rahman's calculations began to crumble.

He saw another familiar face. It was none but Captain Masud. With a soft, gentle smile, the captain asked, "Is ka warn? Waktah ji kama arang." That meant, "How are you? Long time no see."

Abdul Rahman couldn't utter any words. He sensed his emotions, his heartbeat getting lower. With innocent-looking laughter on his face, he stared at Masud and uttered, "Captain sir, speak my language!"

He himself had taught that language to Masud. Yes, Captain Masud learned that language from Abdul. Masud asked, "Where have you been? Where did you disappear for so long? And why were you running like a lunatic? You were lucky your mother came to rehearse the song at the rehearsal session, and if my sentry missed you, you would have been in trouble, maybe already to the afterlife."

His mother's eyes welled up again. She cradled her son's head in her lap and sat down, "In the afterlife? How would you leave this old woman and go? Is God so merciless, so cruel?"

The reply that welled up in Rahman's heart went unheard, "Not in the afterlife; I could have escaped this country at least." Meanwhile, he embraced his mother with both arms. Then he pleaded, "Mother, sing that song once! Somaliye to su."

His mother blushed, "Stop it, my crazy boy. There are so many people here."

Abdul Rahman insisted, "Let it be! No issues."

She began in a soft voice, "Su ma liye to su..." That melody moved everyone in the medical inspection room. As he gazed at his mother's face, Rahman's eyes filled with tears. He felt that his grief-stricken country was spreading the song of awakening everywhere.

She started singing softly, "Su ma liye to su..." That melody touched the hearts of everyone in the medical inspection room. Tears welled up at the corner of his eyes as he kept his eyes on his mother's face. He felt that his wounded sorrowful country was singing the song of awakening to the world. And he knew, at that moment, that he had found his home.

Zone of Confidence

Lieutenant Colonel Shafiqul Islam
(United Nations Mission in Ivory Coast - UNOCI)

On a moonlit night in 2004, amidst our winter training session, we gathered inside the tent to enjoy our dinner. As the youngest officer, I occupied a seat at the far end of the table. Suddenly, out of the blue, the general officer commanding (GOC) made an announcement, "Shafiq, heartfelt congratulations on being selected for the UN mission with your Parent Unit."

At that time, I was serving as the aid-de-camp to the GOC of an infantry division. I distinctly remember that during the winter training session, the division was stationed outside the cantonment in a makeshift encampment alongside the headquarters, where the GOC was also present.

Upon receiving the news of my selection for a United Nations mission, a whirlwind of emotions engulfed me—excitement, apprehension, elation, and even some doubts. Considering the country of deployment was Ivory Coast, it was natural for some underlying concerns and qualms to emerge.

However, the unit I was assigned to, my parent Unit-1, filled me with immense joy and pride. This could explain my heightened excitement and happiness. As the night progressed and dinner came to an end, I found myself immersed in a sea of emotions, oblivious to the passage of time as I absorbed the impact of the good news and received well wishes for the upcoming journey.

The following morning, I called the new unit commander, who warmly congratulated me and provided swift instructions, urging me to join the unit as soon as possible.

Upon joining the unit on time, the chief assigned me to the deployment cell in Dhaka.

As I embarked on this new chapter of my military career and became part of the deployment cell, my responsibilities involved coordinating various logistical tasks. These included flight planning, coordinating every aspect of flights, loading and unloading aircraft, accounting for personnel, issuing necessary supplies from the Central Ordnance Depot, securing and receiving goods, and local procurement, among others. As I engaged with these tasks, my cherished aspiration of being on a United Nations mission was on the verge of realization. Finally, the waiting period ended, and the last battalion flight to Ivory Coast was about to depart from home.

It was July 3, 2005 when I embarked on a United Nations-chartered moderate-sized Boeing from Dhaka. As I bid farewell to my family, including my mother, sister, and dear relatives, at the airport, I couldn't help but feel the weight of my greater responsibilities. Lost in contemplation, I drifted into a light slumber, only to be awakened by the presence of the air hostess delivering a meal tray. After finishing my meal, I engaged in conversation with my comrade seated beside me, discussing the thoughts that had occupied my mind about Ivory Coast for some time.

During our conversation, I learned that a United Nations peacekeeping mission had been initiated in this war-ravaged country through a peace agreement in January 2003. The primary objective was to bring an end to the civil war that had engulfed Ivory Coast. The conflict pitted the government forces, representing Ivoricos or the Ivory Coast, against rebel forces controlling the northern part of the country. Ivory Coast, historically known as 'The Island of Elephants' or 'The Home for Elephants,' derived its name from the ivory trade that once flourished in the region.

However, peace remained elusive in the region until 2017. It was not until that year that an official peacekeeping mission was successfully completed, gradually restoring some semblance of peace. To facilitate the peace agreement in 2003, the country was divided into two parts. The Zone of Confidence (ZOC) was established in the middle, with the Ivorian government controlling the southern part and audaciously fearless rebel forces occupying the northern part.

The main task and primary objective of the United Nations Operation in Cote d'Ivoire (UNOCI), a highly critical peacekeeping mission, was to maintain complete and unwavering control over the ZOC. This was crucial to preserving peace throughout Ivory Coast and preventing direct confrontations or clashes between the two warring factions situated to the north and south of the ZOC. The mission aimed to foster stability and security in the region.

On a rainy autumn afternoon on July 4, I arrived at the international airport of Abidjan, the capital of Ivory Coast. Despite being a country plagued by war, the city's impressive communication systems, intricate infrastructure, and captivating artificial beauty have earned it the moniker "Paris of Africa." Surrounded by towering skyscrapers, the city exudes grandeur and charm. We spent the night at the transit camp in Abidjan before proceeding to the battalion headquarters the following day.

On July 5, we embarked on our journey from Abidjan to the city of Zuénoula. Ivory Coast was nearly eight times larger than Bangladesh. Outside the capital, dense construction sites were scarce. As we ventured beyond the city limits, the landscape transformed into one devoid of towering structures. Instead, we observed small villages or "Fla's" scattered every few kilometers, giving the terrain a picturesque appearance from above.

While the main roads within the city facilitated communication, the roads in the villages were in poor condition, exacerbating the situation during the rainy season. It seemed that both literal and metaphorical turns occurred during these rainy days, affecting those involved in various businesses. Additionally, the main roads, which were wide and straight, served as transportation routes for 18/32-wheeler trucks carrying large pieces of lumber and posts from different areas destined for the capital. The value of this lumber was substantial, as it was highly sought after and essential, attracting the attention of colonial powers seeking to profit from its trade.

As I continued my journey, a veteran driver shared additional insights with me, enriching my understanding of the region. He informed me about the vast stretches of abandoned yet fertile land along the route. Surprisingly, many people had chosen to abandon agriculture altogether or showed little interest in pursuing farming. In contrast, it was the women of the villages who emerged as the true life force, shouldering household chores, nurturing children, and working in the forests and fields with remarkable efficiency. They excelled in various tasks, including domestic responsibilities, outshining their male counterparts in every sphere.

Regrettably, the men of the villages seemed captivated by indulgence, seeking ease and luxury. Night after night, they whiled away their time at local bars, reveling in drinking and merrymaking, seemingly oblivious to the labor and toil undertaken by the women. These trifling acts of natural beauty played out against the backdrop of Ivory Coast, a land of stunning elegance. The lush greenery that enveloped the landscape was simply awe-inspiring. To the south lay the serene Atlantic Ocean, while to the west stretched the rugged peaks of mountains and the verdant plains of the Midwest, boasting abundant resources such as diamonds, gold, uranium, and coconuts. Despite this earthly paradise, the political situation remained volatile, with the government and rebels locked in a tumultuous battle over these very resources, creating an atmosphere of extreme unrest.

As our conversation drew to a close, we arrived at our Bangladesh Battalion (BANBAT) headquarters in Zuénoula. Although considered an important city in the area, Zuénoula could be likened to a small sub-district or Upazila (Thana district) of our own country. It appeared somewhat disorganized and unplanned, its contemporary state tinged with an unfortunate and surreal aura. Perhaps, the ravages of the civil war had left their mark, contributing to the city's current predicament.

Our battalion headquarters were established in an unfinished bungalow that belonged to the local mayor and located on the outskirts of town. A large perimeter with a barbed-wire fence surrounded the area, and a substantial trench had been excavated beneath the fence line for security purposes. The attack on our battalion headquarters by rebels just two years prior, in 2003, prompted the implementation of necessary measures to prevent the enemy's direct entry, including the deployment of armored personnel carriers and fortified vehicles ornamented with weapons. Barricades were strategically positioned around the perimeter, and multiple checkpoints were in place within the Zone to ensure four layers of security.

Upon arriving at the entrance of the battalion headquarters, I observed armed guards diligently carrying out their duties with unwavering vigilance. The complex itself featured four watchtowers positioned at each corner, where sentries kept a constant watch during the night. In the center stood a two-story building, with the office complex situated on the ground floor and senior officers occupying the upper floor. The remaining officers were accommodated in nearby UN CORIMECs, small but well-equipped rooms. Our brave soldiers resided a short distance away in tents located to the north of the playground.

This fortified headquarters served as the primary coordination center for the battalion, overseeing seven or eight different camps. Whenever a soldier arrived in Zuénoula, they had to check in at this coordination center, which served as the central hub for all the encampments in the region. There, they would receive instructions regarding their deployment to the next camp.

As a young and ambitious officer, I was informed after two days of arrival that I would be deployed to a camp called "Bonfla" situated within the ZOC. The term "fla" denoted a village, and Bonfla was one such rural village within the ZOC. I was aware that the ZOC presented numerous challenges and held adventurous moments awaiting every second. Rumors of the region's difficulties abounded, enough to send shivers down one's spine. With a sense of trepidation and unease, I returned to my room, contemplating the forthcoming assignment.

However, I quickly reminded myself that as a peacekeeper in a war-torn land, feeling a modicum of fear was a natural response. I resolved to stay as brave as possible to combat the evil plaguing the world. I questioned myself, "How can a peacekeeper who has come to rescue the innocent from misery be afraid of the consequences?" I answered my own question, "Never, not ever, not anymore."

The following day, I bid farewell to Zuénoula and embarked on a journey toward Bonfla, the village to which I had been assigned. As I traveled through the serene countryside, I encountered vast empty fields, lush forests teeming with majestic trees, lines of herbs, and scattered homesteads every few kilometers. I was fortunate enough to catch a glimpse of the daily culture of the people residing there, witnessing vendors selling their goods. Some offered fresh fish and succulent cuts of meat displayed on tables by the roadside, while others cradled plump rats ready for sale.

As I continued on my journey, I noticed vast stretches of uncultivated land, creating a bucolic scene that extended as far as the eye could see. The road I traveled on was well-maintained, wide, and straight, designed to accommodate heavy-duty vehicles such as 32-wheeler trucks or long-covered vans. It was apparent that the road had been constructed with these purposes in mind.

The speed limit on such roads was high, ranging from 120 to 150 kilometers per hour. However, our driver was obligated to adhere to a speed limit of 60 to 70 kilometers per hour, which had been imposed as an "embargo." This restriction gave me the opportunity to engage in a four- to five-hour conversation with the driver and interpreter where we discussed the contemporary situation in their country.

The ZOC effectively split the country into two parts along its center. It spanned approximately 20 kilometers in some areas and 40 kilometers in others along the east-west axis of the country. Checkpoints were established on both sides of the ZOC, regardless of whether the main road or any other route intersected with it and encountered checkpoints set up by either government forces or rebels.

To cross from the north to the south, when passing through the ZOC, one had to navigate at least four checkpoints: one manned by rebels, another by government forces, and two by the United Nations forces.

During our journey, as we approached the camp, our driver unexpectedly came to a halt near a checkpoint set up by the local government forces. Two armed gendarmes, communicating solely in French, approached our vehicle. Thankfully, I had a bilingual companion with me that day, and their presence proved to be a saving grace. I inquired as to the reason for the extensive questioning and scrutiny, despite our United Nations blue helmet status and the vehicle's clear identification as a United Nations vehicle.

My interpreter explained that the authorities conducted thorough checks on all vehicles and collected tolls or levies from every vehicle except those under United Nations jurisdiction. Since it was my first encounter with this particular checkpoint, all my information was in their records. Overall, the process conducted by the gendarmes was meticulous and dutiful.

However, the reality on the ground was not the same for everyone. This level of thoroughness and duty might have been reserved for the privileged few. For vehicles not affiliated with the United Nations, the authorities would often accept CFAs (Ivorian currency) or other forms of illegal gratification in exchange for allowing their passage through the checkpoints. But since they couldn't extort United Nations vehicles or receive financial benefits from them, they may have expected or relied on receiving biscuits, juice, water, or assistance from other sources. If their expectations were not met, they would intentionally delay vehicles at checkpoints, subjecting those on their journey to unnecessary questioning and harassment, causing agitation and concern.

After crossing the Jendamari Checkpoint, we arrived at the BonFla camp checkpoint within minutes, covering a distance of 10 kilometers. The camp was located within the territory of the ZOC, with United Nations checkpoints on both sides. The camp itself stretched over a vast area measuring 20 kilometers in length and 110 kilometers in width within the expansive ZOC. In addition to our presence, the United Nations Police also resided in the camp, close to the main road.

The camp was situated a little further away from the village market, positioned in a slightly elevated area next to a cocoa plantation. Adjacent to the main road, our camp had its own checkpoint. Additionally, our peacekeepers had initiated a green revolution by cultivating various crops such as calabash, cucurbits, pumpkins, vegetables, and more. Although the Ivorians lacked experience in vegetable production, Ivory Coast boasted a remarkable reputation as the world's number one cocoa producer. A testament to this reputation was our southern checkpoint, Checkpoint No.1.

However, despite Ivory Coast's prominence in cocoa production, cocoa itself was not abundant or inexpensive in the region. Once the cocoa was processed, it was primarily exported to countries like France, where it served as a raw material for chocolate production. The finished chocolate products were then imported back to Ivorian cities like Abidjan, where they were sold at significantly higher prices.

Essentially, the chocolate production process in Ivory Coast was not self-sufficient, as the country heavily relied on exporting its raw cocoa material and then importing the final chocolate products at a premium. Despite these trade challenges, Ivory Coast continued to play a crucial role in the global cocoa industry, making significant contributions to the world's chocolate supply.

The exploitation of Ivory Coast's natural resources, including cocoa, coffee, gold, diamonds, uranium, and more, by colonial powers and neighboring countries was indeed a harsh reality. Unfortunately, the country's abundant wealth did not lead to significant development due to years of war and devastation. This situation further exacerbated discrimination, inequalities, and challenges within the country.

Bonfla, being a simple and humble village, managed to survive the horrors of war and struggled to provide education despite the difficult circumstances. The village had a small market, a few shops, several bars, a handful of homes, and two dilapidated schools. Education was crucial, but the colonial superpower, France, deliberately tried to restrict access to education in remote areas of its colony, including Ivory Coast. Destroying a nation's education and youth is a key factor in destroying the nation itself, and this was achieved by denying education to the people. Evidence of this could be seen in the rundown schools that lacked basic furniture and educational resources. These heinous acts were deceitfully carried out throughout Ivory Coast.

The sorry state of schools, shortage of teachers, and lack of basic educational infrastructure hindered the progress of the nation and impacted its development. However, amidst the limited options available in the villages and markets, the joy of listening to Indian Hindi songs provided some solace. Hindi music, such as the popular song "I am a Disco Dancer," could be heard frequently, and the youth and men would dance and sing along throughout the night. The market offered other goods such as clothing, kasaba (a potato-based food that served as a staple), and low-quality imported medicines. Despite the questionable quality and authenticity of these medicines, the locals had no other choice but to rely on them.

As part of the efforts to support the community and contribute to the pacification of BANBAT, medical camps were organized, schools and colleges were provided with furniture, and food was distributed to those in need with the help of the locals. These humanitarian initiatives aimed to positively impact the lives of those living in the village and provide some relief in their challenging circumstances.

With the onset of my operational duties in the ZOC, a litany of tasks fell under my purview. These responsibilities encompassed patrolling the designated area, vigilantly ensuring the absence of any hostile encounters with either government or rebel groups within the ZOC, and preventing the escalation of any untoward incident from spiraling out of control. Moreover, providing valuable assistance to various human rights associations, fostering amicable relations with the local populace, and safeguarding the soldiers from any internal disciplinary breaches or external threats were also part of my remit.

Amongst these responsibilities, patrolling the area and engaging in delicate negotiations and mediation with the rebels and locals in French proved to be the most arduous task. One particular day, at approximately ten o'clock in the morning, we went out to patrol the area in a show of force. During the course of our patrol, an elderly man suddenly appeared, racing towards our vehicle with fervent urgency and desperately attempting to communicate something to us. Despite our best efforts to decipher his message, we were ultimately unable to comprehend the nature of his plea. He tried with all his heart to say something to us, but we could fathom nothing of what he was saying.

We were accompanied by a 20-year-old local named Eve who lived near the BonFla market. Despite receiving no formal education, he worked as an interpreter for us. He spontaneously helped us (self-motivated) ensure camp cleanliness, chop wood, and fetch and distribute rations. In exchange, he received three meals a day. At the end of the month, he was given victuals, tea, biscuits, and the like.

Dark-skinned Eve, with Usain Bolt's physique, was easygoing, athletic, independent-minded, hardworking, open, and a sports lover. He loved playing football and volleyball, had a good grasp of both sports and excelled in them.

However, his greatest quality that overshadowed all else was his ability to speak Bangla.

He had learned to speak it like a native from two other battalions he had worked with previously. His Bangla had a slight southern accent. He used to say (E…. sir, Khelbe na re? Amake beshi beton debe, ami beshi kaj korbO), "Hey, Sir, will you not play? Sir, give me a raise in salary so that I can avail myself to work more for the camp." When we were preparing to return after the end of the mission, Eve said, "Sir, the last battalion made me a corporal, but you guys will make me a sergeant." (In his guileless mind, he was expecting a raise).

Ah..! How innocent and pure-hearted that boy was! What would our raise/promotion matter to him? His simplicity reminded me of the importance of grounding our pursuits of worldly life.

At times, Eve had been much more helpful than as just an interpreter for us, as he had been instrumental in making our ZOC's intricate operational activities a resounding success. His dedication and hard work were truly commendable.

Anyway, as the elderly man approached us, we talked with him and learned that his home was within the confines of the rebel-controlled zone. Some rebel group members had forcefully taken his school-going girl (15 years old) with them. Everyone knew the consequences of such incidents. First, there was rape, and then there was an extreme level of brutality leading to murder. Upon hearing of this tragic event from the elderly man through Eve, we set out to find her. After half an hour of searching, we were able to locate the girl. We rescued her. Ultimately, at the height of barbarism, they raped her and had taken her possessions but, for unknown reasons, had not taken her life.

We immediately provided her with initial medical care, administering first aid, and handed her over to the local police for further assistance and necessary measures to ensure justice. Her father was relieved to hear that she had been rescued alive. In this war-torn land, where life was precious, instances of molestation and loss of dignity were not exploited or sensationalized. Perhaps, in such a devastated country, the absence of outright cruelty was considered a significant aspiration for the common people.

On another occasion, I found myself trespassing at a rebel checkpoint outside the jurisdiction of the UN (ZOC). Unexpectedly, I came face to face with a child soldier holding a loaded, cocked AK-47, pointing it at me with his finger on the trigger. Having dealt with child soldiers before, I knew how volatile and unpredictable they could be, often reacting instinctively without hesitation.

In an effort to defuse the situation, I spoke a phrase in French, "Monami Bisqui," meaning "Biscuit, my friend" in English. By then, I had learned several French words and phrases, including "Done Moa Kado" (Give me gifts), bonjour (good morning), Bossoa (good evening), and others. I suggested that we both take a moment to calm down.

However, it quickly became apparent that this approach was ineffective, as the child soldier still kept his weapon pointed at me. That's when I asked my young colleague, Eve, to communicate with him.

Eve informed me that the child soldier wanted to strike a deal, saying, "Sir, nothing else will work; he wants money to buy cigarettes." As a non-smoker, offering him a cigarette was out of the question. Moreover, giving him dollars would have been a mistake, as it would have only encouraged him to demand money from others at the checkpoint. United Nations members were strictly prohibited from providing dollars at any checkpoint.

He was prepared to use violence if I refused to give him a dollar or a cigarette. It was evident that he had no qualms about resorting to force to obtain what he desired. However, I was determined to remain calm and find a way to defuse this dangerous situation. Taking control of the moment, I suggested speaking to his chief. The young child hesitated for a moment, his finger dangerously close to the trigger. But as soon as I mentioned his chief, fear washed over him, causing him to withdraw his finger from the trigger momentarily. Gradually, he lowered the gun, his demeanor shifting from aggressive to apprehensive. After a while, the chief arrived at the scene, and I learned from our conversation that he had previously served as a sergeant cook and had now become a rebel leader in the area.

I couldn't help but have doubts about whether any VIP, regardless of their nationality, let alone someone from Bangladesh, would receive the kind of mobile communication privileges he possessed.

The leading vehicle, positioned directly in front of the car in which he sat, was equipped with two loaded rocket launchers (RLs). Similarly, an open van trailing behind carried two rebels adorned with double cross belts filled with ammunition and armed with loaded machine guns, carefully scanning the surroundings. In the rear car, there were two more RLs and two rebels carrying loaded machine guns patiently awaiting orders.

From a distance, I noticed that as soon as he stepped out of the car, the armed guards swiftly disembarked and positioned themselves alongside him. Another individual accompanied him, carrying a tissue box. Despite the absence of scorching heat in this land of perpetual spring, there was no sign of sweat on the rebel leader or his companions. However, I couldn't help but be impressed by the level of security around him and contemplated what to say in the situation.

Suddenly, it dawned on me that he must be a Muslim. Taking into account his religious perspective and his current position, I respectfully greeted him with "Assalamu alaikum."

He smiled and happily extended his hand in response, asking, "'Bonju Kapitan, Sava?' (Good morning, Captain! How are you doing? Is everything going okay?)"

I replied, "'Sava biye, evu?' (Thanks, I am doing good, you?)" With a composed posture and my limited command of French (pidgin-creole), I attempted to convey my message, saying, "'Toa Muslim, moa Muslim, pado problem, but this child soldiers' baqubku problem.' In other words, "We are Muslims, just like you. No issues there. But let me tell you, these child soldiers are very dangerous." (These innocent children are often coerced or forced to fight in wars and conflicts, enduring physical and psychological harm that scars them for life.)

He seemed to grasp the gravity of the situation as he nodded and said something to the child soldier. From that point onward, I never had to confront another child soldier (at least not those who initially appeared problematic). However, they repeatedly requested biscuits and water from us, and we repeatedly provided them with what they asked for, considering their struggles and recognizing their shared humanity. This is how my United Nations mission in the ZOC unfolded. Throughout countless memories, sacrifices, and efforts to establish peace, we succeeded in winning the hearts of the local population. Time flew by, and before we knew it a year had passed, signaling the time for us to bid farewell. Surprisingly, the rebels, whom I had confronted and engaged with during various operations to prevent their acts of violence and oppression, organized a farewell party for me and my team.

The moment of parting was filled with overwhelming emotions and tearful eyes from both rebels and locals. The ZOC camp residents blocked the entrance gate, chanting, "No go, no go, Captain." Their love and respect touched me deeply as I left the ZOC, carrying numerous cherished memories in my heart. The last words from Eve resonated within me and left a lasting impact.

As I prepared for my journey back, Eve embraced me tightly, shedding tears, and pleaded, "Sir, please don't leave me here. And if you must go, please take me with you." Unable to ignore his heartfelt plea, I shared a few moments of tears with him, knowing that his words would forever remain etched in the pages of my treasured memories.

It has been quite some time since I returned from the mission. However, even now, I am profoundly moved when I reflect on the faces of those kind, genuine individuals I encountered in the midst of their hostile and critical circumstances.

Before embarking on the mission, I had no prior acquaintance with them. Yet, upon returning home, I felt a deep sense of longing and loss, as if I had left behind cherished companions. The memory of those simple, genuine people continues to resonate within me, serving as a poignant reminder of the beauty and resilience of the human spirit, even in the most challenging of situations.

Bloodlust at Zuénoula

Major General Md. Sarwar Hossain
(United Nations Mission in Ivory Coast - UNOCI)

'Opponent, contrariety, conflict, and dichotomy'—any of these terms may be fitting as a pervasive prefix for any social relationship, as unanimity on all matters at all times is unlikely. However, it is undeniable that when faced with conflict, adversity, or opposition, we all tend to become jittery and unprepared, whether it be in the workplace or in personal and familial relationships. We strive to transcend these conflicts or oppositions, yet the truth remains that hostility and antagonism do not arise from within. Instead of avoiding adversity, conflict, or opposition, it is imperative to master the art of combatting them, as unresolved conflicts or rivalries can lead to ruined relationships. On the other hand, resolving such conflicts with finesse and precision can foster a new paradigm of growth and progress in the relationship.

The art of conflict resolution is relevant and nearly identical in any context, whether it be personal relationships, families, large organizations, or even at the national level. Extensive conflict resolution efforts are often a part of peacekeeping missions. I had the opportunity to work on one such mission as a captain in the UN Peacekeeping Mission in Kuwait (UNIKOM) in 1993–94 and later as an officer in charge of operations at Paltan in the UN Mission in Cote d'Ivoire (UNOCI), Ivory Coast, a country in Africa, from 2005 to 2006. During my time as an operations officer in the field, I was entrusted with numerous major and critical responsibilities that I had to execute.

Despite lacking prior personal or professional training in conflict resolution, the knowledge I gained from Defence Services Command and Staff College proved to be invaluable in resolving conflicts. While I dealt with conflicts effectively in the moment and resolved them, the true professional evaluation only became apparent in subsequent events, especially during the aftermath phase. It was then that I realized the importance of learning how to professionally evaluate my performance in this field.

Although humans often oppose each other, the term "opposition" or "opponency" has broad meanings and uses. Besides "people-to-people enmity," it encompasses various other meanings and uses. Generally, "opposition" refers to any conflict between parties, such as fighting, war, or, in a broader sense, dissonance, disagreement, or a clash of interests or beliefs. Australian writer-researcher John W. Burton, in his renowned book "Conflict and Communication," stated that "during times of conflict, each party regards the intentions, values, self-interest, or behavior of the other as opposed to their own."

Throughout history, battles have been ignited by people's desire for more wealth or power. History demonstrates that the hunger for property or power sparks wars, and this desire has always been the driving force behind conflicts and casualties. In such cases, each party involved considers themselves to be "right" while viewing others as "wrong." When critically examining the formation process of any state, we find that conflict over the possession of contested property or power by a state, party, or individual forms the basis of antagonism, as they vie for ownership, enjoyment, and the right to use resources.

The situation in Ivory Coast was no exception. Apart from intra-ethnic conflicts and religious riots, the country grappled with complications such as power struggles, unrestrained corruption, the decay of state-run institutions, and the uneven and unfair allocation of power and resources among factions within the government. The country was burdened and entangled with these complex problems, leaving internal forces perplexed and without clear direction, while external forces led the nation toward further complications.

However, after a protracted wait, the situation underwent a dramatic improvement with the ousting of Laurent Bagbo from power in the 2010 general election and the subsequent ascension of Alassane Ouattara as president.

In their quest to establish peace in the volatile Ivory Coast, over six thousand peacekeepers from various nations were deployed. The majority hailed from Bangladesh, France, Pakistan, and several African countries.

Despite the formidable challenges faced by Cote d'Ivoire in its pursuit of peace, the unwavering commitment of its people, leaders, and international partners played a pivotal role in ensuring a future marked by tranquility and prosperity for the nation.

All the peacekeepers from every country operated with a shared purpose—a single, unified objective: peace!

However, due to perceptual disparities among nations regarding training manuals, differing approaches to transforming soldiers into peacekeepers, efforts to enhance the professional standards of respective peacekeeping forces, or the historical relationship between a country and Ivory Coast in conjunction with the Ivorian army, the level of rapport between peacekeepers and the Ivorian public varies. These factors must be taken into account when assessing the reasons for the varying degrees of reliance, sincerity, and purity in their relationship.

Indeed, it is true that the Bangladeshi contingents were highly regarded by the people of Ivory Coast, who fostered a profound sense of camaraderie with them. Nonetheless, locals were somewhat critical of the United Nations' role in the political turmoil that erupted in the town of Giglu, West Ivory Coast, in January 2006. Nevertheless, overall, the Bangladeshi peacekeepers were warmly received by communities and factions throughout Ivory Coast and enjoyed widespread acceptance.

Engaged in peacekeeping operations in the Western Sector, the Bangladesh Battalion (BANBAT 1) dedicated its efforts to Jenula—a site situated at the heart of Ivory Coast. Jenula lies adjacent to the "Zone of Confidence", a protected area serving as a physical buffer between warring factions, effectively separating them geographically. The activities of the BANBAT revolved around this area, ensuring the maintenance of stability and security in Ivory Coast's troubled landscape.

Noon, July 30, 2005.

At the Zuénoula Checkpoint established by Ivorian government forces, tensions escalated as local troops adamantly denied entry to a French patrol team, potentially triggering a conflict.

A war of words ensued when the local troops confronted the French patrol team, refusing their entry into the city. Captain Dempsey Frank, the French coordinating officer in charge of the patrol team, acted promptly and judiciously, attempting to establish contact with the prefect and deputy prefect of Zuénoula in an effort to defuse the situation. Unfortunately, notable progress eluded them. Instead, the decision to block the entrance and subject the French patrol team to an excruciating wait at the checkpoint for over an hour further exacerbated the already deteriorating situation!

According to the Memorandum of Understanding between the United Nations and Ivory Coast concerning freedom of movement, all peacekeepers, including French troops, were granted unrestricted access to every corner of Ivory Coast's territory. The checkpoint was located a mere 400 yards away from the resolute headquarters of the BANBAT. Prioritizing the safety of the area and recognizing the paramount importance of security, I swiftly made my way towards the site of the incident. Even before reaching the checkpoint, I observed Ivorian military personnel and French officers hastily departing for the Ivory Coast Army Headquarters nestled within the city. Hastening my steps, I followed suit.

Upon my arrival, I was determined to locate Lieutenant Uff, the local captain. Stepping into his office, I found myself amidst a group of youthful French officers. Although my command of French was limited and left much to be desired, I persisted in communicating with sincerity, utilizing simple sentences and gestures while relying on body language. Despite my linguistic shortcomings, I discovered that establishing connections, even with the most hesitant individuals, became relatively easier by conversing in their native language, French. Lieutenant Uffe shared the concerns of the town's residents, who harbored fear in response to the presence of French troops within the city. He also disclosed that the frequent movement of French troops within the city had instilled consternation among the local population.

Undoubtedly, their presence often stirred unrest. Consequently, a decision was made to restrict the movement and patrols of all UN forces, including the French, within the city, both during daylight and nighttime hours.

During the discussion, I reminded everyone, including Lieutenant Uffe, that in addition to the French troops, all peacekeeping forces possessed the freedom to independently move throughout any part of Ivory Coast to ensure regional security and peace. It was crucial to seek clarification and explanations from the relevant authorities in case of any misunderstandings. Furthermore, we agreed that the local forces would happily collaborate with all peacekeeping forces by informing the public about the work and contributions of the peacekeepers in the region. Any grievances or concerns should be addressed and resolved through effective communication, negotiation, and positive motivation.

However, it was imperative to emphasize that although occasional challenges and disagreements were not uncommon when multiple peacekeeping forces and the local army worked together towards a unified intention of "peace," any issues or conflicts must be prioritized for swift resolution. A mission for peace carried the utmost responsibility of ensuring safety, well-being, and serenity for all.

I also made a request to be notified if the local forces had any objections or concerns regarding the presence of Bangladeshi peacekeepers. I left no stone unturned in my efforts to bring everyone closer and foster unity in working towards the greater cause of peace.

After hours of discussion that yielded no substantial progress, I sought the assistance of Lieutenant Colonel Kore, the sector liaison officer of the local army. He assured me of a thorough investigation into the issue and the events leading to the situation, along with a reliable and authentic solution after consulting with the local commander of Zuénoula.

Following the meeting, as the lengthy negotiations and debates at headquarters concluded, French and BANBAT patrols returned to their respective camps from the scene.

On a positive note, it was reaffirmed during the return from the scene that the local forces had no objections or complaints regarding the movement of Bangladeshi peacekeepers and had never protested against their deployment in the Ivory Coast region.

Regardless of the circumstances, as evening fell on that day, a French patrol made another attempt to cross the same checkpoint in their pursuit to reach Zeunoula, only to be intercepted once again by the local soldiers. The French liaison officer hurriedly entered the operation room at seven in the evening, conveying this incident with an agitated voice and leaving me with the words, "You might hear gunshots, sir."

Doubting the liaison officer's behavior, I promptly informed the commanding officer about the matter. Lieutenant Uffe of the local Ivory Coast military was also present alongside the intercepted French soldiers at the checkpoint. He maintained continuous communication with Captain Alar, the company commander deployed in Zeunoula.

I urged the French coordinating officer to calmly handle the situation over the phone while simultaneously informing Colonel Kore, the coordinating officer of the Ivory Coast army at the sector headquarters, seeking his intervention. In every situation, maintaining composure and a level-headed approach is vital to advance towards peace.

At the checkpoint, the French and local Ivory Coast military stood locked in a tense standoff, seemingly on the brink of an unavoidable confrontation. Each side was firmly entrenched in their positions. The French displayed resolute determination to proceed with their night patrol into the heart of the city of Zeunoula, while the Ivory Coast military stood unwavering in their resolve to prevent them at any cost. They remained watchful, prepared to repel any French attempts regardless of the consequences.

A strategically positioned truck in the middle of the road served as a formidable barricade, blocking the path and acting as a defense against any attack from the Ivory Coast military on the French APCs. With each passing moment, the pressure intensified as both forces stood poised, ready for a potential clash.

I contacted Lieutenant Colonel Kore again, but he informed me that despite his best efforts, he had been unable to establish communication with Company Commander Captain Alar. A palpable silence hung between the two forces. The stakes were high, and tension permeated the air. The French felt the mounting pressure as they confronted the Ivory Coast military, with both sides prepared for a possible confrontation.

Meanwhile, the French coordinating officer informed me that their battalion headquarters demanded that they enter the city, even if it meant dismantling the checkpoint set by the local army. Such an uncompromising stance in a peacekeeping operation was unlikely to yield desirable outcomes.

Adding to the complexity of the situation, the recent death of an Ivory Coast prisoner named Farman Mahe under the custody of French peacekeepers had sparked controversy. This unfortunate incident led to an investigation into the French peacekeeper's commander, General Henry Posh, and two other soldiers who were accused of murder. Additionally, French Colonel Eric Bugard and two others, including an APC driver, faced charges of aiding and abetting the crime.

Considering the tense backdrop, especially in light of the criticism surrounding the conduct of young French officials, I found myself trapped in a hesitational labyrinth. I earnestly hoped that a similar incident would not occur again involving these young officers. I was determined to ensure that such a recurrence could be prevented. Nevertheless, understanding the intricacies of peacekeeping operations and acting accordingly without proper training, guidance, and motivation is a daunting task, even for the most astute soldiers. Being a soldier and being a peacekeeper are two distinct entities. Recognizing this reality and the distinction between conventional military operations and peacekeeping operations is not a simple matter.

The French liaison officer had diligently informed the capital office in Abidjan about the matter while also inquiring about the BANBAT's intended course of action. I perceived his actions and words as a form of psychological pressure exerted on the BANBAT. Without hesitation, I promptly notified the United Nations sector headquarters about the ongoing situation and urged them to gain insights into the matter.

Simultaneously, the sector headquarters was actively seeking a peaceful resolution to the situation. Anticipating the requirements of the French troops, I had already alerted multiple BANBAT patrols in advance and instructed them to remain on standby, ready to address any emergencies that may arise. When I shared our plans and strategies with the French, they dismissed the notion of additional patrols as unnecessary at that moment. Instead, they requested our assistance in sending an individual with expertise in managing tense moments, someone who could serve as a skilled mediator.

In the meantime, I contacted Colonel Kore to inform him of the latest situation at the checkpoint and urgently requested his intervention in dealing with the deteriorating state of affairs. He assured me of his intention to contact the operations headquarters and seek assistance to resolve the crisis, though he couldn't guarantee a swift response. Finding no immediate solution, I turned to the liaison officer for guidance on how else we could offer support.

Upon discussing the matter further, the liaison officer emphasized the necessity of a skilled mediator and specifically requested that I assume that role. Compelled by his insistence, I, along with one of my patrol teams, arrived at the checkpoint and met with officers from both the French and Ivorian sides. Despite my lack of prior experience in mediation, I drew upon the limited knowledge gained from my training and initiated the negotiations.

Undoubtedly, the task was challenging, but my determination to find a resolution remained unwavering. The primary obstacle as a mediator was to transform the adversaries' perceptions of one another within the conflicting groups. Mindful of this, I aimed to propose a plan that would be acceptable to both parties and that they could embrace.

Meanwhile, Lieutenant Colonel Kore found himself unable to approve a forceful crossing of the Zeunoula checkpoint against the will of the local population. He had conveyed this position to HQ Commander Tangor, making it clear to the French liaison officer that he would not permit any unprecedented intervention without the local public's acceptance. Major General Mangue, the Ivory Coast army chief, had also endorsed this approach. Reconciliation seemed increasingly elusive, presenting a formidable challenge.

Undeterred, we persisted in our efforts, repeatedly warning all parties about the severe consequences of clashes and the overall impact of these events on the establishment of peace in the region. It was only when all other plans failed that a brilliant idea emerged. BANBAT peacekeepers had gained greater acceptance among the local Ivorian soldiers than their French counterparts. Neither the local army nor the people had any objections to the movement of BANBAT in Zeunoula, as they found the BANBAT peacekeepers to be more approachable.

With this in mind, a carefully devised plan took shape. Leveraging the relative popularity of BANBAT soldiers, I proposed a joint patrol consisting of BANBAT and French peacekeepers under my guidance to be sent to Zeunoula town. This plan garnered interest from the locals as well, offering a potential path towards reconciliation.

The devised plan was expected to garner a warm reception from the locals since it allowed both Bangladeshi and French peacekeepers to cross the checkpoints together. By doing so, the locals would have granted permission for the Bangladeshi peacekeepers, and the French would not have crossed the checkpoints independently. This approach would have satisfied the French as well, as their freedom of movement was assured. It seemed that the deadlock could finally come to an end.

However, before taking any action, I deemed it wise to communicate my plans separately to the local and French officials, taking into account the expectations and mindset of both parties. This attention to detail and sensitivity to the situation proved to be crucial in the end. I further emphasized to the local forces that if they did not agree to the proposal, the BANBAT and French forces would consider bypassing the checkpoint and entering the city.

Fortunately, reason prevailed swiftly, and we received promising responses from both sides. My proposal played a pivotal role in facilitating a harmonious resolution to the conflict between the two parties, creating a dignified "ceasefire" situation. With gracious acceptance, the local community embraced the idea of a joint BANBAT-FRENCH patrol, leading to a gradual reduction of tensions on both sides. As the joint patrol triumphantly returned, circling back to the captivating city of Zeunoula, a profound sense of personal accomplishment, akin to that of a victorious conqueror, swelled within me.

ANGELS

MAJOR KAZI ARIF
(UNITED NATIONS MISSION IN SIERRA LEONE - UNAMSIL)

From the tender years of her childhood, Jessica harbored a deep fear as the twilight descended, ushering in the impenetrable darkness of night. The encroaching blackness filled her with dread, conjuring visions of nocturnal witches who would materialize beneath the colossal tree near their humble abode, their presence palpable with a heavy thud and labored breaths. Their exhalations echoed in her ears, akin to the wheezing of her ailing grandmother, who, during the nocturnal hours, transformed into a fearsome creature resembling those very night witches.

Avoiding her grandmother became Jessica's daily endeavor. Once a source of distraction with her numerous tasks and requests, the old woman, even in her relative prosperity, never opened her cloth pouch to offer Jessica any sustenance. Lately, as her grandmother's health deteriorated, she would call out to Jessica from a distance, beckoning her with the promise of food. But what could Jessica provide? Their own meals were scarce, confined to a fleeting moment within the walls of their home.

Her mother, a tireless worker whose employment remained shrouded in mystery, returned each day with provisions in her hands. The nature of her mother's work remained elusive. Jessica's heart swelled with joy at the familiar scent her mother carried, yet she also witnessed the exhaustion etched on her face upon each return. Nevertheless, whenever their gazes met, Jessica detected a glimmer of relief in her mother's eyes. The weariness dissipated momentarily as their eyes locked, leaving Jessica with an inexplicable sense of elation despite her mother's fatigue. Her mother would light the stove upon her arrival, and Jessica, with her wide and beautiful eyes, would watch her in fascination, captivated by her every move. Though her mother's business precluded ample time for joyful interactions, Jessica found solace in those fleeting moments when their eyes connected. The closest bond they shared was on that bed, and lying next to her mother was the sole source of happiness Jessica could grasp. Conversation was limited as her mother's weariness hindered extensive communication. Often,

Jessica would inquire, “Mom, do angels truly exist? Can they witness the suffering endured by humanity?” From infancy, her mother regaled her with tales of celestial beings who would alleviate pain, sorrow, danger, and disaster. Her mother had always assured her, “You shall witness it someday, my dear. God’s angels will arrive and provide all that we need. Abundant food, fine clothing, a car, our own home...” Frequently, her mother would succumb to sleep mid-narration, her voice fading away alongside the stories of angels.

In the depths of night, Jessica would rouse from slumber, tenderly inching closer to her mother and offering gentle caresses with a delicate hand. Amidst those interrupted slumbers, Jessica gravitated towards her mother’s side, reaching out to touch her, sometimes keeping a hand on her throughout the night. Eventually, she would drift back to sleep, nestled against her mother’s arm, comforted by her familiar scent. When sleep eluded her, Jessica would conjure visions of angels in the canvas of her mind, yearning for them to be as beautiful as her dreams. However, even in those waking moments, the dread-filled breaths of the lurking witches beneath the tree permeated her thoughts.

One night, gunshots shattered the stillness, followed by a chilling scream emanating from the vicinity of their dwelling. It felt ominously close. In an instant, her mother scooped Jessica into her lap and dashed into the jungle, as if pursued relentlessly by a malevolent force bellowing, “Where has that harlot vanished to? Today, her existence shall be extinguished!”

Amongst the clamor, a voice interjected with a derisive tone, “Hold on, why should we end her life? If we kill her now, we won’t derive pleasure from her as customers in the future. If we prolong her existence, we stand to benefit. We shall extract all manner of information and clues from her! Come now, let us depart! Tomorrow, we shall return to deliver the news!”

With their abductors departing gleefully, having discharged several more rounds, Mother emerged from the dense foliage. A momentary lamentation gave way to curses hurled at the intruders. Grandma muttered something under her breath. Jessica winced at the words flung in her mother’s direction and felt repulsion toward the individuals who resembled the night witches from her mother’s bedtime stories.

But what of the angels? How resplendent were they? Jessica struggled to recall their faces. Her mother had assured her of their beauty and their ethereal abode in the heavens.

Jessica, day by day, inched towards adulthood, absorbing the sounds of bullets, murder, the harsh whispers of the night witches, hunger, abuse, and her mother's desolate screams. These grotesque facets of existence became more familiar to her than the natural growth of her own being. She witnessed and suffered from the surrounding horrors more intensely than her young soul deserved.

In her time, hunger and anarchy reigned supreme. Every day, she encountered a horde of ferocious individuals armed and devoid of love, compassion, or mercy—creatures resembling the most fearsome beasts. No benevolent souls ever graced them with soothing voices, offering empathy or inquiring about the quality of their days. How did her own days unfold?

The neighborhood where they resided had long been etched into their consciousness—the dusty road, the two towering trees, the small bush nearby, the mango tree next to their humble abode, and the distant jungle. Jessica found solace in these familiar landmarks. Despite immersing herself in dust throughout the day, her body adorned with sandy specks, she always returned home in the evening, cleansed and refreshed after bathing or washing herself at the communal tube well.

She anxiously awaited her mother's return, for it brought her a semblance of peace, a fleeting sense of relief. Grandmother would grumble throughout the evening. Then, one day, she passed away silently, slipping into eternal sleep. Jessica hadn't noticed her grandmother's whining on that particular day. In fact, she hadn't even had the chance...

Lost in the bliss of play among her friends beneath the tree's shade, she returned home in the evening, oblivious to her grandmother's coughs and whimpers. Perhaps Grandma was merely asleep, Jessica thought. When her mother finally arrived, she discovered her mother's stiff figure lying on the floor. Surprisingly, her mother shed no tears, showing no signs of grief. Instead, she held firm to her beliefs—praying earnestly, respectfully burying the body—convinced that her mother would be reborn within a week. Liberation awaited her mother's soul, granted only if the final rites were executed flawlessly.

Jessica informed a few neighbors about the incident and sought the assistance of the church's father. She even ventured to Taylor's to rent black attire for herself and a small group, who would don these garments to conduct the final rites and emancipate her mother's spirit. Fortunately, Jessica's grandmother had saved some money. The following morning, her mother withdrew those savings, utilizing them to arrange for sustenance, transportation, and other necessities. She reached out to those who could potentially aid them in fulfilling their longings.

The morning brought a heavy heart for Jessica. Grandma would no longer call for her, no longer beckon her to come near. She had vanished, evading Jessica's company, even on those occasions when Jessica had ventured to her or had been occupied with outdoor play. From that day forward, her voice would no longer echo. And so, Jessica succumbed to tears, her weeping uncontrollable.

With a heavy heart, she completed the final rites alongside her mother. Inebriated, her mother returned home, shedding countless tears. The melancholy enveloped them both. As her mother clung to Grandma's bed, her cries pierced the night, reaching the moon before she succumbed to slumber. In that moment, Jessica perceived her mother as a little girl akin to herself. She sensed a maturity awakening within her, prompting her to refrain from pestering or nagging her mother that day. Drifting off to sleep independently, she abandoned herself to dreams, while her mother rested on Grandma's bed. After that, Jessica slept alone on the old one, no longer sharing the comfort of her mother's presence.

After Grandma's passing, when Jessica returned home from playing with her friends, it was just Jessica, with nobody else to greet her. The overwhelming fear of solitude gripped her each day after her playtime, as she often dozed off at the doorstep, eagerly awaiting her mother's return. At the touch of her mother, she would spring to life, enveloped in a warm embrace, her mother's voice tenderly calling her "sweet girl."

Yet, something felt amiss about her mother. The familiar scent of motherhood had vanished, replaced by a peculiar amalgamation of odors that alienated Jessica. She would wait for her mother to change clothes, freshen up, and call her for dinner before they retired together for the night. But there were days when Jessica felt helpless, especially when her mother departed before her awakening. All she yearned for was the chance to bid her mother farewell each day as she left their home. She possessed limited knowledge about her mother's profession, never pressing her for answers nor being taken along willingly.

As Jessica matured, growing old enough to comprehend their circumstances and the harsh realities of daily life, she refrained from pleading with her mother to take her somewhere or accompany her during her daily activities.

Her mother never divulged the details willingly, and Jessica respected those boundaries, understanding the importance of shielding her from the harshness they endured. Curiosity regarding her mother's enigmatic occupation persisted, but Jessica learned to cherish the moments they shared, patiently awaiting a brighter future where her mother's efforts would lead them away from hunger, fear, and the prevailing anarchy.

One eve, in the darkest hours, a drunken man stumbled upon her mother's doorstep, his presence an unsettling disruption. He barged forward, his voice slurred and venomous, hurling insults and threats into the night air.

Recognizing the voice as George's, her mother intervened, her words a gentle plea. "George, hush now. My daughter slumbers within. Spare her the disturbance. If something were to transpire between us, it would be tomorrow, never today, not here, not now. Understand?"

George, consumed by his inebriation, refused to relent. "Open this damn door, or I'll break it down. I'll invade every inch of this place. I'll have my way with you, you whore!"

Realizing the futility of resistance, Jessica's mother yielded, opening the door in silence, her resignation palpable. She spoke to George, her voice a whisper, "George, I give you what you desire when we work at the bar. Why bring it here? This environment is not suitable, not as normal as you assume. This is not the place, George."

Unsteady on his feet, George stuttered his plea, "Let me stay tonight. Let us enjoy the night together. Allow me this, oh..."

In the dim light, Jessica's mother cast a fleeting glance toward her daughter's slumbering form, confirming her rest. Leaving the door slightly ajar, she led George to her bed, joining him in whispered exchanges and muffled sounds, a cacophony of roughness that roused Jessica from her sleep, her heart heavy with aching.

Come morning, as Jessica surreptitiously observed her mother, she sensed an unease within her. Unable to decipher her mother's emotions, Jessica herself felt unsettled. That man, he was not an angel or a guardian, despite her mother's embrace. He appeared to her as a specter, a haunting presence. Therefore, she sought to avoid her mother's gaze throughout the day, avoiding the touch of those ghostly beings that roamed their home, their hands violating her innocence.

Pretending to sleep became Jessica's shield, sparing her mother the embarrassment. She longed to escape, to retreat deep into the forest, far away from these phantoms. The jungle beckoned, but her love and compassion for her mother held her back, binding her to this world of torment.

As the days passed, the presence of those ghostly creatures persisted, their calls reaching out to her, their intentions sinister. "Hey, Jessica! Come to me, baby." She yearned to escape to the depths of the forest, far from their reach, but her love for her mother kept her rooted.

As she grew, bit by bit, day by day, the horrors intensified. The witches, ogres, and dreadful ghosts descended upon their lives. The storm of political differences and ethnic conflicts engulfed their town, tearing through the fabric of their existence. Violence became a daily ritual, as opposing factions clashed without mercy.

Samuel, Uncle Robert, and John, their neighbors, bore the brunt of the chaos, their lives cruelly cut short. No one dared to intervene. Teenage girls, women of all ages, anyone deemed female, were rounded up and held captive, subjected to unimaginable horrors, while the rest of the town burned.

The flames rose high, consuming everything in their path. Homes turned to ashes, and possessions were reduced to charred remnants. Beds, pillows, furniture, belongings—all were devoured by the fire's unyielding hunger. Savings, crops, infants, and livestock—all were ravaged by the relentless blaze, leaving destruction and despair in its wake.

The dreams of life were consumed by the flames, and those orchestrating this inferno reveled in their wicked delight amidst the Holee of bloodshed and devastation. The outcome of this raging fire, this vengeful game, remained unknown. Yet, one thing was certain—the fire would not discriminate. It would ultimately consume even those who ignited its wrath.

Amidst this turmoil, Jessica could no longer catch a glimpse of her mother. Instead, she witnessed the tears of a few children on the roadside, surrounded by the lifeless bodies of acquaintances and strangers alike, regardless of age.

There was nowhere to turn, nowhere to seek refuge. The once-thriving houses now lay in ruins, reduced to ashes. Twirling columns of smoke ascended, reaching high above, where the angels were believed to dwell.

In shock and horror, Jessica sought shelter behind a small tree, but she soon realized she needed to ensure her own survival. She chose to feign death, blending in among the lifeless bodies that littered the ground.

The perpetrators of this atrocity showed no mercy, fearing any surviving souls might seek retribution. Unspeakable horrors befell the women, and even the children were not spared from the slaughter and torment. Thus, Jessica had no choice but to conceal herself, her gaze fixed upon the sky, yearning for the arrival of the angels.

Her mother had spoken of heavenly beings with the ability to soar through the air, their wings hidden from mortal view. Could no angel descend upon them that fateful day? Was there no savior to be found?

In the depths of her heart, Jessica fervently prayed for the angels to descend and deliver them from their plight. And her prayer was answered when she beheld a celestial chariot adorned with wings, soaring across the sky. The chariot thundered past the lofty shade tree, descending upon a narrow clearing. From its depths, a group of UN personnel emerged, donning blue helmets inscribed with the letters "UN."

These troops swiftly neutralized or apprehended the assailants present, liberating the captive women. Some of the angels turned their attention to providing remedies and nourishment, tending to the wounded and offering sustenance. They shared food and administered medicine to those in need.

Jessica watched from afar, captivated by the actions of these azure-helmeted angels. She questioned whether this was a dream or a waking reverie.

At that moment, one of them gazed upon her and extended a long arm, bearing the blue helmet with the "UN" inscription. Jessica stood transfixed, contemplating this sight for a long while. Finally, she inquired, "Are you an angel? Have you come to rescue us?"

The angel offered no verbal response, neither confirming nor denying her question. Instead, he cradled her in his embrace, carrying her away, her head resting against his chest.

The Scent of Biscuits and Bullets

Major Mohsin Hasan Imran
(United Nations Mission in Ivory Coast - UNOCI)

After trudging along an unpaved road pockmarked with potholes for over three hours, I arrived at my destination with a tinge of disappointment. Unlike the warm and inviting villages I had encountered in Bangladesh, this place exuded a gloomy and austere atmosphere. Seeking respite from the scorching sun, I sought shelter beneath a tree that resembled a makeshift shack near the school grounds. A group of inquisitive locals had gathered around, their eyes fixed on us with a keen interest.

This village bearing the grandiose name of Bangbo Koadikro was a destitute and secluded area in the northeastern region of Ivory Coast, approximately 400 kilometers away from the renowned city of Abidjan. The nearest city was a three-hour journey from there, accessible only via a treacherous muddy road that became impassable during the rainy season. Only a few affluent villagers owned bicycles or motorcycles, while the rest relied on their own two feet. It was no wonder that the presence of foreign combat-uniformed individuals donning blue helmets stirred curiosity among the villagers.

Having recently arrived in Ivory Coast, I swiftly learned that certain areas were on the verge of a by-election, with a looming risk of clashes. The country had endured a bloody chapter marked by power struggles and wealth distribution, which had abruptly come to an end with a regime change the previous year. This particular village had experienced the cancellation of voting due to violence, making it a crucial center with a significantly higher number of voters compared to neighboring villages. Furthermore, the poor communication system rendered it a highly precarious location.

Little did I know of the misfortune and perils that awaited us when the contingent commander assigned me the responsibility of overseeing this remote area during the morning briefing last week. The chief operations officer had separately warned me to exercise utmost vigilance and then some. It became apparent that the area had witnessed a violent clash during the previous election, with UN troops being welcomed by an attack. I shuddered as I watched a video from that election that captured a gruesome scene of individuals severing the hand of an opposition member with a machete, followed by a horrifying display of burning another opposition member alive while a crowd cheered and jeered. It was a sight that would haunt my memory.

Though the contingent commander assured me that the majority of the locals were peace-loving and innocent, he attributed the situation to a handful of power-hungry individuals fueled by foreign conspiracies. Our mandate from the UN was clear: carry out our assigned duties while prioritizing our own safety and the safety of our assets. Thus, on election day, I took charge around the polling station in that village with two pickup trucks and a contingent of 20 soldiers. The narrow and treacherous road made it impossible to deploy an armored personnel carrier. Following the planned schedule, I left the camp at eight in the morning and arrived at the destination by eleven o'clock without any unforeseen delays.

"Monami, bisqui," I heard the voice of a four-to-five-year-old girl, and I turned around. Those words struck a chord within me. "Monami" meant friend, and the little girl was referring to me as her friend while requesting a biscuit. I had noticed the locals' fondness for biscuits before. With a smile, I lifted the girl up and handed her two packets of biscuits from the car. That simple gesture was enough to forge a bond of friendship. The sweet girl began to share what little she knew in the local language, and even though I couldn't understand a single word, the gleam of happiness and gratitude in her eyes was unmistakable.

Lost in my thoughts with the girl in my arms, I realized that the election held no interest for me. It mattered little who won or lost. My responsibility was to safely transport the ballot boxes, senior UN election officials, and the innocent people who had ventured to this remote and disaster-prone area to perform their duties, back to the main election office in the city. The ordinary villagers were oblivious to the fact that Benia's party was hatching countless schemes to destabilize the country, playing with the fate of the common people.

It all started with a simple plea, "Monami, biscui biscui," and in that moment, I found a young friend in haste. Throughout the day, I enjoyed the company of the locals, aided by Romeo, the interpreter who accompanied me. Upon hearing that I hailed from Bangladesh, the villagers bestowed upon me a new name – the "Bangla Captain."

As the day drew to a close, following the completion of the vote by five in the evening, I embarked on the journey back to the city's election office with 15 unarmed UN staff members and other individuals, including a local police officer, alongside the precious ballot boxes. The villagers bid us farewell in a dignified political manner, lining up to express their gratitude. Their kindness was evident as they continued to approach us, thanking us profusely for the few packets of biscuits and dry food we had provided.

The arduous drive through treacherous mud roads, weaving through hills and forests, took us nearly three and a half hours before we reached the city's election office. Leading the convoy was a vehicle manned by ten soldiers and Senior Warrant Officer Ahmed. Following closely were two civilian vehicles carrying the ballot boxes and civilians, while ten more soldiers accompanied me in the final car.

The first two hours passed uneventfully, save for the bone-rattling jolts that reminded us of the ruggedness of the terrain. Corporal Lal Mia, the pickup driver, vehemently cursed the road, swearing that he had never encountered such wretched conditions, even in his father's time. He wasn't exaggerating—the road was truly dreadful, flanked by an oppressive jungle on either side. The absence of any mobile network compounded our challenges, leaving us isolated. The sweltering heat only added to our discomfort, with the weight of the bulletproof jacket burdening our chests.

We had another hour and a half of travel ahead, after which we would safely deliver the convoy to the election office and await the arrival of other patrol commanders. If all went according to plan, we would return to camp together. However, being cut off from any reliable communication for the majority of the day, we remained unaware of the whereabouts of our fellow officers.

Suddenly, a crackle on the walkie-talkie startled me. It was Warrant Officer Ahmed, and his voice was tense and urgent. He barely managed to utter the words "send over" before revealing the grim situation ahead. Our vehicle had become trapped by a barricade, and the circumstances appeared dire. He implored me to move forward to assess the scene.

Without hesitation, I gripped my trusty SMG and leaped out of the car, accompanied by Romeo, the interpreter, and Iqbal, the runner, who stayed close by my side. We dashed toward the front of the convoy as it came to an abrupt haLieutenant All the soldiers had swiftly abandoned their vehicles, seeking cover along the roadside. Countless rehearsals of counter-ambush drills during our training at the cantonment had prepared us for this moment, but none of us had anticipated putting our training to use so soon in a real-life scenario. It was remarkable how quickly our skills kicked in, helping us navigate through this perilous challenge.

Within seconds, I laid my eyes upon a barricade directly ahead. Behind it stood a group of armed men, numbering around 25 to 30. They created a cacophony of noise, shouting at the top of their lungs. Several imposing figures in the front row brandished their weapons, hurling curses in the local dialect as they advanced toward us. Romeo and I raised our hands in a gesture of surrender, urging them to haLieutenant I spoke to them calmly, "Please, stop there."

Meanwhile, Romeo grabbed a megaphone and recited the phrase he had memorized for this very situation, "We are your friends. We are UN peacekeepers. If there is any misunderstanding, let us talk and resolve it."

Amidst the chaos and the exchange of gunfire, our peaceful intentions fell on deaf ears. The deafening sound of bullets shattered the air, narrowly missing me as I stood there, stunned and terrified. I was in disbelief that I had survived such a close encounter with death. The windshield of a car behind us was riddled with bullet holes, a stark reminder of the gravity of the situation.

I quickly realized that the militants were accustomed to responding with bullets instead of words. Just days ago, they had ambushed and mercilessly slaughtered a patrol group, including a lieutenant from the Nigerian battalion. If we faltered or lost our courage now, we would face the same fate. In that moment, a fragment of my cadet training flashed through my mind: "Dash down, crawl, observe, fire."

Without hesitation, I dropped to the ground and sought cover along the roadside. Darkness enveloped everything, obscuring my vision. The night was a black abyss, filled with the sounds of gunfire, the acrid scent of gunpowder, and the anguished cries of pain. It was a scene reminiscent of hell itself—a truly nightmarish environment. Next to me were Warrant Officer Ahmed and Sainik Iqbal, the interpreter and civilian who had accompanied us. I had no idea if Iqbal had been hit or not. In that chaotic darkness, uncertainty reigned.

For more than 10 minutes, we exchanged gunfire, constantly shifting our positions. Everyone except me had taken cover on the road from the outset, while the terrorists boldly advanced through the middle. It appeared that the militants expected us to retreat in fear after a few rounds of fire, giving them an opportunity to seize the ballot boxes and abduct some of the election officials, just as they had done during the previous election.

I had already emptied two magazines from my SMG. Loading the third, I prepared for further action. Liakat, the LMG man beside me, continued to suppress the enemy with sustained fire. Our combined firepower provided a formidable defense against the terrorists. Gradually, I noticed the sound of the militants' bullets diminishing. They seemed to be retreating towards the jungle on the roadside while still firing. The extent of casualties on both sides remained unknown to me.

I instructed Warrant Officer Ahmed to provide cover while I advanced to clear the barricade and escape the precarious situation. Consequently, I ordered my section to follow me.

As I moved forward, Iqbal's call caught my attention, causing me to turn back. What I saw, even in that perilous moment, amused me. A wireless operator who had accompanied us was sprawled on the ground like an ostrich, trembling with fear. His rifle swung haphazardly from his shoulder as he incessantly mumbled prayers. The civilians in the convoy were scattered, seeking safety in whatever way they could. They resembled headless chickens, lying under trees or taking cover along the roadside.

Cautiously, I approached the barricade. It seemed that the enemy had given up the fight, their batteries drained. But caution urged me to ensure our safety. I secured the area and called for the other section to join me in clearing the barricade.

After moving the barricade aside, I repeatedly called the name of Romeo, the interpreter who had been with us. I couldn't abandon the civilians who had taken cover to protect themselves. It didn't feel right to leave them behind. I needed Romeo's assistance to locate them. However, there was no response. Warrant Officer Ahmed informed me that there was a mobile network in the area and suggested I try calling Romeo's number.

I retrieved my mobile phone and promptly dialed Romeo's number. My first call was to the contingent commander, where I provided a brief update on the situation. The contingent commander inquired about any casualties—whether anyone was wounded or killed. I admitted that I wasn't entirely certain about injuries, but some individuals who had accompanied me were still unaccounted for. I couldn't determine if they were dead or had managed to escape the harrowing ordeal.

After speaking with the contingent commander, he assured me of reinforcements and instructed me to return safely. I then called the interpreter, who answered after several attempts. He informed me that he was hiding in the jungle and planned to leave for Abidjan the next morning, resigning from his job as an interpreter. It took considerable effort to convince him that we were temporarily secure and that we would be leaving soon with the convoy. I explained that the terrorists had been defeated and were hiding in the jungle. I emphasized that if he chose to spend the night in that terrorist-infested area, it would be his personal decision, and he would bear full responsibility for any consequences. I also emphasized the importance of his aid in retrieving the remaining civilians, as most of them did not understand English. Eventually, he agreed and gathered all the civilians near the car.

Once the call ended, I muted my phone and tucked it into my pocket, anticipating a barrage of calls that I decided to ignore. The barricade was almost cleared, and I aimed to avoid any unnecessary risks by employing the fire-and-move technique, focusing on the safety of my team.

Meanwhile, the interpreter had regrouped the scattered people, but the police officer who had accompanied us was nowhere to be found. It was rumored that he had fled when the convoy came to a haLieutenant One civilian had a bullet wound in his leg, and a few others had minor injuries, but fortunately none were life-threatening. It seemed that quick reactions and seeking cover had played a significant role in the survival of everyone in the patrol, despite the bruises and the bullet-ridden vehicles.

We reached the city camp safely around nine-thirty that night. Upon our return, the civilian staff who had accompanied us rejoiced and expressed their gratitude, acknowledging that none of them would have made it through the day without the efforts of our team. The following morning, I discovered that four terrorist bodies had been recovered from the nearby jungle where the incident had occurred.

Two days later, the force commander summoned me to his office and requested a detailed account of the incident. He listened attentively as I recounted the events. Afterward, he commended my actions, highlighting the contrast with a previous incident where peacekeepers had pleaded for their lives while their vehicle was burned. He expressed his pride in our team's performance and awarded me a special commendation.

The force commander also informed me that journalists from the media had come to his office for an interview. The incident would be featured in various newspapers. Feeling somewhat embarrassed, I humbly remarked that I didn't consider myself a hero and believed that any other officer in my position would have acted similarly or even better.

Before I could say more, he interrupted me. Rising from his chair, he approached me, embraced me tightly, and smiled. He reassured me that I had done my best and that everyone was proud of my actions. He even made a lighthearted comment about my bonus life, reminding me to savor each moment. We shook hands, and I bid him farewell. Stepping outside his office, I was greeted by the warmth of the sun, its shimmering rays touching my eyes and face. It was a beautiful world, and the most beautiful feeling was the knowledge that I had survived with that bonus life.

Part II

United Nations (UN) Force Commanders' Reflections

In this part, we have asked the peacekeepers who were in the leadership roles and contributed not only towards policy development also in conducting complex operations with the forces from various nations. Their contributions add a rich tapestry of challenges, hopes, and fears in peacekeeping operations.

CHAPTER I

OPENING DOORS AND CREATING NETWORKS IN UN PEACEKEEPING MISSIONS

MAJOR GENERAL (RETIRED) KRISTIN LUND

Being the first-ever force commander of a UN peacekeeping force led to much curiosity not only in Cyprus, but also worldwide.[1] This curiosity opened many doors and provided me with access to both women's organizations (which often tended to be very critical towards the military) and to traditionally male-dominated organizations. It was my choice to go through these doors because my experience had shown me that taking advantage of these openings would give me access to a much wider part of the society in which I was operating.

During my two years as FC, I was invited every week to do public speaking. I found this opportunity inspiring, and it allowed me to gain valuable insights into different communities, which proved useful for helping me lead my troops. In coordination with the SRSG, I was also invited into networks that played an important role in the support of the peace process, such as the unofficial network of female ambassadors on the island who gave me direct access to the political level.

I also came into contact with the PRIO Cyprus Centre,[2] which was instrumental in giving me both insight into the multiple nuances of the conflict and access to different aspects of civil society. Having a more comprehensive understanding of the society we were operating in made it

---

[1] UN Web TV (2014) 'Ban Ki-moon, appointment of Major General Kristin Lund (Norway) as Force Commander of the UN Peacekeeping Force in Cyprus (UNFICYP)'. 12 May. Available at: webtv.un.org/watch/ban-ki-moon-appointment-of-major-general-kristin-lund-norwayas-force-commander-of-the-un-peacekeepingforce-in-cyprus-unficyp/3559758097001/.

[2] For more information on the PRIO Cyprus Centre (PCC), visit: cyprus.prio.org/

easier to understand the complexity of the problems we were aiming to address. One way I attempted to reach out to the wider society was through "the wives". It was important to get the first ladies from both sides more actively involved – not to support their husbands, but to support the peace process. First ladies act as role models for many women, and they can have an important impact on the societies within which they operate. My goal was to get the first ladies together, establish common ground, and take advantage of their existing networks to promote peace. I set in motion plans to get this process underway, but unfortunately this initiative had to be postponed due to unforeseen circumstances.

When I left a few months later, the new male FC who took over did not recognize the importance of this, and it was not followed up on. Indeed, the importance of gaining access to women was seldom understood by the men on either side of the conflict – until they saw evidence of its success. When I suggested creating a women's military network to the generals from both the North and the South, the idea was not well received. Eventually, the chief of defence in the South agreed to help set up a women's network of officers from Cyprus and female officers deployed by NATO and other troop-contributing countries (TCCs) to UNFICYP. The day after the first meeting took place, I got a phone call from the general and he was thrilled. His female officers were extremely satisfied with the meeting. They found out that they faced the same challenges related to being a minority as their counterparts in many of the nations represented in UNFICYP. Furthermore, by attending the network meeting, they got to know more about the UN in general and the UN operation in Cyprus in particular. UNFICYP staff, on the other hand, learned a great deal about the Republic of Cyprus (ROC) forces. This initial meeting led to several follow-up meetings between the female military from the ROC and UNFICYP. ROC officers were thus able to meet and exchange views and thoughts with people from Argentina, Chile, Paraguay, Canada and European countries. These networking meetings helped the officers gain a better understanding of the UN mission and led to more mutual respect and trust between the ROC officers and UNFICYP officers.

Some Reflections

The establishment of women's military networks allows officers from various nations to find common ground and helps develop mutual respect and trust.

Conflicts have a different impact on women than on men. To gain a more comprehensive awareness of the conflict situation, military and peacekeeping forces addressing the conflict should consist of a diverse group of people.

The involvement of female military and civilian leaders enables peace operations to reach a larger segment of the population and thus to develop a more holistic approach to addressing the conflict.

The diversity of a society should be reflected in the military. The UN should actively continue to encourage nations to deploy more women and list the inclusion of women as a criterion in any agreements they make with TCCs. MG (R) Kristin Lund/Peace Research Institute Oslo/PRIO/GPS PB 20-1

An Interview with the General

The editor interviewed the General in July 2023. Here is an excerpt:

> Interviewer: Thank you, General. The first query I have is related to your article's topic about networking. You've mentioned how the UN mission involves people from diverse backgrounds collaborating. Given your experience as a Force Commander, what challenges do you see in collaboration and networking among these different groups?
>
> General: Of course. One of the most challenging aspects for a leader in such scenarios is effective communication. Because of the variety of cultures and diverse ways of interpreting information, it's essential to ensure that the intended message and objectives are clearly understood. This often requires a higher level of oversight. For instance, during my time as a Force Commander, I had commanders report back to me after receiving orders. We'd review the mission's goals to ensure a shared understanding. This process, although time-consuming, is vital. When dealing with sector commanders from various backgrounds and experiences, it's necessary to verify their grasp of the mission. While this may extend the time taken to communicate instructions, it's crucial for mission success. During crises, there's no room for misunderstandings. So, while issuing orders might take longer, the significance of mutual comprehension cannot be underestimated.
>
> Interviewer: Your insights into the communication challenges are valuable. Could you elaborate on how you managed these issues, especially considering cultural differences and your role as the first female commander?
>
> General: Certainly. Cultural differences played a substantial role, particularly in a mission like Cyprus, where commanders came from diverse cultural backgrounds. As the first female commander, curiosity often worked to my

advantage. People were eager to engage due to the novelty of a female leader. This provided an opportunity to bridge gaps and establish connections. To navigate these challenges, I prioritized close communication with sector commanders. I held weekly meetings with them, in addition to interacting during patrols and other activities. Maintaining a robust and regular line of communication was crucial to promptly clearing up misunderstandings. Additionally, coming from a direct communication culture, I understood that some colleagues might need more time to adjust to my leadership style. By cultivating strong relationships and encouraging open dialogue, I aimed to facilitate effective collaboration.

Interviewer: That's insightful. Your emphasis on communication and relationship-building is clear. Moving on, considering UN Security Council Resolution 1325, which focuses on women, peace, and security, did you undertake any initiatives in line with this resolution?

General: Absolutely. Gender inclusivity was a priority during my time as a Force Commander. We initiated several efforts to promote gender equity and enhance the participation of women peacekeepers. For instance, I advocated for placing gender issues at the forefront when engaging with troop and police-contributing nations. By placing gender discussions at the beginning of the agenda, we managed to double the percentage of female military participation, from 4% to 8%, and raise the percentage for UN police from 10% to 25%. I also took measures to ensure equal access to positions for both genders. In a mission with 146 positions, I ensured that both men and women had equal opportunities. This extended to training facilities and creating an environment that catered to all. I also leveraged networks, such as women's networks in North and South Cyprus, to encourage greater gender parity. Moreover, during discussions with various national contingents, I highlighted the importance of gender-sensitive behavior and the prevention of sexual exploitation and abuse. I even considered factors like whether prostitution was legal in their home countries to address potential issues.

Interviewer: Your dedication to gender inclusivity is evident. Could you share a personal memory from your time as Force Commander that remains close to your heart?

General: Of course. One of the most gratifying aspects was bridging the gap between military and civilian staff. Through initiatives like "Inside the Blue," I aimed to foster a better working environment by bringing both groups closer. This effort required some coaxing to get everyone involved, but the results were rewarding. Witnessing the enhanced camaraderie and mutual understanding between military and civilian colleagues was a highlight for me. I believe that a cohesive team atmosphere is crucial for effective leadership. While challenges did arise, the bonds forged within the team were immensely valuable.

CHAPTER II

WOMEN PEACEKEEPERS: WIDENING THE UN'S VISION IN PEACEKEEPING

PROFESSOR DR. BRIGADIER GENERAL (RETIRED) NAZMA BEGUM

As said, 'Women are the mothers of the world, the queens of every sweet home,' I firmly believe that women possess unique qualities that enable them to do things that men cannot. As the only woman in world history to have served twice as a contingent commander and twice as a country senior in the UN (in Ivory Coast and Central African Republic (CAR)), I have proven that women peacekeepers can effectively fulfill the same roles as their male counterparts in war-affected and conflicted areas. They possess the capability to command and control troops while also connecting with local populations, making it easier for women, men, and children to feel comfortable conversing with them. Women peacekeepers have a heightened ability to discern what people are not saying, which is invaluable in these settings. By providing another perspective, they expand the UN's angle of vision.

In 2016, I made history as the first female military contingent commander in ONUCI. This achievement was a breakthrough in a male-dominated field, and it brought pride not only to me but also to my country. I successfully organized the medical unit (BANMED-12) in my home country and deployed it to ONUCI. The dedication and able leadership of my team led the people of Cote d'Ivoire to praise our medical team, raising the flag of Bangladesh high under the umbrella of ONUCI. Additionally, as the country senior military officer of Bangladesh I was responsible for commanding and controlling the Infantry, Artillery, Engineers, Signal Unit and Bangladeshi MSO and MOs. In any critical situation, I had to make prompt and effective decisions.

I also made a positive impact in advocating for gender issues during my time in Cote d'Ivoire. I set personal examples on gender issues for my unit and the communities around Cote d'Ivoire, becoming a mentor for women and encouraging them to take on challenging roles in society and serve as role models. I used my medical skills to address reproductive health, family planning, child care, breastfeeding, and antenatal care and to prevent HIV, AIDS, and malaria, particularly for female peacekeepers and women in Ivory Coast's communities. I coordinated with NGOs dealing with gender issues and community leaders, teachers, and church leaders in my area of responsibility to incorporate gender issues into various programs in Ivory Coast's society.

My team treated nearly 6,000 members of the local population and provided 300 life-saving surgeries free of cost. We also created an environment where even men who have suffered from sexual violence and sexually transmitted diseases could speak freely by availing of both male and female doctors in a friendly and confidential manner. I participated in national-level meetings aimed at spearheading providing free counselling to victimized women on sexual violence. For our excellent work, I received the force commander's commendation and was nominated for the Military Gender Advocate Award in 2016.

In 2018, I was once again appointed as the contingent commander and country senior in the CAR (MINUSCA), where I led an independent, self-sustained military contingent – the BANMED Level-2 Hospital. Our mandate was to provide medical support to almost 10,000 military and UN civilian staff members and employees as well as the entire civil population within and beyond our area of responsibility. Our duties and responsibilities directly contributed to the agenda of civil-military cooperation protection of civilians.

Daily, my contingent conducted outdoor camps for local patients, regardless of their gender, age, or disease. We provided free medical treatment to nearly 6,000 patients and conducted more than 3,000 pathological, clinical, hematological, and biochemical tests for the local population. Additionally, we conducted 400 X-rays and 275 ultrasounds free of cost. In collaboration with the child protection section, we held a health camp for children in which 300 children were provided with basic health education and 150 children were treated.

Our BANMED Level-2 hospital also supported humanitarian organizations operating in the vicinity, such as the ICRC and WHO, by providing crucial, advanced, and life-saving treatments with every necessary investigation. Under my direct supervision, we treated and rehabilitated cases of women's rights violations, including post-rape stress disorder, assault, and physical violation cases, both mentally and physically.

In recognition of our excellent work, I received the force commander's commendation and was nominated for the Military Gender Advocate Award of the year 2018.

Our outstanding performance was highly appreciated by both the local population and UN authorities, and I was honored with commendations from the SRSG, force commander, Central African defense minister, chief of army staff, medical director as well as local mayors. Additionally, I was once again nominated for the Military Gender Advocate Award for my exceptional work. Our achievements were highlighted in numerous world-renowned magazines in 2016 and 2018. Following my return to Bangladesh, the Government of Bangladesh awarded me one of the highest medals in recognition of my exceptional service.

I was honored to serve as the first female field ambulance commander and assistant director of medical services in the Bangladesh Army at a time when it was uncommon for women to hold such leadership positions. Additionally, I was the first female commander of the medical squadron during my tenure with the Bangladesh Air Force. As a female leader in the military, I sought to be a role model for other women and promote empowerment through each of my appointments, culminating in my promotion to the rank of brigadier general as the first female medical administrator in the Bangladesh Army.

As a dedicated peacekeeper, I am a strong advocate for increased female participation in peacekeeping and peace-building operations, as it has been demonstrated that such participation can significantly enhance the UN's ability and effectiveness. Women peacekeepers have earned the confidence and trust of local populations, including rebels, who are often more receptive to working with women in these capacities. In areas where women's voices may be suppressed, female peacekeepers can serve as positive role models and expand the UN's reach with their unique qualities, experience, knowledge, and abilities.

CHAPTER III

REFLECTIONS OF A FORCE COMMANDER: MY EXPERIENCE WITH THE UNITED NATIONS PEACEKEEPING FORCE IN CYPRUS (UNFICYP)

MAJOR GENERAL (RETIRED) MOHAMMAD HUMAYUN KABIR

From October 5, 2016 to October 6, 2018, I served as the force commander of UNFICYP. During my tenure, unprecedented political advancements within Cyprus resulted in significant progress towards reunification. However, despite all efforts, the settlement dialogue ultimately reached an unsuccessful conclusion, leading to a period of reflection by all parties. Operationally, I oversaw a Security Council-directed strategic review and military capability study of UNFICYP. This study primarily focused on cost savings, which led to a 10 percent reduction in troops and some restructuring within the military component. Specifically, the military liaison structure was enlarged to improve coordination and support different political outcomes.

Throughout my time as force commander, the military component provided a stable platform and supported the political process by maintaining a calm status quo between the opposing forces (OPFORs). One of the most satisfying aspects of my job was the excellent cooperation and integration between the different components within the mission. Although the mission is not integrated by design, it operates as such by necessity and logic – a testament to the leadership and personalities involved at every level. When leaving the mission, I was confident that the military component was in good shape and well-prepared to face future challenges.

As the force commander, my primary responsibility was to command the force, which was comprised of 802 (888 before the strategic review and military capability study in November 2017) officers and soldiers of 12 different nationalities. I also deputized for the special representative of the secretary general (SRSG) and stood in for her during her absence. I set the standards for the military component and ensured they were met through regular inspections and an emphasis on training (both pre-deployment and continuation training). I also had a key liaison role with the senior military command level of the OPFORs and, more widely, with the diplomatic community on the island. This representative role was crucial in ensuring the UNFICYP mandate and mission was understood, and our impartial, yet firm stance was recognized.

UNFICYP was established in 1964 to prevent inter-communal fighting between the Greek Cypriot and Turkish Cypriot communities. The mission had three primary objectives: preventing the recurrence of fighting, contributing to the maintenance and restoration of law and order, and aiding in a return to normalcy. However, in 1974, the situation underwent a significant change when Turkey intervened militarily in Cyprus, introducing an international dimension to the Cyprus problem and resulting in a division of the island. Because of the Turkish intervention, the Mission's responsibilities had to be revised to accommodate these new circumstances, leading to the development of four key tasks:

- Maintaining the military status quo in the buffer zone.
- Contributing to the maintenance of law and order, particularly within the buffer zone.
- Supporting the return to normal conditions by promoting social and economic development.
- Providing humanitarian assistance to communities that were left behind.

In addition to its peacekeeping role, the United Nations also assists both sides in their efforts to find a political solution to the Cyprus problem. This political role is carried out by the Secretary General's Good Offices Mission, which is led by the special adviser to the secretary-general (SASG). Previously, this role was filled by Mr. Espen Barth Eide until September 2017, but the UNFICYP SRSG currently fulfills a dual role until the secretary-general appoints a new SASG. The synergy and consistency between the two distinct missions: UNFICYP and the Good Offices were critical. Throughout my time, there had been a shared understanding and purpose at all levels. The political climate shaped the environment, and UNFICYP supported the Good Offices both directly and indirectly in advancing the political process. From a military perspective, I made every effort to provide maximum possible support to the political process, particularly through direct support at meetings and conferences through confidence-building measures, and indirectly by ensuring that the situation remained as calm as possible during times of progress in the talks.

Key Points on Mandate Implementation from October 2016 to October 2018

Throughout my tenure, I strove to build upon the work of my predecessors and made a positive impact in critical areas. Below are the key points that highlight the implementation of the mandate during my two-year tenure:

The UNFICYP ceasefire lines and buffer zone were still officially undefined, and normalizing the buffer zone remained an unresolved task. Although imaginary lines were drawn between the forward military positions, which became known as the Turkish Forces (TF) in the north and National Guard (NG) in the south ceasefire lines (CFLs), neither side signed a ceasefire agreement. Both sides continued to dispute the exact demarcation of the CFLs, further complicating UNFICYP's de facto buffer zone management task. The increasing civilian activity within the buffer zone made the requirement to "normalize the buffer zone" paradoxical, and there were no immediate solutions to issues relating to access, farming, hunting, construction, and transit.

The UNFICYP Aide Memoire, while formally unaccepted, was generally followed by the OPFORs. The 1989 Aide Memoire provided guidelines for permitted military, police, and civilian actions in or near the buffer zone. In 2018, the 1989 Aide Memoire underwent a thorough review to bring the mission's guiding document up to date. The main focus was on making previously implicit OPFOR violations explicit. Once the updated version was published, I met with both OPFOR commanders, who accepted the guidelines of the Aide Memoire 2018 in principle but were waiting for direction from the respective ministries of foreign affairs and ministries of defense before implementing it formally.

In the context of UNFICYP's contemporary environment, liaison and engagement were considered the primary of the three core military tasks: liaison, monitoring, and maintaining contingency. As a mission with a relatively constrained ability to operationalize and enforce its mandate, the ability to resolve issues swiftly and impartially through assertive persuasion, rather than coercion, was critical to its credibility. However, this was closely supported by the ability to identify issues via monitoring and maintain a credible ability to respond to likely threats, principally risks associated with civil unrest and public disorder. UNFICYP's layered liaison mechanism continued to be based on a principle of identifying and resolving problems at the lowest level, providing the mission with the ability to de- escalate issues via a more senior level of engagement when needed.

Improving liaison processes and maintaining impartial relationships had been my main focus. To achieve this, the military component worked hard to enhance processes for recording liaison events, including their outcomes and decisions taken. In the UNFICYP context, historical precedence and agreements played a vital role in liaising with the OPFORs. However, the mission had previously faced challenges in maintaining coherent corporate knowledge due to frequent personnel turnover and ad hoc recording systems. To address this, the mission implemented SAGE as a recording tool for all letters, meetings, and preannounced activities. This provided transparency of activity from Sector to Force HQ and allowed me to have a better understanding of individual issues, thereby enabling me to provide better direction. Moreover, the sheer number of activities announced and exchanged between the sides demonstrated the significance of the liaison staff's daily intermediary function within the military component.

The UNFICYP military component and the OPFORs – the NG in the south and TF in the north – had maintained excellent relations that required constant effort and vigilance. It was crucial to acknowledge that the two sides had fundamentally different interpretations of the situation and the buffer zone, and these differences needed to be respected. Direct communication between the OPFORs could easily solve many issues. However, despite the UNFICYP military component's continuous efforts to encourage direct communication, no progress was made. During my meeting with the OPFOR commanders, I highlighted the potential benefits of direct communication in keeping the situation calm and stable in the buffer zone. The OPFOR commanders recognized the value of direct communication, but it was evident that no progress could be made without political will and support.

Assessing the effectiveness of a PREVENT mission and determining acceptable risks were dependent on the force capabilities and manning. In 2016, the Security Council authorized an increase in the force to 888 personnel in anticipation of a possible solution to the Cyprus problem. However, with the discontinuation of the settlement talks in July 2017, the Security Council directed a strategic review of the mission and a military capability study. This led to a reduction in military personnel to 802, which involved streamlining some capabilities and restructuring to enhance others, such as continuity liaison. The ability to measure the effectiveness of a PREVENT mission and determine acceptable risks was crucial to this process. Any isolated incidents occurring between the OPFORs or between the OPFORs and civilians could cause significant damage to a potential settlement.

Challenges Faced and Lessons Learnt

The military component of UNFICYP operates continuously to prevent fighting and support civilian and policing tasks, thereby maintaining a peaceful buffer zone. However, activities such as armed hunting, construction, and others near the north CFL often upset the TF and raised tensions. Incidents of Greek Cypriots crossing the north CFL, intentionally or unintentionally, had often led to arrests and allegations of ill treatment. The military component faced its greatest challenge in keeping the buffer zone calm as neither side formally accepted the 1989 or 2018 Aide Memoires, which are the capstone documents. This created difficulties in holding the sides accountable as there was no formal agreement on the exact demarcation of the buffer zone. UNFICYP could only issue a protest letter and include the violation in the mission's biannual report to the UNSG, as there were no meaningful sanctions to apply after OPFOR violations. From a wider mission perspective, dealing with 'illegal' civilian activity in the buffer zone was also challenging. Both the military and UNPOL had extremely limited powers of detention, and there was no mechanism for financial penalty. This demonstrated the constraints of authority without power. The force had strived to maintain the perception of authority within the buffer zone.

UNFICYP's relation with OPFORs was dependent on the issue being discussed and the OPFOR commander's personality. Overall, relations had been good as the OPFORs supported UNFICYP's objective of maintaining low tension within the buffer zone. However, major challenges arose when there were significant differences in belief between the sides and the UN, such as the Strovilia Special Area. In such cases, it was crucial to have a comprehensive understanding of the historical agreements between the OPFORs and UNFICYP, which were often referred to as primary documents during liaison. From a military perspective, the absence of direct communication between the OPFORs remained a significant operational/strategic challenge for the mission. Securing political agreement between the sides on direct communication could easily resolve many challenges faced by UNFICYP in managing the buffer zone.

Although UNFICYP is not a technically integrated multi-dimensional mission, the mandate of UNFICYP necessitated integrated activities by all components, and UNFICYP never overlooked this requirement. Despite all components' efforts directed towards integrated working, there was occasional friction related to conflicting priorities, views, and timelines. The lack of a mission concept for UNFICYP resulted in a lack of clarity over conflicting component priorities. However, the multi-layered structure of integration within the mission allowed issues to be escalated as necessary and resolved at a higher level. At the highest level, the absence of a mission chief of staff and a truly joint operations center and sector tactical operations center complicated integration before the strategic review and military capability study in November 2017.

Given the long-standing nature of the mission, continuity was of utmost importance, particularly at the HQ level and in the liaison network. Establishing smooth and effective liaison depended on several factors, including the duration of deployment for the liaison officers (LOs), their communication skills, and their level of maturity. Some troop-contributing countries (TCCs) deployed their military observer liaison officers (MOLOs) or sector civil affairs military liaison officers (SCAMLOs) for only six months, which presented challenges. Typically, a MOLO/SCAMLO needed about 2–3 months to understand the operational environment and establish effective rapport with their counterparts. Thus, with a six-month deployment, LOs had only 3–4 months to deliver results effectively. In some cases, TCCs deployed young officers and non-commissioned officers (NCOs) as LOs who had difficulties in liaising with the OPFORs and lacked maturity in their conduct and communication skills. Additionally, in some sectors, almost all the MOLOs/SCAMLOs were rotated at the same time, which tremendously affected the continuity of maintaining liaison with the OPFORs and lowered operational efficiency.

The contingent members and staff showed praiseworthy dedication and professionalism, and there was no doubt that they were all accomplishing the mission and tasks required of them. However, the mission presented complex challenges across the length of the buffer zone, and although there was not a high threat of armed conflict, underlying tensions remained, for which we needed to be prepared. The mission required a broad range of skills, including unit cohesion, good judgement, observation skills, conflict resolution skills, negotiating skills, and the need to remain impartial in all circumstances to maintain credibility. Poor language skills in some contingents significantly affected the operational performance of the contingent in the field. F802 was the bare minimum strength required to efficiently deliver the mandated tasks in the buffer zone, given that all peacekeepers were well trained and fit for the mission. However, one TCC sent an ad hoc contingent to the mission comprised of approximately 60–70% Air Force personnel from different units. Another TCC in the past comprised 60% reservists. Naturally, I had apprehension about the capabilities and cohesion of ad hoc units and those based on reserve units.

Recommendations

Peacekeeping operations must have clear, realistic, and up-to-date mandates from the Security Council that are agreed upon by all parties involved. The mandate should have well- defined goals, priorities, adequate sequencing, and flexibility to evolve over time. These goals should be periodically measured by specific indicators, as is sometimes done in the mission concept. As goals are accomplished, the mandate should be continuously reviewed and redefined during mandate renewal, which typically occurs bi-annually or annually, taking into consideration the recommendations of the mission leadership. UNFICYP's mandate and the aide memoire that governed relations with the sides had changed very little over the decades, but the situation on the ground had changed completely. Therefore, it is important to update and revise the mandate periodically to reflect the current realities on the ground and ensure the mission's effectiveness.

The participation of female peacekeepers as troops, police, and civilian staff provides a new opportunity in contemporary multidimensional peacekeeping environments. Gender parity is not only essential for its own sake but also because the involvement of women has been proven to increase the chances of improving efficiency and acquiring sustained peace, while reducing incidences of sexual abuse and exploitation. In UNFICYP, we were proud to have achieved gender parity in the mission leadership. We led by example and hoped that our TCCs would help us build on the progress we had made by sending more female peacekeepers to Cyprus. This would not only benefit UNFICYP but also contribute to a wider international effort to promote gender equality and empower women.

The current logistics system employed within UNFICYP is bureaucratic and not well- suited to contemporary peacekeeping operations. It is dominated by outdated administrative rules, regulations, and financial opaqueness. There is a need for change to make it more responsive to operational needs and appropriately delegated. The SRSG should be able to allocate elements of logistics, force enablers, and funding to the force commander, especially in dynamic operational situations that require quick responses or rapid alterations to plans.

Since 1964, Cyprus has been divided, and the last violent action occurred in 1996. Given the current international context, it is highly unlikely that the parties will engage in any significant confrontations. If no settlement is reached among the stakeholders, the mission's mandate should be modified to an observer mission. However, before that can happen, both sides must accept the Aide Memoire 2018, and they should agree to communicate directly.

Properly employing Force HQ staff officers can be challenging if they do not possess the written and spoken English language skills needed to function effectively. It is crucial for contingents in UNFICYP to have a minimum standard of English, especially for those with command and liaison responsibilities. However, in some cases, this standard was not met. In the politically sensitive mission of UNFICYP, the risks associated with miscommunication are high, and misunderstandings can often be traced back to language barriers. Therefore, it is essential that commanders and LOs possess the required language skills to ensure effective communication and prevent potential conflicts.

The UN command and control (C2) structure is unique and differs significantly from most national structures, which is further complicated by the TCCs' influence over their own troops. This "parallel" chain of command can at times delay and hinder the timely execution of operational tasks. Similarly, the issue of accountability is also hampered by this peculiar C2 structure and TCC caveats. However, employing regional commands with a full understanding of the regional dynamic and specific missions would lead to better overall C2 of peacekeeping operations.

Reflecting on my time with UNFICYP, I am grateful for the fascinating and rewarding experience of working alongside a diverse group of multinational military, police, and civilian personnel. The insights and knowledge gained from this collaborative effort have been invaluable. Although the political progress made during my tenure was promising, I regret not being able to be part of the implementation of a settlement. Sadly, the prospects for a political resolution in the near future seem remote. Some may question the continued relevance of UNFICYP in this context. Nevertheless, I remain hopeful that future incumbents and all those who serve with UNFICYP will persevere in their efforts to find and implement a lasting solution in Cyprus.

CHAPTER IV

THE WAY AND LESSONS LEARNED FROM MINUSTAH IN HAITI (2004–2016)

BRIGADIER GENERAL (RETIRED)
GABRIEL GORGE GUERRORO

> "Moments of crisis carry seeds of opportunity, offering the UN and regional organizations the chance to work together more closely on peacebuilding and peacekeeping".
> —Kofi Annan, 25 Jul 2005.

From chaos to Mission Interim Force and handover to MINUSTAH, the slow start-up was plagued with obstacles – 1st of June 2004

A new crisis had begun in Haiti and President Aristide had to resign his position and leave the country in February 2004. Haiti was on the verge of civil war. The Security Council Resolution 1529 (issued on Feb 29, 2004) immediately provided troops for a Mission Interim Force (MIF) for a term of 90 days with the task of stabilizing Haiti, to later be relieved by a peace operation led by the United Nations. With this, the Latin American force generation process for Haiti had begun. Chile, the only country in the region that had joined the MIF in operation "Secure Tomorrow", was at the forefront of what was a clear intention to deploy the mass of the troops for the future United Nations mission provided by troop-contributing countries (TCCs) from the Latin American region. This appealed to Chapter VIII of the United Nations' Charter, although there was no regional governance mechanism applied to collective security for this case. Immediately, liaison officers from Brazil, Argentina, Chile, and Uruguay who represented and were likely candidates to contribute troops in an upcoming United Nations mission, were straightaway appointed to the United States Army Southern Command (US South Com), from where the operation "Secure Tomorrow" was strategically conducted and operationally executed on the ground by the MIF (made up of troops provided by the United States, Canada, France, and Chile). The resolution authorized this operation for a term of 90 days in Haiti.

When on March 4, 2004, I was assigned to South Com to serve as a liaison officer for the Argentine Republic in the force generation process, my work for MINUSTAH had begun. After this planning, my government / the United Nations appointed me as the first mission CMPO. I worked on the setup of the United Nations' mission (MINUSTAH), which was formalized in a rehatted parade on June 1, 2004 (my tour of duty lasted one year) and returned eight years later on February 24, 2012 as the DFC, a position I held for two years, from the end of 2011 up to March 2014. These two key positions with a buffer of seven years, from the force generation process, going through the first staff in the mission as CMPO and then two years post-earthquake, 2012 and 2013, as a DFC working at the operational and strategic level, allowed me to have a clear vision and perspective of the long and difficult itinerary the UN and Latin America struggled with in their work for Haiti with the 12 years of MINUSTAH.

During the force generation process, diplomatic contacts took place at all levels, with feverish negotiations to obtain Latin American troops as soon as possible, while the MIF fulfilled its mission and the 90-day counter authorized by the UNSC resolution 1529 went on inexorably—the clock was ticking. We were four years away from the Brahimi Report, which had just started to produce some changes in the United Nations' structure, but these were not reflected on the ground yet. At the same time, we were facing and suffering from the caveats of one of the phases where the same report placed great emphasis: "The need of rapid deployment" (Brahimi, 2000, p. Cap III A.).

This was a remarkable challenge, though we still only had a vague idea of what the operation could demand since the UN technical mission was still on the ground to produce their report and recommendations about a new peace operation. That is why it was decided to move the planning team from the Southern Command to the future mission area. This was a way of encouraging our capitals to speed up the decision-making process mechanisms and at least firmly corroborate their commitment by having more accurate information produced for us directly from the ground and the facts happening there.

With this purpose, the liaison officers from Argentina, Brazil, Chile, and Canada, along with members of the Department of Peace Keeping Operations at that time, now known as the Peace Operations Department, and officers from the Southern Command staff, landed on the island on March 18, 2004, to begin a schematic force generation plan and tentative deployment. For two weeks, day and night, and in permanent contact with the Planning Service, Force Generation Department, in New York and the capitals, we worked on a likely distribution of areas of responsibility and candidates to deploy in each one, according to their capacities and the political will; meanwhile, we knew we were running out of time.

After two weeks of forceful work, we were back in the United States Army Southern Command, and, as something largely expected by all of us, on April 30, 2004, the UNSC issued resolution 1542 to establish MINUSTAH as of June 1, 2004, authorizing a ceiling of 6,700 military personnel and 1622 policemen that were not available the first year of the mission. In other words, we were two months away from taking over a mission carried out by one force that was mostly organic, trained jointly and equipped with everything they needed and just over 5,000 troops, while the Latin American force generation process was still nascent and full of uncertainties. We had the formal commitment of Brazil leading the mission and contributing with a corpus of troops (two type A battalions in Port au Prince), 160 troops from Chile rehatted from the MIF in Cape Haitian and, just for two months, 100 troops from Canada, who were also rehatted from the MIF as reserves. As of June 1st, for MINUSTAH we had 2,127 troops in total, roughly a third of what was authorized by the UNSC, and a little less than half the strength of the force on the ground that we were swapping.

A chief administrative officer led the mission, since there was still no special representative of the secretary general in office, and the force commander (FC), General Heleno, had just set foot on the island. The staff began with a little more than ten colonels around May 13, 2004, opening with the mission-orientation course lacking everything, even the most elementary necessities, and worked in a second basement in United Nations for Development Program facilities without ventilation or even water in a country where average temperatures were 35 degrees Celsius in June.

MINUSTAH began amid a great reform for the UN proposed by the Brahimi Report, but which had not yet shown its benefits, and Latin America was facing a great challenge as a whole; this was a mission led by a region that lacked a regional governance mechanism applied to collective security.

Generating the Force and Simultaneously Carrying out Operations

In mid-June 2004, my CMPO board situation was shifting mission military personnel numbers very slowly as operational demand had grown exponentially in all sectors. It is important to remember that the internet and mobile phone communications were not like they are today, and we still used faxes and fixed lines; these were our scarce means of communication that were essential for commanding and operations on the ground. Nevertheless, this fact was something that only gradually improved over the years as Haiti also developed and adapted its communication infrastructure.

This was not a minor detail considering that, since the beginning, MINUSTAH had a portioned deployment, with large empty spaces, little mobility, minimal air resources (not for redeploying troops), and a great deficit in transportation infrastructure such as roads, bridges, etc. in a country where two-thirds of the geography is mountainous.

There was a great need to speed up the arrival of more units since maintaining a fragile stability demanded a constant effort from the few troops on the ground in relation to the task they were entrusted with and the logistical means available.

During this process of generating forces without ever reaching the ceiling during the first year, we experienced encounters such as remnants of ex-military uprisings, hostage-taking, and a hurricane that decimated the Argentine battalion's equipment. Moreover, we had to cope with significant permanent logistical needs while simultaneously organizing summits of TCC´s presidents visiting the island and even the security of a soccer match between Brazil and Haiti as part of a community engagement plan. The initial available troops, who took on these diverse tasks, managed to drive and improve the mission capacity by steadily increasing the number of units on the ground.

The great advantage that sustained us as a regional force was our common cultural roots, even in a circumstance in which we did not have pre-existing regional governance organizations for security and efforts were unnecessarily multiplied with overlaps that could have been avoided, greatly accelerating the block response capacity, not only in terms of time but also in terms of performance.

I witnessed the arrival of each contingent as the mission progressed with its logistics, using its own aircraft, redounding capacities that could have been carried out by a lead nation, saving time, improving logistical efficiency and, most importantly, gaining something essential for the accomplishment of the mission: *operational readiness.*

Most contingents, after overcoming the challenges of enlistment, training, and strategic transportation, faced the urgent challenge of being operationally available in the mission area as soon as possible, which, in many cases, was complicated since the UN developed its logistics very slowly. Similarly, the staff of the mission was only gradually fulfilled when, every six months, there was a rotation that gave us the additional challenge of virtually starting over again and again in a never-ending process that very seriously impacted our operational capability. The worst of it was that the mission's critics noticed this issue.

The FC, General Helleno from the Federative Republic of Brazil, knew how to overcome all kinds of obstacles that arose daily with great leadership and experience, but above all he had a great capacity to handle the most complicated aspect of his mission, which was *The Use of Force,* since we were operating under Chapter VII of the United Nations Charter with great instability in a fragile security environment and scarce resources for exercising command and control.

In those first months, I had the privilege of designing the medal that thousands of peacekeepers who would serve in Haiti would proudly wear on their chests. The ribbon was designed with a green color symbolizing the Latin American forest together with the typical light blue of the Blue Helmets and white symbolizing peace.

The general staff was slowly being completed drop by drop, while the pace of work increased exponentially. Still, having the same culture was a clear advantage in many ways, though it became a great weakness when the vast majority were Spanish speakers but were not able to speak or understand English or French; meanwhile, the key positions in the staff (English and French speakers) did not speak or understand Spanish, generating a serious communication problem in a critical situation.

In this way, we found ourselves in key positions fighting on two fronts: building the organization inward and strength outward; therefore, we needed to provide a stable and secure environment throughout the country as soon as possible. The Brazilian troops, coming from an organic force and being the mass of those available at that moment, were the center of gravity to exercise control over Port-au-Prince, but it became increasingly evident that it was taking an excessive amount of time to assemble the whole force in its entirety and to be able to deploy them throughout the whole country.

The decision-making mechanisms used by the TCC governments in many cases, or the lack of experience in the deployment of troops in others, conspired to shorten the time needed for organization and operational readiness. That is why some very specialized elements such as the Command and Service Company for the Military Headquarters (MHQ´s Com Coy) had to come from a faraway place with all the consequences that this situation implied for the mission itself and the TCC provider, in this case the Philippines.

Throughout this period in which the mission began to develop with a plan that considered all Latin American countries, MINUSTAH was formed with huge national efforts, taking a significant amount of work and time to amalgamate on the ground a force that could function as a system. It was already clear after the Brahimi Report that being quick and efficient was not about piling up units but was more about planning and executing a logical and harmonious sequential deployment with adequate logistics and self-sustainment to be operational as soon as possible. With any delay in time, the cost was human lives, a lesson that many UN missions had learned in the nineties.

It was clear that we were going through a historical moment in which the members of the region were called for the first time as a coalition to pacify a Latin American country. This process was a great learning experience for the United Nations, which was running reforms after the Brahimi Report, moving from the fragmented logistics that were its Achilles heel in the face of the demands of the 1990s, to the Integrated Support Service. It was also an important learning experience for the contributing countries, which had begun to value bilateral exchanges and combined work during periods between deployments, something that had just begun to escalate in the region in the new millennium.

The rotation every six months and the slowness in the organic development of the force conspired against the continuity of support and the establishment of a safe and stable environment. Time began to stretch, and I clearly remember that at that beginning very few people envisioned that this deployment would last close to thirteen years with all the issues and necessary human resources and logistical support to be contributed by the TCCs and police contributing countries (PCCs).

As a region, we were facing a great test of resistance and capacity. From that initial moment, I have always maintained and voiced that this could not be in vain for our nations and that we must take note of what happened, not only in the historical sense of each event, but also in terms of the lessons learned that should serve Latin America in the future, considering everything that happened in Haiti, especially in relation to cooperation and regional collective security.

This initial leg was particularly difficult because, as the saying goes, "what is not trained in peace will be suffered in wartime". I also believe that many of the problems we faced in Haiti could have been standardized and trained for prior to the operation, especially the issue of interoperability, which was critical.

A force of this magnitude in a country far from its own logistics centers with a high level of turnover and a staff in slow formation requires prior standardization agreements to speed up and facilitate its readiness for urgent operations. A force with the mission of providing a secure and stable environment must function as a system since the operational level does not see areas of responsibility as isolated compartments but as one organization that manages to stabilize the entire country at the same time and at the same security level.

This first stage in Haiti was a great experience for Latin America and for the United Nations, but it was clear that the ceiling of the mandate would not be reached without the contribution of troops from regions other than Latin America, something that was not in the initial plan and that would necessitate an additional requirement for interoperability at the mission command level of the military component and the rest of the mission components.

MINUSTAH Towards the Desired State (2012 to 2016)

I arrived as Deputy Force Commander (DFC) in Haiti on February 24, 2012; on that day, Prime Minister Garry Conille had just resigned, and the political climate was tense. However, I did not have much time to adapt since a few hours later I gave a speech at the medal's ceremony for the Philippine contingent. I felt at home, as nothing was strange to me. I was only surprised to see the evolution of troops that made up the numbered type of units.

I then realized how that process of generating forces that began in 2004 had evolved and the great influence that the setup of the organization adopted later because of the consequences of the earthquake and demands from the public. Haiti taught us something very important, which is that neither the hurricanes we faced (four in three years), nor the earthquakes and floodings, nor the cholera epidemic were foreseen in any mandate and were tremendous tragedies for the Haitian people. At the same time, they were colossal challenges that were completely unexpected by the mission and caused setbacks more frequently than we expected.

In addition to the great cost in lives, these events shook the mission to its foundations, and its pace and organization had to be radically altered to overcome each episode and move toward the desired end state. No one at the beginning of 2004 in the general staff considered these contingencies when planning for the long and mid-term, though we as peacekeepers know that we must always be prepared for the planned course to be altered without notice.

In 2012, when I came back, I found a different, renewed country, particularly considering the limitations we had in 2004. At the same time, I confronted a changed mission compared to when I had started in 2004. Now, with more assets and having dealt with great tragedies, we, together with the Haitian people, could definitively move towards the achievement of the goals projected in the strategic planning. We were clearly in the peace-building phase after helping the country gradually recover from the earthquake, which compelled the mission to postpone the end state for two years.

As DFC, I worked directly with six engineering companies under my command, all of which did an extraordinary job in many aspects: vertical construction, roads, canal cleaning, purification and water drilling, debris removal, etc. I felt on the exercise of my command that the battalions skillfully controlled their area of responsibility and there was a great capacity for operations and deterrence against gangs and 'spoilers'. Humanitarian aid, civil and military cooperation action and quick impact projects were great assets for winning Haitian hearts by improving their lifestyle. The new Haitian National Police that I knew in 2004 from its genesis had already reached 12,000 troops and was operating alongside United Nations forces. The Hatian National Coast Guard was recovering and being trained by the Uruguayan Contingent in a tutorial and practice plan by means of boats contributed by the Canadian government together with foreign instructors.

Despite so many difficulties and setbacks, everything seemed to be evolving towards the desired end state envisioned in the strategic plan. Nevertheless, I always had the idea that Latin America was losing energy and resources by not being able to coordinate and make better use of its logistics, human resources, transport, etc. with pre-existing regional training and a proven organization.

In each rotation, I saw the enormous individual and often simultaneous efforts that could have been saved, coordinated, and optimized in advance. Binational efforts such as the Engineer Company from Chile and Ecuador were a sample of this and demonstrated its feasibility on a small scale, though they were only a part of the whole.

Upon analyzing the general configuration of the force, it was clear that despite the great efforts of Brazil, Uruguay, Argentina, and Chile as the main TCC suppliers, Latin America could not reach the ceiling established by the mandate and the military component had to be completed with units from other regions. Although they constituted a component of large cultural diversity, their adaptation to a dissimilar culture generated organic and logistic difficulties and saddled the United Nations Mission Headquarters with additional demands that could have been avoided.

I must admit that my time as G4 of the High Readiness Brigade for the United Nations operation named SHIRBRIG for two years from 1998 to 2000 and my contacts with General Romeo Dallaire (the former force commander in Rwanda during the genocide), along with reading his books, have instilled in me a very particular vision. This insight influences to a great extent my idea of the urgent need to increase cooperation, coordinate, not duplicate efforts, integrate, and standardize to achieve maximum efficiency as quickly as possible at the lowest cost in terms of human and financial resources. Above all, this must be done quickly to save lives, which is what is most important to protect. The MINUSTAH organization ended up being a hybrid between Chapter VIII of the United Nations Charter, Regional Arrangements, and the old-fashioned way of generating force that the Brahimi Report initially advised to supersede and to which the High Independent Panel on Peace Operations report (HIPPO) later refers directly in point III. A. Strengthening global – regional partnerships.

MINUSTAH's Legacy and Lessons Learned

Here are some of the lessons learned from my experience in the mission:

At the operational level, start from scratch when the clock is ticking. Force generation is a fundamental phase in which strategic transportation and logistics deployment are the main challenges to overcome and, at the same time, it is critical to achieve rapid reaction capabilities in a shorter period. This is something the United Nations somewhat resolved in 2019 with the Peace Capability Readiness System (PCRS) and by appealing to Chapter VIII of the Charter. This solution is often not possible for political reasons, regional upheaval or the capacity / will of the states. Afterwards in the field, the operational level suffers because it has a mandate to be accomplished and have human resources assigned, and in almost all cases these are insufficient and arriving late. Resources normally disembark in intermittent waves, while the local situation unfolds and develops with critical events occurring every day in a row; meanwhile, the mission has little or almost no capacity at all to influence the situation.

Since the genocide in Rwanda in 1994, the creation of one United Nations' standby force has been largely discussed, such as the Standby High Readiness Brigade (SHIRBRIG) initiative (1998–2018), which many states systematically resisted for political reasons. Nevertheless, I am fully convinced that SHIRBRIG could have provided an alternative solution in those cases where it was necessary to deploy efficiently. It would perform as an interim force to take over another organization or as intervention forefront force applying the Responsibility to Protect, something not determined yet concerning how and who (what force?) is going to execute the norm with legal authority and, more importantly, legitimacy.

Considering these concepts, the statistics, the theory, and the great experience the Latin American region has in peace operations and cooperation for peace, along with the potential of its training centers for peacekeeping troops that will be assigned to incoming missions, why does the Latin American region not have a regional governance scheme applied to collective security? Why do we engage in individual or at most binational efforts having had an experience like MINUSTAH?

After MINUSTAH, the need to integrate, seek to cooperate and work intensely to add capabilities to be stronger, faster, and more efficient was and still is very clear.

Security is a sensitive issue when sovereignty is at stake and therefore requires a great political consensus. At the same time, thirteen years of MINUSTAH compel us to examine the great effort made by the region to pacify Haiti and the huge experience obtained in generating for Latin America the conditions for a regional governance scheme applied to a collective security system for peace operations. More specifically, the core staff and logistic cooperation for rapid deployment will be the most valuable elements to contribute, as we perceived it in that distant 2004 where the urgency to help a brother country presented us with the need and the challenge of moving towards a mission area in a very short time and as a regional security organization.

The protection of civilians has taken on a centrality and urgency. It emphasizes the need to create and provide forces with very special characteristics, such as a capacity to use force at a midpoint between the lethality and containment of unarmed violence with speed, flexibility, mobility, and the ability to produce intelligence as a multiplier of their aptitude to protect civilians.

At the tactical level, ends and means must converge on the ground

Cultural affinity and international cooperation prior to the conflict of the TCC and PCC has a direct impact on tactical efficiency and mutual support in operations and acts as a capacity multiplier where it is pre-existing, such as the binational engineers' company from Chile-Ecuador.

The rotation and permanence of the troops cannot exceed nine months since it has a direct impact on discipline and causes (as was the case in Haiti) serious misconduct events at the tactical level that endanger the legitimacy of the presence of United Nations' troops in the country and have a great impact at the strategic level. Dealing with the same subject for less than nine months created a problem, especially during the mission setup, causing operational efficiency gaps.

The tactic, despite the recommendations of the Brahimi and HIPPO reports, continues to be impacted by a dissociation between ends and means, which is often due to pressure for political reasons and in most cases budget reasons.

The community engagement process is essential for the fulfillment of the mission and must be deeply internalized in all orders and directives at the operational level to be clearly understood by the tactical level.

Caring for the environment is becoming more and more central within the mission, and the tactical level must decode and execute it effectively, since not doing so endangers the legitimacy of the entire operation, as was the case in Haiti regarding the cholera epidemic, which apparently originated with a bad practice of one battalion at the tactical level.

Tactical mobility using air assets in the mission area must be the priority and the administration by the force commander at the tactical level is a fundamental tool for command and control and essential for exercising leadership by his presence where decisions are made.

The PCRS and pre-deployment visits have been successful attempts by the UN to avoid surprises in the mission area and during operations, when it is too late and there is no way to fix them being thousands of kilometers from the capitals of the TCCs and PCCs.

PART III

ANALYSES OF RESEARCHERS, SCHOLARS, AND OFFICIALS ON UNPKO

In this part, several officials from Bangladesh, scholars from across the globe have shared their insights on the current PKO model and discussed about the future role of the UN in peacekeeping.

Chapter V

Preparedness, Challenges and Opportunities of the Bangladesh Army in UN Peacekeeping

Brigadier General S M Asadul Haque

The Bangladesh Army has actively participated in United Nations peacekeeping missions since the early 1990s. Over the years, it has earned a reputation as a professional and reliable contributor to UN peacekeeping efforts. Since its first deployment in 1988, the Bangladesh Army has sent over 151,930 troops worldwide to participate in UN peacekeeping missions. As of May 2023, Bangladesh is the highest troop-contributing country to UN peacekeeping missions, with over 6,080 soldiers serving in various missions worldwide. The country's significant contribution of personnel and state-of-the-art equipment to the world's most dangerous field missions is a testament to its commitment to maintaining international peace and security. The primary goal of the Bangladeshi Army's participation in UN peacekeeping missions is to help stabilise and rebuild countries that have experienced conflict or instability. It is trained to perform a range of tasks that contribute to this objective, including internal security duties, the protection of civilians, supporting political processes and disarmament, nation-building activities, and facilitating the delivery of humanitarian aid.

The Bangladeshi Army's participation in UN peacekeeping operations has been widely regarded as professional and effective and has garnered international recognition. Having no integral conflict zone, these missions have served the Bangladesh Army for decades as ideal grounds for demonstrating its training and expertise. It has provided invaluable combat experience for Bangladeshi troops and helped the Bangladesh Army strengthen its capabilities in acquiring modern technologies and equipment. By working shoulder to shoulder with different professional armies of the world, the Bangladesh Army has gathered new knowledge and training, enriched itself professionally and culturally, and successfully established a brand value in peacekeeping. In addition, the Bangladesh Army has made a significant contribution to the national economy by earning considerable foreign currency through the reimbursements paid by the UN for its personnel and equipment.

However, the Bangladeshi Army's participation in UN peacekeeping missions has not been without challenges. The complex and rapidly changing nature of modern conflicts and the evolving role of peacekeepers mean that the Bangladesh Army must continuously adapt to new threats and changing circumstances. Additionally, peacekeeping missions often occur in complex environments, and troops must be prepared to operate in unfamiliar terrain and face unforeseen situations. The unique challenges troops face during peacekeeping missions require specific skills, knowledge, and training. They also entail a superior state of preparedness; strong intelligence, surveillance and reconnaissance (ISR) capabilities; mobility; and force protection. Despite these challenges, the Bangladesh Army has also found opportunities for growth and development through its participation in UN peacekeeping missions.

Against this backdrop, this chapter analyses the preparedness of the Bangladesh Army for peacekeeping operations, examines its challenges, explores ways forward for future development, and highlights its opportunities in the domain of peacekeeping. In doing so, the study will first highlight the peacekeeping history, contributions, and current deployments of the Bangladesh Army in UN missions. It will also briefly analyse the current security dynamics of the mission areas with an assessment of the mission's capability requirements and tasks.

The paper has been developed through a qualitative study of the primary and secondary sources. Relevant data projects the most updated information on the peacekeeping missions of the Bangladesh Army as of May 2023. The primary sources also include several policy documents related to peacekeeping training and preparations, relevant literature, academic articles, and reports on the Bangladeshi Army's participation in UN peacekeeping missions.

Blue Helmet History of Bangladesh

The glorious history of Bangladesh in UN peacekeeping and international humanitarian operations started in 1973, just two years after its independence, through the deployment of a group of doctors in Damascus, Syria. In 1988, the Bangladesh Army started its journey with the blue helmet by sending 15 military observers to the United Nations Iran–Iraq Military

Observer Group. Since then, there has been no looking back, and the contributions of Bangladeshi peacekeepers have grown remarkably in the last 35 years. Out of 71 UN peace missions since 1948, Bangladesh has successfully participated in 56 missions in 40 countries on four continents.[1] So far, 187,663 peacekeepers, including 2,569 female peacekeepers, from the Bangladesh Armed Forces and Police, have participated in UN peacekeeping missions.[2]

Table 5-1: Summary of Completed UN Deployments[3]

Army	Navy	Air Force	Total	Police	G/Total
151,930	6,668	8,269	166,867	20,796	187,663

The first Bangladesh Army contingent was deployed in 1992 under the United Nations Transitional Authority in Cambodia.[4] By dint of professionalism, impartiality, and commitment, Bangladesh has become the largest troop-contributing country (TCC) since September 2019. However, its pride has come at the cost of 167 fallen heroes, while 259 suffered combat injuries. Presently, Bangladesh has 7,465 peacekeepers deployed across the world in UN peacekeeping missions.[5]

Table 5-2: Current Force Deployment in UN Missions[6]

Army	Navy	Air Force	Total	Police	G/Total
6,080	364	515	6,959	506	7,465

---

[1] The Daily Star, "United Nations Peace Operations and Bangladesh," March 26, 2021, https://www.thedailystar.net/supplements/celebrating-50-years-bangladesh/news/united-nations-peace-operations-and-bangladesh-2066693.

[2] Overseas Operations Directorate Archive, "UN Peacekeeping Missions of the Bangladesh Army," *UN Archive* (Dhaka: Overseas Operations Directorate, General Staff Branch, Army Headquarters, April 18, 2023).

[3] Ibid.

[4] Armed Forces Division, "Bangladesh in UN Peace Operation," UN Peacekeeping (Dhaka, April 18, 2023), https://afd.gov.bd/un-peacekeeping/bangladesh-in-un-peace-operation.

[5] Overseas Operations Directorate Archive, "UN Peacekeeping Missions of the Bangladesh Army."

[6] Ibid.

Current Bangladesh Armed Forces in UN Peacekeeping

At present, there are 12 ongoing peacekeeping operations deployed across the world.[7] Among these, the Bangladeshi Army's peacekeeping contingents are deployed in six missions in Africa:[8]

- The United Nations Multidimensional Integrated Stabilization Mission in Mali (MINUSMA).
- The United Nations Multidimensional Integrated Stabilization Mission in the Central African Republic (MINUSCA).
- The United Nations Stabilization Mission in the Democratic Republic of the Congo (MONUSCO).
- The United Nations Mission in South Sudan (UNMISS).
- The United Nations Interim Security Force for Abyei (UNISFA).
- The United Nations Mission for the Referendum in Western Sahara (MINURSO).

Twenty (20) Bangladesh Army contingents are presently deployed in six UN missions. These contingents, such as manoeuvre contingents and enabler contingents, are categorised based on their capabilities and tasks. The manoeuvre contingents include mechanised infantry battalions/companies, special forces companies (SFCs), and quick reaction forces (QRFs). SFCs are comprised of mainly army commandos, while QRF companies are a mix of mechanised elements and commandos. By June 2023, the Bangladesh Army will likely deploy a new type of manoeuvre contingent named the mechanized infantry company (base defence) in Mali, which will protect the base camp.

Meanwhile, the enabler contingents include six types of military peacekeeping outfits: engineer contingents, signal (communication) contingents, military police contingents, medical contingents, aero-medical evacuation teams, and explosive ordnance disposal (EOD) platoons. By June 2023, the Bangladesh Army will likely deploy a new type of enabler contingent named armed helicopter units, and by 2025 it is also likely to deploy an unmanned aerial system (UAS) company jointly with the USA in Mali. In addition, the Bangladesh Army has nineteen United Nations Peacekeeping

[7] United Nations, "Where We Operate - United Nations Peacekeeping," 2023, https://peacekeeping.un.org/en/where-we-operate.
[8] Armed Forces Division, "Bangladesh in UN Peace Operation."

Capability Readiness System (UNPCRS) contingents for responding to any further calls from the UN.

With 6,080 peacekeepers deployed in 12 different UN missions and headquarters, the Bangladesh Army accounts for 81% of the total contribution of Bangladesh to UN peacekeeping. Apart from 5,905 contingent members, 175 military staff officers (MSOs) and military observers are deployed in different missions, including the UN Headquarters in New York. Female peacekeepers of the Bangladesh Army were first deployed in 1995 as doctors and in 2005 as MSOs. Ninety-four female officers and 309 female soldiers are deployed in different capacities in UN missions.[9]

Current Security Dynamics in Peacekeeping Missions

Today's peacekeepers are highly vulnerable due to the highly volatile environment and propensity of non-traditional threats. These threats include mercenaries, transnational terrorist groups and technological threat elements; the introduction of improvised explosive devices (IEDs), mines, suicide bombings; and increased abductions, targeted killings, etc., all of which pose increasingly grave threats to peace efforts.[10]

The main terrorist groups operating in the West and Central African regions are Al Qaeda in the Islamic Maghreb, Boko Haram, Movement for Unity, and Jihad in West Africa, Jama'at Nasr al-Islam Wal Muslimin, Islamic State West Africa and Islamic State in the Greater Sahara. There are also self-defence militias, which emerged as a security provider of the affiliated communities. They exploit the poverty, weak governance, and porous borders among conflict-ridden African nations to establish their foothold and carry out attacks and targeted killings against civilians, government agencies, local security forces and peacekeepers.[11]

Armed conflicts and violence in African nations mainly erupt from poverty, competition over resources, political rivalry, ethnic and religious tensions, communal clashes, and the proliferation of armed groups. While the withdrawal of the French from several African countries has created a

---

[9] Overseas Operations Directorate Archive, "UN Peacekeeping Missions of the Bangladesh Army."
[10] Global Terrorism Index, *Global Terrorism Index 2022*, *Global Terrorism Index* (Sydney, Australia, 2022), https://doi.org/10.4324/9780203731321.
[11] Ibid.

security vacuum that has been filled by terrorist groups, the emergence of mercenaries like the Russian Wagner Group has evolved with new conflict dynamics.[12] There are also widespread communal conflicts that are further exacerbated and exploited by terrorist groups by taking sides on ethnic divisions to further their organisational agenda.

The trafficking of humans, narcotics, and weapons; kidnapping for ransom; and extortion are the most notable organised crimes in African countries. Due to corruption, weak state control, and porous borders, it is easy for organised criminal networks to operate. The large scale of poverty and unemployment are also driving the recruitment of criminal gangs.[13]

The impacts of climate change, which is leading to drought, desertification, food and water insecurity, a lack of food for cattle, floods, and other effects, have tremendously affected the livelihood of the local population and aggravated poverty, disease, displacement, and social unrest. In some places, ecological factors stand out as the leading causes of intercommunal conflicts and are exploited by terrorist groups to drive their recruitment.

The current security situation in the mission areas is also marked by political instability and a lack of state control and is rapidly deteriorating due to political unrest and military coups. Because states have failed to provide security in remote areas, non-state actors, such as militias, have seized on opportunities to position themselves as alternatives to governments. Terrorist groups have attracted the local populations to their ideologies by capitalising on their grievances against the government and political system.[14]

Operational Capability Requirements

Besides maintaining peace and stability, the responsibilities of protecting civilians, human rights monitoring, security sector reform, etc. further complicate peacekeepers' tasks. The multidimensional peacekeeping mandate authorises peacekeepers to use force for self-defence and implementation of

[12] Peacekeeping News, "Many Challenges Remain for Sahel Joint Force, Warns Lacroix, Welcoming 'Increased Coordination'," November 16, 2020, https://peacekeeping.un.org/en/un-news/many-challenges-remain-sahel-joint-force-warns-lacroix-welcoming-increased-coordination.

[13] Mark Micallef and Matt Herbert, "Fragile States and Resilient Criminal Ecosystems," *Global Initiative against Transnational Organized Crime*, no. August (2022).

[14] Global Terrorism Index, *Global Terrorism Index 2022*.

the mandate. According to the prevailing trend, the mission environment may become more challenging, entailing more robust, mobile, and technologically adaptive peacekeeping outfits.

Peacekeeping tasks of the Bangladeshi contingents vary according to the security dynamics and mission-specific requirements guided by the mandate, and specified through the Statement of Unit Requirement (SUR) and Memorandum of Understanding (MoU) of a contingent. Generally, the tasks of the contingents encompass the following:

- Protection of civilians.
- Area domination/verification.
- Security/escort of UN members.
- Protection of main supply routes.
- Cordon, search and rescue operations.
- Demobilisation, disarmament and rehabilitation (DDR).
- Civil-military cooperation (CIMIC) – Construction and building works, HRD programs, and pacification activities.
- Provision of medical facilities to military and civilian employees of the UN.
- Provision of communication to the forces and Mission Headquarters.
- Special operations.
- Supporting government institutions and security forces.

To contend with prevailing complex peacekeeping challenges, the Bangladesh Army keeps its peacekeepers updated with the most modern equipment and training. However, with the evolving security dynamics and demands of UN missions, there is a need for an adaptative approach towards training modification, technological improvement, and the regrouping of forces to inculcate a robust peacekeeping attitude. Training for peacekeepers now encompasses a multitude of subjects. Peacekeeping training can be divided into two categories: one includes the 'regular training' that Bangladeshi troops typically receive while participating in various military and UN-related courses, and the other includes 'special training' related to the UN.

- Regular Training
    - Knowledge of the UN and African Union system.
    - General codes of conduct.
    - Humanitarian assistance.
    - Security of men, material, and information.
    - Essential health and hygiene.

  - HIV/AIDS and malaria.
  - Navigation and radio communications.
  - Counter improvised explosive device (CIED) awareness.
  - Sexual exploitation and abuse, child protection and protection of civilians.
  - Community policing.
  - CIMIC.
  - Reform, restructuring, and rebuilding.
  - Disaster management.
- Special Training
  - Mission-specific training.
  - Driving and maintenance of mechanised/plant vehicles and other types of vehicles specific to the mission areas.
  - Laws of armed conflict, international human rights law, and laws of the land in the mission areas.
  - Conflict analysis, communication, mediation, and negotiation skills.
  - Peacekeeping intelligence.
  - Community outreach and engagement.
  - Operational language training (English, French, Arabic, Swahili, Bambara, etc.).
  - Disarmament, demobilisation, and reintegration (DDR).
  - Media management and relations.
  - Gender equality and mainstreaming.

Bangladeshi Army's Peacekeeping Preparedness

Quality-focused and adaptive selection process is the key. The evolving security landscapes in the mission areas demand maximum operational efficiency and robustness. To keep pace with the requirements of the UN and field missions, the selection policies for the UN mission are updated each after two years and amended as and when required. The table of organization and equipment (TO&E) of the contingents are prepared following the SUR received from the UNHQ and updated before the deployment of each contingent based on the recommendation of the already-deployed contingents and any special requirements from the UNHQ or the mission. Operational requirements and robustness are given the highest priority. The officers and men are selected for the missions depending on the mission-specific requirement and individual competency following a strict screening process through a board of officers. Personnel failing to attain the desired benchmark are dropped from the contingents during concentration and are even repatriated from the mission areas at national cost.

Updating the training needs is important for success in PKOs. Training needs for future contingents are regularly updated based on the recommendations of the deployed contingents and from the assessment of current incidents/events in the mission areas. For example, following the increasing trend of IED attacks, the Ordnance Centre and School (OC&S) and Engineer Centre and School of Military Engineering (ECSME) have conducted detailed research and replicated IEDs in their lab to impart training to future contingents. Demographics, environment and climate change are also considered while designing the training. On their return from the mission areas, contingents are always sent to the Bangladesh Institute of Peace Support Operations Training (BIPSOT) to share their experiences and recommend modifications to the pre-deployment training (PDT). A model describing the system of assessing the peacekeeping training gaps and requirements is appended below:

Figure 5-1: Assessment of Training Requirements[15]

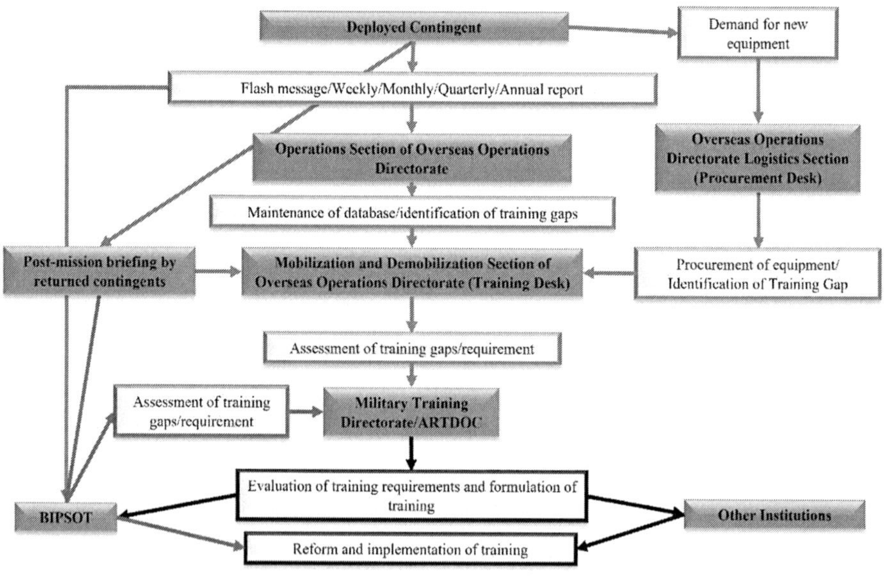

[15] Overseas Operations Directorate Archive, "UN Peacekeeping Missions of the Bangladesh Army."

Operational readiness must be maintained for successful conduct of PKOs. Under the United Nations Peacekeeping Capability Readiness System (UNPCRS), the Bangladesh Army currently has three contingents placed at the rapidly deployable level who are ready to be deployed within 60 days of the deployment order. The Bangladesh Army also has sixteen other contingents placed at different levels of the UNPCRS, substantiating its superior preparedness to respond to any call from the UN. The overseas operations directorate negotiates with the UNDPO about the MOU for each contingent before deployment to ensure they are deployed under favourable terms and conditions. They also monitor the operational employments of the contingents and negotiate with the UNDPO about different deployment parameters of the contingents.

Equipment tables of the contingents are prepared at least six months before deployment to ensure proper logistics preparation. A team of officers from every deployable contingent visits the mission area to carry out logistics reconnaissance and survey three months before the deployment. Moreover, based on thorough stock taking by the deployed contingents, the deployable contingents amend the lists of equipment tables. To ensure the operational effectiveness and readiness of the Bangladesh Army contingents, the overseas operations directorate strives to deliver the contingent owned equipment (COE) in the mission areas in due time. Apart from the regular rotations, it also maintains a buffer stock of self-sustainment items and spare parts for an infantry battalion.

To keep pace with the requirements of modernisation, enhanced mobility and force protection, the Bangladesh Army is gradually replacing all the armoured personnel carriers (APCs) deployed in the missions with light armoured vehicles (LAVs) for greater mobility and mine resistance ambush protected (MRAP) vehicles with built-in jammers for greater protection against IEDs. The contingents are also equipped with state-of-the-art CIED equipment and UASs to ensure better movement protection. The Bangladesh Army pioneered the eco-friendly PKO effort by contributing generators with solar power energy in UNMISS in 2022.

Comprehensive Human Resource Development (HRD) program helps to select PKO members who play crucial role is team building in PKO. Over the years, the Bangladesh Army has evolved with a unique and comprehensive HRD program spanning over 12 months to prepare the contingents' deployment following a highly systematic and meticulous process. Warning orders and deployment instructions are served for almost one year, while contingents' TO&Es are prepared/updated and circulated approximately ten months before. All necessary procedures like the arrangement of government orders, visas, clearance from mission areas, medical screening, and vaccinations are done well ahead under the responsibility of the overseas operations directorate. They also ensure the timely rotation of the contingents through coordination with the UNDPO.

Table 5-3: PDT Model of the Bangladesh Army[16]

PDT Level 1	PDT Level 2		PDT Level 3
Two weeks	Four weeks	Three weeks	Four weeks
Phase 1 (Basic skills training)	Phase 2 (CPTM)	Phase 3 (STM)	Phase 4 (Outdoor Ex)
Refresher/currency training on basic soldiering capabilities and skill development	Training on UN peacekeeping capabilities and skills	Special skill development training and scenario-based training for sub-units	Mission-specific lane training for sub-units and battalion-level mission-specific field exercises

According to UN guidelines,[17] the Bangladesh Army conducts the PDT of contingents for 13 weeks (Table 3) in four phases divided into green and blue training categories. Green training is performed solely under the responsibility of a division military base called a lodger formation. In contrast, blue training is conducted as a shared responsibility among the army headquarters (AHQ) overseas operations directorate, BIPSOT and the lodger formation encompassing core pre-deployment training materials (CPTM), specialized training material (STM), mission-specific training, and operational language training, followed by field training exercise and refresher training at the BIPSOT. This training is conducted over five

16 Overseas Operations Directorate, "Policy on Pre-Deployment Preparation (PDP) and Pre-Deployment Training (PDT) of UN Contingents" (Dhaka: Army Headquarters, November 12, 2017).

17 UN DPO DFS, *Operational Readiness Preparation for Troop Contributing Countries in Peacekeeping Missions*, *United Nations* (Department of Peace Operations, 2018), https://peacekeeping.un.org/sites/default/files/3._military_-_4_guidelines_on_opr.pdf.

months before deployment by designated instructors from the contingents previously nominated and trained at BIPSOT through the Training of Trainers Course of PDT.

Provision of combat-worthy training must be made in time and efficiently. All deployable contingents practice firing with their heavy weapons and mechanised vehicles before deployment. Some contingents undergo jungle lane shooting training that considers the terrain peculiarity of the mission areas. The Special Force contingents undergo special commando training, while mechanised training is conducted for the mechanised infantry contingents, as far as possible, in a customised environment to match the terrain configuration of the mission areas. To keep pace with changes/upgrades and the incorporation of new equipment and technologies, specialist personnel of the contingents are sent to different military and civilian specialised training institutes. This helps the contingents to maintain the technical expertise of the specialist personnel so that operations are not hampered due to a lack of appropriate equipment maintenance.

As IEDs are the most ominous threats for peacekeepers,[18] UN standard CIED training is conducted for all the deployable contingents. For the MINUSMA contingents, the Mobile Training Team of United Nations Mine Action Service (UNMAS) from MINUSMA, Mali, visit Bangladesh annually to conduct mission-specific CIED training. UNMAS teams from UNHQ and MINUSMA have recognised our CIED training facilities, like the IEDD/EOD data centre, simulator, fabrication lab, and CIED fusion centre in OC&S and ECSME, as exemplary for other TCCs.

The PDT is evaluated through a multi-layered evaluation process to maintain the desired training standard. The lodger formation evaluates the green and blue training through a designated board of officers. In addition, a mobile training support team is formed under the responsibility of BIPSOT, including representatives from the overseas operations directorate and sponsor directorates (specific to the arms and service of the deployable contingents) to carry out a periodic evaluation of the training imparted to the contingents.

Continuous and proactive monitoring mechanism plays a vital role in PKO success. 24/7 monitoring of the mission areas is ensured through the Mission Monitoring Centre (MIMOC), the de-facto operational centre of

---

[18] UNODA, "IEDs – a Growing Threat," 2022, https://www.un.org/disarmament/convarms/ieds-a-growing-threat/.

the AHQ Overseas Operations Directorate, to have real-time situation awareness and first-hand information from the mission areas. MIMOC follows a regular reporting chain: all regular reports and impromptu incidents in the mission areas are first reported to MIMOC; from there, it is disseminated to the designated desk officers and circulated in the command channel. This allows the fastest communication with the contingents and enables them to make necessary decisions at the earliest possible time. MIMOC is also utilised for impromptu video tele conferences with the contingents and monitoring of their rotation flights and shipments. It houses physical and digital archives of all the UN missions participated in by the Bangladesh Army.

Grooming officers for mission leadership roles is essential for next generation PKO. Based on its extensive peacekeeping experience, the Bangladesh Army is able to offer its expertise in the different tiers of mission leadership. Officers are prepared for higher UN assignments through services in relevant appointments, e.g., to be eligible for the appointment of contingent commander, an officer must have previous experience as a contingent member. Meanwhile, to become a sector commander, one should have experience as contingent commander.

Challenges for Peacekeeping Preparedness

Maintaining the ratio designated by the uniformed gender parity strategy (UGPS). According to the Uniformed Gender Parity Strategy (UGPS) 2018–2028, [19] the female ratio of each military contingent must be raised to 15% by 2028. By 2023, it is supposed to be 10%, with a 1% increase every year. The Bangladesh Army, however, can contribute only around 6%, on average, of the total military strength deployed in UN missions, while most of the manoeuvre contingents have much less than that. The main challenge in implementing the UGPS is that the overall ratio of females in the Bangladesh Army is only 2%, from which it is challenging to yield 10% for the UN contingents. However, as of 2021, the overall ratio of women among all the TCCs that formed contingents across the UN peacekeeping missions

[19] DPO, "Uniformed Gender Parity Strategy 2018-2028," 2018.

was only 5.8% [20], and, as of 2022, Bangladesh secured the third position in the list of the top contributors of female peacekeepers.[21]

Lack of indigenous defence production vis-à-vis an inadequate supply of equipment spare items. As opposed to the demands of deployable contingents, Bangladesh has a meagre defence production capacity, which is limited to the production of small arms and ammunition, assembly of light military trucks and modification of combat vehicles like APCs/LAVs/MRAPs and other transport and utility vehicles. Again, due to rough terrain and weather conditions and the frequent conducting of operations without sufficient maintenance pauses, equipment wear and tear of the contingents, especially for vehicles' spares, is very high. Due to a lack of indigenous production capability, Bangladesh primarily depends on importing these spares, which are currently at stake due to the ongoing Russia-Ukraine War.

Sometimes, unforeseen situations, like natural disasters, political instability, and conflicts, cause unwanted delays in procuring essential equipment and spares, which poses difficulties for the timely deployment of contingents and affects operational efficiency. COE shipment is also a complex, lengthy, and challenging issue involving exhaustive documentation and clearance processes. Sometimes, difficulties occur with shipments due to delayed receipts of clearance, instability and security issues in mission areas, and natural calamities.

Adaption to the Mission Environment. The terrain and weather in mission areas differ significantly from those of Bangladesh, which makes it challenging to provide training for operational sustenance in a mission-specific environment. Hence, deployable contingents must undergo a long and strenuous PDT to acquire physical and mental robustness and to get acquainted with the equipment used in the mission areas. Moreover, troops also need to get acclimatised to the terrain and weather of the mission areas in the initial days of their deployment.

While French is neither an official nor a commonly used language in Bangladesh, peacekeepers are deployed primarily in francophone countries,

---

[20] United Nations Peacekeeping, "Uniformed Women in Peace Operations (2022)," A4P, 2022, https://peacekeeping.un.org/en/uniformed-women-peace-operations-2022.

[21] SIPRI, "Women in Multilateral Peace Operations in 2020: What's the State of Play?," *SIPRI Multilateral Peace Operations Database*, 2022, https://www.sipri.org/sites/default/files/2020-10/women_in_multilateral_peace_operations_in_2020_small.pdf.

which puts additional pressure on preparedness for operational language training for deployable contingents.

Prospects and Opportunities for the Future

Formulation of peacekeeping doctrines and guidelines. It is high time for the Bangladesh Army to formulate doctrines for concentrating, training, and equipping deployable contingents based on its previous 35 years of experience and all pertinent books, pamphlets, instructions, directives, policies, and guidelines. It should also include the TTP, a mobilisation scheme for the UN deployment and maintenance aspects in the mission areas, and a focus on integrating high-tech and robust equipment.

Development of peacekeeping training. To keep pace with adaptive peacekeeping training due to evolving security dynamics in the mission areas and to address the knowledge gap between officers and men about recent peacekeeping phenomena, a separate study package on peacekeeping should be incorporated into all the training at the formation and institutional levels. This will allow officers and men to quickly pick up the pre-deployment training material once selected for UN missions and undergo the PDT.

Bangladesh received an official invitation through a letter of interest from Elsie Initiative Funding (EIF) as one of the highest troop-contributing countries. The BIPSOT initiated a project proposal on *"Enhancement of Cap of BIPSOT to Train Female Peacekeepers"*, and the proposal was approved by the EIF Steering Committee on 29 Mar 2022. Implementing this proposal will significantly enhance BIPSOT's capabilities in terms of training female peacekeepers.

UN accreditation of the CIED training institutions like OC&S and ECSME is under process. The Bangladesh Army is also eager to share its CIED training expertise with other TCCs under the light coordination mechanism (LCM) initiative of the UNDPO, which was pledged during the 2021 Seoul Peacekeeping Ministerial. Bangladesh also contributed CIED training aids to the UNMAS in 2022.

Induction of modern equipment. In recent years, the Bangladesh Army has procured and deployed much new modern equipment in UN missions, such as MRAPs, UASs, LAVs, and different types of CIED equipment and communication gadgets. Through deployment in peacekeeping operations, its officers and men are being trained on the operation and maintenance of this state-of-the-art equipment. This, in turn, adds to its overall modernisation efforts and is one of the most significant opportunities for the Bangladesh Army through UN deployments.

Diplomatic relationships and investment in African countries. The superior professionalism, dedication, and contribution of the Bangladesh Army's contingents in peacekeeping missions have earned it a bright image and an unwavering friendship with the host African nations. The Bangladesh government may utilise these ties to further deepen diplomatic relationships with those countries and exploit opportunities for investment in agriculture and industries.

Co-deployment/joint pledge with Other TCCs. The USA has proposed to provide UASs for the Bangladesh Army to boost UN peacekeeping operational capabilities under Global Peace Operations Initiative funding. This will enhance the capability and image of the Bangladesh Army as a competent peacekeeping force and enable it to induct 4th-generation UAS technology in collaboration with the USA. Recently, Gambia has also proposed the co-deployment of a contingent with the Bangladesh Army, which reflects the confidence of the UN and other TCCs in Bangladesh as a peacekeeping partner and highlights the superior brand value of the Bangladesh Army as an effective peacekeeping force. It will also open a new horizon of bilateral military cooperation with other nations.

Regional peacekeeping hub in Bangladesh. General Birame Diop, military adviser, DPO, UN HQ, floated the idea of establishing a regional peace operational hub for south and central Asian nations and chose Bangladesh as the leading country. This reflects the UN's confidence in Bangladesh and would open new windows of opportunity for Bangladesh to rebrand itself as a world peacekeeping leader.

With a long and distinguished history of UN peacekeeping, the Bangladesh Army has made a significant contribution and demonstrated its firm commitment to establishing world peace by participating in several UN peacekeeping missions, including the most challenging and dangerous ones in countries such as Mali, South Sudan, the Central African Republic, and D R Congo. Through its contributions over more than three decades, the Bangladesh Army has helped to stabilise conflict-affected areas, protect civilians, and promote the rule of law in the most distressed parts of the world. Since 2019, Bangladesh has been continuously crowned as the world's top troop-contributing country with sustained growth in its peacekeeping contribution. It has also positioned itself as a UN peacekeeping role model with its supreme performance and sacrifices. The ultimate sacrifice of valiant Bangladeshi peacekeepers remains a source of inspiration and strength for total commitment to global peace.

However, challenges remain, including the need to meet the high standards of training and discipline; maintain a continuous troop supply despite huge demands back home; adapt to changing operational environments; keep pace with rapid technological development and the modernisation of equipment; and meet the evolving requirements of the UN. Moreover, the COVID-19 pandemic and current Russia-Ukraine War have added complexity to peacekeeping preparedness. With these challenges come opportunities for the Bangladesh Army to enhance the capacity of its human resources and equipment, utilise and evaluate its training and operational preparedness, and expand its influence in the domain of peacekeeping. The Bangladesh Army can also explore new partnerships and collaborations with other countries and international organisations to leverage its resources and expertise.

References

Armed Forces Division. "Bangladesh in UN Peace Operation." UN Peacekeeping. Dhaka, April 18, 2023. https://afd.gov.bd/un-peacekeeping/Bangladesh-in-un-peace-operation.

DPO. "Uniformed Gender Parity Strategy 2018-2028," 2018.

Global Terrorism Index. *Global Terrorism Index 2022. Global Terrorism Index*. Sydney, Australia, 2022. https://doi.org/10.4324/9780203731321.

Micallef, Mark, and Matt Herbert. "Fragile States and Resilient Criminal Ecosystems." *Global Initiative against Transnational Organized Crime*, no. August (2022).

Overseas Operations Directorate. "Policy on Pre-Deployment Preparation (PDP) and Pre-Deployment Training (PDT) of UN Contingents." Dhaka: Army Headquarters, 12 Nov, 2017.

Overseas Operations Directorate Archive. "UN Peacekeeping Missions of the Bangladesh Army." *UN Archive*. Dhaka: Overseas Operations Directorate, General Staff Branch, Army Headquarters, 18 Apr, 2023.

Peacekeeping News. "Many Challenges Remain for Sahel Joint Force, Warns Lacroix, Welcoming 'Increased Coordination'," 16 Nov, 2020. https://peacekeeping.un.org/en/un-news/many-challenges-remain-sahel-joint-force-warns-lacroix-welcoming-increased-coordination.

SIPRI. "Women in Multilateral Peace Operations in 2020: What's the State of Play?" *SIPRI Multilateral Peace Operations Database*, 2022. https://www.sipri.org/sites/default/files/2020-10/women_in_multilateral_peace_operations_in_2020_small.pdf.

The Daily Star. "United Nations Peace Operations and Bangladesh," 26 Mar, 2021. https://www.thedailystar.net/supplements/celebrating-50-years-Bangladesh/news/united-nations-peace-operations-and-Bangladesh-2066693.

UN DPO DFS. *Operational Readiness Preparation for Troop Contributing Countries in Peacekeeping Missions. United Nations*. Department of Peace Operations, 2018. https://peacekeeping.un.org/sites/default/files/3._military_-_4_guidelines_on_opr.pdf.

United Nations. "Where We Operate - United Nations Peacekeeping," 2023. https://peacekeeping.un.org/en/where-we-operate.

United Nations Peacekeeping. "Uniformed Women in Peace Operations (2022)." A4P, 2022. https://peacekeeping.un.org/en/uniformed-women-peace-operations-2022.

UNODA. "IEDs – a Growing Threat," 2022. https://www.un.org/disarmament/convarms/ieds-a-growing-threat/

CHAPTER VI

FROM DHAKA TO BIHAC (BOSNIA): AN ANALYSIS OF A COMPLEX PEACEKEEPING MISSION

LIEUTENANT COLONEL SYED MD AS-SAZID

Serious inter-communal clash broke out in Croatia in June 1991, when this country and its northern neighbour, Slovenia, declared themselves independent from Yugoslavia, and Serbs living in Croatia, supported by the Yugoslav People's Army (JNA), opposed this movement. Efforts by the European Community to stop the hostilities and resolve the crisis in the framework of the Conference on Yugoslavia had proved unsuccessful. Consequently, the United Nations (UN) became actively involved with the Security Council unanimously adopting its resolution 713 (1991) on 25 September 1991, expressing deep concern at the deteriorating security situation and called on all states to implement immediately a "general and complete embargo on all deliveries of weapons and military equipment to Yugoslavia". While the Council fully supported the efforts already undertaken by the European Community with the support of the Conference on Security and Cooperation in Europe (CSCE), it soon became clear that the most valuable contribution the UN could make at that stage was a peace-keeping operation to create the necessary conditions for the pursuit of political negotiations for a peaceful settlement. Thus, on 21 February, the Security Council, by its resolution 743 (1992), established UNPROFOR for an initial period of 12 months with a condition that the Force should be an interim arrangement to create the conditions of peace and security required for the negotiation of an overall settlement of the Yugoslav crisis within the framework of the European Community's Conference on Yugoslavia.[1]

---

[1] Department of Public Information, "Former Yugoslavia - UNPROFOR," *United Nations Protection Force*, 1996.

UNPROFOR was one of the most challenging and complex missions in the history of UN peacekeeping operations characterized by complex geo-political tensions and a high level of hostility between the warring factions leading to gross human rights violations, including genocide, ethnic cleansing, and rape. Among the many countries that contributed to the UNPROFOR, Bangladeshi peacekeeping contingent played a very significant role in the mission's success. As of 30 November 1994, Bangladesh sent a total of 1,235 military personnel, 40 police personnel and 43 observers from different forces.[2] The only Bangladeshi contingent named Bangladesh Battalion (BANBAT) was deployed in Bihac enclave of Bosnia-Herzegovina and served under tremendous odds within hostile forces' actions and extreme weather conditions. The contingent had been deployed in September - October 1994, and immediately after their deployment, from November, Bosnian forces had laid an impregnable siege all around Bihac. Despite the constant threat of ambushes, bombings, and sniper fire, Bangladeshi peacekeepers remained committed to their mission and carried out their duties with utmost professionalism and dedication under the mandate of UNFPOFOR to protect the population of Bihac from the wrath of Serbian Army. With supplies cut off for about ten consecutive weeks and under relentless attacks by Serbian forces, the heroic saga of Bangladeshi peacekeepers kept the UN flag flying high at the cost of their sweat, blood, and life. They left a saga of heroism and sacrifice that is still cherished by the people of Bihac.

The contribution of Bangladesh to the UNPROFOR in Bosnia and Herzegovina is a testament to the country's commitment to peacekeeping and its willingness to contribute to the global efforts to maintain peace and security. Bangladeshi peacekeepers, being one of the pioneers, not only contributed to the success of the mission but also gained valuable experience in conducting peacekeeping operations in a highly challenging and hostile environment. The country's participation in the UNPROFOR helped to enhance its reputation as a reliable and responsible member of the international community and a leading contributor to UN peacekeeping missions.

Against this backdrop, this paper endeavoured to highlight the glorious and valiant contribution of Bangladeshi peacekeepers in the service of peace in Bosnia-Herzegovina under the mandate of UNPROFOR. In doing so, a detailed historical account of peacekeeping operations by the BANBAT has been given. First, the genesis of the crisis and background of the UN

---

[2] Ibid.

deployment have been discussed, followed by a brief description of its mobilization and deployment. Then operational and administrative challenges faced by the contingent have been discussed, along with the infamous 1994 Siege of Bihac and the stories of survival and fighting of the BANBAT. Finally, the everlasting impression left out by the Bangladeshi peacekeepers and the invaluable lessons carried forward by Bangladesh from this mission has been briefly highlighted.

The article has been developed as a descriptive study following a qualitative analysis of both primary and secondary sources. Several eyewitnesses have been interviewed by the author that substantiated the research findings from the secondary sources and aided the study of relevant documents, primarily available from open sources. The author also had the opportunity to retrieve some of the privileged information from the appropriate archive of the Bangladesh Army.

Genesis of Crisis and Background of the UN Deployment

Religious differences were a key element of the Balkan wars of the early 1990s. Bosnia-Herzegovina, formerly a part of Yugoslavia, had three large ethnic communities with diverse religious entities. Bosnians were Muslims, Croats were Catholic Christians, and Serbs were Orthodox Christians. In the 1980s, Yugoslavia started falling apart as a smaller independent state because of internal political turmoil and nationalist claims along ethno-religious lines filled the power vacuum. Subsequently, in April 1992, the Bosnian war began with the declaration of independence by Bosnia-Herzegovina from Yugoslavia.[3] By December 1995, till the signing of the Dayton Peace Accord, approximately 100,000 Bosnian Muslims and Croatian Catholic Christians were massacred by Bosnian Serb forces backed by the Serb-dominated Yugoslav Army.[4] Failing negotiations, the international community agreed to deploy United Nations (UN) Peacekeepers, and thus the United Nations Protection Force (UNPROFOR) emerged as the first UN peacekeeping force in Croatia and Bosnia-Herzegovina.

Initially, UNPROFOR was established in Croatia as an interim arrangement to create the conditions of peace and security required to negotiate an overall settlement of the Yugoslav crisis. Its mandate was to ensure that Croatia's

---

[3] Eric Patterson, "Bosnia: Ethno-Religious Nationalisms in Conflict," *Religion and Conflict Case Study Series*, August 2013.
[4] Recep Keskin, "The Dispute between Bosnian Muslims and Serbs," *CSUSB Scholar Works* (California State University, 2003).

three UN Protected Areas (UNPAs) were demilitarized and that all persons living there were protected from fear of armed attack. In 1992, UNPROFOR's mandate and strength were enlarged to increase its monitoring and control mechanisms in Croatia and Bosnia-Herzegovina. Later that year, the mandate was further broadened to support the humanitarian efforts by the United Nations High Commissioner for Refugees (UNHCR) and Red Cross. In addition, it monitored the 'no-fly' zone, banning all military flights in Bosnia and Herzegovina, and the United Nations' safe areas' established by the Security Council. In March 1994, UNPROFOR started monitoring the implementation of a ceasefire agreement between the Croatian Government and local Serb authorities.[5]

Figure 6-1: Former Yugoslavia during the 1993 Yugoslav War[6]

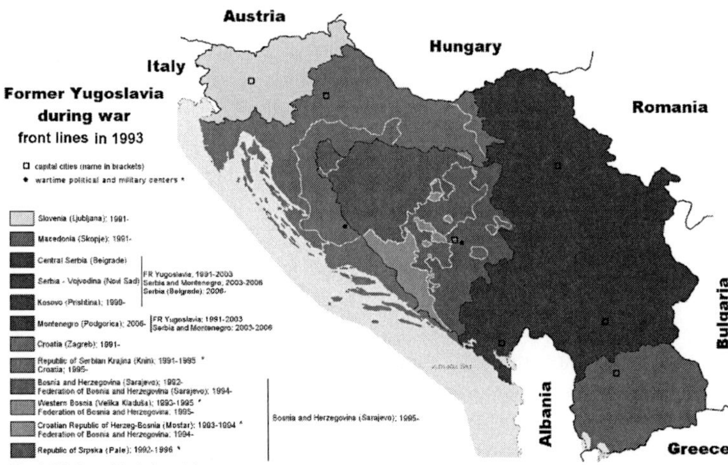

UNPROFOR was one of the few UN missions allowed to use force in self-defence in coordination with the North Atlantic Treaty Organization (NATO) for combat air support. UNPROFOR also monitored the implementation of ceasefire agreements signed by the Bosnian Government with Croat and Serb forces. On 31 March 1995, the Security Council of the UN decided to replace UNPROFOR with three different but interlinked

[5] Department of Public Information, "Former Yugoslavia - UNPROFOR."
[6] Paweł Goleniowski, "Former Yugoslavia during War (1993)," Wikimedia Commons, July 21, 2011, https://commons.wikimedia.org/wiki/ File:Map_of_war_in_Yugoslavia,_1993.png.

missions, such as the United Nations Preventive Deployment Force (UNPREDEP) in Macedonia and the United Nations Confidence Restoration Operation in Croatia (UNCRO) in Croatia, with restructured UNPROFOR operations ongoing in Bosnia and Herzegovina until their replacement by NATO and European Union (EU) missions in December 1995.[7]

The deployment of peacekeepers from Muslim countries in Bosnia-Herzegovina was withheld until 1993 due to complex ethnic, demographic, and geo-political dynamics leading to objections from some European nations. NATO countries failed to muster additional troops, with the situation worsening. Here the Organization of Islamic Cooperation (OIC) played an active diplomatic role in pushing the decision to deploy Peacekeepers from non-European nations, especially from Muslim countries like Bangladesh, Malaysia, and Pakistan. In this context, requested by the UN to deploy one mechanized infantry battalion within 45 days, Bangladesh readily accepted the proposal and started with mobilization and training of the contingent in June 1994 in Savar Cantonment.[8]

Mobilization of the BANBAT

The BANBAT was formed with the nucleus of 10 East Bengal Regiments consisting of 1,220 men.[9] It was assigned to replace a French battalion at Bihac, a small Muslim enclave surrounded by Serbs in Bosnia-Herzegovina.[10] Before the deployment, all Armored Personnel Carrier (APC) commanders, drivers, and firers of the contingent participated in a month-long intensive training organized by the United Nations at the Jovlen Garrison in Slovakia. Since the Force Headquarters was in Zagreb, the capital of Croatia, all kinds of movements and administrative activities were carried out from there.[11]

---

[7] Department of Public Information, "Former Yugoslavia - UNPROFOR."
[8] Salim Akhtar, "Bangladesh UN Contingent Peacekeeping Mission in Bosnia-Herzegovina," *The Daily S*, February 26, 2018.
[9] AHQ Education Directorate, "United Nations Peacekeeping Mission," in *History of Bangladesh Army, 1st Edition*, ed. Lieutenant Colonel Kazi Sajjad Ali Zahir, BP (Retd), vol. 3 (Education Directorate, Bangladesh Army Headquarters, 2015), 161–62.
[10] Akhtar, "Bangladesh UN Contingent Peacekeeping Mission in Bosnia- Herzegovina."
[11] AHQ Education Directorate, "United Nations Peacekeeping Mission."

Arrangements were made with the UN to provide specialized equipment, including APCs, and organize training of the APC crews. The East German Army's equipment, stores, and winter uniforms were supplied from the stock warehouses in Slovakia. A training team of about 100 plus people was sent about a month ahead to Slovakia to train and take over the equipment.[12]

By August 1994, the battalion arrived at Zagreb and stayed for a month at Camp Pleso with soldiers from other NATO countries. The training team from Slovakia, along with BTR-70s, generators, and other equipment, arrived in Zagreb by train. BANBAT received 68 APCs (BTR-70), a few tracked Armoured Recovery Vehicles (ARV), a large number of trailer-mounted generators, and other equipment from the UN. The arrival of Contingent Owned Equipment (COE) from Bangladesh was delayed due to transshipment issues at the port of Alexandria, Egypt.[13]

BANBAT's Deployment in Bihac

Led by Colonel Salim Akhtar (later retired as Brigadier General), the contingent was deployed in four phases on 20, 24, 27 September, and 7 October, respectively, under the mandate of the United Nations Protection Force (UNPROFOR). The contingent started its peacekeeping operational activities by setting up three base camps in the urban areas: the battalion headquarters was established at Choralichi in the centre of the province; a logistics base was set up at Velika Claudusa in the northwest, and a company group was deployed in the southeastern part in Bihac city.[14] There was also one company each at Choralichi and Velika Claudusa camps.[15]

Chief Logistics Officer (CLO) Lieutenant Colonel Liakat Ali Khan (later retired as Colonel) was the Commander of the Velika Claudusa camp, whereas Major Towhid (later retired) was the Commander of the Bihac Camp. However, the operational activities of the Velika Claudusa were controlled by Major Bashir Haider (later retired in the same rank), who was appointed as the Base Operational Officer (BOO). Battalion Quarter Master (QM) Major Shafiqul Islam (later retired) and Ammunition Technician Officer (ATO) Major Ashraf (later retired), among several others, were also

---

[12] Akhtar, "Bangladesh UN Contingent Peacekeeping Mission in Bosnia- Herzegovina."
[13] Ibid.
[14] AHQ Education Directorate, "United Nations Peacekeeping Mission."
[15] Shafiqul Islam, "Individual Experience of Serving in BANBAT (UNPROFOR) in Bosnia-Herzegovina," *Interviewed by Lieutenant Colonel Syed Md As-Sazid, Afwc, Psc* (Dhaka: Overseas Operations Directorate, Army Headquarters, January 1, 2022).

located in the logistics base. Contingent Commander Colonel Salim Akhtar, Chief Operations Officer (COO) Lieutenant Colonel Khaleq Hawlader (later retired), and Base Operations Officer Major Sikder (later retired as Brigadier General) were located at the Battalion Headquarters with several other officers.[16] The BANBAT also had integral subunits within its organogram, like workshop detachment, medical detachment, signals detachment, and logistic support elements.[17]

Operational and Administrative Challenges

From the beginning, the mission was very challenging for the BANBAT. There were Serbian territories around Bihac province, where the short-lived Republic of Serbian Krajina was formed, consisting of a Serb-Croat mixed population. It completely disintegrated Bihac from the states of Bosnia and Herzegovina. There was also deep hatred among different ethnicities and intermittent fighting among the ethnic militias, often deteriorating the security situation and trapping the BANBAT between their fights. The weather was freezing, and the terrain was mountainous,[18] for which Bangladeshi soldiers hardly had any training or experience. The contingent was not equipped to operate in freezing winter conditions, which demanded unique clothing, equipment, and winter survival knowledge. Inadequate driving skills of the APC and other soft-skin vehicles and their lack of driving experience under snow conditions were a big concern. In Bosnia, many fatalities used to occur due to accidents on mountain roads during the winter months.[19]

According to the Los Angeles Times, Bangladeshi peacekeepers were deployed in the Bihac region with an acute shortage of armaments and without adequate winter clothing. Its report also highlighted another critical challenge: the problem had been compounded by Bangladesh's lack of representation in senior leadership positions in the UN. Contrarily, when Bosnian Serbs attacked a French armoured vehicle in September, NATO immediately responded with an airstrike request from the UN.[20] However,

---

[16] Ibid.

[17] Akhtar, "Bangladesh UN Contingent Peacekeeping Mission in Bosnia- Herzegovina."

[18] AHQ Education Directorate, "United Nations Peacekeeping Mission."

[19] Akhtar, "Bangladesh UN Contingent Peacekeeping Mission in Bosnia- Herzegovina."

[20] Dean E Murphy, "Peacekeeper Wounded in Bosnia Dies - Balkans - Bangladeshi Was Opne of Five Injured in Bosnia Attack. UN Officials Denounce It as Most Serious Strike against Their Mission since War Began," *Los Angeles Times*, December

the contingent successfully faced these challenges due to the commanders' high professionalism, morale, and patience.

Siege of Bihac-1994

The Siege of Bihac was laid down for three years, from June 1992 until August 1995, by the Army of the Republic of Serbian Krajina and Bosniak dissenters led by the Fikret Abdic. It ended after the Croatian Army overran the rebel Serbs.[21] Abdich, an influential Muslim businessman, controlled a large area of Bihac and, in September 1993, declared the autonomy of the province of Western Bosnia after the Sarajevo government rejected the UN-EU peace plan for Bosnia. EU supported Abdich to pressure Bosnian President Alija Izetbegovic to accept the peace plan. In August 1994, Abdich and his 25,000 supporters were chased out of the Bihac pocket by the Bosnian Army's 5th Corps.[22]

By the third week of September 1994, when Bosnian forces around Sarajevo and central Bosnia were pressed by intense Serbian attack, the Bosnian Army 5th Corps Commander in Bihac decided to counterattack the Serb forces to breakout and relieve pressure from central Bosnia. It resulted in a significant defeat of the Serbs with a substantial loss of their territory and the capture of massive equipment from the retreating Serb forces, including tanks and artillery guns.[23] After one month, Bosnian and Croatian Serbs coordinated a counterattack on the enclave, swiftly regaining the lost territory and threatening the Bosnian government-held enclave and the BANBAT positioned at its centre.[24]

By 27 November 1994, Serb forces retook around one-third of the area and moved closer to the headquarters of the Bosnian Fifth Corps. France and Britain rejected the US plan to relieve the city. Bosnian-Serb forces first set a deadline of 1900 GMT on 26 November for the town's defenders to

---

14, 1994, Online edition, https://www.latimes.com/archives/la-xpm-1994-12-14-mn-8905-story.html.

[21] Gail Harmon, "War in the Former Yugoslavia: Ethnic Conflict or Power Politics?" (Boston College University, 2007), https://dlib.bc.edu/islandora/object/bc-ir:102115/datastream/PDF/view.

[22] John Pomfret, "Serbs Bombard Muslim Positions in Bihac," ed. Sally Buzbee, The Washington Post (Washigton D. C., December 6, 1994), https://www.washingtonpost.com/archive/politics/1994/12/06/serbs-bombard-muslim-positions-in-bihac/4af12b48-94ab-4399-b871-eefdb407775e/.

[23] Akhtar, "Bangladesh UN Contingent Peacekeeping Mission in Bosnia- Herzegovina."

[24] Pomfret, "Serbs Bombard Muslim Positions in Bihac."

surrender. Later they made a new offer for Bosnian-Muslim troops to surrender to Fikret Abdich's forces. However, the mayor of Bihac, Hamdija Kabiljagic, refused to surrender to Abdich's force, fearing that it might turn into a mass killing by the Serbs. Instead, he instructed his people to block the streets with trees and cars. Diplomatic efforts to halt the fighting continued. The UN proposed a three-day ceasefire in Bihac, but Radovan Karadzic, the Bosnian Serb leader, did not respond to this call.[25]

On 05 December, Croatian Serb forces, firing tank and artillery shells, blasted Muslim positions in the Bihac pocket. Besides Rocket Propelled Grenades (RPG), mortars and tanks, Croatian Serb forces continued to fire from artillery guns from nearby Croatian Serb territory on the Muslim positions in Velika Kladusa. In response, lightly equipped Muslims fought back with small arms. The renewed fighting indicated their intention to connect the Croatian Serb territory to the Bosnian Serb areas by capturing the roads and the railway links passing through the enclave. Croatian Serbs put Fikret Abdich in charge of their captured territory at this stage.[26]

BANBAT's Survival and Fight

For the September 1994 victory of the Bosnian Muslim forces in Bihac, Serb forces linked their success to the influence of BANBAT's presence in that area [27]. From November to December 1994, the details of Serbia and the Republic of Serbian Krajina cut off all logistics supplies to the UN and Bangladesh contingent in Bihac for two months and 22 days during severe winter.[28] During the siege period, many young officers displayed extreme courage and determination in holding their positions under intense fire. In one incident, Captain Masroor E Elahi (later retired as Major) risked his life to rescue his men, including the Contingent Commander, under intense Serb shelling on a meeting location in Bihac city. The major brunt of the Serb attacks was born by Velika Claudusa and Bihac camps. Serbians used to keep the BANBAT camps pinned down with continuous firing from small arms and rockets. There was no luxury of reluctant movement even within the camps due to fear of targeted shooting by Serbs. Any convoy moving out was most often welcomed by live firing, and even while remaining

---

[25] Emma Daly and Andrew Marshall, "Bihac Fears Massacre," ed. Christian Broughton, *Independent*, November 27, 1994, Online edition, https://www.independent.co.uk/news/bihac-fears-massacre-1439958.html.

[26] Pomfret, "Serbs Bombard Muslim Positions in Bihac."

[27] Akhtar, "Bangladesh UN Contingent Peacekeeping Mission in Bosnia- Herzegovina."

[28] AHQ Education Directorate, "United Nations Peacekeeping Mission."

inside APC, the crews could correctly guess the firing from its sound of impacts on the APC.[29]

The siege by Serbs started immediately after the departure of the French battalion, and the BANBAT was not even able to draw their first consignment of the rations and logistics collections.[30] At that time, the Chief Liaison Officer (CLO) of the BANBAT, Lieutenant Colonel Saidur Rahman, Bir Protik (later retired as Brigadier General), stayed at the Velika Klaudusa Camp in the absence of its camp commander. He took the initiative and led a patrol with his best officers and soldiers to negotiate with the local brigade commander of Abdich forces to allow passage for their logistics collections convoy. A few of the bravest officers, like Major Shafiqul Islam and Interpreter Lieutenant Commander Abdur Rouf (later retired as Captain) of the Bangladesh Navy (BN), volunteered to accompany him.[31] Under such a hostile situation, the group moved with small arms and grenades by two soft-skin vehicles rather than APCs and without escort to avoid raising their suspicion. After a deliberate search and navigation, they could find the rebel brigade headquarters inside deep jungles heavily guarded by fortified positions and tanks. The negotiation was somehow fruitful. Although BANBAT could not draw fresh rations across the border, they were advised to follow some lightly guarded routes. Following their guidelines, they could finally collect some supplies after several attempts.[32]

After a few days, the ration stocks again depleted, and the situation became so dire that the camp commander had to calculate the remaining life span of his men that could be survived with the leftover supplies. They started rationing everything starting, from food to fuel. After finishing the surplus stock left by the French forces, they started consuming the half-rotten and expired food stocks at one stage. Meanwhile, Battalion Headquarters at Choralichi collected some rations bypassing the siege. On 12 December 1994, in a desperate attempt, a halfway link patrol with eight APCs, led by Major Masud Rezwan (later retired as Major General), was sent from Velika Claudusa to Choralichi following an unconventional route to collect some

---

[29] Islam, "Individual Experience of Serving in BANBAT (UNPROFOR) in Bosnia-Herzegovina."

[30] Ibid.

[31] Saidur Rahman, "Memories of Peacekeeping Operations in Bosnia-Herzegovina," *Interviewed by Lieutenant Colonel Syed Md As-Sazid, Afwc, Psc* (Dhaka: Overseas Operations Directorate, Army Headquarters, January 1, 2022).

[32] Islam, "Individual Experience of Serving in BANBAT (UNPROFOR) in Bosnia-Herzegovina."

rations from their stocks. While returning, once the convoy had been entering the premises of the logistics base, Abdich's force fired two consecutive missiles from the surrounding hills on the last APC of the convoy.[33]

In this incident, APC Operator No. 396951 Sepoy Md. Ismail Hossain died, and four soldiers, including one JCO (Junior Commissioned Officer), were seriously injured. Captain Nahid (later Brigadier General) and two other soldiers (Lance Naik Shafi, East Bengal Regiment and Sepoy Nazrul, Armored Corps) rescued their wounded comrades with utmost bravery risking their own lives.[34] According to the eyewitnesses, the first missile had hit the APC in a way that one hand of Sepoy Ismail was blown away, and a few splinters hit his eyes and other body parts. Captain Nahid had first seen the explosion and rushed to the spot to rescue his comrades from the APC on blast. When he reached there, the second missile hit the APC and blew away the turret. Luckily, a few jerrycans full of fuel did not catch fire, saving the rest of the APC crews.[35] The contingent was offered a NATO airstrike, but the Contingent Commander asked for an immediate ceasefire to evacuate the wounded soldiers to the US Mobile Army Surgical Hospital (MASH) in Zagreb.[36] Later Sepoy Ismail died there on 13 December 1994. The UN outrightly denounced this attack as the most severe strike against the UNPROFOR since the war in Bosnia-Herzegovina.[37]

Los Angeles Times (issue of 14 December 1994) provided a detailed account of the situation. According to its Staff Writer Dean E. Murphy, with already depleted supplies and stocks, 1200 Bangladeshi peacekeepers were pinned down in a bomb shelter near the northwest Bosnian town of Velika Kladusa as Croatian Serbs and Serbian-backed Muslim rebels advanced on the area. Serb forces from the Krajina region of Croatia appeared to have the intention of overrunning the Muslim holdings of the Bihac township, which lay within the UN's declared 'safe area.' With the battle closing in, the BANBAT got trapped in the region with only two weeks of food supply and insufficient weapons and ammunition to repel Serbian attacks. UN

---

[33] Ibid.

[34] AHQ Education Directorate, "United Nations Peacekeeping Mission."

[35] Islam, "Individual Experience of Serving in BANBAT (UNPROFOR) in Bosnia-Herzegovina."

[36] Akhtar, "Bangladesh UN Contingent Peacekeeping Mission in Bosnia- Herzegovina."

[37] Murphy, "Peacekeeper Wounded in Bosnia Dies - Balkans - Bangladeshi Was Opne of Five Injured in Bosnia Attack. UN Officials Denounce It as Most Serious Strike against Their Mission since War Began."

spokesperson Michael C. Williams told the Los Angeles Times, "The attack is one of the most flagrant, deliberate, and calculated assaults on UNPROFOR to have taken place in the history of this peacekeeping operation." While Sepoy Ismail was the 133rd peacekeeper of the UNPROFOR to die, the mission headquarters was still hesitant to respond for fear of provoking new attacks against the peacekeepers. Even the UN's call for NATO to deploy air assets was deemed too risky because the growing number of antiaircraft missile systems controlled by the Bosnian Serbs covered almost 40% of Bosnia, including the airport in the capital Sarajevo.[38]

Recognitions and Remembrance

The recognition of BANBAT was not quite significant compared to the hardship, performance, and sacrifice of its officers and men in safeguarding Bihac and its citizens under the Serbian siege. However, the contingent received numerous letters of appreciation from the Force Commander, Lieutenant General Rupert Smith of the UK, and several others from UNPROFOR headquarters, who sincerely appreciated the professionalism of Bangladesh soldiers.[39] Captain Masroor-e-Elahi received commendations from UNPROFOR's Bihac Area Commander and later from the Force Commander for his bravery and heroic contributions. Two Bangladesh Army soldiers were martyred during the mission's entire duration, and twelve others were injured.[40]

On 20 April 2017, Bosnian Assistant Minister for Multilateral Affairs, Ministry of Foreign Affairs, Mr Almir Sahovic, in a meeting with Bangladeshi Ambassador to the Netherlands, shared a deep sense of gratitude on behalf of the Bosnian people to Bangladeshi peacekeepers for their contribution during the Bosnian Civil war. In his view, because of Bangladeshi peacekeepers, the people of Bihac were better protected than other parts of Bosnia, like Srebrenica. Over 8000 lives were lost in what is now known as the Srebrenica massacre.[41]

---

[38] Ibid.
[39] Akhtar, "Bangladesh UN Contingent Peacekeeping Mission in Bosnia- Herzegovina."
[40] AHQ Education Directorate, "United Nations Peacekeeping Mission."
[41] Embassy of Bangladesh to the Netherlands, "Bosnia Praises Bangladesh Peace Credential and Economic Vigour," April 22, 2017, http://bangladeshembassy.nl/bosnia-praises-bangladesh-peace-credential-and-economic-vigour/.

Dr M Riaz Hamidullah, His Excellency the Ambassador of the Bangladesh Embassy in the Netherlands, visited Bihac City and met its mayor Mr Suhret Fazlic during his visit to Bosnia-Herzegovina from 13 to 19 June 2021. During this meeting, the local leaders deeply recalled the valiant heroism and contribution of the Bangladeshi peacekeepers deployed in Bihac to protect its citizens.[42]

Lessons Learnt from UNPROFOR

Being one of the pioneer deployments of Bangladeshi peacekeeping contingents, as the first UN mission in the European soil and having experienced an extremely rough weather and hostile environment, BANBAT (UNPROFOR) and its valiant peacekeepers brought home not only immense glory, but also invaluable lessons. Some of these lessons that guided Bangladeshi peacekeepers in last three decades to crown Bangladesh as the top contributing country and a brand name in the domain of the UN peacekeeping are appended below:

Importance of Cultural Awareness and Building Trust. The Bangladeshi peacekeepers learned the importance of understanding the local culture, customs, and traditions of the people they were serving. By dint of their supreme professionalism and willingness to sacrifice, they were quick in building trust and establishing good relationships with the local population that was crucial for the success of their mission.

Adapting to the Environment. The peacekeepers learned the importance of adapting to the challenging environment they were operating in. They had to quickly learn how to navigate through difficult terrain, deal with extreme weather conditions, and operate in a hostile environment. Thus, the peacekeepers of the BANBAT (UNPROFOR) left for the future peacekeepers the lessons of unbending moral strength at the face of extreme odds.[43]

Flexibility and Patience. Bangladeshi peacekeepers learned the importance of being flexible and patient in dealing with complex situations. They had to adapt quickly to changing circumstances and remain patient in the face

---

[42] Overseas Operations Directorate Archive, "UN Peacekeeping Missions of the Bangladesh Army," *UN Archive* (Dhaka: Overseas Operations Directorate, General Staff Branch, Army Headquarters, April 18, 2023).

[43] Akhtar, "Bangladesh UN Contingent Peacekeeping Mission in Bosnia- Herzegovina."

of difficult and stressful situations. With the scarce supply of food, the entire contingent sustained by maintaining a harsh but self-imposed austerity.[44]

Courage and Determination. Officers and men of the BANBAT (UNPROFOR) made a glaring example of unwavering courage and determination for the peacekeepers of all generations. The Contingent Commander Brigadier General Salim Akhtar took immense pride out of his contingents and mentioned, "Credit goes to the absolute tenacity and courage of the Bangladesh soldiers and young officers who served under my leadership and delivered what I asked for under any condition".[45] Officers and men of this contingent, even under hostile fire, did not hesitate to risk their life for evacuating their fallen and injured comrades.[46]

Importance of Humanitarian Assistance. The peacekeepers learned the critical role of humanitarian assistance in maintaining peace and security in conflict zones. They witnessed firsthand the impact of aiding vulnerable communities, and the positive impact it had on building trust and establishing good relationships with the local population. This is a lesson that has always been carried forwarded by Bangladeshi peacekeepers anywhere they have been deployed in last three decades.

Importance of Media. Media always acts as a very important and strong tool for building public and international perception of a peacekeeping mission's successes and failures. Contingent Commander of BANBAT (UNPROFOR) understood the importance of media and went for several interviews with electronic medias like BBC, SKY TV UK, Peter Jennings Show, ABC TV USA, and a few print medias like Los Angeles Times and Philadelphia Times to project the truth and counter any kind of negative propaganda.[47]

Importance of Leadership Positions. Being a nascent peacekeeping nation, Bangladesh had very few staff officers and no senior officer in the leadership positions in the UNPROFOR headquarters, which deprived Bangladeshi peacekeepers from a strong voice in the decision-making areas.[48] Presently, Bangladesh has one force commander, three sector commanders, and many important staffs at the UN headquarters and the

[44] AHQ Education Directorate, "United Nations Peacekeeping Mission."
[45] Akhtar, "Bangladesh UN Contingent Peacekeeping Mission in Bosnia- Herzegovina."
[46] AHQ Education Directorate, "United Nations Peacekeeping Mission."
[47] Akhtar, "Bangladesh UN Contingent Peacekeeping Mission in Bosnia- Herzegovina."
[48] Murphy, "Peacekeeper Wounded in Bosnia Dies - Balkans - Bangladeshi Was Opne of Five Injured in Bosnia Attack. UN Officials Denounce It as Most Serious Strike against Their Mission since War Began."

headquarters of different missions including the Chief of Staff of the United Nations Department of Peace Operations (UNDPO) at New York.[49]

Commencing its journey of UN peacekeeping in 1988 with United Nations Iran-Iraq Military Observer Group (UNIMOG), today Bangladesh is a brand name with more than three decades of experience in peacekeeping affairs and has been crowned as the top Troops Contributing Country (TCC). Out of 69 peacekeeping missions since 1948, Bangladesh has participated in 54 missions in 40 countries. So far, 1,63,887 peacekeepers from Bangladesh have participated in UN missions.[50] Today, Bangladesh Army has around 6000 personnel, both men and women, deployed in various contingents and Military Staff Officers/ Observers in 08 peacekeeping missions.

In the 1990s, Bangladesh Army was nascent in peacekeeping with the sole experience of a peacekeeping field mission under the United Nations Iraq Kuwait Observation Mission (UNIKOM) in 1991 and United Nations Transitional Auth in Cambodia (UNTAC) in 1992.[51] In this context, UNPROFOR was the firsthand experience of Bangladesh deploying a contingent on European soil. The outstanding leadership and unwavering determination of commanders at all levels, superior training, staunch discipline, and unmatched courage of Bangladeshi soldiers combinedly fought against all odds. Inferior equipment, compounded by the severe shortage of rations and stocks, had put a little barrier to their valour and spirit.

The BANBAT (UNPROFOR) in Bosnia-Herzegovina always kept the UN flag flying high under highly adverse situations. Under similar conditions, some contemporary European battalions, including those under NATO, surrendered the areas to the advancing Serbs. Serb forces carried out a massive killing in the enclaves of Zepa and Srebrenica after the responsible UN battalions failed to sustain them. Nevertheless, BANBAT stood against every odd with meagre supplies and soldiers with no fighting experience amidst European winter.[52] They left a footprint of courage and examples of determination that led their successors to fight and sacrifice for the cause of

---

49 Overseas Operations Directorate Archive, "UN Peacekeeping Missions of the Bangladesh Army."

50 Bangladesh Army, "Our Contribution in Global Peacekeeping," Peacekeeping (General Staff Branch, April 2023), https://www.army.mil.bd/PeaceKeeping-Work.

51 Overseas Operations Directorate Archive, "UN Peacekeeping Missions of the Bangladesh Army."

52 Akhtar, "Bangladesh UN Contingent Peacekeeping Mission in Bosnia- Herzegovina."

global peace till today. After all these years, their heroism is still remembered by the Bosnians.

The glorious contribution of BANBAT under the mandate of UNPROFOR is a rare example of professionalism in peacekeeping and a testimony of humanity and kindness by Bangladeshi peacekeepers that kept them alive in the heart of those Bosnian people in their stories and memoirs.

References

AHQ Education Directorate. "United Nations Peacekeeping Mission." In *History of Bangladesh Army, 1st Edition*, edited by BP (Retd) Lieutenant Colonel Kazi Sajjad Ali Zahir, 3:161–62. Education Directorate, Bangladesh Army Headquarters, 2015.

Akhtar, Salim. "Bangladesh UN Contingent Peacekeeping Mission in Bosnia- Herzegovina." *The Daily S*, February 26, 2018.

Bangladesh Army. "Our Contribution in Global Peacekeeping." Peacekeeping. General Staff Branch, April 2023. https://www.army.mil.bd/PeaceKeeping-Work.

Daly, Emma, and Andrew Marshall. "Bihac Fears Massacre." Edited by Christian Broughton. *Independent*, November 27, 1994, Online edition. https://www.independent.co.uk/news/bihac-fears-massacre-1439958.html.

Department of Public Information. "Former Yugoslavia - UNPROFOR." *United Nations Protection Force*, 1996.

Embassy of Bangladesh to the Netherlands. "Bosnia Praises Bangladesh Peace Credential and Economic Vigour," April 22, 2017. http://bangladeshembassy.nl/bosnia-praises-bangladesh-peace-credential-and-economic-vigour/.

Goleniowski, Paweł. "Former Yugoslavia during War (1993)." Wikimedia Commons, July 21, 2011. https://commons.wikimedia.org/wiki/File:Map_of_war_in_Yugoslavia,_1993.png.

Harmon, Gail. "War in the Former Yugoslavia: Ethnic Conflict or Power Politics?" Boston College University, 2007. chrome-extension://efaidnbmnnnibpcajpcglclefindmkaj/https://dlib.bc.edu/islandora/object/bc-ir:102115/datastream/PDF/view.

Islam, Shafiqul. "Individual Experience of Serving in BANBAT (UNPROFOR) in Bosnia-Herzegovina." *Interviewed by Lieutenant Colonel Syed Md As-Sazid, Afwc, Psc*. Dhaka: Overseas Operations Directorate, Army Headquarters, January 1, 2022.

Keskin, Recep. "The Dispute between Bosnian Muslims and Serbs." *CSUSB Scholar Works*. California State University, 2003.

Murphy, Dean E. "Peacekeeper Wounded in Bosnia Dies - Balkans - Bangladeshi Was Opne of Five Injured in Bosnia Attack. UN Officials Denounce It as Most Serious Strike against Their Mission since War Began." *Los Angeles Times*, December 14, 1994, Online edition. https://www.latimes.com/archives/la-xpm-1994-12-14-mn-8905-story.html.

Overseas Operations Directorate Archive. "UN Peacekeeping Missions of the Bangladesh Army." *UN Archive*. Dhaka: Overseas Operations Directorate, General Staff Branch, Army Headquarters, April 18, 2023.

Patterson, Eric. "Bosnia: Ethno-Religious Nationalisms in Conflict." *Religion and Conflict Case Study Series*, August 2013.

Pomfret, John. "Serbs Bombard Muslim Positions in Bihac." Edited by Sally Buzbee. The Washington Post. Washigton D. C., December 6, 1994. https://www.washingtonpost.com/archive/politics/1994/12/06/serbs-bombard-muslim-positions-in-bihac/4af12b48-94ab-4399-b871-eefdb407775e/.

Rahman, Saidur. "Memories of Peacekeeping Operations in Bosnia-Herzegovina." *Interviewed by Lieutenant Colonel Syed Md As-Sazid, Afwc, Psc*. Dhaka: Overseas Operations Directorate, Army Headquarters, January 1, 2022.

United States Central Intelligence Agency. "Balkan Battlegrounds: A Military History of the Yugoslav Conflict, 1990-1995." Washington D. C., USA: Office of Russian and European Analysis, 2013. https://www.loc.gov/resource/g6841sm.gct00210/?st=gallery.

Chapter VII

Proposed Withdrawal of MINUSMA: Regional Security Implications and the Future of UN Peacekeeping in Africa

Lieutenant Colonel Syed Md. As-Sazid

Mali, a country in the heart of West Africa, has long been grappling with a complex web of security challenges. Over the years, these challenges have manifested in the form of armed insurgencies, terrorist activities, and inter-communal conflicts, posing a significant threat to both the stability of Mali and the wider Sahel region. The United Nations Multidimensional Integrated Stabilization Mission in Mali (MINUSMA) was deployed in 2013 when the Malian state was on the verge of collapse following the disintegration of its northern regions by Tuareg rebels and terrorist groups. In response to the evolving security concerns, MINUSMA's mandate focused on restoring peace, protecting civilians, facilitating the political transition, and, most importantly, supporting the implementation of the 2015 Algiers Peace Agreement.[1] Malians also welcomed United Nations (UN) troops in the hopes of attaining security and stability, which has not been achieved over the last ten years. Thousands of people have been killed, and the continued violence has displaced hundreds of thousands.[2]

With more than 300 fatalities of peacekeepers and a budget of $1.2bn, MINUSMA has been the deadliest and costliest UN peacekeeping operation in history. Despite this, Mali's government has accused the mission of worsening the situation, leaving people distrustful of the UN. *"The government of Mali calls for the withdrawal of MINUSMA without delay,"* Malian Foreign Minister Abdoulaye Diop had said in June, stressing that Bamako *"is willing to cooperate with the United Nations on this issue"*.

---

[1] As-Sazid, "Emerging Security Challenges in the Sahel and the Need for an Adaptative Approach towards Peacebuilding."

[2] Mohamed, "Analysis: What's next for Mali after MINUSMA Withdrawal?"

Subsequently, on 30 June 2023, the United Nations Security Council (UNSC) unanimously decided not to renew the MINUSMA mandate and to end this decade-long peacekeeping mission in Mali.[3] However, the proposed withdrawal of MINUSMA has raised serious concerns and sparked intense debate regarding its consequences and the possible impacts it may have on regional security dynamics.

The proposed withdrawal of MINUSMA from Mali, following the induction of the Russian Wagner group and the withdrawal of French forces, is likely to have far-reaching impacts on the political, strategic, and security dynamics of Mali and its neighbouring countries. The possible escalation of transnational Jihadism, terrorism, inter-communal violence, drug/human trafficking, population displacement, and other organized crime in Mali may affect the fragile security environment of its neighbours. Its resultant regional instability may subsequently affect other large UN peacekeeping missions deployed in the Central African Republic (CAR), the Democratic Republic of the Congo (DRC), and South Sudan. Russia and China's covert military and economic influences will likely become more dominant, diminishing the influence of the US and other Western powers. The decision for the premature withdrawal of MINUSMA, without a comprehensive transition plan, could potentially leave a security vacuum in Mali, exacerbating existing challenges and creating new ones.

In these circumstances, understanding the potential consequences of this decision is crucial for formulating effective strategies to mitigate risks, enhance stability, and foster sustainable peace in Mali and the Sahel region. It is also essential to carefully assess the potential risks of escalation and spread of instability into its neighbouring countries, which could affect the regional security architecture. By shedding light on these possible impacts of MINUSMA's withdrawal, policymakers, regional actors, and international stakeholders can make informed choices and take necessary measures to ensure the long-term stability and security of Mali and the wider West African region.

Against this backdrop, this article explores the probable security implications of MINUSMA's withdrawal with comprehensive and predictive analyses of present and future events. In doing so, this paper will analyze the current security landscape in Mali; the role of MINUSMA, including its successes and challenges; factors driving MINUSMA's withdrawal; and the role of

---

[3] Ibid.

major regional and international stakeholders in comprehending the multifaceted impacts of such a decision.

Genesis of the Malian Crises

The fall of the Libyan regime in 2011 and the return of hundreds of combatants to Mali and Niger significantly contributed to the destabilization of the Sahel region.[4] Returning from Gaddafi's Islamic Legion, battle-hardened Tuareg fighters instigated the 2012 Malian civil war, which Al-Qaeda in the Islamic Maghreb quickly picked up and exploited to further their agenda. Subsequently, numerous armed groups emerged in northern and central Mali based on their differences and conflicting interests. While many of the contemporary security crises in Mali are rooted in competition for control over profitable trade and trafficking routes,[5] they are primarily driven by poverty, inter-communal tensions, weak governance, and the influence of jihadist groups.

In response to the Malian crisis, Operation Serval and the Armed Forces for the Chadian Intervention in Northern Mali were launched in 2013. Later, Operation Barkhane, the G5 Sahel Joint Force (G5S-JF), and MINUSMA were mobilized in 2014. Nevertheless, the situation deteriorated, creating potentials of regional instability through terrorism an militancy.[6] The failure of Operation Barkhane to curb militancy, continued political instability in the aftermath of the May 2021 coup, the departure of French military forces, and the activities of Russia's Wagner Group severely impacted stability and security in Mali.[7] In this circumstance, the withdrawal of MINUSMA by the end of 2023 may leave the security situation even more volatile.

Emerging Security Challenges in Mali

The current security landscape of Mali has a complex mix of challenges including climate change, transnational terrorism, the proliferation of illegal arms, communal conflict, unprotected borders, human and drug trafficking, and organized criminal activities that pose severe impediments to peace

---

[4] Chauzal, "From the Malian Crisis to the Sahel Breakdown."

[5] Marshall, "Africa's Crime-Terror Nexus: Transnational Organised Crime, Illicit Economic Networks and Violent Extremism in the Sahel."

[6] Chauzal, "From the Malian Crisis to the Sahel Breakdown."

[7] Institute for Economics and Peace, *Global Terrorism Index 2023: Measuring the Impact of Terrorism*.

operations.[8] These challenges are often interconnected and exacerbate each other with significant impacts on national and regional development, stability, and security. Characterized by asymmetric warfare and the emergence of local militias, non-traditional security threats have increased in scale, intensity, and complexity in recent years. Peace operations have become extremely challenging, with peacekeepers suffering increasing fatalities.[9]

In 2022, Mali recorded its highest number of deaths due to terrorism in the last decade. IS overtook JNIM as the most prominent group in Mali, with 19 attacks and 243 deaths. The tri-border area encompassing Mali's border along with Niger and Burkina Faso remains the most affected.[10] The entry of Wagner mercenaries and the exit of French forces substantially changed the conflict ecosystem. IS quickly exploited this security void, as evidenced by their deadly onslaught against Tuareg armed groups in north-eastern Mali. JNIM launched around 50 attacks on MINUSMA peacekeepers, killing at least 20.[11]

The increasing use of mines and improvised explosive Devices (IEDs) by terrorists remains one of the most significant challenges facing peacekeepers, the Malian Army (FAMa), and civilians.[12] From July 2013 until the end of April 2023, the United Nations Mine Action Service (UNMAS) recorded 1,592 IEDs/mines utilized by non-state armed groups, leaving 1,116 people dead and injuring 2,510 more.[13]

The leading causes of armed conflicts and violence include poverty, resource competition, political rivalry, ethnic and religious tensions, and the proliferation of armed groups.[14] In 2022, armed conflicts in the Sahel caused around 9,000 fatalities,[15] with the highest number recorded in Mali. In the Sahel, notably Mali and Burkina Faso, atrocities by government security

---

[8] Amani Africa, "Peacekeeping in Africa: Emerging Challenges and Critical Lessons for Sustainable Peacekeeping Operations."

[9] As-Sazid, "Emerging Security Challenges in the Sahel and the Need for an Adaptative Approach towards Peacebuilding."

[10] Institute for Economics and Peace, *Global Terrorism Index 2023: Measuring the Impact of Terrorism.*

[11] Nsaibia, "The Sahel Mid-Year Update - Persistent, Expanding, and Escalating Instability."

[12] MINUSMA, "IED Attacks Continue to Take Their Toll on Civilians in Mali."

[13] UNMAS, "UNMAS Report on Mali."

[14] Dieng, "The Sahel: Challenges and Opportunities."

[15] Wilkins, "Deaths, Instability Increase Across the Western Sahel in 2022."

forces significantly increased in 2022, resulting in 1,058 civilian casualties. In Mali, this number was roughly the same as those killed in terrorist attacks.[16]

Inter-communal violence is rooted in ethnic and tribal differences, political rivalry, and competition for control over natural resources and trade/trafficking routes. The already-precarious security situation has been worsened by terrorist groups' attempts to exploit these ethnic fault lines.[17] JNIM has embedded itself as the defender of the Fulani ethnic group against their rival Dogon. Similarly, the Islamic State in West Africa (ISWA) has exploited the intercommunal conflicts between the Fulani and Tuaregs in the Mali-Niger border area. By leveraging these tensions, both groups have successfully increased their recruiting. The snowball effect of the Malian crises has exacerbated Fulani-Dogon intercommunal violence in Central Mali, which has further spread inside Burkina Faso, Chad, Mauritania, and Niger.[18] While claiming legitimacy as local security providers, the self-defence groups have been parties to several deadly communal conflicts in north and central Mali.[19]

Mali's political instability and lack of state control are driven by poor governance and public services, insecurity, corruption, weak law enforcement, and unemployment, which has allowed non-state actors, including jihadist groups and self-defence militias, to represent themselves as alternatives to the government in areas beyond urban centres.[20]

Mali ranks among the top 50 countries in the GI-TOC Organized Crime Index,[21] with increased trafficking of humans, narcotics, and weapons; extortion from trade routes; and kidnapping for ransom. Its vast and uncontrolled desert offers opportunities for criminal networks to operate and evade law enforcement agencies. Since few formal employment alternatives exist, illegal economic activity is considered socially acceptable and necessary for survival.[22] Artisanal gold mining is also responsible for

---

[16] Institute for Economics and Peace, *Global Terrorism Index 2023: Measuring the Impact of Terrorism*.
[17] As-Sazid, "Emerging Security Challenges in the Sahel and the Need for an Adaptative Approach towards Peacebuilding."
[18] Institute of Economics and Peace, *Global Terrorism Index 2022*.
[19] Dieng, "The Sahel: Challenges and Opportunities."
[20] Institute of Economics and Peace, *Global Terrorism Index 2022*.
[21] Micallef and Herbert, "Fragile States and Resilient Criminal Ecosystems."
[22] Marshall, "Africa's Crime-Terror Nexus: Transnational Organised Crime, Illicit Economic Networks and Violent Extremism in the Sahel."

intra-regional migration, fuelling other criminal activities.[23] Weak governance and law enforcement, corruption, porous borders, and poverty in the region exacerbate the problem.[24]

The humanitarian crisis in Mali continues to be characterized by a significant increase in the number of displaced people throughout the country due to non-state armed groups, protracted violence, and conflict with armed forces. According to the Displacement Tracking Matrix, as of September 2022, Mali accounts for 440,436 internally displaced persons, mainly in the regions of Mopti and Timbuktu,[25] while approximately 198,000 Malian refugees are in Burkina Faso, Mauritania, and Niger.[26]

Mali is vulnerable to long-term climate change, which is exacerbated by high population growth, diminished resilience, and violent conflicts.[27] The changing climate has caused droughts, floods, and desertification, leading to food insecurity, increased poverty levels, displacement, and social unrest in the north and centre.[28] In the central regions, drastic ecological change is the primary driver of violence between Dogon, the sedentary farmers, and Fulani, the seminomadic cattle herders.[29]

The '*Agreement for Peace and Reconciliation in Mali Resulting from the Algiers Process*' failed to fulfil its promise of national reconciliation, apparently due to interrelated problems such as a failure to integrate all relevant actors, limiting peace initiative only in the North of Mali, and a lack of political will and commitment on the part of stakeholders. Moreover, it solely focused on the 2011–2012 rebellion in Northern Mali and failed to address the escalating Islamist violence, deadly interethnic conflict, and ongoing insecurity in Mali's central regions. It also suffers from a lack of wide-ranging inclusivity and will of the stakeholders for political decentralization and implementation of the disarmament, demobilization, and reintegration (DDR) of ex-combatants.[30]

---

[23] Micallef and Herbert, "Fragile States and Resilient Criminal Ecosystems."
[24] As-Sazid, "Emerging Security Challenges in the Sahel and the Need for an Adaptative Approach towards Peacebuilding."
[25] Displacement Tracking Matrix, "Mali Crisis Response Plan 2023-2024."
[26] European Civil Protection and Humanitarian Aid Operations, "Mali Factsheet."
[27] De Coning and Krampe, "Climate, Peace and Security Fact Sheet - Mali."
[28] Ghins et al., "Environmental Fragility in the Sahel."
[29] Institute of Economics and Peace, *Global Terrorism Index 2022*.
[30] Devermont and Harris, "Why Mali Needs a New Peace Deal."

MINUSMA's Contributions to Addressing the Security Challenges

Following the political and security crisis that began in 2012, MINUSMA was deployed in Mali with a robust mandate to stabilize the country, support the implementation of peace agreements, protect civilians, and facilitate the delivery of humanitarian assistance. In April 2013, MINUSMA's mandate was authorized by the UNSC under Resolution 2100 to support the implementation of Mali's peace and reconciliation process.

Mandate and Key Objectives of MINUSMA

On 29 June 2022, the UNSC approved Resolution 2640 (2022), renewing MINUSMA's latest mandate for another year until it ended on 30 June without further renewal. MINUSMA's main strategic priority has remained unchanged: *to support the implementation of the Peace and Reconciliation Agreement and political transition in Mali.* The renewed mandate also focused on addressing the root causes of conflicts and inter-communal violence, protecting civilians, re-establishing state authority, and ensuring essential social services in central Mali by implementing a *'comprehensive, inclusive, politically focused strategy'* by the Malian government.[31] It expressed grave concern about the continued deterioration of Mali's political, security and humanitarian situation through persistent terrorist attacks and violence[32] and demanded extra protection for distressed women, children and marginalized groups.[33] The UNSC further reiterated its support for the mediation efforts by ECOWAS and encouraged negotiations to reach an agreement on credible and acceptable timelines for holding inclusive, free, fair, and transparent elections and the swift return to constitutional order. However, it expressed concern about the lack of engagement by the Transition Government of Mali and the signatory armed groups.[34]

MINUSMA's Successes and Challenges

MINUSMA's efforts to promote security and stability in Mali have been significant and multifaceted. It has played an instrumental role in implementing

---

[31] UNSC, "MINUSMA Mandate."

[32] Security Council, "Security Council Resolution 2640 (2022)."

[33] UNSC Press Release, "Security Council Renews Multidimensional Integrated Stabilization Mission in Mali, Adopting Resolution 2640 (2022)."

[34] UNSC, "MINUSMA Mandate."

the 2015 Algiers Peace Agreement; promoting dialogue, reconciliation, and good governance; and supporting the political process in Mali. The mission closely monitored, documented, and investigated human rights violations and abuses while advocating for accountability for these crimes. It has also actively supported humanitarian assistance programs and DDR. Until 2016, MINUSMA was relatively successful in improving stability in northern Mali and reducing violence. It also supported the organization of the 2013 elections and assisted the process, which culminated in the 2015 Algiers Agreement for Peace and Reconciliation in Mali.[35]

Since 2016, however, MINUSMA's effectiveness in stabilization and the protection of civilians (POC) has been questioned due to increased violence and jihadist attacks, especially in Central Mali. While mandated to assist the Malian government in addressing the situation, MINUSMA never received the resources needed to be effective.[36] As Islamist insurgents started gaining hold in northern and central Mali, MINUSMA faced increasing terrorist attacks and, constrained by its mandate, struggled to adapt. Despite restrictions on offensive engagements against the militants, peacekeeping forces were expected to effectively respond to these threats. Although successive updates have gradually made MINUSMA's mandate more robust, they never went so far as to include a clear offensive counter-insurgency strategy. Hence, MINUSMA was compelled to depend upon and operate alongside FAMa and Barkhane forces and assist their counterterrorism drive, making themselves targets for the terrorists. Meanwhile, the withdrawal of Barkhane and Takuba forces and the Malian government's growing reluctance to cooperate with MINUSMA have rendered that task even more difficuLieutenant[37]

Moreover, several troop-contributing countries (TCCs) started withdrawing their troops, leading to a tremendous shortage of forces and equipment. Increasing threats of IEDs and other forms of terrorist attacks compelled the peacekeeping forces to dedicate much of their resources to force protection rather than to the POC. While most UN troops are stationed in communication centres and logistics nodes, armed groups and jihadists maintain their dominance in the remote areas, where, in the absence of any state security apparatus, mass numbers of people are forced to either ally themselves with one of the armed groups or organize self-defence militia for protection against terrorists. As such, most Malians in the north and centre consider

[35] Lijn, "The UN Peace Operation in Mali: A Troubled Yet Needed Mission."
[36] Ibid.
[37] International Crisis Group, "MINUSMA at a Crossroads."

MINUSMA ineffective in providing security.[38] A research team from the Effectiveness in Peace Operations Network found massive criticism and mistrust against MINUSMA among the population.[39]

Despite these limitations, MINUSMA has brought stability to areas where it is deployed. Its presence has helped deter insurgents from securing cities and larger towns. MINUSMA also helped the Malian government maintain its presence in the geostrategically essential locations and remote areas and supported the activities of humanitarian assistance groups.[40] In the north and centre, it significantly contributed to developing public services and healthcare facilities and worked to ensure the social security of women and children.[41]

However, a MINUSMA with a robust counter-insurgency mandate, more capable of ensuring POC and human rights, and willing to directly support the counterterrorism efforts of FAMa would have been better welcomed in Mali. The relationship between MINUSMA and the Malian government worsened after the coup, mainly due to the constraint of MINUSMA's mandate in fighting militants. In July 2021, the Malian government opposed raising MINUSMA's troop ceiling, arguing it was pointless without a more robust counter-insurgency mandate. However, both parties tolerated the status quo as long as Barkhane contributed to counterterrorism operations. With French troops withdrawn following the induction of the Wagner Group and Russia becoming Mali's preferred military partner, coordinating MINUSMA's activities with counter-insurgency efforts became more complicated.[42] Meanwhile, the departure of several resourceful TCCs and French forces further constrained the mission's air and logistics support.[43]

MINUSMA is thus evolving in a challenging context amidst growing anti-MINUSMA sentiment among mass Malians, nourished by years of military presence without improvements in the security situation. Rumours and accusations of favouritism toward northern populations, a secret agenda for dividing Mali and collaborating with terrorist groups, and disinformation campaigns branding MINUSMA as an agent of 'Western colonial forces'

---

[38] Ibid.

[39] Lijn, "The UN Peace Operation in Mali: A Troubled Yet Needed Mission."

[40] International Crisis Group, "MINUSMA at a Crossroads."

[41] Cascais, "Mali's MINUSMA Peacekeeping Mission: A Foreseeable Disaster."

[42] International Crisis Group, "MINUSMA at a Crossroads."

[43] Sharland, "MINUSMA and Protection of Civilians: Implications for Future Peacekeeping Missions."

have exacerbated the hostility.[44] MINUSMA was overstretched geographically, with over 13,000 troops scattered and operating over a vast territory, while Malian authorities were increasingly obstructing the use of MINUSMA's air and intelligence, surveillance and reconnaissance assets, which are essential for movement, force protection and intelligence gathering.[45]

Evolving Geopolitical Dynamics Behind MINUSMA's Impending Withdrawal

According to Sahel expert Niagale Bagayoko, Bamako's decision about MINUSMA's withdrawal was a result of 'exceedingly strained' Malian-French relations.[46] In August 2020, a coup pushed out Mali's last elected president, Ibrahim Boubacar Keita, as frustration with his government's inability to curtail the rebels reached a breaking point. A second military takeover in May 2021 due to internal political turmoil and the lack of a definite timeline for a return to civilian rule infuriated former colonial power France and the Economic Community of West African States (ECOWAS), prompting sanctions against Mali. Bamako insinuated that Paris was pressuring its neighbours in the ECOWAS bloc to take a hardline stance.[47] Due to such boiling tensions with ECOWAS, Mali pulled out of the G5S-JF in May 2022, followed by the departure of French-led Barkhane and Takuba forces.[48]

Moreover, the mission has evolved in the context of the Russian invasion of Ukraine and escalating rivalries between Russia and the West.[49] The UNSC, deeply divided over Russia's war on Ukraine, failed to offer MINUSMA enough political backing and strategic guidance. The US criticized Russia's alleged use of Mali to bypass its sanctions on arms imports and designated Wagner as a transnational criminal organization.[50] Moscow, on the other hand, has increasingly taken the side of Bamako against the West. Since

[44] Gauthier Vela, "MINUSMA in Mali: The Challenges of Multidimensional Peacekeeping."
[45] Sharland, "MINUSMA and Protection of Civilians: Implications for Future Peacekeeping Missions."
[46] Cascais, "Mali's MINUSMA Peacekeeping Mission: A Foreseeable Disaster."
[47] Mohamed, "Analysis: What's next for Mali after MINUSMA Withdrawal?"
[48] Koné, Sow, and Koné, "MINUSMA Leaves Mali - Will Regional Leaders Step up on Security?"
[49] Gauthier Vela, "MINUSMA in Mali: The Challenges of Multidimensional Peacekeeping."
[50] Jezequel and Maiga, "Ensuring MINUSMA's Smooth Departure from Mali."

Russia's all-out invasion of Ukraine, there has been a new trend for China and Russia to abstain on resolutions concerning peace operations, as they complain that the US and its allies lack respect for the views of host governments. Although Russia did not wield its veto power against MINUSMA's mandate, other UNSC members assumed it was aware of and possibly encouraged Bamako's decision to expel MINUSMA. Early in 2023, Mali denounced France's role as steward of MINUSMA's affairs at the UNSC.[51] Thus, in addition to Mali's political challenges and transnational insecurity, MINUSMA is caught between several geopolitical issues.

In January 2023, three different options for MINUSMA's future were discussed at the UNSC: firstly, an increase in uniformed personnel; secondly, reconfiguration of the mission assets to target specific priorities; and, finally, the withdrawal of uniformed personnel and conversion of MINUSMA into a political mission.[52] Meanwhile, in May 2023, the UN High Commissioner for Human Rights published a report accusing the FAMa and Wagner of killing more than 500 people during a military operation in Moura town of the Mopti region in March 2022. This report, published only a few weeks before the vote on the mission's renewal, infuriated the Malian authorities. Bamako vehemently denied these accusations and denounced the UN investigations as 'espionage'.[53]

On 16 June, to the surprise of most UNSC members, Mali's transitional government asked MINUSMA to leave *"without delay"*, citing a *"crisis of confidence"* between Malian authorities and MINUSMA.[54] Mali's Foreign Minister Abdoulaye Diop also said, *"MINUSMA seems to have become part of the problem by fueling community tensions exacerbated by extremely serious allegations which are highly detrimental to peace, reconciliation and national cohesion in Mali"*. Issa Konfourou, Mali's ambassador to the UN, commented that MINUSMA "*has certainly not achieved its fundamental goal of supporting the government's efforts in securing the country"*. With 31 December 2023 as the deadline for departure, the Malian government allowed MINUSMA to respond to the *"imminent threat of violence to civilians"* and to help in the delivery of humanitarian aid.[55]

---

[51] Gowan and Forti, "What Future for UN Peacekeeping in Africa after Mali Shutters Its Mission?"
[52] Gauthier Vela, "MINUSMA in Mali: The Challenges of Multidimensional Peacekeeping."
[53] Jezequel and Maiga, "Ensuring MINUSMA's Smooth Departure from Mali."
[54] Nichols, "Mali Asks United Nations to Withdraw Peacekeeping Force."
[55] Mohamed, "Analysis: What's next for Mali after MINUSMA Withdrawal?"

Implications for Other Peacekeeping Missions

According to some analysts, MINUSMA's withdrawal will initiate the end of an era of large, complex, and multidimensional UN peacekeeping missions in Africa because it has several factors in common with other large UN peacekeeping missions. The most obvious of these is MINUSMA's failure to project as a deterrent counterterrorism force and inability to firmly ensure the POC, even though it is an essential element of its mandate. In 2022, the disillusioned population of the DRC launched a series of violent protests accusing the United Nations Organization Stabilization Mission in the Democratic Republic of the Congo (MONUSCO) of a failure to fight rebel groups. Similar charges have been laid against the United Nations Mission in South Sudan (UNMISS), where peacekeepers rarely intervene in the affairs of militia groups. The United Nations Multidimensional Integrated Stabilization Mission in the Central African Republic (MINUSCA) has also made limited contributions to the POC beyond Bangui and allegedly avoids confronting insurgents. Thus, all four missions have observed diminishing popular support and influence with host governments. Speculation and rumours have gravitated towards the question of whether the expulsion of MINUSMA by the Malian government could inspire other states to follow suit.[56]

In the case of the DRC, the Congolese government had already indicated that it wants to see MONUSCO leave shortly after the 2023 national elections. Contrarily, despite confronting issues between MINUSMA and the Central African Army, the government of the CAR still views MINUSCA positively and is less likely to follow Bamako's example in the immediate future. The government of South Sudan, still immune to Wagner's influence, is also unlikely to take any drastic measures against UNMISS considering its reliance on the mission's logistics and humanitarian support. The UN is also set to support national elections in both South Sudan and CAR in 2024 and 2025, respectively, which makes the closure of UN peacekeeping missions unlikely for these countries in a short-term timeframe. However, New York's tepid reaction to Bamako's drastic measures against MINUSMA may have a mid-term effect considering the apparent weakness of the divided UNSC in opposing such decisions by the host nations.[57]

---

[56] Gowan and Forti, "What Future for UN Peacekeeping in Africa after Mali Shutters Its Mission?"
[57] Ibid.

At a more strategic level, MINUSMA's closure will fuel discussions about alternatives to blue helmet operations in Africa. The UNSC has long debated about whether the UN could instead bear a percentage of the costs of the African Union (AU) led peace operations. Last year, the UN Secretary-General António Guterres strongly voiced this idea, calling for a new AU-led peace enforcement operation in the Sahel to tackle jihadist threats. In contrast to its earlier position under the Trump administration, the US supported it.[58]

Effect on Mali's Stability

While Bamako seems confident in its ability to guarantee the safety of its citizens, in the absence of a clearly defined security strategy, analysts speculate that the military government would primarily rely on Wagner, which neither has the military resources of the UN nor the same commitment to community mediation, human rights and pacification.[59] As such, the Malian government will face more significant challenges in extending its authority across the country to replace MINUSMA's efforts in the POC, especially after withdrawing its critical air assets and logistic support. Given the current insecurity across the country, its implications will likely be dire for civilians without additional international or regional assistance.[60]

Despite a few military victories of the current Malian regime against terrorists, the situation is worsening. Jihadist attacks are increasing, with groups active throughout the country. FAMa and Wagner, in the security provider role, have been accused of killing hundreds of civilians.[61] Wagner's deployment has also strained Mali's relations with its Western partners.[62] According to Sahel expert Abdoulaye Sounaye, Mali had proven incapable of ensuring the security of its citizens, especially in the north. Ulf Laessing, head of the German Konrad Adenauer Foundation in Bamako, said, *"While the Malian government will attempt to replace MINUSMA with*

---

[58] Ibid.

[59] Mohamed, "Analysis: What's next for Mali after MINUSMA Withdrawal?"

[60] Sharland, "MINUSMA and Protection of Civilians: Implications for Future Peacekeeping Missions."

[61] Gauthier Vela, "MINUSMA in Mali: The Challenges of Multidimensional Peacekeeping."

[62] Institute for Economics and Peace, *Global Terrorism Index 2023: Measuring the Impact of Terrorism*.

its Russian partners, the armed rebels and jihadist groups will try to exploit the security vacuum." [63]

Common Malians have mixed feelings regarding MINUSMA's departure. Many think the presence of MINUSMA did not improve the situation and were happy with ending its mandate. According to them, jihadist attacks in the centre and north of Mali did not stop, despite, or perhaps because of, the presence of international troops. However, some expressed concern about the future development projects and aid initiatives organized by MINUSMA in the centre and north of Mali.[64]

MINUSMA's departure will likely worsen Mali's international standing, potentially leading to further reductions in international donor engagement and the immediate loss of a $1.2-billion mission that employs thousands of Malians.[65] Finally, as Bamako distances itself from certain Western and African partners, it risks relying solely on military solutions that cannot bring stability to the country. It is not easy to envision how Mali will finance the $644 million action plan for implementing its stabilization strategy in the central regions.[66]

Potential Impacts on Regional Security and Stability

The withdrawal of MINUSMA will likely strain regional security dynamics. Its withdrawal could create a security vacuum, exposing local forces and vulnerable populations to the resurgence of armed extremist groups. It may escalate tensions among various militant factions vying for territory control, further destabilizing the region. With porous borders and a lack of state control, the escalation of instability in Mali could spill over into neighbouring countries like Niger, Burkina Faso, Mauritania, and Algeria, exacerbating existing security challenges and creating a broader regional crisis. Armed groups, signatories to the 2015 Algiers Peace Agreement, have already referred to MINUSMA's departure as a 'fatal blow' to the peace process.[67] According to Niagale Bagayoko, the withdrawal of the French forces and MINUSMA has highlighted the growing complexities and challenges in

---

[63] Cascais, "Mali's MINUSMA Peacekeeping Mission: A Foreseeable Disaster."
[64] Ibid.
[65] Day, "As Peacekeeping Exits Mali, the Transition Is a New Opportunity for the UN."
[66] Jezequel and Maiga, "Ensuring MINUSMA's Smooth Departure from Mali."
[67] Day, "As Peacekeeping Exits Mali, the Transition Is a New Opportunity for the UN."

resolving conflicts in Africa.[68] The recent coup and military takeover in Niger on 26 July, followed by a strong reaction from ECOWAS and an even stronger counter-reaction from the military regimes of Mali and Burkina Faso, manifests the rising instability and divide in the region.[69]

Since 2017, MINUSMA has provided logistical support to the G5S-JF. In 2022, after Mali withdrew from this force, MINUSMA sub-contracted the provisioning of its logistics support to private companies, which will no longer be possible after MINUSMA's withdrawal, paralyzing the G5S-JF's counterterrorism drive in other countries in the region and rendering them more vulnerable to terrorist attacks. Moreover, considering Mali's geostrategic position in the Sahel and the transnational nature of the security challenges, it will be challenging for other Sahel countries to launch any joint counterterrorism drive without Mali's active participation.[70]

With MINUSMA's departure, the Tuaregs, scattered over several countries such as Mali, Niger, Algeria, Burkina Faso, and Libya, will likely lose confidence in implementing the 2015 Algiers Peace Agreement. Consequently, the ethnic hegemony of the Bambara ruling class may reignite their grievances against decades of discrimination and exclusion from the political and economic processes by successive Bamako-based governments. It may also prompt the rebel armed groups to reunite and take up arms with a pre-peace agreement stance for independent Azawad, which might lead to the disintegration of Mali and threaten its territorial integrity as well as its neighbouring countries sharing the Tuareg homeland.

Some analysts argue that the pre-emptive French military intervention in Mali, followed by the establishment of MINUSMA, was a strategy to protect French economic and mining interests in Niger by preventing the spillover of the Malian crises.[71] Aiming to reduce its military footprint in Africa,[72] following MINUSMA's withdrawal, France is likely to support a regional peacekeeping mission in the Sahel in collaboration with its African partners.

[68] Cascais, "Mali's MINUSMA Peacekeeping Mission: A Foreseeable Disaster."
[69] Adetayo, "A Test of Wills: Can ECOWAS Reverse Niger Coup and Establish a New Order?"
[70] Dieng and Mfondi, "UN Troops to Withdraw from Mali: What Will Change in Terms of Security."
[71] Francis, "The Regional Impact of the Armed Conflict and French Intervention in Mali."
[72] Vincent, "France Will Further Cut Back Military Presence in Africa."

Geopolitical Race Among the Big Powers

While Western actors have dominated the security environment in the Sahel region for over a decade, other countries such as China, Russia, Turkey, and the Gulf states have maintained warm relations with Sahelian states through trade, aid, financial support, infrastructure building and cultural exchange.[73] However, Russia has significantly increased its focus and involvement in Africa over the last two years. Through Wagner mercenaries, Moscow is allegedly gaining a solid foothold in Mali and Burkina Faso by taking advantage of Western policy missteps, growing anti-European sentiment, and longstanding failures of international and local actors to address the root causes of regional instability.[74]

Wagner reportedly operates in eight African countries and is most active in Libya, Sudan, Mali, and the CAR. Hired by host governments for training security forces, it allegedly protects those regimes and combats terrorism alongside their regular forces.[75] Despite the recent mutiny of Wagner and its resultant tensions with the Russian government, Moscow has reassured the retention of Wagner in Mali and CAR, which are deemed as a source of lucrative natural resources for the Kremlin and a springboard for broader influence over Africa.[76] In January, U.S. Undersecretary of State for Political Affairs Victoria Nuland testified to Congress that Wagner's access to gold mines in Mali and the CAR "directly funds" operations in Ukraine.[77] Like in Mali, with deep-seated anti-French sentiment, the new military leadership of Burkina Faso also expelled the French troops in January and turned towards Moscow.[78] The recent coup in Niger also sparked speculation among analysts that Africa is turning into an extended battlefield between Russia and the Western power bloc,[79] with Kremlin's eyes on Niger's vast Uranium deposit.[80]

With minimal funding and investments, Russia's success in Africa is partly driven by tapping into domestic instead of global politics, where the Western powers ignored state-sponsored atrocities, driving the marginalized

---

[73] Sandnes and Carrozza, "Russia, China and New Power Dynamics in the Sahel Region."
[74] Stronski, "Russia's Growing Footprint in Africa's Sahel Region."
[75] Gbadamosi, "How Russia Expands Its Influence in Africa."
[76] Burke, "Wagner Mercenaries Will Not Be Withdrawn from Africa, Says Russia."
[77] Gbadamosi, "How Russia Expands Its Influence in Africa."
[78] Stronski, "Russia's Growing Footprint in Africa's Sahel Region."
[79] Sharma, "Putin vs West - Battlefield Africa."
[80] Blas, "The Long Arm of the Kremlin and the Politics of Uranium."

populations to ally with jihadists, rebels, and self-defence militias as local security providers.[81] Wagner's presence in Africa acts as good publicity for Russia, thereby undermining Western attempts to isolate it. Moscow has also proven its ability to exploit Africa's governance shortfalls, instability, and security vacuums. It has embraced the military regimes in both Mali and Burkina Faso, providing them with security assistance, diplomatic backing, and information operations support. Wagner is also reportedly expanding into Chad, which is positioned strategically in the Central Sahel with borders with countries where Wagner is most active.[82] Contrarily, following the coups, the US suspended free trade access for Mali and CAR to its markets and recently imposed sanctions against top Malian junta officials.[83] However, apart from expressing repeated concerns about Wagner's presence, the US has not come forward with any alternative security strategy for the Sahel to replace Wagner and reduce Russian influence.

Presumably, considering its recent strategic debacle in Afghanistan and its current involvement in the Ukraine War, the US is unlikely to expand its military footprint in Africa. Although the Western power bloc is now implicating Russia in emerging instability in the Sahel, it must be remembered that the erstwhile presence of Western forces and the outgoing MINUSMA also failed in alleviating the main drivers behind Sahel's security quagmire. As such, squeezing Russia out of Africa alone might not solve the problems unless a comprehensive security strategy and peace enforcement plan that takes all the regional and international stakeholders on board is put in place.

On the other hand, China's engagements in the Sahel are centred on its economic interests, in particular sourcing natural resources and minerals, which drives its involvement in security affairs mainly through military exchange programmes and training, contributions to UN missions, donations to the G5S-JF, and the export of arms.[84] As such, with Western influence diminishing and the US maintaining a passive role, China and Russia are gradually expanding all over Africa using their soft and hard power.

---

[81] Gbadamosi, "How Russia Expands Its Influence in Africa."

[82] Stronski, "Russia's Growing Footprint in Africa's Sahel Region."

[83] Al Jazeera Staff, "US Sanctions Mali's Defence Minister, Officials over Wagner Ties."

[84] Sandnes and Carrozza, "Russia, China and New Power Dynamics in the Sahel Region."

Probable Mitigation Strategies

The withdrawal of MINUSMA is paving the way for more assertive African leadership in addressing the Sahel's insecurity. It provides an opportunity to revive the bonding while setting aside institutional rivalries between the UN, AU, ECOWAS, G5 Sahel, and Accra Initiative. The peculiarity and transborder nature of the Sahel and West African regional security risks call on the AU and ECOWAS to play a more significant role and implement African solutions to maintain stability.[85]

The AU and ECOWAS may engage with Mali and strengthen the AU-led Mission for Mali and the Sahel (MISAHEL) to prioritize a political approach aimed at short-term stabilization. They should also support the Algeria-led international mediation for resolving Mali's internal political and security issues. It should aim to revive the 2015 Algiers Peace Agreement's implementation, which might reconcile the possible disintegration between the north and south of Mali.

Besides, ECOWAS may formulate an appropriate framework for minimizing the root causes of regional security challenges and guide the transitional regimes to suitable political approaches and security response mechanisms while considering the Accra Initiative.[86]

Western powers, taking China, Russia and all the regional stakeholders on board, should formulate a comprehensive security and peace-enforcement strategy, as well as its implementation plan to combine the efforts of the AU, ECOWAS and G5 Sahel in either strengthening MISAHEL or establishing another AU-led peace enforcement mission with a strong counterterrorism mandate and a holistic approach towards stabilization.

According to Adam Day, in the current security context, MINUSMA's departure will leave Mali more reliant on Wagner. As such, the UN may propose that the transitional authority of Mali should allow an AU or ECOWAS-led force to deploy in the most troubled areas in the north and

---

[85] As-Sazid, "Emerging Security Challenges in the Sahel and the Need for an Adaptative Approach towards Peacebuilding."
[86] Koné, Sow, and Koné, "MINUSMA Leaves Mali - Will Regional Leaders Step up on Security?"

centre of Mali to fill the security void and assist in conducting a credible election in 2024.[87]

The UN may also establish small-scale political and observer missions in conflict-ridden countries, including Mali, with suitable mandates to render diplomatic and financial support to the AU-led missions.[88] With terrorism and insurgency being significant impediments to peace and stability in Africa, the UN charter may be revised to endorse more robust peace-enforcement mandates that allow both direct involvement and supporting roles of UN field missions in counterterrorism and counter-insurgency.

A greater focus should be placed on the disarmament and rehabilitation of ex-combatants. The highest emphasis should be put on applying soft powers by international and regional actors to mitigate the root causes of insecurity by adopting measures such as poverty alleviation, the delivery of essential public services, creating employment opportunities, women empowerment, mandatory education, local mediation and intercommunal reconciliation, free and easily accessible medical facilities, and reducing the impact of climate change. Armed groups should be hit at their most vulnerable points by disrupting their supplies and funding chains and preventing their recruitment by improving the security and living standard of marginalized populations.

The transitional government of Mali should also revive the peace and reconciliation process in the north and centre[89] and create favourable conditions for a smooth political transition. It may be prudent for the UNSC to condition MINUSMA's exit on the 2024 elections and a stable political settlement beyond 2024 by setting clear benchmarks, like establishing independent electoral institutions, ensuring political inclusivity, preventing human rights abuses, and facilitating a credible national election. Financial packages by the UN, AU, World Bank, and major donors may also be offered with these conditions to assist their implementation.[90]

Despite MINUSMA's departure, the UN may maintain its presence in Mali through different programs and agencies, such as the UN Development

---

[87] Day, "As Peacekeeping Exits Mali, the Transition Is a New Opportunity for the UN."

[88] International Crisis Group, "MINUSMA at a Crossroads."

[89] Dieng and Mfondi, "UN Troops to Withdraw from Mali: What Will Change in Terms of Security."

[90] Day, "As Peacekeeping Exits Mali, the Transition Is a New Opportunity for the UN."

Programme and World Food Programme, which may have critical roles in social development and humanitarian assistance. Bamako should also seek to support these agencies' operations in MINUSMA's absence, especially in remote areas.[91] Shoring up the resilience of neighbouring states through economic and diplomatic support to the regional bodies to manage the impacts of ever-spreading instability must also be a priority.[92]

Following the Libyan Civil War, a vast ungoverned space emerged across the Sahel and Sub-Saharan Africa, where violent non-state actors filled up political and security voids by seizing economic opportunities, both legal and illegal, while maintaining a semblance of political order. While violent extremism, insurgency and inter-communal violence are the worst threats to peace and stability in the region, poverty, insecurity, socio-economic disparity, unemployment, poor governance, and climate change are some of the root causes of violence and terrorism. The state's inability to provide security and the failure of the international forces to effectively suppress the militancy, violence, and organized crime have brought the relevance of these forces into question. Despite tremendous support towards the development and humanitarian sectors and the implementation of the 2015 Algiers Peace Accord, the decade-long deployment of MINUSMA could not offer the desired stability and protection, seemingly due to its mandate's constraints in regard to counterterrorism operations. Hence, with the induction of Russian Wagner Group and the closure of other regional and international missions and counter terrorism operations, as well as the deterioration of the Bamako-MINUSMA relationship, the mere existence of MINUSMA became undesirable to Mali's transitional military regime.

While the end of MINUSMA may not precipitate the collapse of other UN missions, it may point toward the end of an era for large-scale blue helmet missions in Africa. With MONUSCO also likely to draw down soon, the UN's military footprint on the continent will be much smaller in the future. As an alternative to UN peacekeeping operations, if the UNSC decides on supporting AU-led peace enforcement missions in Africa, the space for the UN to deploy its blue helmet forces will shrink further. Several such successful and less successful UN missions have been closed or converted to other missions. In contrast, the UN has experimented with new mission models in recent years, like the United Nations Verification Mission in Colombia, which has involved far fewer personnel and resources. An era of UN peacekeeping may end little by little in Africa, hastened by the departure

[91] Jezequel and Maiga, "Ensuring MINUSMA's Smooth Departure from Mali."
[92] Stronski, "Russia's Growing Footprint in Africa's Sahel Region."

of MINUSMA. But the organization will likely face new and unpredictable calls for help in an era of rising conflicts and geopolitical uncertainty.[93] As such, a fresh approach by the UN that is less centred around military operations and has a broader political strategy and inclusivity might be suitable for the Sahel region.[94]

References

Adetayo, Ope. "A Test of Wills: Can ECOWAS Reverse Niger Coup and Establish a New Order?" *Al Jazeera News*, August 1, 2023.

Amani Africa. "Peacekeeping in Africa: Emerging Challenges and Critical Lessons for Sustainable Peacekeeping Operations." Peace and Security Council, March 18, 2021.

As-Sazid, Syed Md. "Emerging Security Challenges in the Sahel and the Need for an Adaptative Approach towards Peacebuilding." *International Day of UN Peacekeepers Journal* 9, no. 9 (May 29, 2023): 17–35. https://doi.org/10.12968/eyed.2019.20.11.viii.

Blas, Javier. "The Long Arm of the Kremlin and the Politics of Uranium." *Bloomberg*, August 3, 2023.

Burke, Jason. "Wagner Mercenaries Will Not Be Withdrawn from Africa, Says Russia." *The Guardian*, June 26, 2023. https://www.theguardian.com/world/2023/jun/26/wagner-soldiers-will-not-be-withdrawn-from-africa-says-russia-foreign-minister.

Cascais, Antonio. "Mali's MINUSMA Peacekeeping Mission: A Foreseeable Disaster." Deutche Welle, April 7, 2023. https://www.dw.com/en/malis-minusma-peacekeeping-mission-a-foreseeable-disaster/a-66110603.

Chauzal, Grégory. "From the Malian Crisis to the Sahel Breakdown." Stockholm International Peace Research Institute. Stockholm, Sweden, April 7, 2020.

Coning, Cedric De, and Florian Krampe. "Climate, Peace and Security Fact Sheet - Mali." *Stockholm International Peace Research Institute*, no. May (May 2021).

Day, Adam. "As Peacekeeping Exits Mali, the Transition Is a New Opportunity for the UN." IPI Global Observatory. International Peace Institute, July 6, 2023.

---

[93] Gowan and Forti, "What Future for UN Peacekeeping in Africa after Mali Shutters Its Mission?"

[94] International Crisis Group, "MINUSMA at a Crossroads."

Devermont, Judd, and Michelle Marielle Harris. "Why Mali Needs a New Peace Deal," April 15, 2020, 1–7. https://www.csis.org/analysis/why-mali-needs-new-peace-dea.

Dieng, Adama. "The Sahel: Challenges and Opportunities." *International Review of the Red Cross* 103, no. 918 (2021): 765–79. https://doi.org/10.1017/S1816383122000339.

Dieng, Moda, and Amadou Ghouenzen Mfondi. "UN Troops to Withdraw from Mali: What Will Change in Terms of Security." The Conversation, July 17, 2023.

Displacement Tracking Matrix. "Mali Crisis Response Plan 2023-2024." International Organization for Migration, 2023.

European Civil Protection and Humanitarian Aid Operations. "Mali Factsheet." European Commission. Directorate-General for European Civil Protection and Humanitarian Aid Operations (ECHO), 2023.

Francis, David J. "The Regional Impact of the Armed Conflict and French Intervention in Mali." *NOREF Report*. Norway: Norweigian Peacebuilding Resource Centre, April 2013. https://doi.org/10.1080/13629395.2013.799351.

Gauthier Vela, Vanessa. "MINUSMA in Mali: The Challenges of Multidimensional Peacekeeping." *Australian Outlook*, March 8, 2023.

Gbadamosi, Nosmot. "How Russia Expands Its Influence in Africa." *Africa Brief*, Foreign Policy, May 17, 2023.

Ghins, Léopold, Ignacio Madurga-Lopez, Kazuma Yabe, and Jonathan Marley. "Environmental Fragility in the Sahel." OECD, September 2022.

Gowan, Richard, and Daniel Forti. "What Future for UN Peacekeeping in Africa after Mali Shutters Its Mission?" International Crisis Group, July 10, 2023.

Institute for Economics and Peace. *Global Terrorism Index 2023: Measuring the Impact of Terrorism. Institute for Economics and Peace*. Sydney, 2023. http://visionofhumanity.org/resources.

Institute of Economics and Peace. *Global Terrorism Index 2022. Global Terrorism Index*. Sydney, Australia, 2022. https://doi.org/10.4324/9780203731321.

International Crisis Group. "MINUSMA at a Crossroads." *International Crisis Group*, December 1, 2022. https://reliefweb.int/report/mali/minusma-crossroads.

Jazeera Staff, Al. "US Sanctions Mali's Defence Minister, Officials over Wagner Ties." *Al Jazeera News*, July 24, 2023.

Jezequel, Jean-Herve, and Ibrahim Maiga. "Ensuring MINUSMA's Smooth Departure from Mali." *International Crisis Group*. International Crisis

Group, June 27, 2023.
Koné, Fahiraman Rodrigue, Djiby Sow, and Hassane Koné. "MINUSMA Leaves Mali - Will Regional Leaders Step up on Security?" Institute for Security Studies, July 6, 2023. https://issafrica.org/iss-today/minusma-leaves-mali-will-regional-leaders-step-up-on-security.
Lijn, Jair van der. "The UN Peace Operation in Mali: A Troubled Yet Needed Mission." *IPI Global Observatory*, November 26, 2019. https://reliefweb.int/report/mali/un-peace-operation-mali-troubled-yet-needed-mission.
Marshall, Will. "Africa's Crime-Terror Nexus: Transnational Organised Crime, Illicit Economic Networks and Violent Extremism in the Sahel." *Global Risk Insights*, February 16, 2022. https://globalriskinsights.com/2022/02/africas-crime-terror-nexus-transnational-organised-crime-illicit-economic-networks-and-violent-extremism-in-the-sahel/.
Micallef, Mark, and Matt Herbert. "Fragile States and Resilient Criminal Ecosystems." *Global Initiative against Transnational Organized Crime*, no. August (2022).
MINUSMA. "IED Attacks Continue to Take Their Toll on Civilians in Mali." MINUSMA - United Nations Peacekeeping. UN Peacekeeping, 2023. https://doi.org/10.1177/004711787100301002.
Mohamed, Hamza. "Analysis: What's next for Mali after MINUSMA Withdrawal?" Al Jazeera. Al Jazeera News, July 3, 2023. https://www.aljazeera.com/news/2023/7/3/what-next-for-mali-after-minusma-withdrawal.
Nichols, Michelle. "Mali Asks United Nations to Withdraw Peacekeeping Force." *Reuters*, June 17, 2023. https://www.reuters.com/world/africa/mali-asks-united-nations-withdraw-peacekeeping-force-minusma-foreign-minister-2023-06-16/.
Nsaibia, Héni. "The Sahel Mid-Year Update - Persistent, Expanding, and Escalating Instability." The Armed Conflict Location & Event Data. USA, 2022.
Sandnes, Marie, and Ilaria Carrozza. "Russia, China and New Power Dynamics in the Sahel Region." *The Peace Research Institute Oslo (PRIO)*, January 16, 2023.
Security Council. "Security Council Resolution 2640 (2022)." Vol. 2640. New York: UN Security Council, June 29, 2022.
Sharland, Lisa. "MINUSMA and Protection of Civilians: Implications for Future Peacekeeping Missions." IPI Global Observatory. International Peace Institute, July 14, 2023.

Sharma, Palki. "Putin vs West - Battlefield Africa." *Vantage*. Firstpost, July 31, 2023.

Stronski, Paul. "Russia's Growing Footprint in Africa's Sahel Region." *Carnegie Endowment for International Peace*, February 28, 2023.

UNMAS. "UNMAS Report on Mali." United Nations Mine Action Service, April 2023. https://www.unmas.org/en/programmes/mali.

UNSC. "MINUSMA Mandate," June 29, 2022. https://minusma.unmissions.org/en/mandate-0.

UNSC Press Release. "Security Council Renews Multidimensional Integrated Stabilization Mission in Mali, Adopting Resolution 2640 (2022)." *UN Meetings Coverage and Press Releases*. Vol. 9082nd Mee. UN Press, June 29, 2022. https://press.un.org/en/2022/sc14953.doc.htm.

Vincent, Elsie. "France Will Further Cut Back Military Presence in Africa." Le Monde, June 6, 2023.

Wilkins, Henry. "Deaths, Instability Increase Across the Western Sahel in 2022." Voice of America News, December 29, 2022.

CHAPTER VIII

HARNESSING THE SKILLS AND EXPERIENCES OF FORMER UNITED NATIONS PEACEKEEPERS IN HUMANITARIAN INTERVENTIONS

MAJOR (RETIRED) DEL H KHAN

Introduction

Amidst the dust and debris of conflict-ravaged lands, a blue-helmeted figure emerges as a beacon of hope. The iconic symbol of United Nations (UN) peacekeepers, their blue helmets symbolize a commitment that transcends national boundaries – a commitment to upholding global peace and security through the execution of UN-mandated peacekeeping missions. These missions, far more than a mere display of blue, stand as guardians of ceasefires, protectors of civilians, facilitators of aid, and architects of political processes in nations ensnared by the clutches of conflict.

Imagine a world where nations send their sons and daughters to serve together under the UN banner, each bearing their own uniform, speaking their own language, yet united by a singular purpose – to restore order and humanity in the face of chaos. These brave souls, drawn from diverse backgrounds as military, police, and civilians, carry out their mission under the watchful gaze of the UN Security Council. Their mandate: to use force only when necessary for self-defense, their expertise encompassing logistics, engineering, medical support, and communication. These are not mere soldiers; they are agents of peace, architects of stability, and guardians of impartiality.

In the realm of conflict, where allegiances shift like desert sands and trust is scarcer than water, these peacekeepers wield a powerful tool – neutrality. Their impartiality serves as a bridge between warring factions, a conduit for dialogue amidst the cacophony of violence. A testament to their significance, their commitment isn't just to ending conflict; it's to fostering trust in communities shattered by strife.

Yet, as the shadows of conflict fade, another challenge looms on the horizon – the challenge of transition. The journey from the battlegrounds to the humanitarian frontlines is a road less traveled, a road paved with the experiences, skills, and adaptability of former peacekeepers. As we delve into the intricate interplay between UN peacekeeping and humanitarian response, the tapestry of their stories reveals a potential yet to be fully harnessed, challenges yet to be overcome, and a transformation waiting to unfold.

In this exploration, we voyage through the confluence of roles and realities between UN peacekeeping and humanitarian interventions. We unearth the untapped potential lying dormant within the ranks of ex-peacekeepers – individuals forged in the crucible of conflict zones, now poised to reshape the landscape of humanitarian action. By examining the bridge between these two domains, we illuminate the path that leads to leveraging the unique skills and experiences of former UN peacekeepers in the complex, often unpredictable realm of humanitarian interventions."

The Traditional Role of UN Peacekeepers in Humanitarian Interventions

The integration of UN peacekeepers into humanitarian interventions is a nuanced topic marked by a range of advantages and complexities. On one side, UN peacekeepers bring a wealth of extensive competencies, experiential insights, and resources to crisis-stricken regions. Their overarching mission revolves around protecting civilians, preventing violence, and advocating for human rights. This multifaceted engagement includes activities like cease-fire monitoring, patrolling, constructive dialogues with local leaders, providing vital humanitarian aid, and safeguarding civilians at risk of human rights abuses. Another critical dimension of their role is their pivotal contribution to peacebuilding processes, involving negotiation facilitation, mediation between conflicting parties, training local security forces, and investigating human rights violations.

However, integrating UN peacekeepers into humanitarian contexts comes with challenges. One major challenge is defining their mandate clearly. While their primary aim is to protect civilians, their obligation to provide humanitarian assistance is open to interpretation. This ambiguity can lead to confusion and disagreements with humanitarian stakeholders, causing uncertainty in accountability channels. Furthermore, the potential involvement of peacekeepers in using force during humanitarian situations and the risk of entanglement in political or security issues beyond their scope raise concerns. These complexities raise questions about the possibility of militarizing humanitarian aid, which threatens the core principles of impartiality and neutrality that underpin humanitarian efforts.

Historical examples highlight the dynamic interaction between UN peacekeepers and humanitarian interventions. The United Nations Mission in Sierra Leone (UNAMSIL) is a case in point. Deployed during the Sierra Leone Civil War in 1999, UNAMSIL went beyond its conventional role, contributing not only to stability restoration but also to disarmament efforts and the establishment of a democratic order, underscoring the transformative influence of peacekeepers.

The integration of UN peacekeepers into humanitarian contexts requires careful navigation. While their multidisciplinary expertise and operational capabilities bode well for civilian protection and well-being, the intricate balance between their mandate and the unique domain of humanitarian assistance necessitates thoughtful consideration. This exploration, rooted in empirical evidence and intellectual examination, serves as a guide for assessing the feasibility and effectiveness of such collaborations.

Successes of UN Peacekeepers in the Field

The impact of UN peacekeepers within humanitarian interventions cannot be overstated, with their role often leading to significant outcomes that reverberate on multiple occasions. Their significant footprint covers a range of achievements, including providing security to vulnerable populations, facilitating access to vital aid and resources, and crucial support in reinstating essential services in conflict-affected or crisis-ridden areas.

A poignant example of their effectiveness is seen in the Rwandan genocide of 1994. During this tragic period, UN peacekeepers established safe havens that protected thousands of civilians, potentially saving countless lives. Similar narratives emerge from the annals of South Sudan, where peacekeepers played an active role in shielding civilians from violence during the 2013 civil turmoil. Their efficiency also manifested in the effective distribution of aid and resources, demonstrated through their adeptness in providing relief after the Haiti earthquake in 2010. These peacekeepers also established and maintained havens for internally displaced individuals and refugees, showcasing their dedication to providing crucial assistance during dire circumstances.

Their impact expands further, as peacekeepers contribute to the restoration of fundamental services such as education, healthcare, and sanitation in conflict-ridden or crisis-stricken areas. In the crucible of the Democratic Republic of Congo (DRC), the security umbrella provided by peacekeepers created a conducive environment for aid workers to operate, ensuring the provision of essential services even in remote and perilous regions.

However, it's important to acknowledge that the trajectory of peacekeepers' successes isn't without limitations. Instances exist where their protection has faltered, failing to shield civilians from violence – a stark reality illustrated by the tragic Srebrenica massacre in Bosnia and Herzegovina in 1995. Additionally, there have been cases where peacekeepers' efforts fell short, notably the critique of inadequate action to prevent the Rwandan genocide in 1994. Unfortunately, their reputation has also been marred by allegations of human rights abuses, underscoring the complexity of their mission.

The interplay of peacekeepers' achievements and limitations underscores the multifaceted nature of their engagement. While history attests to their effectiveness as stewards of stability, it also documents instances where they grappled with the intricacies of their role. Analyzing this intricate narrative requires acknowledging these nuances and remaining open to the evolving landscape that shapes their contributions in humanitarian interventions.

The Potential Value of Former UN Peacekeepers' Experience and Skills

The array of skills and experiences acquired by former UN peacekeepers during their tenure in peacekeeping missions equips them with a formidable toolkit applicable across a diverse spectrum, including the realm of humanitarian interventions. This reservoir of knowledge includes competencies spanning leadership, decision-making, cultural sensitivity, conflict resolution, and logistics management. These skills are honed through exposure to complex and challenging environments such as conflict zones and disaster-stricken regions. After their service, former peacekeepers possess a comprehensive understanding of working with various stakeholders, from government officials to civil society groups and local communities.

These experiences underscore their teamwork, adaptability, and ability to make sound judgments even in high-stress situations.

In the realm of humanitarian emergencies, ex-peacekeepers become repositories of unique insights and competencies, especially in areas like conflict mitigation, intercultural communication, and leadership. Their deep understanding of local dynamics becomes crucial in responding to the needs of affected populations. This understanding positions them as sources of guidance for response personnel, enhancing their actions with valuable direction. Moreover, the comprehensive training ex-peacekeepers receive in areas like cease-fire monitoring, patrolling, data collection, constructive dialogues with local authorities, humanitarian aid provision, and support for indigenous security forces and peacebuilding initiatives equips them with a reservoir of expertise. This expertise collaborates with their ability to enhance the effectiveness of local law enforcement and the military in terms of safeguarding civilians and mitigating conflicts. Additionally, ex-peacekeepers act as advocates for transparency, reporting on human rights violations and contributing to the development of international jurisprudence and standards.

Their experience in decision-making and leadership plays a crucial role in orchestrating and guiding the efforts of humanitarian agencies in crisis situations. The ability to make swift and effective decisions in high-pressure scenarios becomes pivotal for timely aid distribution. Their foundation in cultural sensitivity improves communication between aid organizations and the communities they aim to assist. Proficiency in conflict resolution is pivotal in negotiating safe passage for humanitarian workers and facilitating aid delivery, a dimension that is particularly relevant in conflict or instability-ridden regions.

In the realm of logistics, the resource management expertise of ex-peacekeepers enhances the efficiency of aid delivery, optimizing the transit and distribution of essential assistance to those in need. Their management of aid transportation and allocation becomes transformational in ensuring the effectiveness of aid delivery. Beyond their range of competencies, the employment of ex-peacekeepers represents both fiscal prudence and the enhancement of the efficacy of humanitarian operations. The harmonization of skill sets and experiences with the rigors of humanitarian emergencies is a testament to the potential harnessed through the integration of former peacekeepers into humanitarian efforts.

Transferable Experience and Skills

Transferring the competencies and experiences held by former UN peacekeepers to the realm of humanitarian interventions holds the promise of invaluable contributions across several pivotal domains:

1. Conflict Resolution: A compelling example of conflict resolution competencies inherent in UN peacekeepers' skillset arises in the UN Mission in Ethiopia and Eritrea (UNMEE). Established in 2000 to oversee the ceasefire between these two nations, UN peacekeepers played a key role in defusing a border dispute and fostering dialogues between these antagonistic nations. Through high-level meetings and the establishment of buffer zones, peacekeepers demonstrated their capacity to resolve conflicts through diplomatic means, ultimately leading to the peaceful resolution of the Ethiopia-Eritrea conflict.
2. Intercultural Communication: The UN Interim Force in Lebanon (UNIFIL) exemplifies the significance of intercultural communication proficiency within peacekeeping endeavors. In the aftermath of the 2006 Hezbollah-Israel conflict, diverse contingents of UN peacekeepers were deployed to Lebanon. Their mission centered on maintaining peace and security amidst the nation's diverse ethnic and religious landscape. Through adept intercultural communication, they fostered trust and cooperation within the local community, thereby contributing to an environment characterized by tranquility and security.
3. Leadership: Leadership competencies wielded by UN peacekeepers resonate in the efforts of the UN Mission in the Democratic Republic of Congo (MONUC), initiated in 1999 to support the nation's peace process. Local communities and government institutions benefitted from their guidance and leadership, instrumental in restoring stability and upholding the rule of law. The nurturing of indigenous security forces, bolstering their ability for civilian protection and peacekeeping, exemplifies the positive impact of their leadership acumen. The culmination of these efforts was evident in the successful 2006 elections in the Democratic Republic of Congo.
4. Local Knowledge: The UN Mission in Sierra Leone (UNAMSIL), founded in 1999, serves as a poignant testament to the significance of local knowledge in peacekeeping endeavors. The nation's civil war led to the deployment of UN peacekeepers to restore peace and protect civilians. Their deep understanding of Sierra Leone's socio-cultural and political dynamics fostered trust with the local population,

instrumental in addressing the root causes of conflict and illustrating the potency of local knowledge in conflict resolution.

5. Humanitarian Assistance: The UN Mission in South Sudan (UNMISS), established in 2011, epitomizes the role of UN peacekeepers in humanitarian assistance. Amidst the aftermath of conflict in South Sudan, they emerged as champions of humanitarian aid delivery. Their impact reached vulnerable populations, particularly women and children. These peacekeepers ensured their access to essential aid, bolstering stability and improving the lives of local inhabitants.
6. Training and Capacity Building: Within the scope of the UN Mission in Kosovo (UNMIK), UN peacekeepers assumed a transformative role in training and enhancing the capabilities of local security forces. Following its establishment in 1999, this mission provided training initiatives for local police and military units. This capacity-building translated into enhanced civilian protection and conflict resolution proficiency. Peacebuilding efforts were further supported through reconciliation endeavors and dialogue facilitation between diverse ethnic and religious groups. The aftermath of these efforts manifested in a more secure and stable environment, nurturing the growth of long-term peace and prosperity in Kosovo.

These specific skills and experiences, combined with former peacekeepers' leadership and decision-making finesse, cultural sensitivity, conflict resolution and mediation abilities, and logistical and resource management skills, position them as valuable assets in humanitarian emergency response teams. Nonetheless, the process of transferring these skills comes with its own set of challenges and limitations.

Challenges and Limitations in Skill Transfer

While former UN peacekeepers hold a treasure trove of experience and competencies, several challenges hinder their immediate availability as aid workers and alignment with the requirements of humanitarian entities:

1. Scarcity and Availability: Former peacekeepers are not readily available due to the limited nature of their engagements. Military and police peacekeepers usually return to their home countries after their missions, and the option to join humanitarian efforts might only be open post-retirement from their parent units. This limits the pool of eligible ex-peacekeepers for humanitarian deployments.

2. Specific Recruitment Process: The recruitment of serving military or police peacekeepers for civilian humanitarian roles is exclusively the prerogative of the UN. Some military bodies may impose constraints on the participation of their personnel in civilian humanitarian activities, further limiting the pool of potential ex-peacekeepers for humanitarian missions.
3. Misalignment of Skillsets: Humanitarian organizations might have distinct recruitment and deployment processes, along with specific technical requirements and training prerequisites that may not perfectly align with the competencies gained by ex-UN peacekeepers. For example, fields like public health interventions may demand technical competencies that may not correspond to those of former peacekeepers.
4. Financial Constraints: Humanitarian organizations may have budget constraints that prevent them from hiring individuals who require higher remuneration due to their exposure to dangerous environments during UN service. This financial implication could strain their budgets.
5. Militarization Concerns: Some humanitarian organizations might be cautious about recruiting former military personnel due to concerns about the militarization of humanitarian aid. Collaboration with military personnel may lead certain entities to favor other candidates who do not have military backgrounds.
6. Preference for Humanitarian Experience: Humanitarian organizations may lean toward candidates with direct experience in the field of humanitarian action, as their familiarity with the organization's principles and practices can facilitate smoother integration.

To address these challenges and facilitate the transfer of experience and skills from ex-UN peacekeepers to humanitarian contexts, a comprehensive approach is necessary.

Effective Strategies for Transferring Experience and Skills

Maximizing the potential contributions of ex-UN peacekeepers in humanitarian contexts requires a well-coordinated strategy that encompasses various elements:

1. Information Repository: Establish a comprehensive database that catalogs the range of skills and experiences accumulated by ex-UN

peacekeepers. This repository should be easily accessible to humanitarian organizations, facilitating a seamless alignment between their expertise and relevant roles.

2. International Peacekeeper Exchange Program: Develop an international exchange program for peacekeepers, facilitating the immersion of former peacekeepers into humanitarian contexts by pairing them with established non-governmental organizations (NGOs) and aid entities. Such partnerships encourage the exchange of insights and skills.
3. Workshops and Seminars: Organize targeted workshops and seminars to enhance ex-peacekeepers' understanding of the humanitarian sector. Topics could include the UN's humanitarian response framework and core humanitarian principles, fostering cooperation and communication within a humanitarian environment.
4. Certification Protocols: Create a certification program that validates the competencies and experiences of ex-UN peacekeepers, ensuring their alignment with the specific requirements of humanitarian roles. Certification could cover areas like conflict resolution and peacebuilding, ensuring a fit with particular positions.
5. Mentorship Platforms: Establish a mentorship network where ex-UN peacekeepers act as mentors, providing support and guidance to humanitarian response personnel, especially newcomers. This mutually beneficial arrangement facilitates the transfer of wisdom and experiences to the next generation of humanitarian workers.
6. Collaboration and Synergy: Enhance collaboration between UN peacekeeping missions and humanitarian organizations through joint training exercises and the inclusion of ex-peacekeepers in leadership roles within humanitarian entities. This cooperation fosters a seamless continuity of expertise.
7. Performance Monitoring and Evaluation: Regularly assess and evaluate the effectiveness of ex-peacekeepers in their new roles, both by humanitarian organizations and UN peacekeeping missions. This evaluation ensures the smooth transfer of their competencies and attributes, reaffirming their impact within humanitarian contexts.
8. Advocacy and Visibility: Elevate the role of ex-UN peacekeepers in humanitarian settings through collaboration with national and regional governments and UN agencies. Engage in media activities, press releases, speaking opportunities, and other communication channels to emphasize their relevance and potential contributions.
9. Tailored Training Modules: Develop specialized training modules designed to align the competencies of ex-peacekeepers with the

requisites of humanitarian contexts. These modules should focus on core skills such as cultural acumen, conflict mitigation, and logistical coordination.

10. Network of Expertise: Foster a network that connects former UN peacekeepers with experienced humanitarian professionals. Utilize virtual platforms and regular gatherings to facilitate the exchange of best practices, knowledge sharing, and professional growth.

Conclusion

The strategic integration of former UN peacekeepers into the fabric of humanitarian interventions represents a potent avenue for enhancing the global response to humanitarian crises. The comprehensive training and diverse experiences acquired by these individuals during their engagement in peacekeeping missions have nurtured a capacity to navigate complex and high-pressure situations. This aptitude equips them to effectively address the multifaceted demands inherent in humanitarian contexts.

While challenges and limitations exist in seamlessly transferring the competencies of former peacekeepers to humanitarian roles, a well-coordinated approach can mitigate these obstacles. By establishing targeted programs, fostering collaboration, and creating platforms for mentorship and training, the potential of ex-UN peacekeepers as valuable contributors to humanitarian efforts can be fully realized.

In essence, the integration of former UN peacekeepers into the realm of humanitarian interventions is not only a practical endeavor but also a testament to the continued commitment of these individuals to the cause of global peace and security. By harnessing their skills, experiences, and insights, the humanitarian sector can benefit from their unique contributions, ultimately contributing to the alleviation of human suffering and the promotion of stability and well-being in crisis-affected regions.

CHAPTER IX

FROM KEEPING THE PEACE TO ENFORCING IT: CHALLENGES OF CONFLICT EVOLUTION IN UNITED NATIONS PEACEKEEPING OPERATIONS

CAPTAIN (NAVY) (RETIRED) OMAR MARTIN TEJADA PEREZ

> "Long are gone the days of just monitoring ceasefires. Now we are working in crisis settings. Now it is time to do more."
> —Mr. Herve Ladsous
> Former Under-Secretary-General for Peacekeeping Operations (Ladsous 2015)

The United Nations (UN) was established in 1945 after the end of World War II (WWII) with the primary objective of avoiding another massacre committed by those promoting the scourge of war. The UN Charter, the foundational document of this international organization, states as one of its main purposes the following:

> To maintain international peace and security, and to that end: to take effective collective measures for the prevention and removal of threats to the peace, and the suppression of acts of aggression or other breaches of the peace, and to bring about by peaceful means, and in conformity with the principles of justice and international law, adjustment or settlement of international disputes or situations which might lead to a breach of the peace. (United Nations 1945)

Chapter V of the Charter establishes a legal framework that entitles the Security Council (SC) to decide where and when to authorize intervention to enforce this mandate.

After WWII ended, the world entered a new conflict dynamic framed by the Cold War. Consequently, the two major powers: the United States and the then-Union of Soviet Socialist Republics and their respective allies struggled to spread their global influence. In this scenario, where a conventional warfare (inter-state) mindset still predominated, peacekeeping forces, incepted initially in 1948, were used to separate regular military forces and monitor inter-state peace agreements by patrolling contention areas known as buffer zones. Therefore, this standardized procedure was the norm at the beginning of the implementations of the first peacekeeping operations (PKOs), which some call "Classic PKOs" or "First-Generation" ones (Hempson 2011).

The superpowers' race for strategic influence also brought up and fueled internal conflicts in several countries worldwide. This situation started to impact the classic model of PKOs and presented new challenges for the troops on the ground. Military and police peacekeepers, also known as *blue berets* or *blue helmets*, faced new challenges in this mixed and increasingly complex environment where conventional and unconventional warfare features started to mingle. The constant change required drastic mindset, equipment, and training updates, so drastic that if not implemented, the reality would consistently outpace the necessary actions on the ground.

Consequently, as the Berlin Wall's fall decreased the intensity of the Cold War, a new warfare paradigm would rapidly arise into what British General Rupert Smith calls the "war amongst the people" (Smith 2006, 5, 28). These global geostrategic and political changes again outpaced the classic military training and doctrine under the conventional view. As a result, several new challenges emerged in the middle of this harsh transition that severely impacted the actions on the ground.

This shift from an inter-state and conventional approach to an increasingly intra-state and unconventional approach (Hempson 2011) suddenly caught military and police peacekeeping forces between the crossfire caused by conflicts whose roots were mainly in local social grievances rather than international interests, though the latter have always been present. Freedom fighters, guerrillas, terrorist groups, and organized crime increasingly mingled with the exploitation of natural resources, smuggling routes, civil protests, and private and state international interests inside national soil. This state of chaos and uncertainty produced by the *fog of war*, as defined by Clausewitz (Kornberger 2013), ironically needed the support of the civilian component if it wanted to be cleared. New types of peacekeeping missions were established that focused not only on stopping conflicts but also on helping to transition from war to peace, adding the concept of peacebuilding to the military approach. This scenario included actions such as electoral support; disarmament, demobilization, and reintegration programs; humanitarian assistance; government capacity-building; and other actions that would enhance conflict prevention, resolution, and transformation measures (Kenkel 2013; Hempson 2011). Here is when the multidimensional concept appeared and overlapped with the purely military approach to the original PKOs resulting in the civilians, police, and military components working hand in hand to fulfill the requirements of a broader and more complex approach where warring parties and the roots of conflicts were not easy to identify.

After this, PKOs' evolution, rather than settling, sped up together with the unfolding of the post-Cold War era. The information revolution introduced new technology, giving more and more people access to real-time news from around the globe. This posed further and more complex challenges for PKOs where technology rapidly increased its essential role in improving mission performance (Portmess and Romaya 2015; Duursma and Karlsrud 2021). Missions' multidimensional character had to deal with increasing local social turmoil and different kinds of non-state actors, while adaptation always seemed to be one step back. The original character of PKOs, namely the separation of two conventional belligerent groups, had changed long ago. This scenario seemed to be designed to challenge every Westphalian-based doctrine, concept, or strategy. In this new setting, Chapter VI of the UN Charter, which authorized the deployment of only consent-based missions and was paramount for the original PKOs, was not enough to protect the force and the local population and, to a lesser extent, to accomplish the mission, resulting in grave failures such as in the cases of Rwanda, Bosnia, and Somalia. A drastic change in the legal framework of the missions was needed; therefore, the time had come for the implementation of Chapter VII, which allowed peacekeepers to be deployed without the host nation's consent and permitted the use of force, if required, to accomplish the mission.[1]

Subsequently, new developments made PKOs come onto the scene by reinforcing the application of Chapter VII of the UN Charter, transitioning from the defensive approach previously described into an offensive character. This situation granted greater power to the use of force option and enhanced the participation of the civilian component to the point of

---

[1] Article 42, United Nations Charter, Chapter VII: Action with Respect to Threats to the Peace, Breaches of the Peace, and Acts of Aggression.

establishing transitional governments while working on the re-institution of law and order. At this time, the task was not only protecting the force but using the force to protect civilians, rebuild political and governmental institutions, and accomplish the mission's mandate. This type of PKO finds theoretical support in human security and positive peace concepts (Kenkel 2013), intensely challenging the traditional Westphalian sovereignty of non-intervention (Pitty and Smith 2011). However, current PKOs may lack international legal support due to the possibility of compromising peacekeepers' neutrality when performing offensive operations for purposes other than self-defense and the risk of becoming another belligerent force instead (Derejko 2023).

Current missions also seem to evolve the multidimensional character into a hybrid one where the UN and different international actors gather to form a mission in which high standards of coordination, cooperation, command, and control must be implemented to succeed (Tardy 2014). This is not an easy task considering the variety of backgrounds, organizational culture, and interests at stake.

By analyzing different theoretical approaches to the ever-changing conflict environment UNPKOs are facing today and describing and linking the author's personal experiences as a UN peacekeeper with this theoretical framework, this book chapter will provide some recommendations focused on improving the quality, legitimacy, and legality of PKOs as a form of international military intervention seeking to reduce the casualties of peacekeepers and the local population.

From Dallaire to Brahimi to Dos Santos Cruz

As seen, the end of the Cold War can be described as a significant event that shifted the conventional warfare approach, or industrial war paradigm as called by Smith, into a more unconventional one where war amongst the people predominates (Smith 2006, 269). Of course, this is not to say that inter-state conflicts are gone, as we can see with the current war between Russia and Ukraine or the tensions between the United States and China, but most of the conflicts are fought within states' territories, where most of the victims are local civilians. In the same vein, the conventional military mindset and doctrine were not ready to face the sharp shift generated after the Berlin Wall fell. As a result, the UN could not adapt to the new paradigm that reshaped political, diplomatic, social, and military thinking in a way few could have foreseen.

Consequently, one of the first acid tests for the UN, and also for the international community as a whole, came soon in 1994 when the United Nations Assistance Mission to Rwanda was not able to stop a mass killing of Tutsis and moderate Hutus in the hands of extremists Hutus. Despite all the evidence collected months before the genocide and the UN force commander's plea for more troops and authorization to secure the protection of civilians, the international community was caught in a deadlock imposed by the SC due to political reasons (LeBor 2008; Dallaire 2009). Meanwhile, a genocide was rapidly taking place, killing from eight hundred thousand (Adebajo 2011, 231) to more than one million people between April and July 1994 (Hernan 2010, 82) in what Waller calls "the most rapid and efficient genocide in the twentieth century" (Waller 2016, 164).

Unfortunately, Rwanda was not the only fiasco, as there were also other UN failures in the 1990s. For instance, Somalia and Bosnia and Herzegovina also challenged the UN's ability, efficiency, capacity, and ultimately legitimacy around the world, making it urgent to assess the situation and revisit the UN's approach, which resulted in the issuance of the Brahimi Report in 2000 (De Coning and Peter 2019, 5). Under the request of the former UN Secretary-General Kofi Annan, Mr. Lakhdar Brahimi, former Algerian Foreign Minister, chaired the Panel of Experts on PKOs. The panel acknowledged that the UN had failed to avoid the scourge of war and wrote a report providing several recommendations for the UN and for state members seeking to improve and adapt PKOs to the then-new challenges and threats to world peace. It emphasized the need for significant institutional change and increasing budgets to support peacekeeping and peacebuilding tasks, issuing recommendations focused on *improving strategic direction, decision-making, rapid deployment, operational planning and support, and usage of modern information technology* (United Nations 2000).

Time went by, and, despite the awareness generated and, to some extent, the implementation of changes recommended by the Brahimi Report, the situation on the ground kept evolving, as previously explained. *Blue Helmets* were challenged by even more complex scenarios resulting in more casualties in the population and also on the peacekeepers' side. UN peacekeeper personnel fatalities during the first- and second-generation mission types amounted to 851 from 1948 to 1990, less than 20 per year on average.

Unfortunately, those numbers skyrocketed, reaching 1711 at the beginning of the new century and rising to 4249 at the time of writing. That means that around 900 peacekeeper fatalities were recorded in the 1990s, 90 on average per year, while another 2538 fatalities have been counted in the 21st century, an average of 110 per year (United Nations Peacekeeping 2022).

Meanwhile, the number of civilian casualties in current armed conflicts may represent up to 90% of the total, a dramatic increase considering that in World War I, it represented between 15% and 25%, and in World War II, it accounted for up to 65% of the estimated deaths (Burkle 2019, 838; Thakur 2013, 168).

Therefore, a significant assessment was still needed; the international community required the UN to adapt to the pace the world's conflicts were setting. If a diplomatic officer chaired the Brahimi Report panel, maybe this time the call might be made to a military expert on PKOs. Consequently, on November 2017, the UN Secretary-General Antonio Guterres appointed Brazilian General Carlos Alberto dos Santos Cruz to lead a high-level review of peacekeeping fatalities and injuries due to violent acts (United Nations Secretary-General 2017). The report issued by this high-level review was called *"Improving Security of United Nations Peacekeepers: We need to change the way we are doing business",* and it presented a set of recommendations advocating for a drastic change in approach to topics such as troop training, logistics, doctrine, intelligence, leadership, and budgetary and even legal considerations in a robust framework. This document supports a more offensive stance as a countermeasure to increasing peacekeepers' casualty reports. It strongly criticizes current UN peacekeeping policies, standards, and troop-contributing countries' flaws. The report seeks to improve deployment standards, troops' training quality, and appropriate logistical support, which have been identified as constant weaknesses on the ground ("Improving Security of United Nations Peacekeepers (Independent Report)," 2017). Unfortunately, as of today, casualty numbers have not decreased, and peacekeepers on the ground are still struggling to face many challenges for which they need to be more appropriately trained and supported. For example, from 2020 to 2021, during the COVID pandemic, an average of 130 peacekeepers died, mostly in the most dangerous missions, namely the United Nations Multidimensional Integrated Stabilization Mission in the Central African Republic (MINUSCA), the United Nations Organization Stabilization Mission in the Democratic Republic of Congo (MONUSCO), and the United Nations Multidimensional Integrated Stabilization Mission in Mali (MINUSMA). Interestingly, all these missions are located in a region of massive illicit goods smuggling, which sheds light on the current dynamics of conflicts, as shown in the *World Atlas of Illicit Flows* (Nellemann et al. 2018).

Sovereignty vs. Intervention

At this point, this book chapter has elaborated on conflict developments before and after the fall of the Berlin Wall. As a result, the turbulent political outcomes of the end of the Cold War impacted PKOs' performance on the ground in a meaningful manner.

However, PKOs, as a form of international intervention, have mainly passed from classic ceasefire agreement monitoring missions to multidimensional and integrated missions seeking to adapt to new challenges. This situation pushed the international community to fall into a dilemma of whether or not to intervene in cases where the state could not or was not willing to protect its own citizens' lives, especially after the horrors witnessed in Rwanda, Somalia, and Bosnia and Herzegovina in the 1990s. Moreover, sovereignty as authority had been jeopardized by the state's use of genocidal forces against its population, where the pursuit of national security considerations clashed with those of human rights protection (Thakur 2013, 75).

In this complex scenario, between 1999 and 2000, former UN Secretary-General Kofi Annan asked the General Assembly to assess and submit recommendations on the international community's failure to meet the challenges posed by internal conflict dynamics. As a result, the Government of Canada and other significant actors announced the implementation of the *International Commission on Intervention and State Sovereignty* (ICISS) in September 2000. The commission issued a report establishing the new concept of the *Responsibility to Protect* (R2P), which highlighted the need to make states accountable for their lack of capacity or unwillingness to protect their own population. This new concept, although not a binding norm, is a set of rules on how the international community is expected to behave to halt violence that would lead to mass atrocities due to internal political struggle. It is about the international community's right to perform a humanitarian intervention in a nation-state's territory (Global Centre for the Responsibility to Protect 2001).

However, the R2P concept is seen by some as a new form of imperialism, a tool to justify intervention in internal affairs due to superpower interests (Herman and Paterson 2010, 19). It has also lacked political support in cases where needed, like the mass killings in Darfur, Syria, and South Sudan during the last decade. Others, however, justify the R2P concept and highlight its prevention, reaction, and rebuilding principles as a mechanism of the international community to stop mass killings. This political and diplomatic toolbox reshapes the traditional concept of sovereignty as authority into sovereignty as responsibility, which entitles obligations toward citizens (Krishnan, 147). In short, R2P is a citizen-based rather than a state-based concept.

It is also essential to highlight the concept of *Responsibility While Protecting* (RWP) proposed by the Brazilian government in 2011 after the intervention in Libya to ensure the use of force only as a last resort, noting that any intervention should carry accountability for the interveners' actions on the host country. The RWP concept does not contradict the R2P approach, but it complements the latter, advocating for the compliance of its novel principles (Hamann and Gabrielsen 2020). Brazilian military actions to pacify various hotspots in Port-au-Prince, Haiti between 2005 and 2007 may be considered an example of an RWP approach that prioritized prevention and negotiation before any enforcement action was taken. However, RWP still needs to be promoted more in the academic literature and international forums considering its potential to create bridges between the supporters and the critics of the R2P concept (Kenkel and Cunliffe 2016, as cited in Hamann and Gabrielsen 2020).

From Theory to Practice

After more than twenty-one years of service in the Peruvian Navy, four of them in the Peruvian Jungle and three and a half as a UN Peacekeeper in Sudan, Haiti, and the UN Headquarters in New York, I have witnessed the dynamics that affect today's conflicts.

Working in the Putumayo River Basin, which marks the border between Peru and Colombia for around one thousand kilometers, gave me the opportunity to learn how different the conventional and the irregular warfare mindsets are. In a place where insurgency, border-organized crime, and social grievances, including poverty and a lack of state presence, mingled equally, the time I walked (and sailed) the area showed me a reality I had ignored beforehand. Here, I learned about counterinsurgency, low-intensity warfare, intelligence, and, to summarize, the war amongst the people, as Smith calls it.

From English to Arabic, to Dinka to Swahili… and All the Way Around

A few years later, after my experience on the border between Peru and Colombia, already back in my home city of Lima, a huge metropolis where one-third of the Peruvian population lives, I received an unexpected phone call. A person on the other side of the line, a senior officer, told me that I had been selected to be a UN military observer in the United Nations Mission in Sudan (UNMIS) and had a couple of weeks to get ready and fly to the other side of the world. I had already received a bit of UNPKO training experience while participating in a multinational peacekeeping operations exercise in Chile in 2002, but I knew this challenge would be different—this would be for real, not training anymore.

Fortunately, my departure to Sudan was delayed for a couple of months, so I had the opportunity to better prepare myself and my family for a one-year tour to Africa with little or no chance of coming back home for a visit during that time. I would be traveling with three other Peruvian military officers whom I did not know previously. After a thirty-two-hour flight that stopped in Amsterdam and Nairobi before landing in Khartoum, we faced our first taste of what our mission would be like. As a normal procedure, we, the four Peruvian officers representing the UN in Sudan, went to check with the Sudanese migration officers at the airport, who decided that, although we showed them all the necessary documents, our visas were not authorized and that we all would be put on an airplane back to Peru. At first, because none of us were able to speak English as well as we would have liked, we could not understand what they were telling us, but it was clear that we were in trouble; sometimes, body language expresses more than words, and the Sudanese officers made themselves clear very soon. At this point, it was obvious that the Government of Sudan was not happy with the UN presence in its territory, sniffing into the country's business, and, as we learned later, this was the standard welcoming procedure to new UN military personnel, providing a bit of softening before our deployment to the different war-torn communities in the south.

After a week in Khartoum full of administrative tasks and finalizing my induction training, I was ready to travel to Rumbek in South Sudan. That morning, I took a plane that would take me from Khartoum to Wau, my first stopover. The plane was ready to take off, and as usual I was stressed and fully alert, waiting for the plane to ascend while it constantly increased its speed on the airstrip. Suddenly, before reaching the end of the strip, the plane's propellers' stopped making noise, and the craft's brakes were immediately activated. It was a failed departure attempt. A mechanical failure prevented us from taking off at the last moment. *Was that another welcoming procedure for UN personnel?* … I thought. Other mechanical and operational challenges would come later during that year, some due to negligence and others due to weather conditions, though nothing bad ever happened.

Rumbek, a small town destroyed by the long war between the South Sudanese Forces and the Sudanese ones, was my destination and the place where I would first learn about real peacekeeping. Soon after I arrived, I realized that the UN was not what I had thought, but it was the only force we had for establishing peace. Very professional people committed to making peace were mingling with others who only wanted to make money out of others' suffering and the conflict status quo. I also soon realized that in the same way my English was not good enough to apply my professional skills the way I wished, I found that I did not have a clue how to communicate with the local population, whose languages were Arabic or Dinka and, in a few cases, Swahili. A few years before, I had learned in the Putumayo that if I wanted to survive, I had better connect with the local population. Now, in the middle of Africa, I knew that I would be in a critical and vulnerable position if any emergency happened and I had to fix it soon.

I then discovered that learning key words and phrases in those languages, such as "Thank you," "My name is," "How are you," Nice to meet you," and "Good morning," among others, would be a good start to connect with the locals; this was not a perfect situation but a good start. A couple of incidents in the local market where I was helped by some unknown locals who never hesitated to protect me against a physical attack confirmed this later on. After more than fifteen years, I still remember these "good Samaritans."

Connection with the local population can undoubtedly save one's life, but connecting with the conflict parties, when possible, will also do so. I also learned this, in one way or another, in the Putumayo, where I had to deal with locals involved in illegal economic activities because they did not have another option; they were part of the conflicting parties. Now, in Sudan, I had to build confidence with the South Sudanese and the Sudanese representatives and militaries alike, a difficult but not impossible task. This connection once saved my life when a South Sudanese military officer attached to my patrol had to deal with a soldier who aimed his rifle at my head and threatened to shoot me for no reason. It was a bizarre situation where an officer ordered a lower-ranked soldier, who was completely out of control, to surrender his weapon and, noticing that his authority was not working, ended up begging him not to shoot at me. By then, I had already spent a few months with this officer and had built a strong connection by exchanging conversation and sharing good moments at the base and on patrol.

After a year of moving around in what is today Sudan and South Sudan, two countries that share only a few things and a history of conflict, I learned that although the UNPKOs' challenges described in previous paragraphs were all present, the smile of a person in need as a result of our actions is enough for understanding that our mission as peace soldiers has been accomplished. I experienced that several times but never more strikingly than in Shambe, where one day I arrived leading a long-range aerial patrol. The Rumbek Team Site Leadership sent us to verify the humanitarian and security situation in this little village located by the White Nile River, one hundred twenty-six kilometers east of Rumbek and two hundred sixty-four kilometers north of Juba. There, a group of around three hundred displaced people gathered on top of a floating pontoon had been abandoned to their luck. When I entered the pontoon, I saw people lying on the floor asking for food and medicines. Elders, adults, and children were starving and had been escaping militia forces that had been chasing them for weeks. This situation was impacting not only the displaced group but also the Shambe's villagers, with the potential to become a humanitarian and security disaster. I will never forget the people's smiles when I started to collect all the information to prepare my patrol report. For me, this would be another report that may or may not have solved the situation, but it seems that for them, a foreign military officer writing about what was happening to them was the only hope they had to be rescued at that moment. After a few days, that group was rescued and transported to Juba, where they were provided with a safe haven. I would never know if our visit had something to do with that, but those smiles are among the most precious memories I still have from that time.

The Land of Mountains and the Little Cardboard House

After another deployment in the Peruvian jungle and then having worked with the United States Marine Corps in North Carolina, I was deployed to Haiti as a contingent company commander in the United Nations Stabilization Mission in Haiti (MINUSTAH). This experience put me in a position where I could witness the quality of training, leadership, and other challenges already described in the previous paragraphs. Here, I realized that not all the units were appropriately equipped or trained, while others always kept the best combat standards required for this kind of mission. Again, the UN bureaucracy in Haiti taught me what I had experienced years before in UNMIS when I realized that the UN is not perfect, but it was the best thing we had to make peace, at least for those who commit themselves to do so. The following story summarizes one of the most rewarding experiences I ever had as a peacekeeper:

> Sarah was an American girl who worked for a private company in Port-au-Prince in 2014. When I met her, she asked me if I would like to go with her to visit an orphanage and deliver anything that may bring some happiness to the eighty little kids who lived there. I had recently arrived in Haiti and was commanding an infantry company of more than three hundred men and women as part of the MINUSTAH.
>
> After accepting Sarah's invitation, I met my staff officers and planned the visit. We organized the convoy's security, formed teams to perform amusement shows, delivered groceries and clothing, and provided basic dental and medical care for the kids and the orphanage staff. The activities were performed as per our plan until I entered the main two-story building that hosted all the kids and had a couple of classrooms on the first floor. Unfortunately, the structure was made of humid and damaged wood, making the building unsafe for occupancy and posing a significant risk to the kids. At that moment, I thought I had to do more for the well-being and safety of those children and told the orphanage staff that I would try to find a way to improve the building conditions, but the only thing I could promise was to do my best, and this would not mean anything but that.
>
> When I arrived at my base, I passed by the Paraguayan military engineering company, asked its commandant if he could assign me two engineers to assess the building situation, and try to make a plan; he agreed without hesitation. The next day, I returned to the orphanage with the engineers to do the assessment and returned to our base, waiting for the final report to be submitted to me in a week.
>
> The next day, early in the morning, I received a phone call from an unknown number. A lady with a Spanish accent on the other side of the line told me that she had learned about my idea of helping the orphanage. Her name was Patricia, and she ran an NGO at the time. She was willing to join me on the project and offered money to fund it. I asked her to visit me a week after the engineering report was submitted to my office to discuss the issue.
>
> The report arrived on time, and I met the engineers and Patricia to decide on the best course of action and define the roles. We made a quick impact project (QIP) that had to be approved by the MINUSTAH. I would provide the workforce, Patricia's NGO would fund the project and be in charge of the logistics, and the Paraguayan engineers would lead the work on the field. Everything was coordinated, so we all thought it was only a matter of time before getting the approval. We were wrong.

After four months of waiting for the UN approval, the QIP was rejected by the very last office in the MINUSTAH headquarters due to a minor administrative flaw. It would be difficult for me to try to fix it because I was about to leave the mission in a couple of weeks and had other significant commitments before my departure. I was devastated because I had put so much effort into seeing the project become a reality. After a six-month tour of duty, I finished my mission and returned home at the end of July 2014, wholly disenchanted and demotivated, thinking that my mission in Haiti had been in vain. Once again, I was wrong.

In the first week of December 2014, Patricia sent me the following email:

From: Patricia

To: Omar

Subject: Orphanage

Dear Omar,

How are you? Did you stay in Peru or have you traveled again? I hope you are well.

I have wanted to write to you for a long time to tell you that we have finally made the orphanage. My organization found extra money and started to work on that. We inaugurated it on October 29, 2014.

The next day I returned to Spain, and now the UN has hired me to work at MONUSCO as a civilian. I arrived a week ago and will be based in Lubumbashi. So, if you come through the Congo, please let me know.

I am sending you the photos of the orphanage. Thank you for your support and enthusiasm. Without your encouragement, I would have left it as impossible from the beginning. You have contributed enormously to making it finally come true. I hope it fills you with joy and pride. A big hug,

Patricia

The World Could Be as Big as an Apple

After my experience in MINUSTAH, I returned to Peru and was again deployed to the Valley of the Rivers Apurimac, Ene, and Mantaro, an immense area where most of the Peruvian cocaine production takes place. Here, I was the second in command of a unit composed of seven hundred Marines and Navy personnel. Counterinsurgency and counternarcotic operations were our main mission, and my experience in PKOs and other deployments within Peru had given me enough understanding of how to deal with the challenges of a combat zone. After a successful year in that zone, I was ready to apply for a position at the Department of Peace Operations in the UN Headquarters in New York.

The following year, after a long recruitment process, I received a message that I was selected to work as a peacekeeping affairs officer in the Office of Military Affairs at the Department of Peace Operations. It took me fifteen years from the first time I knew there were military positions available in the UN HQ, two UN missions, three deployments to the jungle, one year working with a foreign military, one year studying abroad, and a master's degree before I was selected. Now, I knew I would be part of the strategic level of the UNPKOs, that high-level stage where decision-making impacts the people on the ground. Now, I could change all the systematic flaws I had experienced on my missions, those things that negatively impacted the performance and lives of the peace soldiers every single day. Now, I could teach these people in New York about peacekeeping and make them aware of things they would not be able to see from their comfort in their offices far from the combat zones. Well, once again, I was wrong.

In New York, I met up with some of my former colleagues from UNMIS and MINUSTAH and others with lots of experiences from the military, police, and civilian components. I also met diplomats, politicians, and ambassadors from the different permanent missions (PMs) to the United Nations. It was as if the whole world was directed from the UN HQ building and the few blocks around where the PMs are located. It felt like the world was as big as an apple. While some of them were trying to do their best to work for peace, yet more could be done to transform conflicts.

Here, I learned again that the UN is not a perfect organization. Ideally, it should work for the member states' interests, however, the SC Permanent Members-the P5-have greater influence in decision making when it comes to world peace. In New York, I understood that in politics, there are no friends but allies, and no enemies but adversaries, and that peace and war are driven by those dynamics. In New York, I also learned about national caveats, international military intervention, vetoes, lobbies, influence, politics, diplomacy, and the balance of power, among other issues that sometimes have nothing to do with good training, leadership, language skills, proper equipment, or even peace. However, in New York, the Big Apple, I also learned that all that I had experienced was worth it and that there are many more giving much more of themselves to establish peace. I learned that, despite all the flaws, there is still hope.

The evolution of conflict in the 20th and 21st centuries has represented a significant challenge in PKOs' performance. Unfortunately, the lack of adaptation has been reflected on the ground, leading to genocidal practices that were not stopped by the international community, where millions of civilians have suffered the horrors of these actions. Additionally, UN peacekeepers, military, police, and civilian personnel have also suffered from the impacts of this situation, which has shown increases in fatality records despite several UN attempts to fix the adaptability gap after the Cold War ended. In this regard, the issuance of the Brahimi, ICISS, and Gen. Dos Santos Cruz reports seems unable to solve major issues affecting the actions on the ground. Although they reflect the conflicts' roots, all those reports lack implementation speed, with the UN's reaction to current challenges always one step behind.

Moreover, concepts like the R2P need more international community support to make them a binding norm rather than a simple set of moral ideas that, in practice, have no desired impact on today's conflicts. One of the significant obstacles to implementing such a norm comes from the SC's lack of transparency and commitment, which has led the world to face genocides like the one in Rwanda with its hands tied. Another major obstacle to not finding an effective international mechanism to avoid human atrocities is the lack of an international law framework to cover peacekeepers' necessary actions to address current challenges.

Furthermore, PKOs are strongly linked to the SC's decision-making, so it is not surprising to say that any wrong decision issued by the latter will negatively impact the former's performance and the lives of peacekeepers and, ultimately, those of the local population. Better, and sometimes faster, mechanisms of decision-making processes to effectively protect civilians in war-torn scenarios have to be developed by the SC.

Finally, the current robust character of the PKOs should stay within the legal framework of the R2P and should also take into account the RWP stance, at least while a more appropriate international legal framework is developed. Today, high standards of leadership performance are required. Leadership in complex scenarios like the current ones needs a global view, a deep local cultural understanding, and a flexible approach to interact across the spectrum of peace operations as a whole. Before implementing any mission, a thorough assessment has to be issued to weigh the impacts of coercive measures on the local population's life. Peacekeeping force protection should not clash with the protection of civilians and their human rights, but the international community has to explicitly support these mandates by setting clear rules and taking accountability.

References

Adebajo, Adekeye. 2011. *UN Peacekeeping in Africa: From the Suez Crisis to the Sudan Conflicts.* London: Lynne Rienner.

Burkle, Frederick M. 2019. "Revisiting the Battle of Solferino: The Worsening Plight of Civilian Casualties in War and Conflict." Disaster Medicine and Public Health Preparedness 13 (5-6): 837–41. https://doi.org/10.1017/dmp.2019.77.

Cedric, de Coning, and Peter Mateja. 2019. "United Nations Peace Operations in a Changing Global Order." https://library.oapen.org/handle/20.500.12657/22919?show=full.

Dallaire, Roméo and Brent Beardsley. 2004. *Shake Hands with the Devil: The Failure of Humanity in Rwanda.* New York: Carrol & Graf. Kindle.

Derejko, Nathan. "The Attack of the Peacekeepers." Mauro Institute for Peace and Justice. 2023. YouTube Video, 1:02:19. https://www.YouTube.com/watch?v=fuVur_RIYhY&ab_channel=MauroInstituteforPeace%26Justice

Duursma, Allard, and John Karlsrud, 2021. 'Technologies of Peace,' in Oliver P. Richmond, and Gëzim Visoka (eds), The Oxford Handbook of Peacebuilding, Statebuilding, and Peace Formation. New York. Oxford University Press. https://doi.org/10.1093/oxfordhb/9780190904418.013.28, accessed December 7, 2022.

Global Centre for the Responsibility to Protect. 2001. "The Responsibility to Protect: Report of the International Commission on Intervention and State Sovereignty, 2001." Global Centre for the Responsibility to Protect. https://www.globalr2p.org/resources/the-responsibility-to-protect-report-of-the-international-commission-on-intervention-and-state-sovereignty-2001/.

Hamann, Eduarda, and Maria Gabrielsen. 2020. "Brazil's Evolving "Balancing Act" on the Use of Force in Multilateral Operations: From Robust Peacekeeping to "Responsibility While Protecting." Status and the rise of Brazil. Palgrave Macmillan, Cham. 153-173.

Herman, Edward S., and David Peterson. 2010. The politics of genocide. New York. NYU Press.

Hempson, Donald A. 2011. "Avoiding the Scourge of War: The Challenges of United Nations Peacekeeping." Origins. Accessed December 3, 2022.

https://origins.osu.edu/article/avoiding-scourge-war-challenges-united-nations-peacekeeping

"Improving Security of United Nations Peacekeepers (Independent Report)." 2017. United Nations Peacekeeping. https://peacekeeping.un.org/en/improving-security-of-united-nations-peacekeepers-independent-report.

Kenkel, Kai Michael. 2013. "Five Generations of Peace Operations: From the 'Thin Blue Line' to 'Painting a Country Blue.'" Revista Brasileira de Política Internacional 56 (1): 122–43. https://doi.org/10.1590/s0034-73292013000100007.

Kornberger, Martin. 2013. Clausewitz: On strategy, Business History, 55:7, 1058-1073, DOI: 10.1080/00076791.2013.838035

Krishnan, S. 2020. "UN Peacekeeping, Responsibility to Protect and Humanitarian Intervention." India Quarterly 76.1: 120–135.

Ladsous, Herve. "UN Peacekeeping: Challenges from the field today and tomorrow." United Nations. 2015. YouTube Video, 1:01. https://www.YouTube.com/watch?v=OyJULAN7-so&ab_channel=UnitedNations

LeBor, Adam. 2008. "VII A Rwandan reprise." In "Complicity with Evil": The United Nations in the Age of Modern Genocide, New Haven: Yale University Press. https://doi.org/10.12987/9780300135145

Nellemann, Christian, Jürgen Stock, and Mark Shaw. 2018. World atlas of illicit flows. Global Initiative. Accessed November 25, 2022. https://globalinitiative.net/analysis/world-atlas-of-illicit-flows/

Pitty, Roderic, and Shannara Smith. 2011. "The Indigenous Challenge to Westphalian Sovereignty." Australian Journal of Political Science 46 (1): 121–39. https://doi.org/10.1080/10361146.2010.546336.

Portmess, Lisa, and Bassam Romaya. 2015. "Digital Peacekeepers, Drone Surveillance and Information Fusion: A Philosophical Analysis of New Peacekeeping." Theoria 62 (145). https://doi.org/10.3167/th.2015.6214502.

Smith, Rupert. 2006. *"The utility of force." The art of war in the modern world.* New York. A. Knopf. Kindle.

Tardy, Thierry. 2014. "Hybrid Peace Operations: Rationale and Challenges." Global Governance: A Review of Multilateralism and International Organizations 20 (1): 95–118. https://doi.org/10.1163/19426720-02001007.

Tejada, Omar. Haiti. Una gota de Esperanza en medio de la adversidad y la burocracia: La historia de la reconstrucción de un orfanato. Sociedad Argentina de Estudios Estratégicos y Globales. May 16, 2020.

https://saeeg.org/index.php/2020/05/16/haiti-una-gota-de-esperanza-en-medio-del-desastre-la-adversidad-la-burocracia-la-historia-de-la-reconstruccion-de-orfanato/
Thakur, Ramesh. 2013. The People vs. the State: Reflections on UN Authority, US Power and the Responsibility to Protect. www.un-ilibrary.org. United Nations. https://www.un-ilibrary.org/content/books/9789210563413.
United Nations. 1945. "United Nations Charter." United Nations. 1945. https://www.un.org/en/about-us/un-charter/full-text.
United Nations. 2000. "Brahimi Report: Report of the Panel on United Nations Peace Operations." UN Doc., A/55/305-S/2000/809.
United Nations Peacekeeping. n.d. Fatalities. Accessed on November 7, 2022. https://peacekeeping.un.org/sites/default/files/stats_by_year_1_79_September_2022.pdf
United Nations Peacekeeping. n.d. Our history. Accessed on November 7, 2022. https://peacekeeping.un.org/en/our-history
United Nations Secretary-General. 2017. Lieutenant Gen. Carlos Alberto dos Santos Cruz of Brazil - to lead a high-level review of peacekeeping fatalities and injuries due to violent acts. November 17, 2017. Accessed November 7, 2022. https://www.un.org/sg/en/content/sg/personnel-appointments/2017-11-17/lt-gen-carlos-alberto-dos-santos-cruz-Brazil-lead-high
Waller, James. 2016. Confronting Evil: Engaging Our Responsibility to Prevent Genocide. Oxford; New York: Oxford University Press.

CHAPTER X

DECIPHERING THE DILEMMA: PERSIST OR CEASE? UNVEILING THE FOUR PROLONGED UN MISSIONS

KAWSER AHMED, PHD

When the then-Egyptian president Gamal Abdel Nasser nationalized the Suez Canal, it triggered tensions among the colonial powers, since it is the quickest sea route between Asia and Europe and about 15% of global shipping traffic moves through it.[1] This event came on the heels of the already-explosive Israel-Arab situation that led to the Israel Army's push towards the Suez Canal on October 29, 1956. This move threatened to escalate into a full-blown war between Egypt, Israel, France, and the United Kingdom. In response to this crisis, Canada proposed the idea of sending an international peacekeeping force to the area under UN supervision (The Canadian Encyclopedia, 2022). Within days, Canadian diplomat Lester B. Pearson presented a resolution at the UN General Assembly that established the first-ever United Nations Emergency Force aimed at monitoring and resolving conflicts in volatile regions (Canadian War Museum 2022).

Today, UN peacekeeping operations (PKOs) are critical for maintaining global peace and security by providing essential assistance to countries affected by conflict. According to statistical evidence, despite occasional highly visible failures of some peacekeeping missions, UN PKOs have made significant contributions to global peace and security (Bratt 1996). These missions have not only helped prevent the escalation of conflicts, as they have also facilitated a transition towards stable and effective post-conflict governments. Furthermore, the UN has expanded its diplomatic initiatives for conflict resolution with an increasing number of political

---

[1] https://www.reuters.com/article/egypt-suezcanal-oil-idINKBN2BI26C

missions since 2005 (Hellmüller, Tan, & Bara 2023). Although we previously referred to the Suez crisis, which gave birth to the idea of UN peacekeeping, in January 1949 the first-ever UN mission was established in Kashmir and was called the United Nations Military Observer Group in India and Pakistan (UNMOGIP) with the aim of supervising the ceasefire between India and Pakistan in the state of Jammu and Kashmir (UNITED NATIONS 2023e).

Over the seven decades of the UN's PKOs, it has undergone transformations not only in the methods of deployment of its personnel (both civil and military) on the ground but also in terms of the types of conflicts it has faced. For example, the first-generation missions involved establishing an interposition force between belligerent groups that had given their consent to the establishment of a peacekeeping mission. The second-generation missions began when peacekeepers were interposed as a buffer between opposing troops to prevent imminent warfare (Mackinlay & Chopra 1992). These tasks have grown in complexity, and peacekeepers today are required to undertake multidimensional tasks that include monitoring ceasefires, providing humanitarian aid, restoring the rule of law, promoting human rights and supporting the development of democratic institutions. Thus, the evolving nature of peacekeeping duties suggests that contemporary peacekeeping missions now face various challenges ranging from maintaining law and order to peacebuilding (here, peacebuilding refers to former UNSG Kofi Annan's concept) (Annan 2016).

Additionally, UN PKOs also face challenges related to funding and troop contributions. So far, the UN has deployed 113 missions and 120,000 military, police and civilian personnel at various missions across the world to maintain world peace (Hellmüller, Tan, and Bara 2023). Despite the number of personnel deployed, there is often insufficient funding to properly equip and train peacekeepers. That begs the question of whether there is sufficient political and financial commitment from member states to enable effective UN PKOs. UN peacekeeping missions are further constrained by the limited number of troop-contributing countries willing and able to participate in such operations.

The United Nations Security Council (UNSC) authorizes UN peace operations and their specific tasks vary depending on the nature of the conflict and the challenges they present. The planners at the United Nations Department of Peacekeeping Operations, when planning for a mission consider the following criteria: the legitimacy of the mission, including appropriate mandates and achievable objectives; ensuring impartiality and consent from all parties involved while also advocating for the protection of human rights and gender equality; ensuring the safety and security of UN personnel as well as those under their protection; promoting sustainable peace through community engagement and capacity building; and continued evaluation of the mission's effectiveness in achieving its objectives (UNITED NATIONS 2023a). Accordingly, the UN mandates are set with the goal of promoting and protecting human rights, restoring stability to conflict-ridden areas and providing support for sustainable peacebuilding efforts. Although the mandates are not set in stone, some challenges may arise during the mission's implementation. The UNSC often reviews mission mandates based on the evolving situation on the ground and revises/modifies mandates accordingly to ensure the legitimacy of the mission (Pushkina 2006). However, the extension of mission mandates often depends on the willingness of member states to continue contributing troops and resources as well as the availability of funds to sustain and support the operations.

Some of the key influencing factors in achieving UN mandates in a particular mission lie with the logistics and operational challenges that UN PKOs face. This includes limited resources in terms of transportation, communication equipment, and infrastructure, which can hinder the ability to effectively deploy and manage peacekeeping personnel. Furthermore, attacks on peacekeepers pose a significant challenge to UN PKOs. These attacks, whether non-state actors or state forces carry them out, can threaten the safety and lives of peacekeepers, undermine public trust in the UN's ability to maintain peace and security, and hinder the mission's effectiveness.

With the overall scope in mind, the author selected four long-standing UN missions (Kashmir, Lebanon, Cyprus, and Western Sahara) as case studies. The central question driving the analysis is the effectiveness of these missions in fulfilling their mandates and tackling challenges. Subsidiary questions include assessing the feasibility of achieving their mandates in today's geo-political contexts and considering whether it is time to reconsider these missions' activities and reallocate valuable resources. One might argue that given the mandates, it appears that the selected UN PKOs are fulfilling some parts of the mandates; however, the bigger question is whether the partial fulfillment of mandates is cost effective when there is a UN budget crunch. Others might further argue that if the missions are withdrawn, hostility would break out; thus, in this chapter a detailed geo-political analysis is provided to explain that the conditions on the ground would not actually be worse and could potentially remain status quo if not improved.

In this chapter, the author only focuses on the selected PKO cases and not on the special political missions (SPMs) and special envoy/special advisor (SE/SA) (Hellmüller, Tan, & Bara 2023). While each UN peace operation is unique, there is a level of consistency in the types of tasks mandated by the UNSC. Depending on their specific mandate, peace operations may be deployed to prevent the outbreak of conflict or its cross-border implications. They can also play a role in stabilizing conflict situations after a ceasefire, creating an environment conducive to achieving a lasting peace agreement. Furthermore, these operations may assist in implementing comprehensive peace agreements and guide states or territories through a transition towards stable governance based on democratic principles, good governance, and economic development. The expansion of mandated tasks reflects the need to adapt to different conflict scenarios and address emerging threats effectively.

The analysis of these case studies aims to provide insights into the strategies employed by the UN in addressing logistical and operational challenges and the need for analyzing mandate achievement capacity. Additionally, the case studies provide an opportunity to assess how changing mandates and adjustments made by the UN have affected mission effectiveness. Using key UN documents and academic literature, this chapter seeks to contribute to the wider discourse on the reform of UN PKOs and shed light on the complex realities of implementing such missions in challenging environments.

A few studies have discussed ways to evaluate the success and failure of UN PKOs. Other than solely relying on the decisions of the UN SC purely viewed from a 'violence' or lack thereof perspective, there has hardly been any discussion in academia about why the UN HQ continues to fund some missions for decades while others are defunded. However, some scholars have attempted to establish some criteria to measure the success of PKOs as below:

> In summary, one may judge the success of peacekeeping troops by their ability to prevent the onset or renewal of warfare, as well as to limit the death and destruction of violent incidents short of war in their area of deployment... [A] second criterion on which to judge the success of peacekeeping is its ability to facilitate the resolution of the disagreements underlying the conflict (Diehl 1993: 36-37 cited in Bures, 2007).

The UN Peace Mission Mandates (UNPMM) study looked at 113 UN peace missions, including 54 PKOs, 50 SPMs, and 9 SE/SAs, that were either founded or had their mandates changed between January 1, 1991 and December 31, 2020. It reveals that the majority of missions (more than 60%) do not have any mandate modifications that occur after the founding year (Hellmüller, Tan, & Bara 2023, 6). This is usually because amendments to mandates must be approved by the UNSC, thus UN peace operations may take a while to respond to rapidly changing ground conditions. But periodically new mandate duties are assigned to the remaining missions, which is why we present the data in a timeseries format. For example, the UN Mission in the DRC (MONUC), which had new missions assigned to it in seven out of its eleven years of operation, is an extreme example. Researchers categorized all the necessary duties over a period of three decades into three categories in the same essay: minimalist, moderate, and maximalist. When the job table from the article was examined more closely, it became clear that the majority of the missions could only deliver minimalist tasks, not maximalist ones. There were 18 maximalist[2] jobs included in the table, which was more than the other categories (ibid, p, 7).

[2] **Table 10-1: Mandate Tasks in UN Peace Missions, 1991–2020.**

Minimalist tasks	Moderate tasks	Maximalist task
Co-ordination of donors, partners, and UN agencies	Conflict assessment and early warning	Civil society capacity building
Demilitarization	Disarmament, demobilization and reintegration (DDR)	Dialogue and reconciliation (local)
Demining	Electoral assistance	Dialogue and reconciliation (national)
Elimination of chemical weapons program	Electoral security	Dialogue and reconciliation (regional)
Good offices and mediation (track 1)	Information campaigns	Economic reforms
Humanitarian assistance	Monitoring/ investigating international humanitarian law (IHL)/international human rights law (IHRL) violations	Good governance
Observing, monitoring, or reporting (OMR) military	Recovery, rehabilitation, and reconstruction	Human rights promotion
OMR police	Support to international criminal justice	Institution-building (IB) transitional state institutions

This chapter addresses the aforementioned central and subsidiary questions through a critical analysis of four long-duration UN PKOs. Some of the factors contributing to the selection of cases are: 1) the duration of mission since deployment, 2) changing geo-political contexts and mission objectives, 3) mission expenditure, 4) fatalities, and 5) an inability to address root causes of conflicts and rather exacerbating the conflict – encouraging 'Cold Peace'. This concept denotes the presence of the mission as a 'conflict transformation blocker' due to it being ever present in the region. The chapter is organized into two sections. In the first section, the historical context of the conflict based on which the missions were deployed including timelines is presented, followed by mission mandates and key actors involved in the conflict and peacekeeping efforts. The second section contains analysis.

The author does not intend to undermine the achievement of these missions so far; rather, questions are raised about the practicality of these missions based on the geo-political situation and resource constraints to argue whether these are still justified given the unachievable mandates. Such an approach enables us to consider a few issues in PKOs. Firstly, it provides a better understanding of the complex realities of implementing peacekeeping

---

Refugee/Internally displaced person (IDP) assistance	Protection of civilians (POC) children	Promotion of independent media
Small arms and light weapons (SALW)	POC conflict-related sexual violence	Rule of law (ROL) judicial reform
Secure environment for delivery of aid	POC general	ROL legal reform
		ROL penal system reform
		Sexual and gender-based violence
		Security sector reform (SSR) military
		Security sector reform (SSR) military
		SSR police
		Support to permanent state institutions
		Transitional justice
		Women's rights and participation

Source: adapted from 3 Mandate Tasks in UN Peace Missions, 1991–2020, p, 7

missions in challenging environments, which can inform future decisions on mission design and mandate formulation. Secondly, it ensures accountability and transparency in the use of resources and taxpayer money of the member states that provide troops and equipment to the UN. Furthermore, it is vital to acknowledge that the success of UN peacekeeping missions cannot be measured solely based on violent conflict or a lack thereof. Rather, success should be evaluated based on various factors such as promoting human rights, facilitating political processes, the disarmament and demobilization of armed groups, and ensuring the protection of civilians.

A Brief Introduction to the Four Protracted UN Missions

United Nations Military Observer Group in India and Pakistan - UNMOGIP (UNITED NATIONS, 2023e)

On January 24, 1949, the first batch of United Nations military observers arrived in the designated mission area with the purpose of overseeing the ceasefire between India and Pakistan in the state of Jammu and Kashmir (J&K). These observers, led by the military adviser appointed by the UN secretary-general, constituted the core of the UNMOGIP. After the resumption of hostilities in 1971, UNMOGIP has remained present in the region to monitor and report on developments related to the strict adherence to the ceasefire agreement established on December 17, 1971. Their primary responsibility is to observe the situation and provide updates to the secretary-general.

United Nations Interim Force in Lebanon (UNIFIL)(UNITED NATIONS 2023d)

Initially, the Security Council established UNIFIL in March 1978 to verify the withdrawal of Israeli forces from Lebanon, restore global peace and security, and aid the Lebanese government in regaining control in the region. The mandate of the mission had to be modified twice: in 1982 and 2000 due to changing circumstances.

Following the crisis in July/August 2006, the council bolstered the capabilities of the force and decided that in addition to its original mandate, UNIFIL would also be responsible for tasks such as monitoring the cessation of hostilities, supporting and accompanying the Lebanese armed forces during their deployment in southern Lebanon, and providing assistance to ensure humanitarian access to civilian populations and the voluntary and safe return of displaced individuals.

The United Nations Peacekeeping Force in Cyprus (UNFICYP) (UNITED NATIONS 2023c)

The Security Council established this force in 1964 with the aim of preventing further conflict between the Greek Cypriot and Turkish Cypriot communities. Following the hostilities in 1974, the council assigned additional responsibilities to the force. In the absence of a political resolution to the Cyprus problem, UNFICYP has continued its presence on the island to oversee ceasefire lines, establish a buffer zone, engage in humanitarian efforts, and assist the secretary-general's good offices mission.

The United Nations Mission for the Referendum in Western Sahara (MINURSO) (UNITED NATIONS 2023b)

This mission was established in 1991 based on settlement proposals accepted by Morocco and the Frente Popular para la Liberación de Saguia el-Hamra y de Río de Oro (Frente POLISARIO). The mission's purpose was to oversee a transitional period leading to a referendum where the people of Western Sahara would choose between independence and integration with Morocco. The special representative of the secretary-general was given the responsibility for the referendum and was supported by a group of civilian, military, and civilian police personnel known as the United Nations Mission for the Referendum in Western Sahara. In 2016, the Security Council adopted a resolution urging the parties involved in the conflict to demonstrate political will for more intensive and substantive negotiations.

In Table 10-2 below, data pertaining to these four missions are compiled:

Table 10-2: Mission Data

Name of the mission	First deployed	Geographical location	Number of personnel participated	Fatalities	Budget (One fiscal year as a unit of analysis)
United Nations Military Observer Group in India and Pakistan (UNMOGIP)[3]	January 1949 **73 years**	India - Pakistan	110 (68 civilians) and 42 experts on Mission	13	Approved budget: (01/2021–12/2021): $10,519,800
United Nations Peacekeeping Force in Cyprus (UNFICYP)[4]	March 1964 **58 years**	Cyprus	151 civilians, 68 police 52 staff officer, 740 troops	183	Approved budget: (07/2021–06/2022): $57,567,300
United Nations Interim Force in Lebanon (UNIFIL)[5]	March 1978 **44 years**	Lebanon	10,365 (809 civilians) 209 staff officer, 9347 troops)	329	Approved budget: (07/2021–06/2022): $ 510,251,500
United Nations Mission for the Referendum in Western Sahara (MINURSO)[6]	April 1991 **32 years**	Western Sahara	230 civilians, 198 mission experts, 2 police, 7 staff officer, 20 troops	20	Approved budget: (07/2021–06/2022): $60,908,900

From Table 1, we find that the UNIFIL and UNFICYP missions have suffered the most casualties since their deployment.

Also, in the table, the budget for one fiscal year is taken as a basis of calculation, which makes UNIFIL the most expensive mission among all.

In Table 10-3 below, the mandates of the missions are elaborated.

[3] https://peacekeeping.un.org/en/mission/unmogip
[4] https://peacekeeping.un.org/en/mission/unficyp
[5] https://peacekeeping.un.org/en/mission/unifil
[6] https://peacekeeping.un.org/en/mission/minurso

Table 10-3: Mission mandates at a glance[7]

Name of the mission	Mission mandates and responsibilities	Remarks
UNMOGIP[8]	• The area of responsibility of the mission is along the Line of Control (separating the Pakistan- and Indian-administered regions of Kashmir) and the Working Boundary (separating the Indian-administered Kashmir and Pakistan). • As part of the 1949 Karachi Agreement,[9] the Mission also has the mandate to investigate and report on alleged ceasefire violation complaints lodged by one of the parties. • It maintains the UN presence in the area of responsibility and monitors the situation through operational tasks, including conducting investigations of alleged ceasefire violation complaints, field trips to military units, area reconnaissance and manning of observation posts.	Through UN intervention, the rights to self-determination of Kashmiri people are forever lost. The mission cannot effectively investigate ceasefire violations because of the nation states' direct involvement in the conflict. Minimalist tasks - 2 Moderate tasks - 0 Maximalist tasks - 0
UNIFIL	According to United Nations Security Council Resolutions 425 and 426[10] in 1978, UNIFIL is tasked with the following objectives: • Confirm the withdrawal of Israeli forces from southern Lebanon • Restore international peace and security	How is UNIFIL expected to restore international peace and security amidst state failure and tensions between Hamas and other actors? Minimalist tasks - 1

[7] Data compiled in the table were compiled from the mission websites

[8] https://unmogip.unmissions.org/

[9] The Karachi Agreement of 1949 was signed by the military representatives of India and Pakistan, supervised by the United Nations Commission for India and Pakistan, establishing a cease-fire line in Kashmir following the Indo-Pakistani War of 1947. It established a cease-fire line, which has been monitored by United Nations observers from the United Nations since then. https://peacemaker.un.org/sites/peacemaker.un.org/files/IN%20PK_490729_%20Karachi%20Agreement.pdf

[10] https://unsco.unmissions.org/sites/default/files/s_res_4261978.pdf

	• Assist the Government of Lebanon in ensuring the return of its effective authority in the area. In addition, Security Council resolutions 1655, 1697, and 1701 have further reaffirmed the mission's mandate	Moderate tasks - 0 Maximalist tasks - 2
UNFICYP[11]	As given in resolution 186 (1964), UNFICYP's mandate is to use its best efforts to: • Prevent the recurrence of fighting; • Contribute to the maintenance and restoration of law and order; • Contribute to a return to normal conditions.[12]	How do we define 'normal conditions'? And how does one expect the UNFICYP to establish peaceful conditions and to what end? Minimalist tasks - 0 Moderate tasks - 1 Maximalist tasks - 1
MINURSO	MINURSO was originally mandated in accordance with the settlement plan to: • monitor the ceasefire; • verify the reduction of Moroccan troops in the territory; • monitor the confinement of Moroccan and Frente POLISARIO troops to designated locations; • take steps with the parties to ensure the release of all Western Saharan political prisoners or detainees; • oversee the exchange of prisoners of war to be implemented by the International Committee of the Red Cross; • repatriate the refugees of Western Sahara, a task to be carried out by the	The fundamental responsibility to identify eligible voters could not be accomplished so far. UN's presence encouraged 'Cold Peace' and almost extinguished Sahrawis' right to self-determination. Minimalist tasks - 5 Moderate tasks - 2 Maximalist tasks – 0

[11] https://betterworldcampaign.org/mission/cyprus-unficyp#:~:text=The%20United%20Nations%20Peacekeeping%20Force,ethnically%20divided%20island%20of%20Cyprus.

[12] https://unficyp.unmissions.org/unficyp-mandate

	United Nations High Commissioner for Refugees; • identify and register qualified voters; • organise and ensure a free and fair referendum and proclaim the results; • reduce the threat of unexploded ordnances and mines.[13]	

In the above table, using the UNPMM data set, the author shows the number of minimalist, moderate, and maximalist tasks assigned to the four cases.

III. Analysis

Challenges of achieving mission mandates of PKOs are numerous. A detail description of these is appended subsequently.

Despite the long duration of the UNMOGIP, it has not been able to fully achieve its mandates, and the conflict between India and Pakistan over Kashmir remains unresolved. Below are some of the reasons why it has struggled to achieve its mandates:

A lack of political willingness of the parties to resolve conflict matters most. The failure of the UN mission in Kashmir is largely due to the involved parties' inadequate willingness to implement a negotiated settlement for peace. A proxy war over Kashmir has endured since neither India nor Pakistan have demonstrated enough desire in ending the conflict. The conflict is still unsolved, as evidenced by the subsequent proliferation of terrorism emerging from the region. India's control over J&K is now total as it brought constitutional change regarding its "special status"; Article 370 of the Indian Constitution that grants the state of J&K a special status. When the British colonial government in India ended in 1947, Maharaja Hari Singh of the then-colonized state of J&K signed a Treaty of Accession to bring the state of J&K over to the Indian side. Additionally, having a special status limits the parliament of India's ability to pass laws that benefit the whole country. With the exception of choices relating to defense and international relations, the state of J&K is effectively granted a "temporary" special status that allows it to make those decisions and have its own flag and constitution. On August 9, 2019, Indian Interior Minister Amit Shah notified India's parliament that Article 370, a constitutional clause that

[13] https://minurso.unmissions.org/mandate

accords J&K special status, needed to be repealed. The central government subsequently approved and repealed it.

A non-binding mandate makes it difficult to enforce recommendations. It has a non-binding mandate, which limits its ability to take decisive actions. The mission can only observe and report violations of the ceasefire agreement, but it does not have the authority to enforce compliance or make any binding recommendations. Take, for example, the issue pertaining to access. The mission has faced significant challenges in accessing the disputed region due to restrictions imposed by both India and Pakistan. This has limited the mission's ability to monitor and investigate violations, and it has not been able to fully fulfill its mandate.

The mission operates in a prevailing hostile environment since India and Pakistan have become archrivals that have both exercised their rights to control respective areas of J&K. The mission has faced several security incidents, and militants have targeted its personnel, which has further limited its ability to carry out its work effectively. For example, violence in the mission area is widespread. According to the Indian Daily newspaper, 41,000 people have died in just 27 years; there have been 69,820 militancy-related occurrences over the specified period, which equates to the state experiencing 2586 militancy incidents a year, which India attributes to cross-border terrorism from Pakistan (Jacob and Naqshbandi 2017). In addition, the mission has struggled to address the underlying issues and root causes of the conflict, which has contributed to its inability to achieve its mandates.

In addition, this historic dispute over Kashmir continues to pose a significant threat to regional stability as both nuclear-armed countries lay claim to this region. In addition, there have been several wars fought between these two nations such as the Indo-Pakistani War of 1947, the 1965 war, and the Kargil conflict of 1999, which escalated tensions even further.

In sum, the UN mission in Kashmir has struggled to achieve its mandates due to a combination of factors, including a lack of political will, a non-binding mandate, a lack of access, a hostile environment, and the complex nature of the conflict. While the UN mission in Kashmir has played an important role in monitoring the ceasefire and investigating violations, its limited mandate and the lack of cooperation from the parties involved have prevented it from making any significant progress towards resolving the conflict. Research shows that such challenges will be difficult to surmount in the near future.

UNIFIL was deployed to help restore peace and security in southern Lebanon and to monitor the border between Israel and Lebanon. UNIFIL's mission is to monitor the cessation of hostilities, accompany and support the Lebanese Armed Forces as they deploy throughout southern Lebanon, and ensure that there are no unauthorized weapons or armed personnel in the area between south of the Litani River and the Blue Line in southern Lebanon. Despite being one of the longest-running UN peacekeeping missions, UNIFIL has struggled to achieve its mandates as well. Below are some of the reasons why the UN mission in Lebanon has faced challenges in achieving its mandates.

The multi-ethnic, multi-party political and governance structure in Lebanon exacerbates conflict. In the words of some scholars, "Lebanon's politics is characterized by a corrupt, sectarian structure, constant interference from neighbouring states like Israel, Syria and Iran, and many years of war" (Khatib and Wallace 2021). One of the primary reasons for the failure of UNIFIL to achieve its mandates is the lack of political will from the parties involved to implement its recommendations. The conflict between Israel and Lebanon over border disputes and the presence of Hezbollah have persisted for decades, and both parties have not shown enough interest in resolving the conflict. Moreover, Lebanon's inactive government and its political fragmentation have added to the challenges facing UNIFIL. The country has a history of internal conflicts and sectarian divisions that further impede progress towards peace and stability.

A limited mandate and resources impact a mission's performance. UNIFIL has a limited mandate that does not give it the authority to enforce compliance or take decisive action. Additionally, the mission has faced resource constraints, including limited personnel and equipment, which has limited its ability to carry out its work effectively. When pursuing terrorists or taking other actions in the mission's area of responsibility that result in extrajudicial killings and human rights breaches, UNIFIL also encounters legal obstacles. Hezbollah, which is presently operating in Lebanon with full force, is not included in UN Resolution 2373 so as to maintain the UN's neutrality. The resolution condemns intimidation used to prevent UNIFIL from doing its work, but it makes no mention of those responsible for such affairs. This ensures that the Lebanese Army enters south Lebanon and is the only armed force there and more generally that the government of Lebanon and not Hezbollah exercises sovereignty and control over all of Lebanese territory, and Resolution 1701 outlines the specific actions UNIFIL should take (Abrams **2018b)**. UN Resolution 2373, adopted in August 2017, extended the mandate. It also added that the Security Council "recalls its authorization to UNIFIL to take all necessary action in areas of deployment of its forces and as it deems within its capabilities, to ensure that its area of operations is not utilized for hostile activities of any kind…." (Abrams, 2018a). However, this begs the question of whether UNIFIL has the capacity to implement the stated tasks assigned by the UN HQ.

Inadequate support from the international community and complex geo-political conditions impedes mission objectives. UNIFIL has not received adequate support from the international community, including financial resources and political backing. This has limited the mission's ability to carry out its mandate effectively and to address the underlying issues that contribute to the conflict. The conflict in Lebanon is undoubtedly multifaceted and complex with various actors involved. UNIFIL's mandates ensure stability and security in southern Lebanon, which is constantly undermined by incidents such as border violations, violence, and ongoing tensions between Hezbollah's forces and Israel (Arab 2012). Despite over a decade of presence in Lebanon, UNIFIL has struggled to address the underlying issues contributing to the conflict due to its multifaceted nature. Hezbollah's involvement in regional conflicts has an impact on Lebanon's politics and socioeconomic situation, making it difficult for Lebanese authorities to maintain peace effectively.

Operating in a hostile environment compounds mission risk. The mission operates in a hostile environment where violence and tension are high, and militants have targeted its personnel. This has limited the mission's ability to carry out its work effectively and to maintain the safety and security of its personnel. Israel's involvement in Lebanon has included Operation Litani in 1978 and its subsequent invasion and occupation of parts of southern Lebanon. Its occupation of southern Lebanon lasted for nearly 18 years, during which it faced constant attacks and guerrilla warfare from Hezbollah. International pressure and domestic opposition eventually led to Israel's withdrawal from Lebanon in May 2000, fulfilling UN SC Resolution 425.[14] However, Israel's post-1978 invasion resulted in significant human suffering, displacement, and political tensions. The conflict had long-lasting implications for the region, contributing to the rise of Hezbollah as a major player in Lebanese politics and armed resistance against Israel.

Lebanon's state fragility affects the mission's effectiveness. The UN`s office of human rights warned in 2022 that Lebanon is in danger of becoming a failed state amidst widespread poverty (UNITED NATIONS

[14] https://unsco.unmissions.org/sites/default/files/s_res_4261978.pdf

2022). Its economy is in ruins and its political parties are vying for power. The situation in Lebanon has become increasingly complicated due to other factors such as waves of refugees, political instability and economic collapse. The war in Iraq and civil war in Syria have led to a large number of refugees seeking shelter in Lebanon. A study funded by the United Nations High Commissioner for Refugees showed that only 20% of Iraqi refugees were food secure, while more than one million Syrians were displaced to Lebanon after their own civil war broke out. Lebanon's fragile health system is also struggling under the weight of hosting the world's highest number of refugees per capita following the Syrian crisis. It has been impacted by regional conflicts.

The Armed Conflict Location & Event Data Project (ACLED) in its mid-year 2022 update listed Lebanon as one of the top 10 conflicts to worry about; it observed that, despite a decline in demonstrations, political violence intensified (ACLED 2022).

In sum, the UN mission in Lebanon has faced significant challenges in achieving its mandates due to a combination of factors, including a lack of political will, a limited mandate and resources, inadequate support from the international community, the complex nature of the conflict, and the hostile environment. While UNIFIL has played an important role in maintaining stability and monitoring the border between Israel and Lebanon, its limited mandate and the lack of cooperation from the parties involved have prevented it from making any significant progress towards resolving the conflict.

UNFICYP was established in 1964, and despite being one of the longest-running UN peacekeeping missions it has faced several challenges in achieving its mandates. Below are some of the reasons:

A simmering conflict situation is preventing the mission from achieving its objectives. UNFICYP's efforts to fulfill its duties have been significantly hampered by the inability to resolve the conflict between the Greek and Turkish Cypriots. Both sides have accused the UN of being biased towards one community, which has made it difficult for the mission to maintain a balance between the two (Zahedi 2001). Officials have warned Al Jazeera that Turkish-Cypriot authorities might be getting ready to remove United Nations peacekeepers from their posts in Northern Cyprus, sparking a fresh political and security crisis on the split island. Tahsin Ertugruloglu, who oversees Northern Cyprus' foreign affairs, told Al Jazeera that "[the United Nations Peacekeeping Force in Cyprus (UNFICYP)] needs to enter into a formal agreement with the Turkish Republic of Northern Cyprus (TRNC) in order to continue its presence and operations there" (Psaropoulos 2022).

A lack of political will of the conflict parties has hindered the mission's effectiveness. The conflicting parties have not demonstrated adequate political will to end the conflict, and some have even actively opposed UNFICYP's attempts to uphold peace and stability. The UN mission's recommendations were not followed, which has made the situation on the island much more difficuLieutenant Since the collapse of the most recent peace agreement in 2017, efforts to resume finding a peaceful resolution to the issue between Greek and Turkish Cypriots have failed. The UN expressed its disappointment over the inability of its most recent conference in Geneva to mend the divide between the two factions. According to Stephane Dujarric, a spokesman for UN Secretary-General Antonio Guterres, the meeting's goal was to "determine whether common grounds exist for the parties to negotiate a last solution to the problem in Cyprus within a foreseeable horizon" (Smith 2021).

Historic baggage and EU involvement have compounded the situation further in Cyprus. The conflict began in 1974 when Turkey launched a military intervention in response to a coup in Cyprus, which aimed to unite the island with Greece. Turkish forces invaded Cyprus, resulting in the occupation of the northern part of the island. This led to a displacement of populations, with Greek Cypriots being displaced from the north and Turkish Cypriots from the south. The war in Cyprus is also linked to previous hostilities between Greece and Turkey, notably the Sèvres Treaty's[15] legacy.

Greece, Turkey, and the European Union are just a few of the numerous parties involved in the complicated dispute in Cyprus. The mission has struggled to address the core issues and the causes of the conflict, which has made it more difficult for it to effectively perform its tasks. Examining some of the underlying factors that have stymied efforts to resolve the dispute is

---

[15] Treaty of Sèvres, (August 10, 1920), post-World War I pact between the victorious Allied powers and representatives of the government of Ottoman Turkey. The treaty abolished the Ottoman Empire and obliged Turkey to renounce all rights over Arab Asia and North Africa. The pact also provided for an independent Armenia, for an autonomous Kurdistan, and for a Greek presence in eastern Thrace and on the Anatolian west coast, as well as Greek control over the Aegean islands commanding the Dardanelles. Rejected by the new Turkish nationalist regime, the Treaty of Sèvres was replaced by the Treaty of Lausanne in 1923.
https://www.britannica.com/event/Treaty-of-Sevres

crucial to better explore the complexity of this problem. One crucial issue that has had an indirect impact on how the Cyprus crisis has been handled is the EU's conditional policy approach toward Turkey (Baştürk 2013). Scholarly research suggests that Tayyip Erdogan received a number of benefits from European politicians in exchange for his backing of The Annan Plan, a proposed settlement strategy meant to end Cyprus' protracted territorial conflict (Michael 2007). Turkey and Greece come under significant strain when foreign parties impose restrictions like these, which makes it difficult for them to engage in productive negotiations. The Commission also noted democratic shortcomings in Turkey with regard to minorities' rights and freedom of speech.

Due to a number of issues, including continuous war, a lack of political will, resource limitations, the conflict's complexity, and its shifting character, the UN mission in Cyprus has had a difficult time fulfilling its duties. While UNFICYP has been crucial in preserving peace and security on the island, its capacity to properly carry out its responsibilities has been diminished by the lack of movement toward a peace agreement.

MINURSO was established in 1991 to help resolve the conflict between Morocco and the Polisario Front over the sovereignty of the region. It has experienced a number of difficulties in carrying out its objectives while being one of the longest-running UN peacekeeping missions. Here are some of the factors that have made it difficult for the UN mission in Western Sahara to fulfill its duties:

The mission was established with the goal of holding a vote to determine the region's destiny, but this has not happened yet, so the dispute has not advanced toward a settlement. Since the Polisario Front favors independence and Morocco insists on its proposal for autonomy, the disputing parties have been unable to come to an agreement on the referendum's rules. The parties' disagreement over the qualifications for referendum voters' eligibility, especially between Morocco and the Frente POLISARIO, has been a serious barrier (Ahmed 2017). Specifically, there has been a debate on who qualifies as a valid voter. The referendum voter identification and registration procedure has proven to be a difficult and controversial process.

Persistence of a lack of inter-group collaboration since the inception of the mission has crippled the mandate achievement capacity. MINURSO has had trouble completing its mission because of the parties' lack of cooperation. Both the Polisario Front and Morocco are accused of impeding the mission's operations by limiting its ability to move freely and access specific areas. During the referendum, the Sahrawis had the option of selecting between independence or integration with Morocco. Differences over how these options should be interpreted and implemented, particularly with regard to Sahrawis' rights to self-determination, continue to exist, further complicating the referendum process. Despite efforts by the United Nations and a number of mediators, the parties have been unable to reach a mutually accepted settlement. The peace process has come to a halt due to a lack of political will and commitment to reaching a settlement. Security issues have consistently hampered holding the referendum. It has been challenging to establish measures that will foster confidence and trust between the parties, including assuring participant safety and addressing worries about potential results.

The mission's success is hampered by the shifting political landscape and human rights violations. Since MINURSO was established in 1991, the local political landscape has seen a significant shift. It has become increasingly challenging for the mission to accomplish its goals as a result of several countries, including the United States, recognizing Morocco's sovereignty over the Western Sahara. Furthermore, the Frente POLISARIO works in regions where Morocco currently exercises de facto control over most of the nation. This division makes it challenging to logistically organize for and protect the referendum process. Human rights violations, including the detention of political prisoners and the dispersal of peaceful protests, continue to occur in the Western Sahara (Amnesty International 2022). The lack of action by MINURSO to address these violations has come under fire, deepening the distrust between the warring parties.

The UN mission in Western Sahara has faced significant challenges in achieving its mandates due to a combination of factors, including the lack of progress towards a resolution, the lack of cooperation from the parties, ongoing human rights violations, the changing political landscape, and the lack of international support. While MINURSO has played an essential role in maintaining peace and stability in the region, the lack of progress towards a resolution of the conflict has limited its ability to fully achieve its mandates.

Lessons Learned

The case studies of four protracted UN missions have revealed valuable lessons that can inform future peacekeeping efforts. One important lesson learned from the four protracted UN missions is that political will plays a fundamental role in achieving sustainable peace. The involvement and cooperation of all parties involved in the conflict, as well as external stakeholders such as international organizations and neighboring countries, greatly impact the outcome of any PKO. Additionally, it has been found that adopting a comprehensive approach that complements military provisions with civil elements is necessary for long-term sustainability. While the implementation of military provisions may proceed smoothly, progress on civil elements can often lag behind. Therefore, there needs to be a concerted effort towards creating a safe and secure environment that enables the implementation of other aspects of the peace process.

The efficiency of United Nations operations in numerous crisis zones across the world is hampered by a lack of resources. Four lengthy UN operations have provided us with important lessons about the difficulties in preserving peace and security. Global interest in UN-led operations declined during the 1990s, especially following the end of the Cold War. This resulted in a decrease in the amount of money and military assistance given to the UN. Unsustainably low financing continues to be a major problem for UN-led operations everywhere. Additionally, because of countries' post-Cold War downsizing, there are now more people with military training available. This has made it challenging for the UN to find and keep skilled workers, which has had a major influence on mission success. For instance, UNTB10 identified financial constraints as its primary challenge while creating its operational plan. States that provide resources for UN-led missions may support the deployment of such resources for some operations but not for others owing to political concerns or the availability of finances at any particular moment. This results in excessively large and sometimes unsupportable mission expenditures, which ultimately reduce mission effectiveness.

One might wonder why the UN does not wind up the above-mentioned protracted missions because they could not achieve their mandates. This is not the only factor considered when deciding whether to terminate a UN PKO. The level of progress made toward achieving the mission's goals, the security situation, the political climate on the ground, and the willingness of the parties involved to work with the UN are all taken into consideration when determining whether to end a mission.

The UNSC typically decides what the mandates of peacekeeping missions should be, and it can change or extend those mandates through a resolution. A peacekeeping mission's mandate may be changed in reaction to new events that may call for a different strategy or in response to shifts in the circumstances on the ground. The mission's goals, the mission's size and makeup, or the rules of engagement may all be revised as part of the mandate. The precise number of times their mandates have been altered vary depending on the mission and the circumstances in the area.

The choice to carry out the four lengthy missions previously stated, notwithstanding their failure to fulfill their objectives, was probably motivated by a number of different considerations. These may consist of:

1. The need to preserve stability: Often, even when a UN PKO has not completed its goal, its presence serves to preserve stability in a region. A recurrence of war and instability brought on by the mission's cancellation might have serious repercussions for the local populace as well as the larger international community.
2. The significance of the conflict: For the international community, the conflicts in the areas where the extended missions are taking place are frequently of crucial importance. It is pivotal for the UN to keep a presence in the region to monitor the situation and stop the conflict from worsening since these conflicts may have larger ramifications for regional and global stability.
3. The parties' desire to continue the mission: The UN cannot carry out a peacekeeping deployment without the parties' agreement. Occasionally, the parties may agree to extend the mission even though it has not fulfilled its objectives because they value the UN's continued presence in the area.

Deciding whether to continue a UN PKO even after it has failed to accomplish its mission is a difficult decision that takes into account a number of factors. While the UN strives to carry out its mandates and objectives, there are times when the mission's continued presence is deemed crucial to uphold peace, monitor developments, and prevent a conflict from getting worse. It is difficult to state unequivocally whether it is wise to enable these missions to continue or to transfer resources elsewhere from a cost-benefit viewpoint because it depends on a number of factors.

1. On the one hand, maintaining these missions may be costly given that PKOs cost billions of dollars every year. The continuation of military operations without achieving their objectives can be seen as a waste of resources, and it may be preferable to use that money elsewhere to settle other international conflicts. On the other hand, if these operations are stopped without the underlying problems being resolved, costs can be incurred.
2. In addition, when deciding whether to continue with these missions or to discontinue them, political and diplomatic considerations

should be taken into account. The UN and member states may feel morally obligated to continue PKOs in some areas, even if progress has been slow or difficult to achieve.

3. Cost-benefit analysis is an essential tool for decision-making, but in complex geopolitical situations like these UN missions, it may not be sufficient. Additional factors, such as potential withdrawal costs, political and diplomatic considerations, and moral obligations may need to be taken into account when deciding whether to continue or divert resources from these operations.

In light of the UN peacekeeping system reform, upcoming peacekeeping deployments might be re-evaluated. The UN has admitted that a variety of challenges confront PKOs, including a lack of finance, poor coordination, and problems in reacting to fresh and complex combat scenarios. To address these problems, the UN has been attempting to enhance the peacekeeping system. By improving peacekeepers' ability to protect people, interact with local populations, and manage difficult conflict scenarios, for instance, the UN has been attempting to improve the effectiveness of peacekeeping missions. Along with other UN agencies, regional organizations, and non-governmental organizations, the UN has been attempting to improve the coordination of PKOs. This is part of an attempt to more successfully integrate political, humanitarian, and development agendas into PKOs.

In conclusion, the above-mentioned UN PKOs have significantly contributed to international peace and security, though they have encountered various challenges along the way.

The decision to continue or cease these operations is a complex one, as it involves various factors such as financial cost, political and diplomatic considerations, and moral obligations. However, the main argument made in this chapter is that effective UN PKOs require constant evaluation and improvement as well as close coordination with various organizations. The other factor discussed at length in this chapter is about the geo-politics within which these missions are carried out, which are beyond the UN's control to influence. Further, the long-term presence of a mission also encourages conflict parties to rely on the presence of UN forces rather than taking ownership of their own peace process.

At this juncture, when the UN's expenses are under scrutiny from member states, it is important to assess each situation on a case-by-case basis and determine if the benefits of continued peacekeeping outweigh the costs. Furthermore, the UN needs to continue exploring innovative approaches, such as engaging regional organizations or transitioning from peacekeeping to peacebuilding. This chapter suggests that the UN should conduct evaluation to consider these long-standing missions' mandates once again and take practical steps to divert more resources to peacebuilding and the prevention of conflicts instead of just managing the status quo.

References

Abrams, Elliott. 2018. "UNIFIL and the Hezbollah Tunnels." Accessed December 7, 2018. https://www.cfr.org/blog/unifil-and-hezbollah-tunnels.

Abrams, Elliott. 2018b. "What is the Role of UNIFIL?". Accessed August 30,2018. https://www.cfr.org/blog/what-role-unifil

ACLED. (2022). Political Violence Intensifies Despite Decline in Demonstrations. Retrieved from https://acleddata.com/10-conflicts-to-worry-about-in-2022/lebanon/mid-year-update/

Ahmed, Kawser. 2017. "Social conflict and peace-building: the perceptions, experiences, and contributions of leaders of selected community-based organizations in Winnipeg, Manitoba." PHD diss., University of Manitoba

Ahmed, Kawser. 2017. "THE WESTERN SAHARA CONFLICT: A PERSPECTIVE". In *Peace Issues in the 21St Century*, edited by Shreesh Juyal and John Duncan, 203-222. (Newcastle upon Tyne: Cambridge Scholars Publishing,2017).

Amnesty International.2022. "Morocco and Western Sahaara. accessed n.d. https://www.amnesty.org/en/location/middle-east-and-north-africa/Morocco-and-western-sahara/report-Morocco-and-western-sahara/

Annan, Kofi. 2016. "Learning the Lessons of Peacebuilding." In *Manchester University Press EBooks*. https://doi.org/10.7765/9781526111692.00013.

Arab, Fawaz, and ARMY WAR COLL CARLISLE BARRACKS PA CENTER FOR STRATEGIC LEADERSHIP. 2012. *UNIFIL peacekeeping and Lebanese National Security*. US Army War College.

Baştürk, Melike. 2011. "The Issue of Cyprus in the EU Accession of Turkey." *Claremont-UC Undergraduate Research Conference on the European Union* 2011 (1): 15–22. https://doi.org/10.5642/urceu.201101.04.

Bratt, Duane. 1996. "Assessing the Success of UN Peacekeeping Operations." *International Peacekeeping* 3 (4): 64–81. https://doi.org/10.1080/13533319608413640.

Bures, Oldrich. 2007. "Wanted: A Mid-Range Theory of International Peacekeeping." *International Studies Review*, September. https://doi.org/10.1111/j.1468-2486.2007.00697.x.

"Canada and Peacekeeping Operations." Canadian War Museum. Accessed 2022. https://www.warmuseum.ca/learn/Canada-and-peacekeeping-operations/

Hellmüller, Sara, Xiang-Yun Rosalind Tan, and Corinne Bara. 2023. "What Is in a Mandate? Introducing the UN Peace Mission Mandates Dataset." *Journal of Conflict Resolution*, March, 002200272311598. https://doi.org/10.1177/00220027231159830.

Jacob, Jayanth and Naqshbandi,Aurangzeb. 2017. "41,000 Deaths in 27 Years: The Anatomy of Kashmir Militancy in Numbers." Accessed September 25, 2017. https://www.hindustantimes.com/india-news/the-anatomy-of-kashmir-militancy-in-numbers/story-UncrzPTGhN22Uf1HHe64JJ.html.

Khatib,Lina and Wallace, Jon. 2021. "Lebanon's politics and politicians". Accessed August 11, 2021. https://www.chathamhouse.org/2021/08/lebanons-politics

Mackinlay, John Chopra, Jarat. 1992. "Second generation multinational operations". *Washington Quarterly, 15*(3), 113-131.

Michael, Michális S. 2007. "The Cyprus Peace Talks: A Critical Appraisal." *Journal of Peace Research* 44 (5): 587–604. https://doi.org/10.1177/0022343307080857.

Psaropoulos, John. 2022. "New crisis brewing on Cyprus after US lifts arms embargo". Accessed October 27,2022. https://www.aljazeera.com/news/2022/10/27/new-crisis-brewing-on-cyprus-after-us-lifts-arms-embargo

Pushkina, Darya. 2006. "A Recipe for Success? Ingredients of a Successful Peacekeeping Mission." *International Peacekeeping* 13 (2): 133–49. https://doi.org/10.1080/13533310500436508.

Smith, Madison. 2021. "Recent Peace Talks Fail To Resolve Ongoing Cyprus Conflict". Accesed May 10, 2021.https://theowp.org/recent-peace-talks-fail-to-resolve-ongoing-cyprus-conflict/

The Canadian Encyclopedia. 2022. "Suez Crisis". Accessed n.d. https://www.theCanadianencyclopedia.ca/en/article/suez-crisis

United Nations. 2022. "Lebanon: UN expert warns of 'failing State' amid widespread poverty". Accessed May 11,2022. https://www.ohchr.org/en/press-releases/2022/05/lebanon-un-expert-warns-failing-state-amid-widespread-poverty

United Nations. 2023a. "Forming a New Operation". Accessed n.d. https://peacekeeping.un.org/en/forming-new-operation

United Nations. 2023b. "MINURSO". Accessed n.d. https://minurso.unmissions.org/

United Nations. 2023c. "UNFICYP Fact Sheet". Accessed n.d. https://peacekeeping.un.org/en/mission/unficyp

United Nations. 2023d. "UNIFIL Fact Sheet". Accessed n.d. https://peacekeeping.un.org/en/mission/unifil

United Nations. 2023e. "UNMOGIP". Accessed n.d. https://unmogip.unmissions.org/

Zahedi, Leila Berge.2001. "Mission impossible?: peacekeeping and peacemaking in Cyprus". Master's thesis, University of Oslo

CHAPTER XI

UNITED NATIONS PEACEKEEPING AND SOFT POWER TOOLS: THE CASE OF BANGLADESH

RAIAN HOSSAIN AND NAHIAN SALSABEEL

Bangladesh has become one of the largest troop- and police-contributing countries to United Nations peacekeeping operations (UNPKOs) since 1988. It has earned recognition in the international arena for its contributions to global peace. Since then, Bangladesh's personnel in UNPKOs have showcased the small South Asian state's commitment to maintaining international peace and security while upholding international rules, norms, and principles. Bangladesh's determination and active participation in UNPKOs are driven by its constitutional commitment, foreign policy objectives and historical experiences from the independence war of 1971. Bangladeshi blue helmets have adapted to the new demands of multi-dimensional UNPKO missions with their professional skills, hard work and dedication.

Much of the academic literature that discusses Bangladesh and UNPKOs has examined historical developments (Zaman and Biswas 2017), reasons for participation (Zaman 2016), challenges (Biswas 2022), policy discussions (Zaman and Biswas 2017) gender dynamics (Kumar 2022), four generations of peacekeeping and Bangladesh (Hossain 2018), and UN peacekeeping and small states, with the case example of Bangladesh (Krishnasamy 2003). Much of the scholarly literature has also shed light on how contributions to UNPKOs can act as a soft power mechanism for different states, including powerful and small states. Soft power and Bangladesh's participation in UN peacekeeping has been a prominent topic of debate among scholars (Mostaque and Sultana, 2017; Pulami, 2022). However, this study goes beyond this topic by unpacking the soft power tools used by Bangladesh through its contributions to UNPKOs. In filling such a gap, this study, firstly, examined the conceptual base of soft power and UNPKOs; secondly, it analyses Bangladesh's participation in UNPKOs and foreign policy; thirdly, it discusses the soft power tool of Bangladeshi peacekeepers in UNPKO missions rather than a mainstream narration of how Bangladesh's contribution to UN peacekeeping is a soft power mechanism; and, fourthly, the study provides policy recommendations about areas in which Bangladesh can explore in terms in soft power and UNPKOs. This study is based on seven semi-structured elite interviews by the two authors (virtually) and is triangulated with secondary academic (journal, book chapters) and non-academic sources (websites, official reports, newspaper articles).

Methodology

This study is a qualitative study. It incorporates information from semi-structured interviews and data available in the public domain to analyse and discuss the various soft power tools used by Bangladesh in regard to UNPKOs. The authors conducted seven semi-structured key informant interviews collected via the snowball sampling method. The aim of the interviews was to gain an in-depth understanding of real-life developments in Bangladesh's role in UNPKOs and the ways in which Bangladeshi peacekeepers strive to stand out. The authors maintained a balance between the interviewees' backgrounds and experiences, with a specific focus on their roles in UNPKO missions and their leadership ranks[1]. Specifically, the study has included narratives of former UNPKO force commanders, medical staff, administrative staff, and instructors in Bangladesh. Lastly, secondary sources were used to triangulate the narratives the authors located from the interviewees while ensuring a balance between national and international sources and perspectives.

Soft Power & United Nations Peacekeeping Operations: The Conceptual Base

Power is a fundamental concept in the realm of the global political world. Until the end of the Cold War, the dominant power narrative concerned material components and resources like military, economic and natural resources. These parameters have traditionally been the barometers for the use of hard power in international politics (Mukherji 2015). New developments in globalisation, economic interdependence, the rise of non-state actors, the spread of technology, non-traditional security threats, and changing political issues paved the way for the concept of soft power to

---

[1] See Appendix 1

emerge in international relations (Mukherji 2015). Although Joseph Nye Jr. coined the term "soft power," the concept of soft power is not new, as it can be found in the writings of both Western and non-Western scholars (Mustaque and Sultana 2017).

With the change in the global political world of reducing the usage of traditional hard power instruments, Nye (1990b) noted the rise of the power of private actors and small states and characterised this change as having five core trends: economic interdependence, transnational actors, nationalism in weak states, the spread of technology and changing political issues. He has defined soft power in a variety of ways over the years. In his 1990 book 'Bound to Lead', he introduced the concept of the ability of a state to influence another through the utilisation of its cultural allure, shared values, societal structure, political system and economic prowess (Nye 1990a). More than a decade later, Nye illustrated soft power as "the ability to shape the preferences of others" or "the ability to get others to want the outcome you want because of your cultural or ideological appeal" (Nye 2004). In 2011, he again defined the concept as the ability of a state to use co-option means of agenda-setting, persuasion and attraction to achieve preferred outcomes (Nye 2011). The common theme of all these definitions is the use of persuasion and attraction, rather than just military means, to shape the policies of other states.

Countries that are identified as smaller states, including Bangladesh, are perceived to lack resources and the ability to influence international politics. Contrarily, scholars have also contended that small states may possess intellectual and propagandistic capacities far larger than their size, which in turn allows them to have a far larger influence over and standing in international relations than expected (Fox 1959; Vukadinovic 1971; Commonwealth Secretariat 1997; Chong 2010). While Long (2017) indicates that small states face power asymmetries in international politics, he dedicates much of his work to illustrating the realities of power, and that despite being considered small in material dimension, smaller states can use soft power to paint a positive global image while achieving their national interests and objectives.

One of the newer discussions around soft power has revolved around the possibility of the military being used as a soft power instrument. Even though the military is often associated with hard power mechanisms, it is suggested that the military also has softer sides (Pulami 2014). Military diplomacy has become one of the central ways countries can increase their soft power capacities (Atkinson 2014). Similarly, contributions to UNPKOs and missions have become a new way for states large and small to exercise soft power. For example, despite opposition to UNPKOs in the early 1990s, China emerged as one of the key personnel contributors in the 2000s and 2010s (Mohan and Gippner 2013). Although the West has frowned on its investment initiatives in Africa and Asia, this is one of China's better-known investments in the multilateral system (Gowan 2020). The Chinese contribution to UNPKOs has helped the emerging global power to build an international image of an emerging responsible power (Mohan and Gippner 2013). In Adebajo's (2018) assessment of Nigeria's participation in UNPK missions, she discusses how the large-scale contribution of troops and police to many different UNPK missions within Africa and beyond has allowed the country to be viewed as a major regional and, to a lesser extent, global peacemaker. Likewise, Pakistan has used UN peacekeeping as a tool in its foreign policy to demonstrate its commitment to international norms of peace, security, and stability. It places a high value on peacekeeping operations, both for its contribution to maintaining global security and peace and for its role in carrying out this duty, and it has established itself as one of the top-ranking international peacekeepers and peace builders, earning admiration from the international community, projecting an attractive face of the country, and broadening the base of cooperation with the international community (Farwa and Garewal 2018). This study unpacks Bangladesh's version of the story and illustrates Bangladesh's key soft power tools in UNPK missions and their utility. Given its commitment to international peace, this research narrates the major soft power tools used by Bangladesh in its participation in UNPKO missions. The next segment provides an overview of Bangladesh's participation in UNPKOs and its relations to Bangladesh's foreign policy objectives.

UN Peacekeeping & Bangladesh's Foreign Policy: An Overview

Bangladesh has already made a considerable footprint in UNPKOs. Dhaka's first deployment with 15 military observers to the UN Iran–Iraq Military Observation Group mission was in 1988, just 17 years following the South Asian state's bloody war of independence. This is a considerable number given that the state's military capacity was just starting to grow (Heitzman and Worden 1988) and the new state's annual experiences with natural disasters (Rahman 1989). Since then, Bangladesh has become one of the top contributors of troops and police in UNPKOs and currently stands as the largest contributor of military and police personnel. Since 1988, Bangladesh has contributed more than 175,000 troops, including more than 1,800 female peacekeepers in 54 peacekeeping missions across five continents (Chowdhury 2022), including the Middle East, Africa, Europe, and Asia (Zaman and Biswas 2017). Bangladesh is also one of the highest contributors of female troops and police in UNPKO missions (United Nations Peacekeeping 2022a). Additionally, the small South Asian state has sacrificed more than 165 lives in fulfilling its commitments to UNPKOs (Daily Star 2022). There has been a massive increase in Bangladesh's personnel contribution to UNPKOs from 1988 to 2022.

Bangladesh goes above and beyond in terms of its contribution of troops and police to UNPKOs. Additionally, UNPKO personnel from Bangladesh have captured the hearts of the locals in many states through their care and empathy (Akhter 2022). While there has been much recognition in terms of formal commendations of Bangladeshi peacekeepers by past UNPKO leaders, including former secretary-generals Ban Ki Moon and Kofi Annan as well as the current UN Secretary-General Antonio Guterres, the expressions of gratitude from local populations in conflict-affected areas indicate how Bangladeshi military and police personnel effectively impact peace within these local communities. Together with its military successes, Bangladesh's proactive and constructive stance has helped the country gain political and diplomatic clout internationally and has made it credible to the UN and other international actors (Zaman and Biswas 2014). Experts and academics have debated the implications of and the justifications for Bangladesh's role as one of the top troop-contributing nations to UN peacekeeping. Banerjee (2008), for instance, cites several reasons why Bangladesh supports UN peacekeeping, including the need to fulfil international obligations, the desire to present a positive image of the nation, the need to divert the armed forces' attention away from any praetorian desires, the subsidising of the army with the reimbursements obtained from peacekeeping missions, and the financial benefits derived by specific members of the armed forces. According to Zaman and Biswas (2014), Bangladesh's participation in UNPKOs has been justified by three core reasons:

- First, participation in UNPKOs satisfies the country's constitutional and international commitments.
- Second, participation in such operations enables military personnel to meet with members of other armed forces and hone their professional abilities. Such global exposure aids in their operational experience and first-hand knowledge of the newest military techniques and tools.

- Third, receiving cash compensation is a significant inducement for Bangladesh to participate in UN operations. UN payments allow the armed services to boost their soldiers' pay and maintain equipment. In other words, peacekeeping missions support Bangladesh's military forces.

Bangladesh's increase in peacekeepers and firm contribution to upholding international peace and security is driven by two primary foreign policy objectives and motives. First, it is guided by the famous quote by Bangabandhu Sheikh Mujibur Rahman, "Friendship to all; malice to none". Secondly, it is driven by the concept of promoting international peace, security and solidarity based upon Article 25 of the constitution of the People's Republic of Bangladesh, which states:

"The State shall base its international relations on the principles of respect for national sovereignty and equality, non-interference in the internal affairs of other countries, peaceful settlement of international disputes, and respect for international law and the principles enunciated in the United Nations Charter, and on the basis of those principles shall –

1. strive for the renunciation of the use of force in international relations and for general and complete disarmament.
2. uphold the right of every people freely to determine and build up its own social, economic and political system by ways and means of its own free choice; and
3. support oppressed peoples throughout the world waging a just struggle against imperialism, colonialism or racialism."

Bangladesh has long been guided by the principle of 'Friendship to all; malice to none'. This has been the fundamental norm that Bangladesh has continually sought to uphold with its foreign policy since its independence in 1971. It also reveres and follows the essence of the philosophy through the promotion of national equality and sovereignty, refraining from meddling in the internal affairs of other countries, peacefully mediating international conflicts, and upholding international law and principles (Karim and Uddin 2016). When deconstructed, Bangladesh's foreign policy renounces the use of force in international relations and for general and complete disarmament, supports the cause of world peace, opposes all forms of colonial oppression and discrimination, and supports people fighting for liberation and human rights (Afroze 2005). These important principles reflect the connection of Bangladesh's foreign policy with its motivation for active involvement in UNPKOs. As such, contributing to UNPKOs allows

Bangladesh to achieve its normative foreign policy agendas while maintaining peace as a core driving feature. Bangladesh's contribution of military troops and police to UNPKOs allows the effective use of its capabilities in promoting international peace, security and solidarity while fulfilling its constitutional duty and commitment. Additionally, this is a meaningful way it extends its friendship to states in conflict through the use of military and police via peacekeeping missions based upon its foreign policy motivation and objectives.

The 1971 Liberation War is another major factor that has motivated Bangladesh to participate in international peacekeeping (Armed Forces Division, n.d.). The state's experience with genocide, persecution, suffering, displacement, and torture in 1971 allows it to understand the horrors victims face in such humanitarian crises and has enabled the people of Bangladesh to understand the plight of a distressed population and act accordingly. As such, Bangladeshi peacekeepers can actively contribute to peace operations and international peace so that people in danger of conflict do not face such dire humanitarian crises.

Soft Power Tools

Peacebuilding and Education. Bangladeshi troops and police under the banner of the blue helmets have significantly contributed to peacebuilding initiatives in different UNPKO missions. Troops and police of the small South Asian state have pursued peacebuilding through different activities, projects and programs. All of our interviewees were quick to point out that much of the success of different peacebuilding programs and activities put forward, planned, and executed by Bangladeshi peacekeepers is rooted in past successful experiences of peacebuilding in the Chittagong Hill Tracts (CHT) region. Bangladesh's military was at the forefront of the peace making and peacebuilding initiatives in the CHT region. As a result, they could use their previous experience and know-how to help plan UN peace-making, peacebuilding and peacekeeping programs directed at the local community levels.

According to Interviewee I, Bangladeshi blue helmets have brought together different members of multiple fighting groups, which often have contentious relations with one another, through activities such as friendly sports matches and cultural programs. To ensure peace in such activities and programs, participating members of different fighting groups are temporarily disarmed for the duration of said activities and programs. Such initiatives allow members of fighting groups who would otherwise refuse to come into peaceful contact to have some common peaceful experiences. "At this stage, the goal is to bring the parties together before formal peace talks can take place", Interviewee I added. Additionally, such contact also encourages informal meetings that pave the way for conversations about the struggles of different communities that the members represent and can encourage initial sentiments of the need for peace. Subsequently, these activities can also lead to more formal peace talks in many cases. Besides arranging community sports, Bangladeshi peacekeepers also distribute sports equipment such as footballs, jerseys, rackets and bats among the local population (interviewee VI). Involvement in sports activities helps gain the support and confidence of the local population.

Interviewee III discussed how blue helmets also provide skill-based training to members of different conflicting parties to bring these parties together. This was reiterated by an instructor at BIPSOT as well. "Skill-based training programs like welding, computing, agriculture, accounting, etc., also provide many members of hostile groups to learn new skills that can be conducive for rebuilding the economy for future stability", she stated when asked about how Bangladeshi peacekeepers may help in rebuilding the economy. These skill-based training programs can also sometimes help develop new hobbies among the members, which can preoccupy their time, as it may otherwise be used up in fighting. All the former and present peacekeepers interviewed stressed that Bangladeshi blue helmets are at the forefront of such initiatives as instructors and facilitators of different skill-based training programs.

Interviewee II mentioned that, "Blue helmets also contribute to peace and peace education through infrastructure development of educational institutions, paying fees of tutors if the local government is unable to, and ensuring awareness for peace." Additionally, Bangladeshi peacekeepers have also supported the education system of different local communities through donations of educational materials for impoverished children and in remote areas (UNPK, 2020b; Ludanga, 2022). It has almost become a trend for Bangladeshi peacekeepers to buy educational materials (schoolbooks and bags) and stationary materials (pens, pencils, erasers, staplers) from Bangladesh and distribute these among local schools during their deployment (Interviewee VI). As local village schools in Africa often only have two or three sets of books per class, Bangladeshi peacekeepers often provide book printing and photocopy support to local schools and educational institutions to increase the quantity of educational resources (Interviewee VI). Interviewee VI, while sharing his own experience, referred to how his battalion in Sierra Leone established five schools and supplied them with necessary fixtures and fittings. Bangladeshi peacekeepers also arrange training and exhibitions for income-generation activities for the local community, such as agriculture, farming and herding (Interviewees IV & VI). Some of the funds for these activities come as direct contributions from the Government of Bangladesh and donations made by Bangladeshi peacekeeper units or brigade-level fund-pools.

Bangladeshi peacekeepers also teach in different schools as tutors where necessary. Interviewee IV discussed how peacekeepers have had to invigilate board exams and transport exam transcripts in different missions because of the lack of capacity in conflicting states. She stated that UN peacekeepers, especially Bangladeshi peacekeepers, participate in such activities because "education is an important supporting pillar for peace and for rebuilding the economy." Bangladeshi peacekeepers also help promote and uphold democratic practices in their electoral duties in peacekeeping missions. Interviewee VI mentioned that Bangladesh peacekeepers perform electoral duties and facilitate national and local elections by providing peace and security for candidates and the local population. These electoral duties showcase Bangladesh's firm commitment to its values and principles of being a constitutionally democratic state.

Bangladeshi peacekeepers have also been known for their contributions to the public health of local communities. According to Interviewee VI, they have addressed the healthcare needs of other UNPKO personnel involved in the missions as well as those of local people. Bangladeshi medics under the banner of blue helmets have delivered healthcare services for treatable diseases to locals in remote communities of Africa. Their engagement in such communities also raises awareness around nutrition, hygiene and the dangers of untreated health issues. Furthermore, Interviewee I mentioned how, "In addition to sending medical personnel, Bangladesh also provides medicine to post-conflict nations and communities where Bangladeshi troops and police are deployed for UNPKO missions." The Bangladeshi medical team under the banner of UNPKOs support the local and vulnerable communities in remote areas as well (UN 2018). Field-level hospitals provide X-ray support, pathology testing and diagnosis services to the local community upon their need (interviewee VI). Bangladeshi medical peacekeepers use Bangladeshi medicines to provide treatment to the local community, which has created room for pharmaceutical companies to expand their operations in those countries. Bangladeshi medics deployed in UNPKO missions can diagnose and help treat maladies and illnesses in local communities. However, despite such great achievements, Bangladesh is yet to maximise such achievements and convert them into trade and commerce usage and benefits (Interviewee VII).

Bengali Language and Culture. Sierra Leone adopted Bengali as one of its official languages in 2002 in honour of Bangladesh's role in the UNPKO mission in the West African state. Our interviewees gave differing but comprehensive perspectives explaining why the Bengali language and culture are revered and sometimes adopted among local communities of different states. According to a Bangladeshi UN peacekeeper (Interviewee V) currently serving in South Sudan, Bangladeshi peacekeepers achieve much of their respect and adoration in the local communities because of their professionalism and their low involvement in exploitative misdeeds that many peacekeepers from other states are accused of. Bangladeshi blue helmets have relatively fewer allegations of sexual exploitation and abuse (Kumar 2022, p.195). According to Interviewee I, much of the respect that Bangladeshi UNPKO personnel achieve is a reflection of the respect they have for the local community members, especially women and girls. This sense of respect for local women and girls, which is nurtured in all Bangladeshi children through religious, cultural, and moral lessons, prevents them from exploiting them as well as any other members of local communities.

Bangladeshi troops and police are also provided with special training before their deployment to any UNPKO missions. "One of the important messages of the training is that they are representatives of Bangladesh at UNPKO missions abroad, and their conduct plays a deterministic role in whether Bangladesh will be asked to contribute soldiers in the future," said Interviewee IV. Therefore, troops and police deployed to UNPKO missions are made aware of the role they play in ensuring a positive image for Bangladesh as a peacekeeper-contributing state at the international as well as at local levels. Other interviewees also brought up factors such as the empathetic, caring, and selfless attitude of Bangladeshi peacekeepers. A Bangladeshi peacekeeper deployed at UNMISS mentioned how Bangladeshi troops and police want to ensure as much as possible the survival of the local people in destitute conditions, including by sharing their own meals, clean water, and other handy resources. Such acts help build greater rapport between the local people and Bangladeshi peacekeepers.

Another important explanation for the mass acceptance of the Bengali language and culture by local communities is rooted in their willingness to give back to the people that helped them rebuild their communities. Interviewee I highlighted that, "Learning the Bengali language is not only a way to communicate with Bangladeshi peacekeepers, but also a way to symbolically give back to the people they are extremely thankful to." Not only that, but school children and the local population often learn to sing traditional and patriotic songs of Bangladesh. Bangladeshi peacekeepers also involve and invite local community leaders and fellow peacekeepers from different countries in cultural and festival celebrations like Bengali New Year (Pohela Boishakh) or by observing International Mother Language Day (Interviewee VII). These not only help showcase Bangladeshi culture but also promote our cuisine abroad (Interviewee VII).

This is further showcased in naming conventions undertaken for important infrastructure in states where UNPKO missions have previously taken place. For example, the people of Ivory Coast have named a major road in Danane town 'Bangladesh Road' in recognition of the contributions made by Bangladeshi peacekeeping forces (The Daily Star, 2007). Similarly, a major highway that connects three provinces in Sierra Leone has also been named after Bangladesh (The Daily Star, 2003). More recently, the government in South Sudan has also named a road in its capital, Juba, 'Bangladesh Road' (The Business Standard, 2021). A Bangladeshi engineering contingent under the flag of UNPKO was directly involved in the construction of many of these major infrastructure projects. They are also heavily involved in reconstruction efforts following natural disasters. Therefore, the local-level engagement of Bangladeshi peacekeepers has successfully won the hearts of the local population.

Bangladeshi Female Peacekeeper. Women, Peace and Security (WPS) played an important role in focusing on female participation in UNPKOs by appearing on the Security Council agenda in 2000 and consequently leading to the adoption of the Security Council Resolution 1325. This resolution, along with Security Council Resolutions 2242 and 2538, stresses the importance of the inclusion of female peacekeepers in UNPKO missions and encouraging female peacekeepers at decision-making levels.

Female participation in UNPKO missions helps access the voices of women and girls in local communities, especially in conservative communities where women are either not allowed or do not want to speak to male peacekeepers. Bangladesh's contribution of female troops also consists of medics; according to Interviewee IV, "Bangladeshi female medics were among the first to be able to access local women and girls, especially in remote communities, and easily draw out their narratives, hear their needs, and communicate them with the rest of the mission." As Interviewee III stressed, "It is well-established now that if we want sustainable peace in conflicting areas, we need to incorporate all people in the area. Therefore, you cannot have sustainable peace without incorporating local females and without knowing their narratives and possible contributions. A comprehensive solution requires everybody's perspectives".

Furthermore, Interviewee V stated that "Bangladeshi female medics provide medical support to the personnel involved in the UNPKO mission, as well as the local women, who would not have expressed their healthcare needs to male medics, especially their needs around menstrual hygiene, prenatal and postnatal health awareness and care." Bangladeshi female peacekeepers are also trained as midwives to help new mothers in post-conflict communities (Interviewee IV).

One of the challenges Bangladesh faces in terms of contributing female peacekeepers to UNPKO is keeping up with the proportionate demand. Multiple UNSC resolutions have attempted to coax greater numbers of female peacekeepers from different contributor states. Interviewee IV explained how greater demands for female Bangladeshi peacekeepers stand as a challenge because Bangladesh cannot fulfil the required numbers. She further stated that, "If Bangladesh cannot fulfil the proportionate demands of female peacekeepers, we have to end up surrendering the same proportionate number of peacekeepers we send." Both Interviewees II and III addressed that the comparatively fewer numbers of female military troops, officers, and police lie at the root of the issue.

Nevertheless, Bangladeshi female peacekeepers have been championing the WPS agenda by playing dynamic roles in UNPKOs. For example, in 2017, Bangladesh contributed two female combat pilots to MONUSCO (UN News, n.d.). In 2023, an all-female 180-member Bangladesh police contingent participating in MONUSCO under the blue helmet banner were awarded UN medals for their contributions (Dhaka Tribune, 2023). Bangladesh also contributed an all-female police unit to MINUSTAH in 2010 (UN News, n.d.). Bangladeshi female peacekeepers have also been an inspiration for women in Wau, South Sudan to consider a career and join the country's security force (United Nations Peacekeeping, 2020a). Therefore, female Bangladeshi peacekeepers have played a crucial role in contributing towards the WPS agenda. Furthermore, Bangladesh has also been able to incorporate the lessons learnt by female UNPKO personnel internationally in its own National Action Plan on Women, Peace and Security (Government of Bangladesh 2019).

The Future of Bangladesh in Peacekeeping Operations

The following recommendations are based on the empirical findings and areas Bangladesh can further utilise as a soft power tool in UN peacekeeping missions.

Championing gendered peacekeeping is a necessity of the present time. Bangladesh's participation in peacekeeping missions does involve female troops and police officers, but the numbers are still minimal and insignificant. Despite UN efforts to address the gender imbalance via the UN Security Council Resolution in Women, Peace and Security, most contributing members, including those of the West, have been unable to provide adequate female peacekeepers as required (Interviewee II). While Bangladesh has turned out to be an example of breaking gender biases with its firm political commitment domestically, it has room to break similar ground in UN peacekeeping missions. Bangladesh's increased contribution of female peacekeepers would significantly improve its contribution to peace-making and peacebuilding mechanisms. Increasing female peacekeepers would further improve Bangladesh's international prestige in achieving gender mainstreaming agendas (Interviewee VII). Focusing on the gender aspect of peacekeeping and overcoming gender imbalances can itself turn out to be a soft power tool and create more future opportunities for Bangladesh in UN peacekeeping missions.

Green peacekeeping & disaster management is an important part of contemporary peacekeeping in the age of climate change. Bangladesh is a vulnerable country due to climate change, and mitigating such risks has been part of Bangladesh's development story. Bangladesh's military and police readiness and frequent deployment in national emergencies such as natural disasters and calamities domestically have provided Bangladeshi peacekeepers with experience in disaster management (Interviewee II). Lobbying for including such criteria in mission mandates and using such unique skills in natural disaster management would likely put Bangladeshi peacekeepers among the top choices in climate-vulnerable missions in Africa. Including a climate security mechanism as part of pre-deployment training for commanding officers and leadership role holders would increase the climate and environmental considerations in missions' operations. Bangladeshi peacekeepers have room to play a greater role in reducing their carbon footprint (waste management) and greening efforts (increased usage of solar panels) in their peacekeeping missions and become role models (Interviewee VII). Gearing up with green peacekeeping also would allow further engagement with the host community to promote sustainable and renewable energy, ultimately helping to achieve Sustainable Development Goals.

The need for digital peacekeeping is one of the many requirements for effective peacekeeping nowadays. 'Digital Bangladesh to Digital Peacekeeping' – The motto of 'Digital Bangladesh' has been one of the transformative driving features of Bangladesh's rapid socio-economic development over the last decade. Bangladeshi peacekeepers can set a good example of using digitilisation and information and communication technology initiatives and peace-making and peacebuilding activities (Interviewee IV). Such a tech-enabled initiative would help develop the host country and communities' economic and social lifestyles in the long run. Bangladeshi peacekeepers can train the local security forces on GPS and satellite images and provide training workshops or courses (computer and software training) for locals and youths.

Engagement with non-state actors in the host country is crucial for sustainable peace. International NGOs and civil society organizations play a critical role as humanitarian assistance providers on the ground in post-conflict societies. UNPKO personnel provide them with additional assistance they may need to operate, especially when it comes to security and infrastructure establishment. Bangladesh's peacekeepers play an active and important role in engaging and working with international non-governmental actors. For example, Bangladeshi blue helmets deployed from the engineering core of the Bangladesh Armed Forces have collaborated with telecommunication and IT companies in conflict-driven host countries and helped establish telecommunication towers and other communication mechanisms in numerous post-conflict states (Interviewee IV). Bangladesh's military and police have vast experience in working with different non-state actors in community development activities in their own nation-state. Similarly, Bangladeshi peacekeepers working along with UN civilian staff and non-state actors (NGOs, human rights organisations and civil society) can positively impact the livelihood and peacebuilding mechanism in post-conflict reconstruction efforts besides maintaining peace and stability (Interviewee VII).

Bangladesh plays a key role in international peacekeeping as a troop- and police-contributing country to UNPKOs. Bangladeshi blue helmets have successfully tapped into the hearts and minds of local community members in different UNPKO missions. Their success has not only been recognised locally in the communities they have helped, as they have also received international commendations. This is perhaps materially best displayed by the growing numbers of Bangladeshi troops and police accepted to UNPKO missions. Beyond local and international recognition of the success of Bangladeshi peacekeepers, Bangladesh has also been able to leverage this success into developing a positive image of itself in the international arena. This has allowed the small South Asian state to use its role as a top troop- and police-contributing state to UNPKOs as a soft power mechanism. The roles and activities of Bangladeshi peacekeepers on the ground have acted as important soft power tools for Bangladesh, as they have implicitly molded the appreciation of Bangladeshi blue helmets. Much of this is thanks to the special medical, gender-based and professional training as well as cultural and normative lessons they receive. Bangladesh now needs to look beyond the economic aspect from UNPKOs; rather, it can be a major branding tool for Bangladesh for its peacemaking and peacebuilding initiatives if utilised properly.

References

Adebajo, Adekeye. 2013. “11 Nigeria”. In *Providing Peacekeepers: The Politics, Challenges, and Future of United Nations Peacekeeping Contributions*, edited by Alex J. Bellemy and Paul D. Williams. Oxford: Oxford Publications

Afroze, Shaheen. 2005. ‘Small States in Global Perspective: In Search of a Role Model in Regional Stability’. In Small States and Regional Stability in South Asia edited by Mohammad Humayun Kabir. Dhaka: The University Press Limited. P. 27

Akhter, Samina. 2022. How Bangladesh is Playing significant role in UN peace keeping mission. Accessed March 29, 2023. https://moderndiplomacy.eu/2022/11/08/how-Bangladesh-is-playing-significant-role-in-un-peace-keeping-mission/.

Armed Forces Division. n.d. Bangladesh In UN Peace Operations. Accessed March 30. https://afd.gov.bd/un-peacekeeping/Bangladesh-in-un-peace-operation

Atkinson, Carol. 2014. “Military Soft Power in 21st Century: Military Exchnages and Partner Development”. In *Military Soft Power: Public Diplomacy through Military Educational Exchanges*, by Carl Atkinson. Lanham: Rowman & Littlefield Publishers.

Banerjee, Dipankar. 2008. ‘South Asia: Contributors of Global Significance’, Peace Operations edited by in Donald C.F. Daniel, Patricia Taft and Sharon Wiharta. Washington, DC: Georgetown University Press, pp.195–6.

Biswas, Niloy Ranjan. 2022. Bangladesh in the United Nations Peace Operations: Maintaining Sustained Engagement within Space of Contestation, *Journal of Bangladesh and Global Affairs,* Vol.1, No.1, pp. 77-90.

Buerk, Roland. 2006. "The cream of UN Peacekeepers" Accessed March 14, 2023. http://news.bbc.co.uk/2/hi/south_asia/3763640.st.

Chowdhury, Masud. 2021. "Bangladesh in UN Peacekeeping Missions: Contributions and Global Recognitions". Accessed March 10, 2023. https://www.observerbd.com/news.php?id=389720

Commonwealth Secretariat. 1997. "A future for small states: overcoming vulnerability: report by a Commonwealth advisory group." London: Commonwealth Secretariat.

Dhaka Tribune. 2023. "Bangladesh women peacekeepers in Congo awarded UN medal." Accessed 14 June 2023. https://www.dhakatribune.com/Bangladesh/2023/01/13/Bangladesh-female-police-contingent-in-congo-awarded-un-medal

Farwa, Ume and Garewal, Ghazanfar Ali. 2018. "An Analysis of UN Peacekeeping as Pakistan's Soft Power Asset". *NUST Journal of International Peace and Stability*, vol. 1, no. 2, 91-102

Fox, Annette Baker. 1959. *The power of small states: diplomacy in World War II*. Chicago: University of Chicago Press.

Government of Bangladesh. 2019. National Action Plan on Women, Peace and Security. Accessed June 10, 2023. http://1325naps.peacewomen.org/wp-content/uploads/2021/04/NAP-WPS-Bangladesh-2019_2022-1.pdf

Gowan, Richard. 2020. "China's pragmatic approach to UN peacekeeping". Accessed March 20, 2023. https://www.brookings.edu/articles/chinas-pragmatic-approach-to-un-peacekeeping/

Heitzman, James and Worden, Robert (eds.). 1988. Bangladesh: a country study. Washington: Library of Congress

Hong, M. 2015. *The Soft Power of Singapore. In The Rise of Singapore*, edited by Amy Lugg, 2:101–22. New Jersey: World Scientific Publishing Co. Available from https://doi.org/10.1142/9789814704946_0019.

Hossain, Raian. 2018. Peace Support Operation in Transition: Fourth generation UN peacekeeping. Accessed April 14, 2023. https://bipss.org.bd/pdf/Peace%20Support%20Operation%20in%20Transition%20Fourth%20generation%20UN%20peacekeeping.pdf

Karim, Sajjad and Uddin, Mohammad Jasmin. 2016. 'Foreign Policy of Bangladesh'. BIISS Journal, vol.37, no.4, pp. 339-362.

Keohane, Robert O. and Nye, Joseph S. 1977. Power and Interdependence: World Politics in Transition, Boston: Little Brown and Company.

Krishnamay, Kabilan. 2010. Bangladesh and UN Peacekeeping: The Participation of a 'Small' State *Commonwealth & Comparative Politics*, vol.41, no.1, pp. 24-47

Kumar, Anand (2022) Bangladeshi Women in UN Peacekeeping: A Case Study. *Journal of Defence Studies*, vol. 16, no. 3, July–September 2022, pp. 191–208

Long, Tom. 2017. "Small States, Great Power? Gaining Influence Through Intrinsic, Derivative, and Collective Power". International Studies Review, doi: 10.1093/isr/viw040

Ludanga, Jimmy. 2022. "Bangladeshi Peacekeepers Handover Books and Sports Items to Primary School Children." Accessed 4 April, 2023. https://unmiss.unmissions.org/Bangladeshi-peacekeepers-hand-over-books-and-sports-items-primary-school-children

Manik, Julfikar Ali. 2007. "Ivory Coast town proud of its Bangladesh Road." Accessed April 1, 2023. https://archive.thedailystar.net/2007/07/24/d70724011511.htm

Ministry of Foreign Affairs, People's Republic of Bangladesh. "Constitution-Foreign Policy of Bangladesh" Accessed April 19, 2023. http://www.mofa.gov.bd/site/page/0498e3d1-9bb7-45f0-988c-cb360e9949e2/Foreign-Policy-of-Bangladesh.

Mohan, Garima and Gippner, Olivia. 2013. Theoretical Approaches to Understanding Chinese and Indian Participation in Peacekeeping. *NFG Working paper No.5/2013*. http://dx.doi.org/10.17169/refubium-22950.

Morgenthau, Hans J. 1954. *Politics among Nations: The Struggle for Power and Peace*. New York: Knopf.

Mukherji, Mithlesh Jayas. 2015. "India's Rising Soft Power: Implications for India's Relations with Its Southern Asian Neighbours". *World Focus*, vol.36, no.8, pp. 102-110.

Murthy, C. S. R. 2007. 'Unintended Consequences of Peace Operations for Troop-Contributing Countries from South Asia', in Chiyuki Aoi, Cedric de Coning and Ramesh Thakur (eds), Unintended Consequences of Peacekeeping Operations, Tokyo: UN University Press, p.160.

Nye Jr., Joseph. 1990a. *Bound to Lead: The Changing Nature of American Power*. New York: Basic Books.

Nye Jr., Joseph. 1990b. "Soft Power". *Foreign Policy*, vol. 80, pp. 153-171

Nye Jr., Joseph. 2004. *Soft Power: The Means to Success in the World Politics*. New York: Sage Publications, Inc.

Pulami, Manish Jung. 2022. "Bangladesh in Global Peacekeeping: Military as a Soft Power in Portraying the Country as a Responsible Power".

Journal of Bangladesh and Global Affairs, vol. 1, no. 3, pp.5-21. https://doi.org/10.55875/jbga.bd.dec22.001

Rachman, Gideon. 2022. "Qatar's Soft Power Play". Accessed March 19, 2023. https://www.ft.com/content/43832508-d3ff-400b-a51f-3ba8c65d562a

Rahman, Syedur. 1989. "Bangladesh in 1988: Precarious Institution Building Amid Crisis Management." Asian Survey, vol. 29, no. 2, pp. 216–22. JSTOR, https://doi.org/10.2307/2644582. Accessed 14 Mar. 2023.

Sotomayor, Arturo C. 2018. "13 Nepal". In *Providing Peacekeepers: The Politics, Challenges, and Future of United Nations Peacekeeping Contributions*, edited by Alex J. Bellemy and Paul D. Williams. Oxford: Oxford Publications

The Business Standard. 2021. "South Sudan names road after Bangladesh." Accessed April 3, 2023. https://www.tbsnews.net/foreign-policy/south-Sudan-names-road-after-Bangladesh-291190.

The Daily Star. 2003. "Road in Sierra Leone named Bangladesh." Accessed April 2, 2023. https://www.thedailystar.net/news/road-in-sierra-leone-named-Bangladesh

The Daily Star. 2022. "165 Bangladeshi peacekeepers killed in 3 decades." Accessed March 29, 2023. https://www.thedailystar.net/news/Bangladesh/news/165-Bangladeshi-peacekeepers-killed-3-decades-3149091

United Nations Peacekeeping, 2020a. "Female Bangladeshi peacekeepers inspire the women of Wau to join security forces." Accessed June 10, 2023. https://peacekeeping.un.org/en/female-Bangladeshi-peacekeepers-inspire-women-of-wau-to-join-security-forces

United Nations Peacekeeping, 2020b. "Bangladeshi peacekeepers boost local education in Wau." Accessed April 5, 2023. https://peacekeeping.un.org/en/Bangladeshi-peacekeepers-boost-local-education-wau.

United Nations Peacekeeping, 2022b. "Troops and Police Contributors". Accessed March 13, 2023. https://peacekeeping.un.org/en/troop-and-police-contributors

United Nations Peacekeeping. 2022a. "Contribution of Uniformed Personnel to UN by Country and Personnel Type". Accessed March 13, 2023. https://peacekeeping.un.org/sites/default/files/01_contributions_to_un_peacekeeping_operations_by_country_and_post_57_December_22_rev1.pdf.

United Nations. 2018. "Bangladeshi UN Peacekeepers Deliver Critical Medical Aid to Remote Communities." YouTube. February 5, 2018.

UN News. "Photo story: Women on the front line for peace." Accessed 14 June 2023. https://news.un.org/en/gallery/1128737#:~:text=MONUSCO%2FForce%20%7C%20In%202017%2C,Lieutenant%20Tamanna%2DE%2DLutfi.

Vukadinovic, Radovan. 1971. 'Small states and the policy of non-alignment'. In *Small states in International Relations*, edited by August Schou and Arne Olav Brundtland. New York: Wiley Interscience, pp.99–114

Wigner, Gregory H. 2017. "Soft power by other means: defense diplomacy as a tool of international statecraft". Phd diss. Boston University

Zaman, Rashed and Biswas, Niloy Ranjan. 2017. Bangladesh and United Nations Peacekeeping Missions: The Quest For A National Policy To Meet The Challenge of Uncertainty, *Journal of the Asiatic Society of Bangladesh (Hum.)* Vol. 62, no. 2, pp. 155-181.

Zaman, Rashed and Biswas, Niloy Ranjan. 2017. The Contribution of Commonwealth Armed Forces in UN Peacekeeping: The Case of Bangladesh, The Round Table, Vol. 106, no. 4, pp. 437-452.

Zaman, Rashed and Biswas, Niloy Ranjan. 2014. 'Bangladesh's Participation in UN Peacekeeping Missions and Challenges for Civil–Military Relations: A Case for Concordance Theory' International Peacekeeping, vol. 21, no.3, pp. 324-344. 10.1080/13533312.2014.938913. Accessed March 14, 2023.

Zaman, Rashed. 2016. UN peacekeeping mission. In *Routledge Handbook of Contemporary Bangladesh* Routledge, pp. 414-424.

Appendix

List of Elite Interviews:

Interviewee I - Former Deputy Force Commander, United Nations Organization Stabilization Mission in the Democratic Republic of the Congo (MONUSCO). Interview conducted in 18/03/23.

Interviewee II - Former Commandant, Bangladesh Institute of Peace Support Operations Training (BIPSOT). Interview conducted in 22/03/23.

Interviewee III - Former Force Commander, United Nations Peacekeeping Force in Cyprus, UNFICYP. Interview conducted in 28/03/23.

Interviewee IV - Instructor, BIPSOT. Interview conducted in 05/04/23.

Interviewee V - Medical Officer, United Nations Mission in South Sudan, UNMISS. Interview conducted in 06/04/23.

Interviewee VI - UN Peacekeeper, United Nations Mission in Sierra Leone (UNAMSIL), United Nations Mission in Liberia (UNMIL), MONUSCO, United Nations Multidimensional Integrated Stabilization Mission in the Central African Republic (MINUSCA). Interview conducted in 08/04/23.

Interviewee VII - Head of a leading security think tank in Bangladesh. Interview conducted in 14/04/23.

CHAPTER XII

COPING WITH CHANGE: ADAPTING PEACEKEEPING TRAINING TO A CHALLENGING WORLD

CARLOS GODOY

United Nations (UN) peacekeepers have been delivering security and stability to war-torn zones for more than seven decades. Peacekeeping forces were conceived to support ceasefires after the inter-state conflicts that erupted after World War II (Day 2020, 1). They consisted of unarmed military officers deployed in buffer zones between opposing states to supervise peace agreements (Goulding 1993, 452). The end of the Cold War and the widespread intra-state conflicts demanded modern peace operations to adapt to new asymmetric conflict settings involving state actors and non-state actors, such as violent extremist groups, terrorist organizations, and organized crime networks (Abdenur, Kuele, and Francisco 2018, 3). The challenges brought by an evolving scenario required the redesign of UN peacekeeping to make it more effective in fulfilling increasingly broad demands. This need became clear after the Rwanda and Bosnia massacres when peacekeepers helpless stood by (Autesserre 2019, 104). Amongst the many initiatives aimed at improving the effectiveness of UN peacekeeping, the Brahimi Report, presented in 2000, was the turning point in framing peace operations in a challenging contemporary world (UN Peacekeeping 2000). The report, which offered explicit and implicit recommendations for making peacekeeping successful (Banerjee 2005, 24), was followed by other high-level initiatives that gave rise to documents, such as the Capstone Doctrine (UN Peacekeeping 2008), and reports, such as the High-Level Independent Panel on Peacekeeping Operations (2015) and Improving the Security of UN Peacekeepers (2018a), which aimed at making UN peacekeepers able to effectively and safely carry out their tasks in the field.

The history of UN peacekeeping has shown, and the Brahimi Report emphasized, that a lack of training and preparedness can jeopardize missions and cost the lives of those whom peacekeepers should protect. Peacekeeping, particularly in the third millennium, requires skills other than those that soldiers typically develop through basic military training (Cutillo 2013, 2). Therefore, more comprehensive and more specific training are needed for UN peacekeepers who, despite the critiques of their failures (Autesserre 2019, 102, 104; Fortna 2004, 269), provide peace and security throughout the world with the second largest military force deployed abroad. Nevertheless, as the correlation between training and efficacy in the field is often underestimated (Cutillo 2013, 1), many UN peacekeeping military units still lack adequate capabilities (Smith, and Boutellis 2013, 3; Jett 2019, 142) to face the complex security and protection tasks posed by the contemporary world.

The chapter explores the evolution of the peacekeeping training delivered by member states to their uniformed personnel, how it has been adapted to the changing world scenario and how training could adapt to follow peacekeeping trends and deal with the increasing challenges UN peace operations have been facing in the new millennium.

Training For Peace: Adapting to Evolving UN Peacekeeping

Established in 1948, the UN Truce Supervision Organization, which was comprised of a group of unarmed military observers who were deployed in Palestine to help monitor a truce between Israel and the surrounding Arab countries (Goulding 1993, 452), marked the origin of UN peacekeeping. This peacekeeping model, often called traditional peacekeeping, remained in place until the end of the Cold War and was characterized by observers or lightly armed forces deployed in buffer zones without an interventionist intent (Wedgewood 1995, 635). During this phase, specific training for UN peacekeeping was almost non-existent due to the limited scope and narrow security tasks of the mandates (Gledhill, Caplan, and Meiske 2021, 207; Cutillo 2013, 3) and the scenario's moderate levels of complexity and risk that required only regular military-like skills from the peacekeepers. The UN did not have a specific body that established training standards for troop-contributing countries (TCCs) to comply with (UN Peacekeeping, Department of Peace Operations); therefore, what to train on and how to train peacekeepers was at the discretion of the TCCs. With the end of the Cold War, the world saw the emergence of a new peacekeeping operations phase known as the second generation (Banerjee 2005, 18, 20), which distinguished the 1990s with a significant increase in the number of UN peacekeeping operations and the scope and complexity of the missions assigned to peacekeepers, who became engaged, for example, in the protection of human rights and state-building tasks (Kondoch, and Howe 2017, 2).

The Srebrenica genocide in 1995 and other peacekeeping setbacks in Africa demonstrated that the "one-size-fits-all peacekeeping" approach would not work to end regional conflicts, which were unique and involved multiple non-state actors (Banerjee 2005, 18, 20-21). This understanding gave rise to what may be called the third phase of peacekeeping (21) or "third-generation peacekeeping," when there was an attempt to take advantage of the expertise of local organizations that would possibly be better suited to keep peace in their respective territories. An overlapping fourth phase characterized by many interventions launched in failing states until the end of the decade, mainly for humanitarian reasons (21), might be called the fourth generation of peacekeeping. A new model of peacekeeping operations, called multidimensional, was launched in Kosovo, East Timor, Sierra Leone, the Democratic Republic of Congo (21), and Eritrea/Ethiopia with a broader scope that required peacekeepers to tackle war crimes and massive human rights violations, support disarmament and demobilization campaigns (Mackay 2003, 218; Briscoe et al. 2015, 24), and protect mass movements of people, all while coping with state reconstruction efforts. More importantly, these operations broadened the array of skills required from peacekeepers (Cutillo 2013, 2) and stimulated ways to correct the limitations of the earlier approaches and improve the efficacy of UN peacekeeping. Although all the continents had seen peacekeepers' boots on the ground during this period, the UN still had not formally recognized the importance of training for successful peacekeeping (1) nor established a structure to develop training standards and guide the TCCs on their implementation.

In 1995 (UN 1995), the UN and member states recognized their responsibility to provide pre-deployment training to peacekeepers (Abdenur, Kuele, and Francisco 2018, 4) and the need for UN-approved procedures and materials to guide the TCCs to undertake them. Also, the presentation of the "Brahimi Report" (Curran 2017, 12) at the Millennium Assembly in 2000 marked the start of changes in UN peacekeeping training structures. Nevertheless, even though the report stressed the importance of peacekeeping training and offered recommendations for the production of training material, the implementation of practical measures to standardize peacekeeping training only started in 2007 when a formal structure for training was implemented as part of the restructuring process of the Department of Peacekeeping

Operations (DPKO) and inception of the Department of Field Support. Moreover, the Integrated Training Service (ITS) was created under a new Division for Policy, Evaluation and Training designed to centralize all peacekeeping training activities (Abdenur, Kuele, and Francisco 2018, 4). Since then, ITS has overseen peacekeeping training activities; developed training standard operations procedures, policies, and materials; and supported TCCs in the peacekeeping training process (4). Following these initiatives, in 2008 the DPKO issued the document known as the "Capstone Doctrine," which, among other objectives, would serve as a guide to personnel training to integrate UN peacekeeping operations (UN Peacekeeping 2008; Cutillo 2013, 1). The creation of a specific structure dedicated to training represented a significant step towards the improvement of the performance of the peacekeepers by establishing minimum individual and collective capacity standards and providing guidance and assistance to TCCs to conduct training for peacekeeping operations.

UN Peacekeeping Training and Preparation

Since the creation of the ITS and the advent of the Capstone Doctrine, the UN has undertaken many initiatives to improve the performance and preparation of peacekeepers for peace operations. The organization recognized that training for all the field personnel involved with peacekeeping operations is critical for the mission's success (UN 2010, 4). Training for peacekeeping was defined as any activity aimed at providing peacekeeping personnel, individually and collectively, whether military, police, or civilian, the knowledge, skills, and attitudes needed to effectively and professionally meet the challenges of the operations and demonstrate the values and competencies cultivated by the UN. Further, a methodology was designed, and the responsibilities for the peacekeeping training and preparation were clearly defined (5-14). Training refers to the process whereby personnel achieve or improve the necessary skills and behaviours to properly perform their duties in the field. In contrast, preparation refers to being ready for deployment in a peacekeeping operation. Although the concepts intersect, the Department of Peace Operations ITS prepared a comprehensive set of policies, guidance documents, and instruction manuals, which are constantly updated (UN Peacekeeping "Policy") and guide the delivery of training to future peacekeepers (Curran 2017, 10).

Training for peacekeeping targets military, police and civilian personnel and is divided into three main phases: pre-deployment training (PDT), induction training, and ongoing training, depending on whether it respectively takes place prior to the deployment in operation, upon arrival in the field or during the operation (Cutillo 2013, 5). However, as military peacekeepers represent 88% of the UN personnel deployed in peacekeeping operations (UN Peacekeeping 2023), this chapter focuses on the PDT of military personnel preparing for UN peacekeeping.

PDT is the critical component of peacekeeping training (5) as it prepares field personnel to be "fully operational and ready to face their assignments from day one" (5). This training is delivered in the TCCs by their trainers and based on the Core Pre-deployment Training Materials (CPTM) elaborated on by the ITS. It has generic and specialized components and aims to complement military training with mission-specific skills (Curran 2017, 11), internalizing UN principles and standards. Indeed, even though the ITS provides extensive guidance materials, the vast array of tasks the complex contemporary conflicts impose on peacekeepers, which often encompass the protection of civilians, monitoring of elections (UN Peacekeeping "Military"), and dealing with asymmetrical warfare, require TCCs to go beyond military training (Cin 2001, 3) to prevent flaws that can jeopardize the operation and put at risk the lives of civilians or their peacekeepers (De Sylva 2017, 47, 54, 61, 70; UN Peacekeeping "Improving Security of United Nations Peacekeepers" (Independent Report), 2018). Yet, the UN guidance is limited to skills specific to the UN, as military training is the TCCs' responsibility (Cutillo 2013, 7).

Preparation, in turn, comprises the process whereby the military, trained under national standards and regulations, reaches operational readiness to apply the knowledge and skills acquired and exercised during the training phase. At the end of the preparation, they will be capacitated to accomplish their tasks according to mission-specific concepts, rules, and operations orders (UN Peacekeeping 2018b, 2-3). The guidelines for operational readiness describe in detail the knowledge, skills and respective standards the various ranks of a military contingent must achieve (6-10). Further, they present a sample of individual and collective PDT to guide the TCCs through the preparation for deployment (11-15).

The high number of peacekeeper fatalities stemmed from aggression, their general underperformance in the field, deficiencies in protecting civilians, and misconduct related to sexual exploitation and abuse. These factors compelled the UN Security Council to adopt a resolution urging TCCs to provide more and better PDT for the military personnel deployed (Wilén 2018, 1) to meet the standards set by the UN (UN 2018, 2-4, 6).

Cooperation To Better Prepare Peacekeepers

The importance of training and preparation to equip peacekeepers with a diversified range of military and non-military skills required by the increasingly complex conflict scenarios the UN is called to stabilize is evident. The expansion of peacekeeping and the increasing participation of TCCs, even with reduced contingents, led to the creation of peacekeeping training institutions and centers worldwide (Cutillo 2013, 9). This trend continues to this date, and many training institutions have gone further, creating associations to establish cooperation strategies (Abdenur, Kuele, and Francisco 2018, 10), which usually involve the exchange of instructors and students, lessons learned, information, and teaching practices (10).

The associations of training centers are structured at regional and world levels and are comprised of military, police, and civilian organizations (Cutillo 2013, 9). In addition to the International Association of Peacekeeping Training Centers, regional associations (Cutillo 2013, 9) comprised of military, police and civilian providers of peacekeeping training have been created: the African Peace Support Training Association in Africa, with 15 members and partners; Association of Asia-Pacific Peace Operation training Centers, with 27 members and observers; European Association of Peace Operations Training Centers, with 33 members; and Latin America Association for Peacekeeping Training Centers (ALCOPAZ), with 32 members and observers. Regional partnerships are not limited to exchanges of personnel. They are effective mechanisms to promote synergy between centers, for example, through mobile training teams (MTTs), which are small groups of training experts from one country sent to a partner training center in the initial structuring phase or are newly created to improve their expertise in a specific area. In Latin America, Brazil has sent MTTs to its ALCOPAZ partners, Mexico and Colombia (Abdenur, Kuele, and Francisco 2018, 10). In the same token, since the 1990s, Scandinavian countries have continually revised a common catalogue of peacekeeping training courses under the Arctic Cooperation Agreement so that their peacekeeping training institutions offer different courses, better allocate resources and avoid the unnecessary duplication of efforts (NORDEFCO 2023). Further, training institutions can focus on certain themes to become regional centers of excellence in specific subjects (Cutillo 2013, 2; BIPSOT 2023; KAIPTC 2023).

Training For Peace: Matching Military and Peacekeeper Skills

To become a good peacekeeper, a regular soldier, irrespective of their origin, must acquire essential knowledge and skills that only specific peacekeeping training provides (Cutillo 2013, 1). Therefore, peacekeeping training generally follows similar curricula and levels of performance worldwide, which aligns with the objective the UN has pursued since the creation of specific training bodies at the DPKO: that all TCCs meet minimum training and preparation standards before deployment in peacekeeping operations. Nevertheless, the focus each TCC applies to the training depends on national methods of instruction, past experiences, and the specific mission context (Cutillo 2013, 7). The array of courses offered, whether in peacekeeping training organizations in Africa, Asia, Europe, or Latin America, encompasses subjects far beyond the CPTM. They often involve the protection of civilians, training for staff officers and military observers, disarmament, demobilization and reintegration, negotiation, conflict management, languages training, civil-military coordination (Curran 2017, 104-105), humanitarian assistance, mine awareness training, UN logistics, conflict management, and women, peace and security (ALCOPAZ 2023; BIPSOT 2023; KAIPTC 2023). The changes in the world context, with local and regional impacts, are increasingly demanding peacekeepers to develop enough empathy to protect unassisted populations and understand the unique needs of women and children (Mackay 2003, 218). Future peacekeepers achieve this level of understanding when they internalize training and skills beyond those of military combat.

Although training schools incorporate UN-provided training materials, training and preparation for peacekeeping are constantly updated according to the contextual evolution of the mission area (Cin 2001, 5) to keep it dynamic and realistic. To this end, teachings include political, social, economic, ethnic and cultural features of the mission zone and conflict background (5). Further, to be vibrant and stimulating, they use methodologies that create simulated scenarios based on real cases (Curran 2017, 66-67, 98, 106) and role-playing drills. This allows trainees to match their combat skills with the peacekeeping knowledge they acquired during the pre-mission training stage, putting into practice conflict resolution and negotiation techniques, UN-specific expertise, lessons learned, and language skills (5). Training, as it is a TCC responsibility, can be undertaken by national peacekeeping training centres, foreign MTTs that offer specific types of support, or by divisions of national Armed Forces institutions (Abdenur

2018, 8). Preparation, however, is usually carried out through collective exercises planned and conducted by the TCCs' defence ministries. Also, training and preparation are executed in the months preceding the deployment in the mission as member states typically allocate enough time (often around six months) to prepare their contingents to begin their participation in the mission or to replace rotating contingents (Cutillo 2013, 7).

The Critical Role of Peacekeeping Training: Lessons from the Field

UN peacekeeping, despite the criticisms it has received for several painful debacles, particularly in Bosnia and Rwanda (Nsia-Pepra 2019, 470), has been, since its inception, an effective mechanism for bringing security and stability to conflict zones and encouraging warring parties to negotiate an agreement. Despite some failures, peacekeeping has had a history of successes for more than 70 years, of which training is a significant component. During the Yugoslav wars, the use of roadblocks by groups of armed men was the principal impediment to the UN Protection Force's (UNPROFOR) effective delivery of humanitarian aid (Curran 2017, 68). As the mission's mandate relied on the consent of all parties in conflict, peacekeeper convoys usually got their way through the roadblocks after applying negotiation techniques (68). Negotiation, indeed, is one of the backbones of modern peacekeeping training, which UNPROFOR peacekeepers learn and internalize during PDT. Another compelling example of the relevance of peacekeeping training comes from the UN Organization Mission in the Democratic Republic of Congo. One of the mission's reports highlighted training as the most effective way to prevent sexual exploitation and abuse (SEA) (96). Although SEA prevention is included in the mandatory peacekeeping training, the UN secretariat has been supporting TCCs in improving the delivery of SEA training for peacekeepers through train-the-trainers courses and the creation of new mandatory SEA training programmes for all uniformed personnel (UN "Preventing Sexual Exploitation and Abuse").

Another example of the positive impact of peacekeeping training comes from the UN Multidimensional Integrated Stabilization Mission in the Central African Republic (MINUSCA). As Howard (2019) maintains, "Despite many challenges, MINUSCA has undoubtedly helped to protect civilians." As previously discussed, the curricula adopted to train peacekeepers worldwide includes the protection of civilians, which became one of their critical tasks and is therefore significantly emphasized throughout the training. Considering this, the effective protection of vulnerable populations is expected to be the outcome of a modern peacekeeping operation. Also, as Bernath (2003) elaborates, training is also connected to the success of the UN Mission in Sierra Leone: "UNAMSIL was successful because... the mission had the strong support of the UN Security Council and the international community.... translated into a strong mandate and a force with sufficient strength, logistics, and training to carry out its mandate" (136).

Peacekeeping Trends and Challenges: The Impact on Training

The changes in conflict over the past 30 years have dictated the trends and challenges of future peacekeeping. To be effective and continue to deliver peace to war-torn zones worldwide, peacekeepers have to adapt to constantly evolving scenarios. They must be prepared to tackle armed groups connected to sophisticated transnational crime networks, human rights abuses targeting vulnerable people, terrorism, and unconventional threats posed by small-scale groups capable of inflicting unthinkable collateral damage to civilians and peacekeepers (Mackay 2003, 218; Day 2020, 4-6). Indeed, the high number of UN peacekeeper fatalities in the past 11 years after asymmetric warfare attacks has emphasized the need to train peacekeepers for conflict scenarios that are increasingly different from past conventional warfare missions where opposite parties were well-defined and easily identifiable (UN Peacekeeping 2018a; Tuvdendarjaa 2022, 4). In such a context, it is essential to redesign peacekeeping training by adding innovative approaches and new technologies to protect civilians who are often victims of terrorist groups, criminal gangs, and extremism (Challenges Annual Forum 2022, 12; Tuvdendarjaa 2022, 4). More importantly, to enable peacekeepers to protect civilians better, which is one of the foundations of the UN's credibility and thus should continue to be a central theme of their peacekeeping operations' mandates in the years to come, training must teach soldiers how to develop people-centred strategies that rely on the community's knowledge (Autesserre 2019, 103; Challenges Annual Forum 2022, 11-12). In other words, peacekeepers must train to "place people at the center" of their initiatives (Challenges Annual Forum 2022, 11-12).

Safety and stability are not enough to stop conflict. The success of peacekeeping entails the sustainable transition to peacebuilding, which implies that development activities will probably remain a significant mechanism of UN peacekeeping (Glendhill 2021, 214). Consequently, peacekeepers will be required even more to act as "early peacebuilders" by engaging in small-scale, fast-implemented development activities known as quick impact projects (Curran 2017, 42-43). In this regard, training must enable peacekeepers to plan and implement small-scale projects in collaboration with local and international agencies to benefit the community. Civilian actors have had their roles significantly expanded since the advent of multidimensional peacekeeping operations, and their presence will increasingly grow in importance in UN peacekeeping in the long term, requiring peacekeepers to communicate effectively and cooperate with diverse actors (55, 62). As a result, training for peacekeeping operations must emphasize the development of varying levels of communication skills.

The proliferation of armed groups in complex settings where peacekeeping operations are established, besides contributing to weakening state authority, also increases the risks for vulnerable populations and undermines the path to sustainable peace. The resulting threat to civilians and peacekeepers and the predictable protraction of conflict has required the UN to begin the trend of robust peacekeeping mandates. Endorsing this trend, the "Force Intervention Brigade," an offensive military force that was added to the UN Organization Stabilization Mission in the Democratic Republic of the Congo to undertake attack operations to neutralize and disarm groups considered a threat to the state and civilians, was established in 2013 (Curran 2017, 138; Day 2020, 2-3). Peacekeeping training must enable troops to use a more extensive and robust range of military tactics during field operations. Another growing trend in contemporary peacekeeping is the weaponization by armed groups of modern technologies that are often accessible and controlled from handheld devices. This allows these groups to activate hidden explosive artifacts, disrupt digital systems, recruit extremists, spread mis- and disinformation, and stimulate societal division and polarization, making it highly difficult for peacekeepers to address these challenges (de Coning 2021, 9; Day 2020, 4; Challenges Forum Annual Report 2022, 11). Peacekeepers need to train to use intelligence and adequate technological tools to be more effective in deterring, identifying, and neutralizing threats that stem from electronic and digital devices and to be capable of matching the physical and cyber domains during operations (de Coning 2021, 9; Challenges Forum Annual Report 2022, 11).

Climate change, which is characterized by long-term shifts in weather patterns and their harmful environmental impacts, is expected to increasingly affect human life. Although it is not a direct ignitor of conflict, climate change side effects like floods and the destruction of arable land can give rise to large-scale displacement, exacerbation of the competition for natural resources, and stimulate people to join armed groups, thereby aggravating conditions favourable to conflict (Day 2020, 4). Hence, peacekeepers must accurately adapt their planning and analysis processes and operational and logistics standards to risks and threats related to climate change effects. Moreover, they will have to be extremely careful in reducing their environmental footprint to demonstrate to the host nation the peacekeepers' commitment to practices that minimize the human impact on the environment (de Coning 2021, 9). Peacekeepers must adopt practices to reduce gas emissions and adequately manage water and energy use and waste disposal from the training and preparation stages. In addition, they have to conduct simulated drills of environmental disasters.

Lastly, women's participation in peacekeeping operations is a vital goal that the UN has been pursuing by urging TCCs to increase female personnel in their contingents (Bridges, and Horsfall 2009, 1). The rationale for expanding the rate of women in peacekeeping, a trend for the future of such missions, is the increase in the success and effectiveness of these operations when there are more females in the contingents (8). Increasing the number of women in peacekeeping inhibits sexual exploitation-related misconduct on the part of peacekeepers, helping to build trust with the host nation (2). Also, female peacekeepers have better access to people who do not trust or do not feel comfortable with men for cultural or religious reasons, amplifying access to community members and consequently helping to minimize risks and get better intelligence in support of operations (Wilén 2018, 7). For that reason, training must extensively use female peacekeepers to enable them to develop intelligence, negotiation, and psychological skills and to practice case exercises that require females for specific tasks.

Despite criticisms for past failures, UN peacekeeping is the best resource available to bring stability to war-torn zones worldwide and protect vulnerable groups against abuses. The better peacekeepers are trained, the more effective their performance and the peacekeeping operation will be, minimizing risks to the safety of the populations and UN troops. The changing contemporary world, however, presents challenges that make more flexible and adaptable peacekeepers increasingly more relevant and needed in the future. Peacekeepers must train to tackle asymmetric combatants and implement early development projects simultaneously. Training that teaches peacekeepers to place locals at the core and provides instruction on how to use new technologies to protect people is more important than ever to increase trust with the host nation and facilitate the successful implementation of mandates. Further, peacekeeping training must teach peacekeepers to comply with rigorous environmental standards and extensively use females to increase the efficacy of the mission.

References

Abdenur, Adriana Erthal, Giovanna Kuele, and Ariane Francisco. 2018. "Enhancing Peacekeeping Training Through Cooperation: Lessons from Latin America." Igarape Institute. http://www.jstor.org/stable/resrep20626

ALCOPAZ. 2023. "Oferta Acadêmica." Accessed May 31, 2023. https://www.alcopaz.org/oferta-academica/

Autesserre, Séverine. 2019. "The Crisis of Peacekeeping: Why the UN Can't End Wars." *Foreign Affairs (New York, N.Y.)* 98, no. 1: 101–118.

Banerjee, Dipankar. 2005. "Current Trends in UN Peacekeeping: a Perspective from Asia." *International Peacekeeping (London, England)* 12, no. 1: 18–33.

Bernath, Clifford, and Sayre Nyce. 2003. "A peacekeeping success: Lessons learned from UNAMSIL." In *International peacekeeping: The Yearbook of International Peace Operations*, pp. 119-142. Brill Nijhoff.

BIPSOT. 2023. "Course Schedule 2023." Accessed June 1, 2023. https://www.bipsot.net/index.php?option=com_content&view=category&layout=blog&id=59&Itemid=251%29%2C+and+Kofi+Annan+international+Peacekeeping+Training+Centre+%28https%3A%2F%2Fwww.kaiptc.org%2Ftraining%2Fall-training-courses%2F.

Bridges, Donna, and Debbie Horsfall. 2009. "Increasing Operational Effectiveness in UN Peacekeeping: Toward a Gender-Balanced Force." *Armed Forces & Society* 36, no. 1: 120-130.

Briscoe, Ivan, Margriet Drent, Kees Homan, Frans-Paul van der Putten, and Dick Zandee. 2015. "Trends in Peacekeeping Operations." Edited by Jaïr van der Lijn, Luc van de Goor, Jan Rood, and Rosa Dinnissen. *Peacekeeping Operations in a Changing World*. Clingendael Institute. http://www.jstor.org/stable/resrep05385.5

Cin, Carlos, and Alaciel Campos Dugone. 2001. "Heart of Peacekeeping Training: The Challenge of Avoiding Darkness in a Soldier's Mind." *Peacekeeping & International Relations* 30, no. 1-3: 2–.

Challenges Annual Forum. 2022. *Annual Forum Report 2022: The Future of Peace Operations* (New Delhi: Challenges Forum, 2022). https://challengesforum.org/wp-content/uploads/2022/12/CAF22_report.pdf.

Curran, David. 2017. *More Than Fighting for Peace? Conflict Resolution, UN Peacekeeping, and the Role of Training Military Personnel*. Vol. 8. Cham: Springer International Publishing. https://doi.org/10.1007/978-3-319-46305-6

Cutillo, Alberto. 2013. "Deploying the Best: Enhancing Training for United Nations Peacekeepers." International Peace Institute. http://www.jstor.org/stable/resrep09549

Day, Adam. 2020. "The Future of UN Peace Operations in a Changing Conflict Environment." *UN DPO Future of Peace Operations Project*.

de Coning, Cedric. 2021. "The future of UN Peace Operations: Principled Adaptation through Phases of Contraction, Moderation, and Renewal." *Contemporary Security Policy*, DOI: 10.1080/13523260.2021.1894021.

De Sylva, Lesley. 2017. "The Reality of New Forms of UN Peacekeeping: High Risk Missions and Fatalities." Master's Thesis, New York University.

Fortna, Virginia Page. 2004. "Does Peacekeeping Keep Peace? International Intervention and the Duration of Peace After Civil War." *International Studies Quarterly* 48, no. 2: 269–292.

Gledhill, John, Richard Caplan, and Maline Meiske. 2021. "Developing Peace: the Evolution of Development Goals and Activities in United Nations Peacekeeping." *Oxford Development Studies* 49, no. 3: 201-229. DOI: 10.1080/13600818.2021.1924126

Goulding, Marrack. 1993. "The Evolution of United Nations Peacekeeping." *International Affairs (London)* 69, no. 3: 451–464.

Howard, L. 2019. "Assessing the Effectiveness of the UN Mission in the Central African Republic." *IPI Global Observatory*. Accessed June 3, 2023. https://theglobalobservatory.org/2019/11/assessingeffectiveness-un-mission-central-african-republic/#_ftnref3

Jett, Dennis C. 2019. *Why Peacekeeping Fails 20th Anniversary Edition*. 2nd ed. Cham: Springer International Publishing. https://doi.org/10.1057/9780312292744_3

KAIPTC. 2023. "All Training Courses." Accessed June 1, 2023. https://www.kaiptc.org/training/all-training-courses/.

Kondoch, Boris, and Brendan Howe. 2017. "Introduction to the Developments, Trends and Challenges to UN Peacekeeping Operations." In *Peacekeeping and the Asia-Pacific*, 1–12. The Netherlands: BRILL.

Mackay, Angela. 2003. "Training the Uniforms: Gender and Peacekeeping Operations." *Development in Practice* 13, no. 2-3: 217–223.

NORDEFCO. 2023. "Course Catalogue." Accessed May 31, 2023. https://www.nordefco.org/Courses2.

Nsia-Pepra, Kofi. 2019. "Robust Peacekeeping: The Most Appropriate Operational Paradigm to Address Contemporary UN Peacekeeping and Civilian Protection Challenges." In *Routledge Companion to Peace and Conflict Studies*, pp. 470-480. Routledge.

Smith, Adam C., and Arthur Boutellis. 2013. *Rethinking force generation: Filling the capability gaps in UN peacekeeping*. New York: International Peace Institute.

Tuvdendarjaa, Muggi. 2022. "Challenges of the United Nations Peacekeeping Operations." *Security Nexus* 23.

United Nations. 1995. "General Assembly Resolution 49/37 (A/RES/ 49/37)." Accessed May 17, 2023. https://digitallibrary.un.org/record/172280/files/A_RES_49_37-EN.pdf?ln=en

—. 2015. "Report of the High-Level Independent Panel on Peace Operations." Accessed May 16, 2023. https://digitallibrary.un.org/record/795940/files/A_70_95--S_2015_446-EN.pdf?ln=en

—. 2018. "Security Council Resolution 2436 (S/RES/2436)." Accessed May 17, 2023. https://digitallibrary.un.org/record/1643360/files/S_RES_2436_%282018%29-EN.pdf?ln=en

—. "Preventing Sexual Exploitation and Abuse." Accessed June 3, 2023. https://www.un.org/preventing-sexual-exploitation-and-abuse/content/training

United Nations Peacekeeping. 2000. "Report of the Panel on United Nations Peacekeeping Operations." Accessed May 16, 2023. https://peacekeeping.un.org/sites/default/files/a_55_305_e_brahimi_report.pdf

—. 2008. "United Nations Peacekeeping Operations Principles and Guidelines." Accessed May 16, 2023. https://peacekeeping.un.org/sites/default/files/capstone_eng_0.pdf

—. 2010. "Training for all United Nations Peacekeeping Personnel." Accessed May 16, 2023. https://resourcehub01.blob.core.windows.net/training-files/Training%20Materials/002%20Policies/002-001%202010.20%20DPKO_DFS_POL_TrainingforallUNPKPersonnel_1May2010.pdf

—. 2018a. "Improving Security of United Nations Peacekeepers (Independent Report)." Accessed May 17, 2023. https://peacekeeping.un.org/sites/default/files/improving_security_of_united_nations_peacekeepers_report.pdf

—. 2018b. "Operational Readiness Preparation for Troop Contributing Countries." Accessed May 19, 2023. https://peacekeeping.un.org/sites/default/files/3._military_-_4_guidelines_on_opr.pdf
—. 2023. "Troop ad Police Contributors." Accessed May 21, 2023, https://peacekeeping.un.org/en/troop-and-police-contributors
—. "Department of Peace Operations." Accessed May 15, 2023, https://peacekeeping.un.org/en/department-of-peace-operations.
—. "Military." Accessed May 17, 2023, https://peacekeeping.un.org/en/military
—. "Policy." Accessed May 16, 2023, https://peacekeepingresourcehub.un.org/en/policy
Wedgwood, Ruth. 1995. "The Evolution of United Nations Peacekeeping." *Cornell International Law Journal* 28, no. 3: 631–.
Wilén, Nina. 2018. "Improving Peacekeeping Performance — Dilemmas and Goals." Egmont Institute. http://www.jstor.org/stable/resrep21373

CHAPTER XIII

STRENGTHENING DISASTER RESILIENCE IN UN PEACEKEEPING MISSIONS: THE ROLE OF BANGLADESH CONTINGENTS

MOHAMMED SADMAN SAKIB,
AFRA SAYARA RAHMAN,
AND KAWSER AHMED, PHD

The United NaUN has tions (UN) has been at the forefront of promoting international peace and security since its inception in 1945. UN peacekeeping has become a powerful tool for assisting nations or groups in conflict to settle disputes through dialogue and mediation. Over the past 70 years, the UN has sent nearly a million people from 125 nations to 71 peacekeeping missions, some in extremely difficult conditions (United Nations 2023). Bangladesh has emerged as a leading troop-contributing country in UN peacekeeping missions, and its contributions have been widely acknowledged. It has made significant contributions to the promotion of peace and security through its active participation in various missions. This book chapter aims to examine the role of a selected Bangladesh contingent in UN peacekeeping missions where it undertook disaster resilience-related activities within their mission mandate (whether explicitly mentioned or not). These are the United Nations Organization Stabilization Mission in the Democratic Republic of the Congo (MONUSCO), the United Nations Mission in South Sudan (UNMISS), and the United Nations Multidimensional Integrated Stabilization Mission in Mali (MINUSMA). Within the UN system, the Department of Operational Support (DOS) operates under the secretary-general (United Nations n.d.) and provides operational support policies and guidelines about communication and infrastructure for missions according to mandates (United Nations n.d.). All participating missions adhere to guidelines that might include improving communication and infrastructure, leading to better operational support in mission areas. According to the UN Peacekeeping Operations Principles and Guidelines, one of the core business components is post-conflict tasks, which include infrastructure development and assistance. According to the "Guidelines for Operational Readiness Preparation for Troop Contributing Countries in Peacekeeping Missions," engineers to perform logistics tasks must be provisioned as a minimum requirement for the contingents to participate in missions (Lacroix and Khare 2020, United Nations 2008). Thus, Bangladesh contingents in UN peacekeeping operations (UNPKOs) perform the tasks of infrastructure development and maintenance in mission areas as part of their peacekeeping role. For example, in 2020, Bangladesh contingents in UNMISS participated in the rehabilitation of 3200 km of road as a part of the mission mandate (United Nations 2020). This chapter aims to provide an understanding of one of the roles of UN peacekeeping missions in promoting sustainable disaster resilience in host countries. This will be achieved by examining the unique characteristics of UN peacekeeping through selected case studies from Bangladesh's peacekeeping missions. Through this analysis, readers will gain a deeper understanding of the importance of UN peacekeeping in supporting local communities by improving disaster resilience.

UN Peacekeeping Roles and Responsibilities

UN peacekeeping helps countries transition to peace after conflict. Legitimacy, burden sharing, and the requirement to deploy soldiers, police, and civilian experts from around the world to implement UN Security Council and General Assembly tasks are among the matters found in a mission mandate. UN peacekeepers support the political process, protect civilians, and support human rights. However, a question often asked is: "Are UNPKOs cost-effective in reducing conflict conditions?" To answer this question, it is not only important to compare the cost of dealing with the consequences of conflict (such as human displacement and infrastructure damage in the conflict zone), but also to compare the UN's spending on peacekeeping with global military spending (Guterres 2017). Over the past 72 years, over one million people from 125 nations have served in 71 peacekeeping operations. Currently, 12 peacekeeping operations employ 90,000 people (United Nations 2021). It can be argued that, in war zones, peacekeepers' presence reduces overt inter-group violence and increases the chance of lasting peace (United Nations 2021). UN peacekeepers are responsible for various activities, including the protection of people, prevention of wars, promotion of human rights, and provision of on-the-ground assistance (United Nations n.d.). Among other UN mission activities, "Providing field assistance" is the activity this chapter will mainly concentrate on (United Nations, n.d.). The effectiveness of peace operations depends on operational assistance. The DOS contributes to the success of peace operations by delivering timely, effective, efficient, and accountable support solutions. It collaborates with UN and non-UN partners to organize, deploy, and support 35 peacekeeping, special political, and other field operations in the most difficult circumstances on earth. Today, the DOS assists in the administration of a combined annual budget of more than $7 billion, 131,000 authorized military troops, and 24,000 approved civilian employees at over 270 duty stations in more than 30 countries. It offers services ranging from budget, finance, human resources, and technology to supply chain and facility management (United Nations, n.d.).

Role of Bangladesh Contingents in UN peacekeeping

Bangladesh's participation in UNPKOs is driven by several factors. Firstly, the country possesses a well-trained and qualified pool of military and police personnel who can carry out peacekeeping duties effectively (Akhter 2022). Secondly, Bangladesh's deep-rooted commitment to global peace as a non-aligned member emanating from its foreign policy (friendship to all and malice to none) supports its involvement in various peacekeeping missions (United Nations HQ 2018). The nation's engagement in peacekeeping efforts has bolstered its reputation as a trustworthy and accountable member of the international community, earning commendations from the United Nations and other organizations (United Nations HQ 2018). Bangladesh's contributions extend beyond personnel, as it also provides financial support and resources to enhance the capabilities of peacekeeping forces (Mazumder 2022). Bangladesh's involvement in UNPKOs fosters both regional peace and security and the development of prosperous and peaceful societies worldwide (Armed Forced Division 2019).

Disaster Resilience in Peacekeeping. According to the United Nations Office for Disaster Risk Reduction (UNDRR), a disaster is a severe disruption that surpasses a community's or society's capacity to handle, resulting in widespread losses and effects on people, property, money, or the environment (UNDRR 2015). Disaster resilience involves the ability of individuals, communities, and systems to anticipate, withstand, and recover from both natural and human-caused disasters (Combaz 2015). The primary goals of disaster resilience are to decrease a community's vulnerability to disasters and enhance its ability to cope with and recover from their impacts. Disaster resilience encompasses activities such as disaster response and recovery, risk reduction, and emergency preparedness (UNDRR 2022). Disasters can severely damage or destroy roads, making them impassable or challenging to access and hindering rescue and aid efforts (Pavement Interactive, n.d.). These roads provide safe access to essential services and should be able to withstand natural disasters like heavy rains and floods (ibid).

Logistics plays a vital role in swiftly and effectively delivering relief to affected communities during the response phase of disaster management. It minimizes waste and inefficiencies, enabling aid workers to focus on assisting those in need (Maghfiroh and Hanaoka 2020). Logistics also plays a vital role in aiding the rebuilding and restoration process by minimizing waste, maximizing efficiency, and ensuring the timely delivery of goods and equipment to the right places, enabling communities to recover and rebuild sustainably (Amna 2013).

One of the UN's peacekeeping roles is to support a host nation in carrying out its reconstruction needs after a disaster (United Nations 2022) and assist with emergency relief, disaster response, and recovery efforts (United Nations 2008). Hence, the condition of roads in disaster-affected areas plays a crucial role in the ability of communities to recover and rebuild. If roads are poorly maintained or damaged, it can hinder the access of aid workers and emergency responders, causing delays in delivering essential relief; bad roads also increase risks for survivors (United Nations 2011). Thus, building roads and maintaining existing ones can enhance a community's preparedness and resilience for future disasters. These roads provide safe and reliable access to crucial markets and services, reducing the community's vulnerability to natural catastrophes (United Nations 2019). Additionally, well-built roads can facilitate the evacuation of towns during disasters, saving lives and minimizing the disasters' overall impact.

When the UN peacekeeping force constructs roads, they also offer technical support and training to the host nation and local authorities for road maintenance and repair (United Nations 2022). In terms of sustainable recovery, preventing the spread of diseases, and mitigating the negative impacts of catastrophes, the UN peacekeeping force plays a significant role in the mission area by ensuring the safety and security of the affected population during the rehabilitation phase. This approach aligns with the knowledge building perspective, which emphasizes the importance of capacity development and knowledge transfer to empower local actors in post-conflict and disaster-affected areas. By providing technical support and training, the UN peacekeeping force enables the local authorities to acquire the necessary skills and knowledge to maintain and repair the roads themselves. Nonetheless, the UN peacekeeping force should function as a facilitator and enabler rather than a dominant authority in dictating the terms of disaster management. By aiding and facilitating the rebuilding and development processes, a peacekeeping force allows a host nation to take the lead. This approach acknowledges the importance of local knowledge and expertise, empowering local authorities to drive their own development agenda while leveraging the support and resources offered by the UN peacekeeping force. Although UN field missions' primary role does not include supporting a host nation's disaster management efforts, some missions do provide technical and financial assistance and coordinate with various international agencies that operate in the disaster area.

To enhance operational support, peacekeeping contingents are required to actively participate in initiatives aimed at improving communication and infrastructure. As per the United Nations Peacekeeping Operations Principles and Guidelines, one of the key components of UN peacekeeping missions is the completion of post-conflict tasks, including infrastructure development and assistance (United Nations 2008). This study shows that the Bangladesh contingents in UNPKOs are often actively engaged in infrastructure development as part of their peacekeeping activities, following the DOS's guidelines. However, UN peacekeeping missions have limited skills and resources, and their primary goal is to help the host nation and its people transition out of conflict (Carnahan, Durch, and Gilmore 2006). As a result, it is expected that the host nations will take the lead in the rebuilding and development process after a natural disaster with the assistance of UN peacekeeping troops in the area.

The Department of Operational Support (DOS) and Disaster Resilience (DR). The DOS is an indispensable component of the UN Secretariat, which provides advisory, operational, and transactional support services to UN Secretariat entities across the world (United Nations n.d.). The DOS endeavors to optimize its operational efficiency and facilitate the attainment of its objectives with greater efficacy and success (United Nations n.d.). Moreover, in cases where it is deemed requisite, the DOS executes actions on behalf of its clientele with delegated authorization, furnishing counsel and support to augment their proficiency. The DOS plays a crucial role in facilitating communication and coordination between the UN Secretariat and a diverse range of affiliated organizations. By providing essential resources and support, the DOS enables these entities to effectively pursue their respective mandates and achieve their organizational goals. The provision of support services is a crucial aspect of organizational management, particularly in the context of operational support. Such support services may encompass a broad range of administrative functions, including but not limited to logistics and transportation (United Nations, n.d.). Additionally, specialized services such as security and protection may also be included within this purview (United Nations, n.d.).

Communication between and among UN agencies and host nations is important. The DOS provides communication and logistics support, which makes the task of improving the infrastructure in the mission areas a part of a peacekeeping mission. The DOS provides support for UN peacekeeping mission activities, and activities such as communication and infrastructure construction help the community achieve disaster resilience. For example, a road built to access remote areas is not used only by the peacekeepers; the community and other organizations use it as well. If a disaster strikes, these roads can be used for better response by responders. For example, if the communication in the mission area is disrupted, then the peacekeeping operations could face difficulties in achieving their mission objectives. In addition, airports and related infrastructure are also maintained as part of UN peacekeeping operational support (United Nations n.d.). Suffice it to say, operational support is not limited to the mission troops; rather, partner organizations and non-partner organizations also get operational support (United Nations n.d.).

As an example, in 2020, engineers from UNMISS rehabilitated 3200 kilometers of roads to prevent conflict and promote peace as a mandate of the peacekeeping missions (United Nations 2020). Floods heavily affect roads and reduce mobility, which is a barrier towards the development of South Sudan, and UNMISS activity has promoted peace and prosperity in the region (United Nations 2020). In December 2021, the Unity region in South Sudan was devastated by floods, which left 70% of the state inundated with water. UNMISS, in collaboration with humanitarian organizations and local authorities, promptly initiated a response, whereby peacekeepers from the engineering corps from Pakistan undertook the construction of 70 kilometers of dikes to safeguard the town, camps for displaced families, the airport, and roads that served as crucial conduits for humanitarian aid and commerce (Kelly 2022).

Methodology

The methodology for this book chapter involved gathering data from the primary source of the AFD of Bangladesh's reports. These reports served as the main repository of information for the study and provided valuable insights into the UN peacekeeping missions conducted by the Bangladeshi armed forces in MONUSCO, UNMISS, and MINUSMA.

The aim of the study is to find links between UN peacekeeping missions and disaster resilience. Within the qualitative approach, the research approach for this chapter was a case study using content analysis and thematic coding to extract meaningful information from the reports. Additionally, relevant secondary sources such as journal articles, reports, and official documents were consulted to supplement the primary data and enrich the analysis. By employing this methodology, the chapter aimed to provide an in-depth examination of the Bangladesh Armed Forces' experiences in these missions, shedding light on their peacekeeping efforts, which helped to build disaster resilience. The timeframe for the study is 2020–2022.

There are limitations to this study, as the authors relied on reports of AFD Bangladesh and no other sources pertaining to the missions that are considered cases in this chapter are available in the public domain. However, this chapter explores a new area of achieving disaster resilience through UNPKOs. Although many studies have focused on different aspects of peacekeeping activities and contributions, none have explored the relation between peacekeeping activities and disaster resilience. By filling this gap in the literature, the chapter has the potential to not only advance theoretical understandings but also offer practical implications for UNPKOs. In addition, a potential impact of this chapter lies in its ability to provide a fresh perspective on disaster resilience through UNPKOs and expand the boundaries of knowledge within the field of disaster management.

The Case Studies

Over the past three decades, Bangladesh's partnership with the United Nations in peacekeeping has increased in scope and depth (Permanent Mission of the People's Republic of Bangladesh to the United Nations 2021); Bangladesh is one of the largest troop contributors as well (Armed Forces Division 2019). MONUSCO, UNMISS, and MINUSMA are some ongoing peacekeeping missions in which peacekeepers from Bangladesh are involved, and these would be the cases to present for the study.

Bangladesh has many deltaic regions and often faces natural disasters in terms of floods and cyclones. It is one of the most risk-prone countries in the world in the area of disaster and climate change risk, and it ranked seventh among risk-prone countries in 2023 (UNDP Bangladesh 2023). The Bangladeshi military has a long history of participating in disaster management activities at home. In 2007, during Cyclone SIDR, Bangladesh's armed forces conducted rescue and relief efforts in the affected areas (Rowe

2007). Bangladesh's armed forces have been a part of disaster management activities, which has contributed towards building skills, knowledge, and expertise in disaster management. These experience, skill, and knowledge have helped Bangladesh's peacekeeping contingents contribute to disaster resilience in UNPKOs.

United Nations Organization Stabilization Mission in the Democratic Republic of the Congo (MONUSCO)

The Democratic Republic of the Congo (DRC) has been a significant site for Bangladesh's peacekeeping endeavors, particularly in the context of MONUSCO. The Bangladesh Armed Forces have played an important role in promoting humanitarian activities by actively engaging in a diverse range of activities. These endeavors include but are not limited to the initiation and execution of undertakings such as the construction of thoroughfares and airfields, the establishment of facilities for the supply of water, and the mending and creation of bridges and culverts. In 2022, the Bangladesh Engineering Contingent, also known as BANENGR-13, undertook a significant initiative to rehabilitate 400 internally displaced persons who were previously residing in Munigi. This rehabilitation effort was carried out with the ultimate goal of relocating these individuals to Bushagara, which is located in the North Kivu region. In the aforementioned year, the battalion undertook the installation of washing points in strategically significant locations of MONUSCO, specifically, the Disarmament, Demobilization, and Reintegration (DDR), Munigi, and Banatu compounds. In addition, the authors have successfully implemented wastewater treatment facilities for UN personnel stationed in Himbi, Sake, Kiwanja, and Goma Airport. The year 2021 witnessed a significant contribution from the Bangladesh military towards the development of Goma Airport. Furthermore, in 2022, the military also played a pivotal role in the reconstruction of the roads leading to the airport. In the year 2020, the concerned authorities undertook the construction of the Kihire Bridge in Sake. Furthermore, the utilization of sea containers for the construction of bridges was carried out in the regions of Rubare and Burai, both of which are situated in the North Kivu province.

United Nations Mission in South Sudan (UNMISS)

The Bangladeshi peacekeeping contingent deployed in UNMISS is currently involved in peacekeeping operations and is making significant contributions towards reconstruction and capacity-building endeavors in the region. Throughout the year 2021, the organization in question engaged in a series of multifaceted endeavors, including but not limited to the provision of emergency water supplies to water bladders, the construction of eastern gate ablution facilities, and the establishment of a solar farm. Furthermore, the individuals in question were actively engaged in the process of rehabilitating the UBA-MVOLO thoroughfare, erecting the perimeter road encircling the IDP-3 encampment and the UN edifice, and fabricating culverts along the UN House Data Center pathway. In the year 2022, the battalion's endeavors were notably intensified, as evidenced by the execution of reparative measures on Spot Road at the IDP3 patrol road, Rokon-Mubdri-Mvolo road, and Mundri-Maridi road. The individuals in question undertook the task of preserving and restoring the external boundary thoroughfare of the UN edifice while simultaneously executing the installation of conduit culverts within the premises.

United Nations Multidimensional Integrated Stabilization Mission in Mali (MINUSMA)

The current study focuses on the duties of Bangladeshi peacekeepers in MINUSMA. This research elucidates the multifaceted nature of the peacekeeping operations undertaken by Bangladeshi troops in Mali. In addition to its peacekeeping endeavors, BANENGR-9 is also involved in non-military developmental support activities. For example, in 2020, in addition to the assigned duties of personnel protection, it assisted with the provision of a water supply that had undergone the process of recycling wastewater in the locality of Village J, situated within the confines of the Kidal Super Camp. Additionally, it was tasked with the construction of an access road leading to Echo Tower 6, which is also located within the aforementioned camp. In 2022, the battalion undertook a series of infrastructure development projects in the Aguelhok Camp and Kidal Super Camp regions. These projects included the construction of wastewater treatment plants in Aguelhok Camp, the resurfacing of all roads and camps within Kidal Super Camp, the repair of the perimeter road in Aguelhok's main camp, the establishment of a proper access road from Kidal Super Camp to the airstrip, and the repair of damaged sections of the airstrip fence at Kidal Super Camp. These initiatives were aimed at enhancing the overall operational efficiency and effectiveness of the camps while also improving the living conditions of the personnel stationed therein.

Findings

Host communities in a conflict zone where a peacekeeping mission is deployed are in a "double jeopardy" situation. On the one hand, they are subjected to hostile armed groups' routine violence, and on the other they are also vulnerable to natural disasters due to the poor support system put in place by the government. Thus, building infrastructure is a task that increases a country's resilience to natural disasters in places like Congo, South Sudan, and Mali. These areas are frequently susceptible to a wide variety of natural hazards, such as floods, droughts, and tropical storms, all of which have the potential to do severe damage to the infrastructure already present and disrupt the provision of important services. Therefore, it is essential to create new infrastructure and maintain existing infrastructure to reduce the effects of disasters. Table 13-1 below shows MONUSCO activities by Bangladesh military contingent.

Table 13-1: MONUSCO activities by Bangladesh military contingent

MONUSCO				
Disaster Resilience Activities	Events conducted by BD contingents	BANENGR -13		
		2020	2021	2022
No	IDP evacuation/rehabilitation operations	N/A	N/A	Rehabilitation of 400 x IDPs from MUNIGI to BUSHAGARA at Nord KIVU
Yes	Water supply facilities erection/establishment program/project	N/A	• Installation of washing points to all MONUSCO axes including DDR, MUNIGI and BANATU compounds • Establishment of a WWTP for UN forces at HIMBI.	Establishment of a waste water treatment plant (WWTP) for UN forces at SAKE, KIWANJA and GOMA Airport

Yes	Road construction/repair project	KABITI – PINGA road rehabilitation at PINGA.	• Rehabilitation of the road from SAKE to NAYABINDO • Rehabilitation of the road in front of the mayor's office in BENI Town. • RUGARI to BUKIMA road rehabilitation.	• Road rehabilitation of KANYARUCHINYA-BUSHAGAR road (4.2 Km) • Road rehabilitation at KIWANJA for a length of 2.5 km
Yes	Airport construction/repair project	N/A	Development work of CCO at GOMA AIRPORT.	Repair of a damaged access road at GOMA Airport
Yes	Bridge/culvert construction/repair project	Construction of KIHIRE bridge at SAKE.	N/A	Construction of an improvised bridge using sea containers at RUBARE and BURAI under Nord KIVU

Source: Authors' own compilation of data from AFD mission reports

Bangladeshi peacekeepers have put effort into building infrastructure, which contributes to resilience. The tables below show the findings from the MONUSCO, UNMISS, and MINUSMA activities that are contributing to resilience. The activities, whether the UN authority to the Bangladesh contingents mandated them, were not mentioned in the report. If these activities are conducted for the operations to operate effectively, they have served the dual purpose of disaster resilience and peacekeeping.

Table 13-2 below shows MINUSMA activities by Bangladesh military contingent

MINUSMA				
Disaster Resilience Activities	Events conducted by BD contingents	BANENGR -9		
		2020	2021	2022
No	Internally displaced population evacuation/rehabilitation operations	N/A	N/A	N/A
Yes	Water supply facilities erection/establishment program/project	Provide recycled water supply to a helipad and Village J of Kidal Super Camp	N/A	Construction of WWTPs at Aguelhok Camp
Yes	Road construction/repair project	Construction access road to the Echo Tower-6 of Kidal Super Camp	Improve internal roads using local laterite/send material in Aguelhok Camp	Resurface all roads and camps of Kidal Super Camp

Yes	Airport construction/repair project	N/A	N/A	1. Position 12 sea containers along perimeter of the 16-acre extension at Kidal Airstrip 2. Repair the damaged part of the airstrip fence at Kidal Super Camp 3. Construction of nine new watch towers and repaired two old watchtowers along the perimeter of the new runway of the airstrip 4. Clearing of the new airfield 150 meters from the runway center line on both sides as well as the apron and gradually sloping towards the drainage system
Yes	Bridge/culvert construction/repair project	N/A.	N/A	N/A

Source: Authors' own compilation of data from AFD mission reports

Table 13-3 below shows UNMISS activities by Bangladesh military contingent

Table 13-3: UNMISS activities by Bangladesh military contingent

UNMISS				
Disaster Resilience Activities	Events conducted by BD contingents	BANENGR -9		
		2020	2021	2022
No	Internally displaced population evacuation/rehabilitation operations	N/A	N/A	N/A
Yes	Water supply facilities erection/establishment program/project	N/A	Water supply to emergency water bladders, eastern gate ablution and solar farm	Water supply to emergency water bladders, eastern gate ablution and solar farm.
Yes	Road construction/repair project	N/A	• MSR Maintenance JUBA-MVOLO Road. • Perimeter road repair around IDP-3 Camp. • UN house water point road repair	• Spot road repairs at IDP3 patrol road. • Maintenance of MSR (Rokon-Mubdri- Mvolo). • Maintenance of the road from UNH to NSS • Maintenance of MSR (Mundri-Maridi).
Yes	Airport construction/repair project	N/A	N/A	N/A
Yes	Bridge/culvert construction/repair project	N/A.	Construction of a culvert at UN House Data Centre Road	Placed a pipe culvert in the outer perimeter road of UNH

Source: Armed Forces Division Bangladesh reports on UN peacekeeping 2023

As shown in the three tables above, the activities carried out by Bangladesh contributed towards disaster resilience. Not all the activities, such as internally displaced population evacuation and rehabilitation operations, are disaster resilience activities, as they are humanitarian responses. But the activities of water supply facility establishment, road construction, bridge and culvert construction and repair projects, and airport construction and repair projects contributed to disaster resilience. During a disaster, it is important to respond and act as fast as possible. If the response is slow and the activities are hampered due to a lack of proper communication, then the losses, both on a financial and human scale, will be high. Thus, the maintenance and construction of infrastructure not only allow peacekeeping operations to function but also increase the disaster resilience of the areas where these projects are taking place.

Efficient disaster response in peacekeeping operations can be done with proper planning and resource mobilization. The need for effective disaster response and recovery activities is one of the primary reasons why the development of physical infrastructure is so important in a conflict zone. The accessibility and connectivity offered by the road networks that have been constructed are necessary for the speedy deployment of the peacekeepers in conflict-affected zones. The infrastructure also serves the purposes of emergency services, the conveyance of relief supplies, and the evacuation of populations impacted by a disaster. In a similar manner, the availability of operational water infrastructure ensures access to safe drinking water, sanitation, and hygiene, all of which are crucial for maintaining public health and preventing the emergence of waterborne diseases in the aftermath of natural disasters.

In the case of Bangladesh contingents, the activities they carry out provide disaster resilience and ensure effective peacekeeping efforts. The roads, bridges, and airports are built for peacekeeping operations, but they have helped build better communication, which is important during disaster response for the responders to carry out their activities. If the resources are allocated properly, then the efficiency of both disaster response and peacekeeping activities can increase, leading to an effective effort.

Peacekeeping operations can promote economic and social stability by incorporating disaster resilience strategies within a mission objective, thereby facilitating the mission's mandates. Robust infrastructure contributes to the general economic and social stability of a nation, which is a positive outcome. Building strategically planned road networks paves the way for increased accessibility to markets, streamlines commercial activity, and encourages overall economic expansion. This, in turn, improves the country's capacity to recover and rebuild after natural disasters, lowering its reliance on aid from other countries. In addition, dependable water facilities support agricultural activity, which in turn ensures food security and allows people to make a living. This is true even in the face of climate uncertainty. Thus, the activities carried out by the Bangladesh contingents in building roads, bridges, water supply facilities, and airports are contributing to promoting economic and social stability by making communications easier, which contributes to the economy. Furthermore, the allocation of resources will increase through increased economic activity, resulting in communities that are more resilient. Thus, the incorporation of disaster resilience strategies within mission mandates can promote economic growth and social stability.

Resilience to future disasters can be achieved through the development and maintenance of physical infrastructure in the host communities. Infrastructure provides resilience towards disasters, either directly (e.g., embankments) or indirectly (e.g., road communications). The development of host communities' physical infrastructure plays a significant part in bolstering their level of readiness and resilience in the face of calamities. For instance, reducing the potential for floods by building roads with elevated surfaces and drainage systems that are up to code can help, while establishing robust water supply systems can help ensure that water is available even during extended droughts. The infrastructure built through MONUSCO, UNMISS, and MINUSMA can help build resilience by maintaining proper communication. Airports, roads, and bridges improve response activities, allowing the community to recover sooner. The proper supply of water through water supply facilities will maintain the water supply, making the disaster response activities less stressful and allowing the responders to put efforts into other activities, which happens when the community is resilient towards disasters.

Certain elements need to be taken into consideration while planning to maintain infrastructure to promote disaster resilience in PKOs. It is important to emphasize that in order to make infrastructures more resistant to the effects of natural disasters, in-depth planning, policy frameworks, and institutional capacities should precede their development. It is imperative to have efficient governance and regulatory processes in place to guarantee the quality of infrastructures as well as their continued upkeep and viability. In addition, the engagement and participation of the community in various decision-making processes is necessary to guarantee that the built environment will cater to the particular requirements and priorities of the local people. In this case, the infrastructure is built to support the PKOs' activities. The infrastructure can provide resilience to the community where it is being built, but proper planning can also save money and promote disaster resilience. For instance, if the roads and bridges are built to sustain themselves for a long period, it will promote economic activity, which will help the community allocate resources. This in turn will contribute towards disaster resilience as resources get exhausted during disasters, and if the resources to mitigate the disasters are present in the community, the disaster's impacts can be reduced, making the community more resilient with time.

The Bangladeshi peacekeepers now stationed in Mali, the DRC, and South Sudan are improving water and communication infrastructure as part of their mission to restore stability in those countries. It makes the communities in which they are creating such infrastructures more resilient to the effects of adverse events. According to the debate, they are essential components for developing resilience. Additionally, they are necessary for actions relating to disaster response and recovery. The activities for maintaining peace are contributing to the disaster resilience of the countries.

In conclusion, UNPKOs have been one of the crucial tools being used to promote peace in war-torn regions for the past seven decades (United Nations 2021). The peacekeeping forces often include engineers, logistics experts, and resources for operational support (United Nations 2008). The purpose of building and maintaining infrastructure in peacekeeping missions is to keep the peacekeeping operations functioning, which contributes to disaster resilience. Roads and infrastructure are important for communication, which helps in various disaster phases in order to conduct relief operations and restoration activities (Amna 2013; Raillani et al. 2020). Being the largest troop contributor in UNPKOs since 1988 (Armed Forces Division 2019), Bangladesh has immensely contributed to disaster resilience in the host communities. In this chapter, the case studies of MONUSCO, UNMISS, and MINUSMA were presented, where peacekeepers from Bangladesh have significantly contributed to the development of roads, bridges, airports, and water supplies as peacekeeping mission activities. Infrastructures such as roads, bridges, and airports make response and recovery efforts easy due to the ease of communication and accessibility. The benefits of the infrastructures are not limited to response and recovery efforts but also include providing resilience from future disasters. This chapter is unique in that it explores the link between UNPKOs and disaster resilience. The future prospects of such studies are vast and have the potential to contribute to academia and have practical implications as well.

References

Akhter, Samina. 2022. "How Bangladesh Is Playing Significant Role in UN Peace Keeping Mission." Modern Diplomacy. November 8. https://moderndiplomacy.eu/2022/11/08/how-bangladesh-is-playing-significant-role-in-un-peace-keeping-mission/.

Amna, Shahzadi. 2013. "Logistics Support and Its Management during Disaster Relief Operations."

Armed Forces Division. 2019. "Bangladesh in UN Peace Operation" Armed Forces Division (AFD).2019. Accessed June 15, 2023. https://afd.gov.bd/un-peacekeeping/bangladesh-in-un-peace-operation

Carnahan, Michael, William Durch, and Scott Gilmore. 2006. "Economic Impact of Peacekeeping Final Report." https://www.stimson.org/wp-content/files/file-attachments/EIP_FINAL_Report_March2006doc_1.pdf.

Combaz, Emilie. 2015. "What Is Disaster Resilience? - GSDRC." GSDRC. 2015. https://gsdrc.org/topic-guides/disaster-resilience/concepts/what-is-disaster-resilience/.

Guterres, António. 2017. "Peacekeeping Is Cost Effective, but Must Adapt to New Reality." United Nations Secretary General. May 30, 2017. Accessed June 15, 2023. https://www.un.org/sg/en/content/sg/articles/2017-05-30/peacekeeping-cost-effective-must-adapt-new-reality.

Kelly, Maya. 2022. "Five Ways That UN Peacekeeping Partnerships Drive Peace and Development." Edited by Urjasi Rudra. United Nations Peacekeeping. May 5, 2022. https://peacekeeping.un.org/en/five-ways-un-peacekeeping-partnerships-drive-peace-and-development.

Lacroix, Jean-Pierre, and Atul Khare. 2020. "Operational Readiness Preparation for Troop Contributing Countries in Peacekeeping Missions." United Nations Peacekeeping. December 1, 2020. https://peacekeeping.un.org/sites/default/files/3._military_-_4_guidelines_on_opr.pdf.

Maghfiroh, Meilinda F. N., and Shinya Hanaoka. 2020. "Multi-Modal Relief Distribution Model for Disaster Response Operations." Progress in Disaster Science 6 (April): 100095. doi:10.1016/j.pdisas.2020.100095.

Mazumder, Kamal Uddin. 2022. "Bangladesh's Involvement in UN Peacekeeping". Pakistan Today. October 28, 2022. Accessed June 15, 2023. https://www.pakistantoday.com.pk/2022/10/28/bangladeshs-involvement-in-un-peacekeeping.

Pavement Interactive. n.d. "Natural Disasters and Highway Infrastructure." Pavement Interactive. Accessed June 15, 2023. https://pavementinteractive.org/natural-disasters-and-highway-infrastructure/.

Permanent Mission of the People's Republic of Bangladesh to the United Nations. 2021. "Peacekeeping." Permanent Mission of the People's Republic of Bangladesh to the United Nations. December 1, 2021. Accessed June 15, 2023 https://bdun.org/bangladesh-priorities-at-the-un/peacekeeping/.

Raillani, Hajar, Lamia Hammadi, Márcia M. Altimari Samed, Abdessamad El Ballouti, and Vlad Stefan Barbu. 2020. "HUMANITARIAN LOGISTICS IN THE DISASTER RELIEF SUPPLY CHAIN: STATE OF THE ART." Pp181–93. doi:10.2495/RISK200161.

Rowe, Salim Mia and Helen. 2007. "Hundreds Dead after Cyclone Rips through Bangladesh." The Sydney Morning Herald. November 16, 2007. Accessed June 15, 2023 https://www.smh.com.au/world/hundreds-dead-after-cyclone-rips-through-bangladesh-20071117-1auw.html.

UNDP Bangladesh. 2023. "Climate Vulnerability Index (Draft)." United Nations Development Progarmme. May 22, 2023. Accessed June 15, 2023 https://www.undp.org/bangladesh/publications/climate-vulnerability-index-draft.

UNDRR. 2015. "Disaster." Undrr.org. 2015. https://www.undrr.org/terminology/disaster.

UNDRR. 2022. "Our World at Risk Transforming Governance for a Resilient Future". United Nations office for Disaster Risk Reduction. May 26, 2022. Accessed June 15, 2023 https://www.undrr.org/gar2022-our-world-risk-gar.

United Nations. 2011. "Dag Hammarskjöld and United Nations Peacekeeping by CARL BILDT." 2011. UNMIK. August 31, 2011. https://unmik.unmissions.org/dag-hammarskj%C3%B6ld-and-united-nations-peacekeeping-carl-bildt

United Nations. 2020. Review of "UNMISS Engineers to Rehabilitate 3200 Kms of Roads to Deter Conflict and Build Peace." United Nations Peacekeeping. November 17, 2020. Accessed June 15, 2023. https://peacekeeping.un.org/en/unmiss-engineers-to-rehabilitate-3200-kms-of-roads-to-deter-conflict-and-build-peace

United Nations HQ. 2018. "Bangladesh "Fully Committed" to UN Peacekeeping as Vital Element of Global Peace and Security – UN Force

Commander “. Department of Operational Support. February 7, 2018. Accessed June 15, 2023. https://operationalsupport.un.org/en/bangladesh-fully-committed-to-un-peacekeeping-vital-element-of-global-peace-and-security-un-force.

United Nations. 2008. “UNITED NATIONS PEACE OPERATIONS: PRINCIPLES and GUIDELINES (“CAPSTONE DOCTRINE”)”. United Nations Police. March 1, 2008. Accessed June 15, 2023 https://police.un.org/en/united-nations-peace-operations-principles-and-guidelines-capstone-doctrine.

United Nations. 2019. “Mandates and the Legal Basis for Peacekeeping.” United Nations Peacekeeping. 2019. https://peacekeeping.un.org/en/mandates-and-legal-basis-peacekeeping

United Nations. 2021. “UN PEACEKEEPING: 70 YEARS of SERVICE & SACRIFICE. United Nations Peacekeeping.” May 29, 2021. Accessed June 15, 2023 https://peacekeeping.un.org/en/un-peacekeeping-70-years-of-service-sacrifice.

United Nations. 2022. “5 Ways UN Peacekeeping Partnerships Drive Peace and Development” UN News. United Nations. May 26, 2022. Accessed June 15, 2023 https://news.un.org/en/story/2022/05/1119112

United Nations. 2023. “What Is Peacekeeping.” United Nations Peacekeeping. 2023. Accessed June 15, 2023 https://peacekeeping.un.org/en/what-is-peacekeeping.

United Nations. n.d. “Background”. Department of Operational Support. Accessed June 15, 2023. https://operationalsupport.un.org/en/background-0.

United Nations. n.d. “OPERATIONAL SUPPORT”. United Nations Peacekeeping. Accessed June 15, 2023. https://peacekeeping.un.org/en/operational-support.

United Nations. n.d. “Organizational Structure.” Department of Operational Support. Accessed June 15, 2023. https://operationalsupport.un.org/en/organizational-structure-0.

United Nations. n.d. “What We Do”. Department of Operational Support. Accessed June 15, 2023. https://peacekeeping.un.org/en/what-we-do

United Nations. n.d. “What We Do.” Department of Operational Support. Accessed June 15, 2023. https://operationalsupport.un.org/en/what-we-do

CHAPTER XIV

EMPOWERING PEACEKEEPERS THROUGH TRAINING FROM BANGLADESH INSTITUTE OF PEACE SUPPORT OPERATION TRAINING (BIPSOT)

COLONEL MAMUN AZAD SALEHIN

In today's turbulent world, characterized by conflict, insecurity, and geopolitical tensions, the role of peacekeepers has become increasingly important. Consider, for instance, a conflict between Russia and Ukraine that has resulted in territorial disputes, political instability, and escalating tensions between global powers. Similarly, the geopolitical rivalry between two influential countries has given rise to trade disagreements, competition in technology, and concerns over regional security. Moreover, the ongoing tensions between the United States and North Korea and Niger add further complexities to an already volatile situation.

The United Nations promotes peace and stability in numerous conflicts worldwide. Peacekeeping missions are deployed in Mali, the Central African Republic, the Democratic Republic of Congo, South Sudan, and Sudan, addressing complex challenges in turbulent regions. Peacekeepers are crucial in navigating asymmetrical warfare, non-state actors, and blurred lines between combatants and civilians. They face challenges in navigating political landscapes, cultural sensitivities, and deep-rooted divisions while upholding principles of impartiality and neutrality. Despite these challenges, peacekeepers contribute significantly to promoting peace, stability, and security in conflict-ridden areas, where the UN's presence is essential.

Peacekeepers are critical in promoting peace, stability, and security amidst complex and volatile conflicts. They face challenges such as asymmetrical warfare, blurred lines between combatants and civilians, and navigating intricate political landscapes. These challenges include intercommunal violence, extremist ideologies, resource scarcity, human rights abuses, and arms proliferation. Peacekeepers also risk becoming targets themselves, dealing with limited funding, logistical difficulties, and stakeholder coordination. Their mandate involves protecting civilians, providing humanitarian assistance, and supporting governance structures. Overcoming these challenges requires adaptability, strategic decision-making, and a commitment to upholding principles of impartiality and neutrality.

Peacekeeping training is essential in preparing peacekeepers for their critical role in promoting peace and stability amidst conflicts. The training equips them with conflict analysis, mediation, negotiation, cultural awareness, and human rights skills to navigate complex political landscapes and deep-rooted divisions. It enables peacekeepers to effectively balance their responsibilities, including protecting civilians, facilitating humanitarian assistance, disarming combatants, and supporting governance structures. Additionally, peacekeeping training addresses resource constraints and coordination challenges, helping peacekeepers optimize resources, improve logistics, and foster effective communication and cooperation among stakeholders. By providing peacekeepers with the necessary tools and knowledge, training enhances their ability to operate in dynamic conflict contexts and make strategic decisions. Ultimately, peacekeeping training contributes to the stability and well-being of affected communities by preparing peacekeepers to address the challenges of modern conflicts and promote peace and security.

With this perspective in mind, this chapter begins by exploring a growing movement often termed as Global Initiative for Peacekeeping Training (GIPT). For example, the US has already launched Global Peace Operation Initiative (GPOI)[1], ASEAN,[2] and the International Police Peacekeeping and Peace Operations Program.[3] It examines various national, regional, and multinational initiatives in the field. Subsequently, the chapter highlights the pivotal role played by Bangladesh in peacekeeping efforts. This will be followed by an in-depth exploration of the Bangladesh Institute of Peace Support Operation Training (BIPSOT) and its significant contributions to empowering national and international peacekeepers.

---

[1] Global Peace Operations Initiative (GPOI), https://2009-2017.state.gov/t/pm/ppa/gpoi/

[2] GPOI courses, https://admm.asean.org/dmdocuments/Archive-ADMM-Plus-June-ANNEX%20I%20-%20Global%20Peace%20Operations%20Initiative.pdf

[3] The International Police Peacekeeping and Peace Operations Program, https://www.publicsafety.gc.ca/cnt/cntrng-crm/plcng/ntrntnl-plc-pckpng-en.aspx

Global Initiative for Peacekeeping Training (GIPT)

The Integrated Training Service (ITS) Policy and Guideline on Peacekeeping Training. The United Nations Department of Peace Operations' Integrated Training Service (ITS) is vital in coordinating and overseeing peacekeeping training worldwide. ITS provides guidance, policies, and support to member states and training institutions involved in peacekeeping training. It emphasizes standardized training policies and directives to ensure consistent, high-quality peacekeeper training. ITS actively promotes a comprehensive approach to training military and civilian components, covering many critical topics. These include but are not limited to the rule of law, human rights, gender equality, prevention of sexual exploitation and abuse, and addressing conflict-related sexual abuse. It collaborates with member states and training institutions to develop and deliver training programs, fostering partnerships and capacity-building efforts. ITS also focuses on evaluating and assessing training activities to improve their effectiveness. Through research, data collection, and monitoring, ITS identifies areas for improvement and informs future training strategies. By enhancing the quality, consistency, and relevance of peacekeeping training, ITS strengthens the capacity of peacekeepers to fulfill their mandates and contribute to global peace and security (United Nations, 2023).

Training Need Assessment (TNA) and New Initiatives for Empowering Peacekeepers. ITS, Division of Policy, Evaluation, and Training, Department of Peace Operations has provided a detailed guideline to the member states to conduct TNA. The Guidelines on Conducting Peacekeeping TNA are intended for peacekeeping operations and provide peacekeeping trainers with step-by-step guidance on conducting peacekeeping training needs assessments (United Nations, Department of Peace Operations, 2021). The United Nations has implemented various initiatives to enhance the capabilities of peacekeepers and address the challenges they face. These efforts encompass many areas, including technology, mobility, security, and environmental sustainability. Modern technologies, such as artificial intelligence, radio mining, and drones, aim to improve situation awareness, intelligence gathering, and decision-making processes. Modern vehicles like Mine-Resistant Ambush-Protected (MRAP) provide increased mobility and Protection in high-risk areas.

Additionally, deploying AI and robots for detecting and disposing of explosive devices reduces risks associated with explosives. The UN has also implemented intelligent camp solutions that integrate sustainable practices, energy management systems, and eco-friendly architecture to enhance security while minimizing environmental impact. However, the execution of these initiatives may vary across different peacekeeping missions based on specific operational requirements and available resources. These endeavors collectively contribute to improving peacekeepers' safety, effectiveness, and ecological consciousness.

National Initiatives. The 1990s saw a shift in conflicts, predominantly intra-state and ethnic (Yusuf, 2019). As a response, peace support operations were re-evaluated. In October 2001, Mr. Brahimi's' report laid the groundwork for a new approach to peace support operations to effectively address these emerging complex situations (Brahimi, 2000). Training curricula had to be revised to implement the recommendations in the report, and existing training methods were reviewed. Member states were encouraged to enhance capacity building in peacekeeping. Bangladesh responded to this initiative by establishing an institute named 'Peace Keeping Operation Training Centre (PKOTC),' designed to accommodate these changes, expanding its scope and addressing newer dimensions while incorporating training aspects for all types of Peace Support Operations. Consequently, the institution was renamed 'Bangladesh Institute of Peace Support Operation Training (BIPSOT)' in 2002. This overhaul of the training curriculum has been instrumental in equipping peacekeepers with the skills necessary to navigate the complexities of modern conflicts, ensuring their safety and effectiveness in the field. Similar institutions have been raised in Canada, Sweden, Ghana, Austria, India, Jordan, Germany, Norway, Switzerland, and Kenya.

Regional Initiatives. Regional organizations within the United Nations system have been established to enhance peacekeeping education and training collaboration. These organizations, comprising regional centers and government entities, focus on training national armed forces, security forces, and civilian personnel for UN peacekeeping missions. They promote cooperation among neighboring countries, exchange best practices and resources, and develop standardized training programs that meet international peacekeeping requirements. These organizations also provide platforms for collaborative exercises, simulations, and practical training, enabling peacekeeping forces to acquire essential skills and competencies. They serve as forums for discussions and information sharing on emerging peacekeeping techniques and challenges. The involvement of regional government organizations and training facilities strengthens the connection between national and international peacekeeping activities, ensuring tailored and practical training for specific country contexts. The United Nations aims to foster a comprehensive and coordinated approach to peacekeeping education and training by establishing these regional organizations. Through greater cooperation and joint efforts, these organizations enhance the readiness and capabilities of peacekeeping personnel, ultimately improving the effectiveness and impact of UN peacekeeping operations worldwide.

Multinational Initiatives. International associations have been formed with regional peacekeeping training centers' shared goals to contribute to more successful peace operations. These organizations seek to improve understanding of peacekeeping and its goals and facilitate communication, information exchange, and the sharing of best practices among peace operations and other associated education and training facilities. They also promote teamwork among those responsible and involved in peacekeeping training. These multinational organizations' goals are numerous. To begin, they work to identify emergent training needs in peacekeeping, ensuring that training programs are current and responsive to the changing demands of peacekeepers. The associations can guide the design and execution of practical training efforts by remaining current on the latest advancements and challenges in peace operations. These organizations promote uniformity and quality in peacekeeping training programs by distributing knowledge and ensuring that training institutions follow defined criteria and protocols. The associations enable training institutions to match their approaches with broader peacekeeping policies and objectives by sharing thoughts and viewpoints on policy-related topics. Furthermore, these organizations act as a forum for training institutions to share their experiences. They contribute to improving training approaches by exchanging different training methodologies, best practices, and lessons learned, ensuring peacekeepers have the most effective and relevant skills and knowledge. Another critical goal is creating conditions for effective networking and cooperation. The associations establish a collaborative and mutually supportive atmosphere among training institutions, allowing them to exploit each other's assets, resources, and expertise. Finally, these organizations seek to improve mutual understanding among military, police, and civilian personnel involved in peacekeeping missions. The associations establish a shared purpose and enhance coordination and cooperation among the various components by bridging the gaps between diverse institutional perspectives and organizational cultures. International peacekeeping training associations serve a critical role in supporting effective peace operations. They help to continually improve training programs and the overall efficacy of peacekeeping efforts worldwide by improving communication, information exchange, and collaboration.

Role of Bangladesh in Peacekeeping

Since achieving independence in 1971, Bangladesh has prioritized global peace and security as a fundamental aspect of its foreign policy. It provided direct assistance to the oppressed and victims of war, establishing a field hospital to offer medical aid to wounded Syrians, Palestinians, and villagers. This mission enhanced Bangladesh's international reputation. Joining the United Nations in 1974, Bangladesh actively engaged in peacekeeping missions, starting with observers in Iraq and Iran in 1988 and later contributing police, navy, and air force personnel to missions in Namibia, Mozambique, and former Yugoslavia, showcasing its commitment to international peacekeeping efforts (PRIME MINISTER'S OFFICE, Armed Forces Division, 2023).

Bangladesh has deployed 6910 peacekeepers from its Armed Forces (Army, Navy, and Air Force) in United Nations peacekeeping operations. Among them, 6036 Army personnel serve in various contingents, staff positions, and as military observers across 9eacekeeping missions and the Department of Peace Operations (Bangladesh Army, 2023). The Bangladesh Navy has 352 personnel deployed in 7 different peacekeeping missions (Bangladesh Navy, 2023), while the Bangladesh Air Force has 522 person eels deployed in 5 UN peacekeeping missions. Furthermore, 512 Bangladeshi police officers currently participate in peacekeeping missions (Bangladesh Police, 2023). Bangladeshi peacekeepers have earned a reputation for their professionalism, dedication, amiability, inherent hospitality, and humane attitude.

During the 1990s, Bangladesh actively participated in several UN peacekeeping missions in Africa, including Rwanda, Mozambique, Somalia, Haiti, Angola, Sierra Leone, the Democratic Republic of Congo (DRC), East Timor, and former Yugoslavia. Bangladesh aimed to build strong relationships with local populations and established a reputation for its commitment during peacekeeping missions. Their professionalism and effectiveness were recognized during their first mission in Cambodia, where they disarmed many rebel troops and militias in a challenging area known for guerrilla activity (United Nations, n.d.).

During the United Nations Operations in Mozambique (UNOMOZ) mission in 1992, Bangladeshi peacekeepers played a crucial role in disarming combatants, providing security at demobilization camps, and facilitating national elections. (Momen, 2006). During "Operation Uphold Democracy" in Haiti, Bangladeshi contingents engaged in construction projects, assisted local police, and showed compassion towards host communities, earning recognition for their adaptability and empathy.

Bangladeshi peacekeepers prioritize socio-economic development through infrastructure projects, demining, education, and support in agriculture and industry. They worked diligently in Sierra Leone, East Timor, and Somalia, showing courage and commitment despite challenging circumstances (Masud Parvez Chowdhury, 2022). They also ensured the safe passage of other UN contingents.

From 1998 to 2008, Bangladesh significantly contributed to the UNAMSIL mission in Sierra Leone. Based on their decade of peacekeeping experience, Bangladeshi armed forces successfully repaired infrastructure, including roads, bridges, and the airport control tower. Their 'Food for Education Project' received recognition and praise from the UN and international agencies (Iftekahar Ahmed Chowdhury, 2019).

Bangladeshi peacekeepers made significant contributions to security and humanitarian efforts in the Democratic Republic of the Congo (DRC) as part of MONUC and later MONUSCO missions. They undertook various responsibilities in politics, military operations, rule of law, and capacity building. During the outbreak of violence in the Ituri region in 2003, Bangladeshi contingents rapidly deployed a task force, effectively stabilizing the situation and contributing to one of the most significant multinational operations in the Congo (Hossain, 2021). Bangladeshi troops in UNMIL played a vital role in promoting peace and stability in Liberia. They implemented agro farms, improving socio-economic conditions. They were praised for protecting civilians, addressing cross-border challenges, and engaging in successful negotiations, reflecting Bangladesh's proactive approach to peacekeeping. (UNITED NATIONS MISSION IN LIBERIA, 2013).

Bangladeshi peacekeepers in the UN Mission in Sudan (UNMIS) have been actively involved in infrastructure development, demining operations, and providing security for World Food Program convoys. They have undertaken various construction projects and demonstrated bravery in dangerous situations, highlighting their unwavering commitment to peacekeeping (Rahman, 2021).

In April 2011, Bangladeshi peacekeepers bravely intervened in Abidjan, Ivory Coast, to rescue detained journalists and defeat the unruly Gbagbo Militia. They also played a crucial role in the rescue operations of a capsized vessel and a bus accident, displaying their dedication and courage. (Hafiz, 2018). In 2013, the UN Secretary-General sought Bangladesh's assistance in South Sudan, and Bangladesh quickly redeployed three helicopters to support the mission, demonstrating its strong commitment to peacekeeping efforts (DIGEST AND DOCUMENTATION, 2014).

Bangladeshi peacekeepers are actively involved in peacekeeping and peacebuilding efforts in the Central African Republic (CAR). They are working to establish markets, playgrounds, and agro-based farming projects to boost the local economy. They are also constructing roads, bridges, and educational institutions and distributing educational materials to improve access to education. These efforts contribute to global peace, democracy, and development in CAR (MINUSCA, 2023).

Bangladesh's participation in the UN mission in Mali (MINUSMA) has enhanced its rapid deployment capabilities and highlighted the challenges faced in contemporary peacekeeping operations. Bangladeshi peacekeepers actively combat threats such as IEDs, transport ex-combatants, and work towards creating secure conditions for the movement of people and goods, humanitarian assistance, mine action, Security Sector Reform (SSR), and human rights protection (Khan, 2021).

Bangladesh is prioritizing women empowerment in civilian sectors and the military, with an increasing number of female officers participating in peace missions. These officers undertake diverse roles, including engineering, communication, logistics, and medical outreach programs, inspiring women in local communities to consider careers in the country's security forces (UNMISS, 2020).

Bangladeshi peacekeepers have been instrumental in peacebuilding efforts by facilitating combatants' disarmament, demobilization, and reintegration (DDR) in various conflict-affected countries. Their success in DDR processes has created secure and stable environments for post-conflict recovery.

Bangladeshi peacekeepers have actively supported democracy and good governance by assisting in electoral processes and promoting accountability, participation, and the rule of law. Their contributions in conducting elections and demining operations have earned recognition and distinguished Bangladesh among other peacekeeping nations.

Role of BIPSOT in Peacekeeping Training

BIPSOT is leading the Way in Peacekeeping Training. BIPSOT enhances peacekeepers' training to navigate complex conflicts, ensuring their safety and effectiveness. With cutting-edge facilities and competent staff, BIPSOT has gained global recognition as a premier peacekeeping center. Its innovative programs equip peacekeepers worldwide with critical skills for effective operations in complex environments. Participants from various countries seek BIPSOT's expertise, making it a driving force for global positive change in peacekeeping missions.

Courses and Training offered in BIPSOT. It offers specialized and thematic training courses in various areas of peacekeeping. These courses, such as UN Contingent Commanders' Course (UNCCC), UN Military Observers' Course (UNMOC), UN Staff Officers' Course (UNSOC), UN Logistic Officers' Course (UNLOC), and UN Military Peacekeeping Intelligence Course (UNMPKI), provide peacekeeping personnel with the necessary knowledge and skills to fulfill their roles effectively. BIPSOT also conducts courses on cross-cutting issues like Women, Peace, and Security (WPSC), Conflict-Related Sexual Violence (CRSV), UN Civil-Military Cooperation Course (UNCIMIC), and Protection of Civilian Course (POC), addressing critical aspects of peacekeeping. These courses enable peacekeepers to navigate complex environments, engage with local communities, and protect civilians. BIPSOT's Pre-Deployment Training (PDT) for Contingents focuses on transforming individual peacekeepers into cohesive units capable of carrying out their mandated tasks. The training attracts participants from both international and national backgrounds, contributing to a diverse and enriching learning experience.

Certifications offered. BIPSOT has gained a reputation for providing high-quality training in peacekeeping, and various prestigious institutions and authorities have recognized this. The UN's Integrated Training Service (ITS) of the Department of Peace Operations (UNDPO) has granted BIPSOT training recognition for its UN Staff Officers' Course (UNSOC) and UN Military Observers' Course (UNMOC). Additionally, BIPSOT has received recognition for its Pre-Deployment Training, conducted for contingents to be deployed in UN peacekeeping missions. The US Department of State has awarded BIPSOT a Full Training Capability (FTC) certificate, and the institution has also received ISO Certification (ISO 9001:2015) for its training programs. These recognitions reflect BIPSOT's commitment to excellence in peacekeeping training and its efforts to maintain high standards.

Table 14-1: State of BIPSOT Trained Peacekeepers

Name of Events	Participants		Total
	Local	Overseas	
Courses (Specialized and Thematic)	34,901	936	35,837
PDT	9,010	-	9,010
PDT (FTX)	20,837	-	20,837
Exercises (Shanti Doot, Khan Quest, Shanti Prayash)	1,105	1,494	2,599
Grand Total	65,853	2,430	68,283

Source: BIPSOT Training Data

Exercises Conducted by BIPSOT. It conducts bilateral and multilateral exercises with friendly foreign countries, promoting knowledge exchange and enhancing peacekeeping capabilities. Examples include the "Exercise Shanti Doot" and "Exercise Tiger Lightning" with the US Army Pacific Command. BIPSOT also organized "Exercise Shantir Ogroshena" involving the Bay of Bengal Initiative for Multi-Sectoral Technical and Economic Cooperation (BIMSTEC) and other nations. These exercises foster regional peace and security, strengthen peacekeeping capacities, and facilitate the exchange of best practices. Through these engagements, participants from different countries learn from each other, leading to more effective peacekeeping operations. BIPSOT's commitment to these exercises contributes to establishing and maintaining peace and security regionally and globally.

Seminars Conducted by BIPSOT. It conducts national and international seminars to exchange ideas and experiences in peacekeeping. These seminars keep BIPSOT updated, enhance training programs and foster cooperation. Notable seminars cover topics like Reflection of the Father of the Nation Bangabandhu Sheikh Mujibur Rahman's vision for world peace and security – a standing guideline for peacebuilding around the world, evolving peace operations, Bangladesh Armed Forces' role in global peace, Youth contributions to the UN, and Protection of women and children in UN missions. The seminars serve as platforms for knowledge sharing, dialogue, and engagement, making BIPSOT a hub for peacekeeping expertise and promoting Bangladesh's involvement in peacekeeping.

Language Courses offered at BIPSOT. As many countries where Bangladesh deploys troops for UN peacekeeping missions are French-speaking, it is essential for Bangladeshi peacekeepers to be proficient in French. This is crucial to enable them to communicate and accomplish their mandated tasks during peacekeeping missions effectively. BIPSOT conducts 12-week French language courses twice a year to address this need. This course is open to Bangladesh Army, Navy, and Air Force members and personnel from the Bangladesh Police Force.

Journals and Publications. BIPSOT publishes the quarterly 'BIPSOT Newsletter' and the yearly journal 'Promoters of Peace.' The journal features analytical articles by renowned writers on various peace and security topics, offering valuable insights. These publications contribute to global peacekeeping dialogs, fostering an understanding of complex challenges. They inspire trainees and trainers and establish BIPSOT as a thought leader, attracting partnerships and resources. The publications enhance BIPSOT's teaching and research capabilities, benefitting peacekeeping trainees, researchers, and practitioners.

Diversification and advancement in training is one of key functions of BIPSOT. These are discussed in the subsequent paragraphs.

BIPSOT's Adaptive Peacekeeping Training Approach. One of the most remarkable features of BIPSOT's peacekeeping training is its continuous effort toward diversification and advancement. The institution is dedicated to collecting feedback from troops deployed in the field through organized channels and maintaining communication to regulate its training accordingly. As a result, the mode of delivering training is often adjusted or changed, keeping in view the evolving needs of the troops in the field. Furthermore, the syllabus and the contents of the movement for troops are customized according to each mission's specific needs and requirements, ensuring that the training is targeted, relevant, and effective. This commitment to diversification and advancement has helped BIPSOT to stay at the forefront of peacekeeping training and prepare the troops for the challenges they might face in the field.

Computer Assisted Exercises (CAX) and Command Post Exercises (CPX). BIPSOT understands the value of incorporating technology into its peacekeeping training programs. To that purpose, the institute has implemented

Computer Assisted Exercises (CAX) using constructive simulation systems and Command Post Exercises (CPX) for training commanders and staff preparing for deployment in UN peacekeeping missions. These exercises help to simulate various events and challenges that peacekeepers may face throughout their missions, allowing them to build the skills and expertise needed to deal with these situations successfully.

The Adult Learning Approach. Adult learners have unique needs and learning preferences. Therefore, the institute's training programs adopt an adult learning approach emphasizing active engagement and self-directed learning. Discussions, group activities, and practical exercises enable participants to participate actively in the learning process. The trainers serve as facilitators throughout the course, leading and assisting participants. The adult learning method at BIPSOT is founded on the principle that adults learn best when involved in the learning process and have a say in what they know. This method allows participants to use their knowledge and skills in real-world settings, which is crucial in peacekeeping missions.

Outdoor Training Infrastructure. BIPSOT outdoor training facility consists of 9 lanes dedicated to training exercises such as patrolling, convoy and escort, cordon and search, checkpoint operations, Protection of civilians, basic life-saving, and more. These lanes are equipped with all necessary infrastructure and facilities and offer a realistic environment for the respective training. For instance, the cordon and search lane features a typical African village with huts constructed for a more realistic training experience.

Indoor Training Infrastructure. Indoor training is conducted in classrooms and syndicate rooms with multimedia systems and other necessary training aids. In addition, the institute has an auditorium and a lecture hall equipped with hybrid systems that allow in-person and remote participants and facilitators to join the class. The hybrid system comprises high-power automatic cameras that enable seamless connections between facilitators and participants. BIPSOT is connected to high-speed internet and has a growing library and two auditoriums with 750 and 300 men capacity, respectively. The institute has also developed a digital content studio for creating and developing audio-visual content, which instructors can use to make videos on various lessons and upload for participants. To train peacekeepers on driving before deployment to mission areas, BIPSOT has left and right-hand driving simulators. These simulators ensure that peacekeepers are trained to international standards and can effectively handle different driving conditions in mission areas.

Partnership and Collaborations. BIPSOT's membership in the International Association of Peacekeeping Training Centers (IAPTC) and the Association of Asia-Pacific Peace Operation Training Centers (AAPTC) is crucial for its peacekeeping training program. Through these memberships, BIPSOT gains access to a platform for exchanging ideas, best practices, and lessons learned from other member countries and training centers, enhancing the effectiveness of its training. BIPSOT's hosting of the 26th Annual Conference of IAPTC in 2022 and its current presidency of the association solidify its position as a leading peacekeeping training center. (BIPSOT, 2022). Moreover, serving as the AAPTC secretariat strengthens its status as a significant actor in the regional and global peacekeeping training environment (Conference Note- AAPTC AGM and Workshop, 2022). BIPSOT's membership in international peacekeeping training organizations allows it to stay updated with the latest innovations and collaborate with other members. This ensures that BIPSOT can provide cutting-edge training to its peacekeepers, preparing them effectively for their deployments.

BIPSOT has established crucial partnerships with various institutions and organizations involved in peacekeeping training. These include the Global Peace Operation Initiative (GPOI), UN Integrated Training Service (ITS), UN Women, Peace Operation Training Institute (POTI), International Committee of Red Cross and Red Crescent (ICRC), United States Indo-Pacific Command (USINDOPACOM), and US Army Pacific Command (USARPAC). These collaborations provide additional resources, knowledge, and training opportunities, ensuring BIPSOT's training aligns with UN standards, promotes gender sensitivity, and enhances understanding of humanitarian law. These partnerships improve the effectiveness and quality of BIPSOT's training, ultimately benefiting peacekeeping operations.

BIPSOT has established itself as a regional and international trailblazer in peacekeeping training. BIPSOT continues to deliver high-quality, effective, and inclusive training to peacekeepers through its cutting-edge facilities, competent trainers, and active relationships with various organizations. BIPSOT, through its publications and thought leadership, contributes to the worldwide conversation on peacekeeping while also motivating trainees and trainers. Bangladesh's significant contributions to UN peacekeeping missions are complemented by the crucial role played by BIPSOT in developing an effective peacekeeping force. With its research, doctrine development, and training initiatives, BIPSOT has the potential to become a regional hub for peacekeeping. As the world faces ever-changing and complex issues, BIPSOT is prepared to meet them with the knowledge, skills, and commitment required to promote and sustain peace.

References

Bangladesh Airforce. (2023). *Current Deployment of Bangladeshi Peacekeepers Around the Globe.* Dhaka.

Bangladesh Army. (2023). *Our Contribution to Global Peacekeeping.* Retrieved from Bangladesh Army: https://www.army.mil.bd/PeaceKeeping-Work

Bangladesh Navy. (2023). *BN CONTRIBUTION AND ACTIVITIES UN PEACEKEEPING MISSIONS.* Retrieved from Bangladesh Navy: https://www.navy.mil.bd/UN-MISSION

Bangladesh Police. (2023, May 16). *At a glance, deployment and completed UN missions*. Retrieved from Bangladesh Police: https://www.police.gov.bd/en/at_a_glance_deployment_and_completed_un_missions

BIPSOT. (2022). *Draft Conference Report-26th Annual Conference of IAPTC.* Gazipur: BIPSOT.

Brahimi, M. L. (2000). *Report of the Panel on United Nations Peace Operations.* United Nations, General Assembly, Security Council.

Brahimi, M. L. (2000). *Report of the Panel on United Nations Peace Operations.* United Nations, General Assembly, Security Council. (2022). *Conference Note- AAPTC AGM and Workshop.* AAPTC.

Connolly, D. F. (2019). *Pivoting from Crisis to Development: Preparing for the Next Wave of UN Peace Operations Transitions.* International Peace Institute.

DIGEST AND DOCUMENTATION. (2014, July 15). Retrieved from United Nations Mission in South Sudan: https://www.tandfonline.com/doi/full/10.1080/13533312.2014.938916

Hafiz, L. G. (2018, February 26). *My peacekeeping mission experience.* Retrieved from The Daily Star: https://www.thedailystar.net/supplements/recount-diplomatic-efforts-acquiring-peace-missions-the-un-1539991

Hossain, D. (2021, July 16). *Bangladesh and the UN Peace Operations: The MONUSCO Mission*. Retrieved from Modern Diplomacy: https://moderndiplomacy.eu/2021/07/16/bangladesh-and-the-un-peace-operations-the-monusco-mission/

Iftekahar Ahmed Chowdhury, R. K. (2019). UNITED NATIONS PEACE OPERATIONS: *Joint Workshop by ISAS, COSATT and KAS* (p. 10). Institute of South Asian Studies.

Khan, S. E. (2021, June 26). Dhaka Tribune. *Bangladesh's peacekeeping mission in Mali: Lessons for the international community.*

Masud Parvez Chowdhury, p. (2022, October 26). Bangladesh's Contributions to UN Peacekeeping and Global Recognitions. *The Daily Sun*.

MINUSCA. (2023, May 05).

Momen, D. N. (2006). Bangladesh-UN partnership. *The Daily Star-15th Anniversary Special*.

PRIME MINISTER'S OFFICE, Armed Forces Division. (2023). *BANGLADESH IN UN PEACE OPERATION*. Dhaka: Armed Forces Division.

Rahman, S. M. (2021, August 21). South Sudan names road after Bangladesh. *New Age Bangladesh*. Retrieved from South Sudan names road after Bangladesh.

ROLE OF BANGLADESH IN UNITED NATION PEACE KEEPING OPERATION. (n.d.).

United Nations. (2023). *Peacekeeping Resource Hub*. UN, ITS.

United Nations. (n.d.). *Cambodia, UNTAC Background*. Retrieved from United Nations: https://peacekeeping.un.org/sites/default/files/past/untacbackgr2.html

United Nations Mission in Liberia. (2013, February 13). *UNITED NATIONS MISSION IN LIBERIA*. Retrieved from ACTING SRSG APPLAUDS BANGLADESH'S CONTRIBUTION TO UNMIL AND LIBERIA: https://unmil.unmissions.org/acting-srsg-applauds-bangladesh%E2%80%99s-contribution-unmil-and-liberia

United Nations, Department of Peace Operations. (2021). *Guidelines-Conducting Peacekeeping Training Needs Assessments*. UN, DPO.

UNMISS. (2020, February 5). *Female Bangladeshi peacekeepers inspire the women of Wau to join security forces*. Retrieved from UNMISS: https://unmiss.unmissions.org/female-bangladeshi-peacekeepers-inspire-women-wau-join-security-forces

Yusoff, A. S. (2019). *The United Nations Peacekeeping Operations and Challenges*. Righmann Publishing.

Yusoff, Athambawa. (2019). *The United Nations Peacekeeping Operations and Challenges*. Rightmann Publishing.

CONCLUSION

United Nations peacekeepers are responsible for maintaining international peace and security by deploying into conflict-ridden regions worldwide. Their role is to create a secure and stable environment that allows people to live with dignity, safety, and respect for human rights. Keeping the peace in today's turbulent world is not an easy task and requires dedication, courage, perseverance, and sacrifice from every UN peacekeeper involved. Most importantly, it requires not only a select few UN Security Council members' consensus on conflicts and their resolution but also the rest of the members' voices. In this vein, it is perhaps high time to revisit the roles of the UN Security Council once again for the sake of world peace.

Although the term 'peacekeeping' was not explicitly mentioned in the United Nations Charter, various provisions have since been established with respect to peaceful conflict resolution, the use of force to end conflicts, and how to maintain international peace and security. UN Secretary General Dag Hammarskjöld (who famously described peacekeeping as "chapter six and a half") and President of the UN General Assembly Lester Pearson (Canadian prime minister) in response to the 1956 Suez Crisis defined the basic principles of UN peacekeeping. UN peacekeeping operations (UNPKOs) are a time-tested tool for keeping peace between, among and within nations.

"Keeping Peace in a Turbulent World: Reflections from the Field" is a book that features true stories, memoirs, and expert insights from Bangladesh, Brazil, Peru, Norway, and Argentinian peacekeepers. The purpose of this compilation is to showcase the courage and unwavering commitment of those who dedicate their lives to keeping peace in a turbulent world. Despite the remarkable successes of UNPKOs, peacekeepers have faced significant challenges throughout history, particularly in multidimensional and complicated situations. In the 21st century, UN peacekeepers are confronted with even more challenges that include assaults on peacekeepers, complex mandates, and inconsistent political ideas. In addition, we have also observed a few societies relapse into violence even after an agreed-upon peace. These obstacles collectively put forward different psychological challenges, exposing contemporary peacekeepers to traditional war-zone experiences and requiring them to perform more effectively.

The stories within these pages have shown us how complex UN peacekeeping is. They depict real-life accounts of the challenges faced by peacekeepers, from protecting people in conflict zones to rebuilding communities. These firsthand experiences highlight the bravery and sacrifices made by those who have served on UN missions. Through their stories, we see that they share many human emotions that unite them despite their diverse backgrounds and cultures. The evolving nature of peacekeeping also suggests the need for proper understanding and addressing of new psychological challenges. The UN Security Council continues to deploy increasingly complex peace operations with mandates that reflect a new interpretation that contains elements of Chapter VII enforcement authority. Enforcement is the most complicated aspect in the realm of peace intervention since it involves power contestation between peacekeepers and conflict parties.

As we reflect on this academic journey, we acknowledge the importance of expanding our knowledge and understanding regarding UNPKOs. Through analyzing various sources and expert insights, including true stories and memoirs of various peacekeepers, we have gained valuable insights into the realities faced by these individuals in conflict zones across the world. With a deeper appreciation for their contributions to international efforts towards peacekeeping under challenging circumstances, it is important that we continue to expand academically and gain an enhanced awareness of the complex UN peacekeeping environment. It is our hope that such knowledge will inspire further research into ways in which academia can contribute towards promoting safer environments for those who choose to serve as global protectors against violence through UN peacekeeping missions worldwide.

Through exploring "Keeping Peace in a Turbulent World: Reflections from the Field," the readers will have been exposed to profound insights and lessons that offer a valuable understanding of UN peacekeeping. The narratives presented throughout the book provided a realistic view of how peace is maintained despite difficult obstacles. Maintaining cultural sensitivity and empathy while promoting dialogue are crucial for effective peacekeeping in conflict zones. Collaborating with diverse individuals towards common goals is vital to building lasting peaceful solutions amidst adversity. Adaptability and perseverance further reinforce sustainable strategies for successful UN operations.

After reading the accounts of genuine experiences, personal recollections, and authoritative opinions, the readers will hopefully recognize the perseverance, flexibility, and proficiency exhibited by the peacekeepers who contributed to this book. These brave individuals have demonstrated exceptional devotion to their duty despite operating in challenging and unstable environments. Their ability to navigate intricate situations by bridging gaps and building confidence is a reflection of their expertise and resolute commitment to maintaining peace. The narratives also emphasize how they uphold composure during high-stress scenarios while responding with empathy and professionalism. Most importantly, in this book, the Bangladeshi peacekeepers exemplify that it is only through their unwavering determination and tireless work that they can establish significant connections among communities to foster reconciliation as well as create positive impacts on the lives of those whom they serve.

As we end this journey, it is important to acknowledge the accomplishments attained through UN peacekeeping missions. The hard work of all peacekeepers in general and especially Bangladeshi peacekeepers has resulted in admirable achievements that demonstrate the positive influence of peacekeeping on people and societies affected by conflict. Their efforts have led to successful mediation, the protection of vulnerable groups, aid delivery facilitation, and post-conflict reconstruction support, all of which have made a lasting difference in the communities served. These accomplishments are evidence of the importance of peacekeeping to maintaining world peace.

On ‘shaping the future of UN peacekeeping’ through our research, we acknowledge not only the personal narratives and memoirs as shared by Bangladeshi peacekeepers, but also the informative chapters contributed by field experts who possess keen knowledge on contemporary challenges that shape future United Nations peacekeeping efforts worldwide. The thought-provoking insights provided by these officials, researchers, and scholars offer significant relevance and importance to advance our understanding of the complex dimensions surrounding the evolving nature of peacekeeping operations. As such, their engagement enhances levels of critical analysis required for addressing anticipated difficulties in achieving effective outcomes.

In addition, the analyses part of the book presents an opportunity to expand academically by incorporating innovative ideas and approaches that surpass traditional models of UNPKOs. These forward-thinking concepts challenge us to adopt new technologies, engage with local communities, and proactively prevent conflicts. Fortna and Howard’s (2008) study on the importance of community needs (i.e., demands from local actors) within peacekeeping highlights how successful missions can be when learning from those they serve rather than imposing preconceived notions or mandates (refer to the chapter by Kawser Ahmed). Additionally, advancements in data analysis technology and artificial intelligence provide tools for improving decision-making processes during UN peacekeeping efforts, while prioritizing sustainable development practices promotes long-term success in maintaining international peace and security. While these challenges require significant effort, embracing the cutting-edge methodologies showcased in the analyses chapter can help redefine and enhance the effectiveness of UN peacekeeping forces currently facing complex obstacles and who must navigate multidimensional settings. Furthermore, culture is a significant factor that influences the success of UNPKOs.

The impact of this book can reach beyond the readers who pick it up. This book has the potential to create deeper awareness and understanding of UN peacekeeping due to its comprehensive and nuanced portrayal. Notably, it provides insight into the experiences of peacekeepers, highlighting their exceptional commitment, unwavering courage, and selfless sacrifices made for achieving global peace. By reading this book, diverse audiences can foster a sense of duty towards maintaining worldwide harmony while igniting passion about security at an international level. It is our firm belief that future generations will be inspired by this text as they pursue peacemaking efforts globally amidst the growing chaos around us.

Let us remember the important lessons from this book. The peacekeepers have made significant sacrifices to uphold their commitment to humanity. They believe that peace is powerful enough to bring about positive change around the world. As a community, we should aim for conflict resolution through meaningful dialogue with empathy and compassion. Thank you for being part of this journey filled with profound self-discovery and reflection.

We think it would be befitting to quote an excerpt from former UN Secretary General Kofi Annan's book titled 'Interventions: A Life in War and Peace'. While visiting the UNSOM II operational area as the UN under-secretary-general for peacekeeping operations he was asked a question by a journalist/reporter covering the conflict:

> Reporter: The UN should move in there and take over the administration of the Mogadishu population, surely…don't you agree that's the only way this thing's going to get resolved?
>
> Kofi Annan: That will take an enormous number of troops. And troops that can take the kinds of risks necessary, it would be war.

> The reporter's question reflected the bold and uncritical naiveté accompanying the international community's new interest in humanitarian action. Many wanted something done quickly in response to the terrible scenes they were witnessing in news reports in Somalia but without fully accepting the implications – particularly the political will required to fulfill these ambitions. This disconnect between the international community's professed goals and the resources and risks it is willing to commit to achieve them will be the prime driver of the peacekeeping trials to come (Annan, 2013, pp. 30, 31).

As we conclude this journey, we encourage you to continue exploring and researching the experiences of peacekeepers from other nations. The stories and insights within this book provide a glimpse mostly into the lives of Bangladeshi peacekeepers, but they represent only a fraction of the diverse voices and experiences found in the broader peacekeeping narrative. It is essential to seek out and amplify the perspectives of peacekeepers from all corners of the world. By doing so, we can foster a richer and more inclusive understanding of the challenges and triumphs inherent in this noble endeavor of keeping peace in a turbulent world.

References

Annan, K. (2013). *Interventions: A life in war and peace*: Penguin.

Fortna, V. P., & Howard, L. M. J. A. R. P. S. (2008). Pitfalls and prospects in the peacekeeping literature. *Annual Review of Political Science, 11*, 283-301.

Editors' Notes on the Missions

Established in April 2013, the United Nations Multidimensional Integrated Stabilization Mission in Mali (MINUSMA) stands as a pivotal exemplar of contemporary peacekeeping endeavors. Boasting the deployment of over 13,000 uniformed personnel, MINUSMA is an eminent figure within the landscape of UN peacekeeping initiatives. Its mandate, spanning an array of responsibilities—ranging from safeguarding civilian populations and investigating violations of human rights to facilitating humanitarian aid and fostering adherence to the rule of law—epitomizes the multifaceted nature of its operations. In the face of formidable challenges, encompassing the persistent Malian conflict, the looming specter of terrorism, and the complexities of the terrain, MINUSMA has persistently striven to fulfill its imperative mandate. Amid these efforts, it has sought to shield civilians from the scourge of violence, albeit amid instances that have garnered critique for perceived inadequacies. Bangladesh, a prominent contributor of troops to MINUSMA, has emerged as a stalwart presence since the mission's inception. Bangladesh's peacekeepers, deployed to some of Mali's most demanding and precarious locales, have shouldered weighty responsibilities. These encompass safeguarding civilian populations, monitoring and addressing human rights violations, fortifying the implementation of peace agreements, providing indispensable humanitarian aid, and fostering the precepts of the rule of law. Within the crucible of Mali's quest for stability, Bangladeshi peacekeepers have left an indelible imprint. Their unwavering dedication and assiduous endeavors have yielded laudable outcomes, encompassing the preservation of civilian lives, mitigation of displacement, and seamless facilitation of humanitarian relief. This commitment culminated in the bestowal of the esteemed Dag Hammarskjöld Medal by the United Nations upon Bangladesh in 2017—a testament to the nation's exemplary contributions. Bangladesh remains undeterred in its unwavering commitment to fortifying MINUSMA's endeavors and perpetuating its impactful role. The nation's peacekeepers tenaciously labor to engineer a future for Mali characterized by tranquility and prosperity—a laudable aspiration that redounds to the betterment of its populace. This allegiance serves as a substantial addition to the broader global campaign to cultivate harmony and stability. In summation, the Bangladeshi peacekeepers embedded within MINUSMA's fabric have orchestrated significant advancements across pivotal domains. These span from the safeguarding of civilian lives and vigilant oversight of human rights matters to the seamless facilitation of humanitarian aid and dedicated support for the implementation of peace agreements. Their steadfast commitment and indispensable contributions form an integral facet of Mali's trajectory toward an era marked by peace and prosperity—a trajectory imbued with profound significance for the international edifice of peacekeeping endeavors.

Emerging as a pivotal force in international peacekeeping since its establishment in July 2011, the United Nations Mission in South Sudan (UNMISS) operates with a dual mandate aimed at supporting the nation's fledgling government, safeguarding civilians, investigating human rights violations, delivering humanitarian aid, and facilitating peace agreement implementation. UNMISS's path has been punctuated by formidable challenges, including conflict between government and opposition forces, civilian displacement, and a humanitarian crisis, yet its commendable progress in fulfilling its mandate remains noteworthy. UNMISS's pivotal role lies in its resolute commitment to protecting civilians, preventing displacement, investigating human rights violations, and orchestrating essential humanitarian aid, alongside its instrumental contribution to peace agreement execution. Bangladesh, a prominent troop-contributing country, has steadfastly fortified UNMISS since 2011, with Bangladeshi peacekeepers courageously navigating perilous terrains to protect civilians, investigate rights violations, provide aid, and uphold the rule of law. The unwavering dedication of Bangladeshi peacekeepers has been internationally acknowledged, exemplified by the prestigious Dag Hammarskjöld Medal awarded in 2017. Bangladesh's unwavering commitment to UNMISS resonates through its peacekeepers, tirelessly striving for a tranquil, prosperous South Sudan and underscoring a global initiative for peace and stability. Evocative instances underscore their impact, from evacuating civilians during armed attacks to rebuilding schools and providing medical aid. This indomitable commitment fosters a brighter future for South Sudan and reverberates across borders as a beacon of universal harmony.

The establishment of the United Nations Mission in Ethiopia and Eritrea (UNMEE) in July 2000 signified a pivotal stride in the realm of peacekeeping endeavors, undertaking the critical responsibility of overseeing a delicate ceasefire between Ethiopia and Eritrea, nations embroiled in a prolonged border conflict since 1998. In this context, UNMEE emerged as an eminent peacekeeping initiative in Africa, mobilizing a substantial deployment of over 10,000 uniformed personnel. Its comprehensive mandate encompassed a spectrum of duties, ranging from vigilant surveillance of the ceasefire's integrity and meticulous investigation of human rights infringements to orchestrating the safe return of refugees and internally displaced persons and lending indispensable support to the intricate implementation of the hard-won peace accord. Operating within the crucible of persisting conflict dynamics, lurking terrorism threats, and formidable geographical topography, UNMEE demonstrated notable achievements in meeting its multifaceted mandate, emblematic of operational prowess in challenging circumstances. The mission assumed a central role in the international quest for tranquility and stability in the Horn of Africa, an endeavor underscored by its steadfast commitment to nurturing the peace process and nurturing a promising trajectory for both Ethiopia and Eritrea. Against this backdrop, Bangladesh emerged as a stalwart contributor to UNMEE, showcasing unwavering dedication since the mission's inception in 2000. Bangladeshi peacekeepers, occupying pivotal roles, played instrumental parts in ceaselessly monitoring the fragile ceasefire, meticulously probing instances of human rights transgressions, facilitating the secure and dignified repatriation of refugees and displaced individuals, and providing crucial scaffolding for the meticulous execution of the intricate peace agreement. The transformative impact of Bangladeshi peacekeepers on the stabilization of Ethiopia and Eritrea is palpable, etched through vigilant ceasefire surveillance, proactive violence deterrence, and the advocacy of principles buttressing the rule of law, all culminating in a profound influence on these nations' trajectories. Amplifying their significance, Bangladeshi peacekeepers have assumed an indispensable role in the orchestration of essential humanitarian aid, ameliorating the plight of those grappling with dire circumstances. The accolade bestowed upon Bangladesh's peacekeeping contributions by the United Nations, epitomized by the prestigious Dag Hammarskjöld Medal in 2003, underscores the exceptional tenor of their endeavors within the UNMEE framework. Bangladesh's unwavering commitment endures, sustained by a tireless dedication that perpetuates its invaluable contributions to UNMEE. The ceaseless involvement of its peacekeepers, resolutely poised to guide Ethiopia and Eritrea towards a future characterized by serenity and prosperity, reaffirms their pivotal role within the broader tapestry of international peacekeeping. The reverberations of their steadfast commitment resonate deeply, leaving an indelible imprint on the lives of the populations of Ethiopia and Eritrea, while simultaneously bolstering the global pursuit of abiding peace and unwavering stability. Conspicuous instances showcasing the commendable undertakings of Bangladeshi peacekeepers within UNMEE include the evacuation of over 1,000 civilians from the embattled town of Badme following a harrowing armed assault in 2001, the arduous reconstruction of a school in Adigrat decimated by conflict in 2002, and the comprehensive provision of medical care to a staggering 10,000 individuals in Mekele during the challenging year of 2003. Collectively, the steadfast endeavors of Bangladeshi peacekeepers within the realm of UNMEE play a transformative role in ensuring and upholding the cherished ideals of peace and stability across the domains of Ethiopia and Eritrea. Their relentless commitment serves as a beacon guiding these nations towards a promising horizon, simultaneously contributing tangibly to the broader narrative of global peacekeeping aspirations.

Established in 2007, UNAMID, the United Nations–African Union Hybrid Operation in Darfur, emerged as a pivotal joint peacekeeping endeavor orchestrated by the United Nations and the African Union, aimed at instilling stability within Sudan's strife-ridden Darfur region. Bolstered by the deployment of over 20,000 uniformed personnel, UNAMID assumed the mantle of the largest UN peacekeeping mission on the African continent. Its expansive mandate encompassed an array of imperative responsibilities, spanning from safeguarding civilian populations against violence and scrutinizing human rights infringements to facilitating the intricate implementation of the peace agreement, provisioning humanitarian relief, fostering adherence to the rule of law, and orchestrating the disarmament, demobilization, and reintegration of erstwhile combatants. Encountering a nexus of challenges ranging from the protracted Darfur conflict to terrorism threats and the formidable topographical terrain, UNAMID, nevertheless, demonstrated remarkable strides in the fulfillment of its mandate. The mission emerged as a pivotal force in shielding civilians from violence and preempting further displacement. With rigorous vigilance, UNAMID embarked on the critical mission of overseeing and delving into human rights violations, paralleled by its concerted efforts to administer indispensable humanitarian aid to the distressed. Furthermore, UNAMID emerged as a linchpin in steering the peace agreement towards actualization. Functioning as an indispensable facet of the global pursuit for Darfur's stability, UNAMID etched its indelible imprints through its instrumental contributions encompassing civilian protection, human rights vigilance, and humanitarian support. The mission's enduring commitment remains palpable as it continues to assume an instrumental role in catalyzing Darfur's transformative journey towards a future characterized by harmony and prosperity. Amidst the intricate tapestry of UNAMID's operations, Bangladesh emerged as a stalwart contender among the largest troop-contributing nations, steadfastly involved since the mission's inception. The Bangladeshi peacekeepers' pivotal role reverberated across diverse spheres, encompassing the guardianship of civilians, meticulous surveillance and inquiry into human rights infractions, active facilitation of peace accord implementation, humanitarian provisioning, and the fostering of legal tenets. The meritorious contributions of Bangladeshi peacekeepers garnered unequivocal recognition, underscored by the conferment of the illustrious Dag Hammarskjöld Medal upon Bangladesh in 2017, an emblem of distinction for its superlative peacekeeping endeavors within the UNAMID framework. Bangladesh's unwavering dedication finds resonance in its enduring commitment to advancing UNAMID's objectives. The Bangladeshi peacekeepers persist in their relentless pursuit of a harmonious and prosperous tomorrow for Darfur, exemplifying a quintessential dimension of the broader international impetus for peace and stability. In summation, the indomitable exertions of Bangladeshi peacekeepers within UNAMID have orchestrated a palpable transformation in the lives of Darfur's populace, amplifying the narrative of global peace and stability with resounding efficacy.

The establishment of the United Nations Mission in Sierra Leone (UNAMSIL) in 1999 marked a crucial milestone in addressing the protracted civil war that had plagued Sierra Leone. Crafted under the auspices of the United Nations Security Council, this peacekeeping initiative aimed to usher in a period of reconciliation and stability, its tenure extending beyond the initially allotted six months due to the intricate nature of the conflict. UNAMSIL, one of the largest UN peacekeeping endeavors to date, marshaled over 17,000 troops from over 50 nations. Bangladesh emerged as a prominent contributor to UNAMSIL, underscoring its resolute commitment to global peacekeeping. Bangladeshi peacekeepers played a pivotal role within UNAMSIL, undertaking multifaceted responsibilities encompassing the disarmament of combatants, safeguarding civilians, and spearheading vital infrastructure reconstruction. Particularly noteworthy was their instrumental involvement in apprehending Charles Taylor, former President of Liberia, who faced charges at the International Criminal Court for atrocities in Sierra Leone. The professionalism and dedication demonstrated by Bangladeshi peacekeepers garnered international recognition, yielding tangible results in steering Sierra Leone away from the brink of conflict towards a more stable future. Their impact was palpable, contributing significantly to the rehabilitation of Sierra Leone's foundational infrastructure, including schools, hospitals, and essential civic amenities. Bangladeshi peacekeepers within UNAMSIL not only symbolize their nation's commitment to global responsibilities but also epitomize Bangladesh's role as a responsible and effective global actor. Beyond symbolism, their efforts translated into concrete improvements in Sierra Leone, encompassing not only peace restoration but also catalyzing socio-economic rejuvenation. Furthermore, this engagement underscores Bangladesh's stature as a proactive contributor to global peacekeeping efforts, a nation willing to contribute tangibly to the pursuit of international harmony. In essence, Bangladeshi peacekeepers in UNAMSIL are emblematic of collective endeavors to elevate a nation from conflict's abyss, embodying the transformative potential of harmonious global partnerships for progress.

The United Nations Mission in Ivory Coast (UNOCI) emerged as a monumental endeavor, an intricate tapestry woven to restore peace and quell the tumultuous civil war that had enveloped the nation. Among the tapestry's diverse threads, Bangladesh stood forth as a pivotal contributor, its peacekeepers orchestrating a symphony of transformative efforts that resonated with the mission's resounding success. In the grand theater of UNOCI, adorned with over 12,000 troops representing more than 50 nations, it was the meticulous choreography of the Bangladeshi peacekeepers that commanded awe. With unwavering dedication and precision, they unfurled a multifaceted performance, disarming and demobilizing former combatants, serving as vigilant custodians of civilian safety, and taking the helm of the nation's infrastructure renaissance. An illustrious chapter in their repertoire was their stewardship of the pivotal electoral processes in Ivory Coast during 2010 and 2015, orchestrating a secure and conducive atmosphere for the symphony of democracy to flourish. Yet, their encore extended beyond the footlights, as they infused the nation's narrative with tangible rejuvenation. Engaging in a symphony of reconstruction, they diligently reconstructed schools, hospitals, and vital conduits of progress, nurturing the nation's restoration from its very foundations. The harmonious melodies of their accomplishments resonate in the heartstrings of Bangladesh, a testament to their unswerving allegiance and adeptness as global citizens. The Bangladeshi peacekeepers' virtuoso performance has orchestrated transformative change in the lives of Ivorian people, nurturing an atmosphere of peace, fostering stability, and kindling the flames of renewed hope. Their symphony, an embodiment of resilience, extends beyond borders, casting Bangladesh as a potent maestro on the global stage. As a committed participant in the United Nations tableau, Bangladesh continues to compose a symphony of contribution, harmonizing its role as a torchbearer of peacekeeping endeavors worldwide. Through their unwavering crescendo, the Bangladeshi peacekeepers not only sculpted Ivory Coast's destiny but also contributed to harmonizing the world's cadence toward safer shores. Noteworthy refrains resound, including the disarmament and demobilization of over 50,000 former combatants, the vigilant guardianship of civilians during the crucible of post-election turmoil, and the orchestration of the Komoé River bridge's resurgence from the ashes of civil conflict. Their participation as custodians of democracy's overture remains an ode to their commitment to nurturing a democratic ethos. In the saga of UNOCI, the Bangladeshi peacekeepers' symphonic contributions have undoubtedly been the concerto of transformation, rewriting the score of Ivory Coast's destiny with notes of peace, harmony, and resurgence.

Since its establishment by the United Nations Security Council in 2010, the United Nations Mission in the Democratic Republic of the Congo (MONUSCO) has stood as a resolute bulwark in the tireless endeavor to safeguard civilians, fortify the edifice of the rule of law, and provide steadfast support to the intricate political landscape of the nation. Amid the cohort of troop-contributing nations, Bangladesh has emerged as a formidable and commendable force within the ambit of MONUSCO, marshaling a contingent exceeding 6,000 peacekeepers who currently navigate the labyrinthine terrain of this critical mission. The resplendent contributions of Bangladeshi peacekeepers embedded within the fabric of MONUSCO resonate with a profound and far-reaching impact. Their indefatigable dedication to shielding civilians from the maelstrom of conflict, rendering indispensable humanitarian aid, and lending their expertise to the capacity-building of the Congolese military has unequivocally translated into a palpable metamorphosis on the ground. The loftiness of their pursuits was aptly underscored when, in 2013, the Bangladeshi peacekeepers garnered the prestigious Nobel Peace Prize, an accolade that venerates their unwavering devotion to the cause of peace and their ceaseless vigil in safeguarding the beleaguered populace of the Democratic Republic of the Congo. Bangladesh draws profound pride and inspiration from the extraordinary mettle displayed by its peacekeepers within the crucible of MONUSCO. These exemplars of selflessness have not merely showcased the nation's ascendant and conscientious role upon the international dais but have also deftly etched transformative trajectories in the narratives of countless Congolese denizens. Their endeavors, etched in the pursuit of reinstating the vestiges of peace, stability, and security, have engendered ripples of positive transformation that reverberate across embattled communities, engendering resurgent aspirations and emboldened sanctuaries of safety. The vanguard role essayed by Bangladeshi peacekeepers stands as an eloquent testimony to the nation's unwavering fidelity as a cherished constituent of the United Nations fraternity. Bangladesh's steadfast commitment to the vanguard of international peacekeeping undertakings serves as a clarion call of its unwavering disposition to contributing to the sanctity of global security endeavours. Within the Democratic Republic of the Congo, the Bangladeshi peacekeepers have not only indelibly imprinted the sands of transformation but have also unerringly chiseled the contours of a more secure world, one that casts its protective mantle over all strata of humanity. Of notable import is the tenacious involvement of Bangladeshi peacekeepers in safeguarding civilians, facilitating the disarmament of insurgent cadres, and laying the foundational bedrock of secure precincts within the precincts of the Democratic Republic of the Congo. Augmenting their noble pursuits is their munificent provision of life-sustaining humanitarian provisions, spanning sustenance, potable water, and indispensable medical succor, an embodiment of compassion that reaches the crest of its resounding impact in the lives of the most vulnerable. Concomitantly, their dexterous efforts in augmenting the capabilities of the Congolese military resonate as a symphony of empowerment, nurturing an ecosystem where civilian protection harmoniously coalesces with the principles of a steadfastly upheld rule of law. The contributions cast forth by the Bangladeshi peacekeepers, wrought with unswerving dedication, are a narrative interwoven with life salvaged, livelihoods rejuvenated, and the very fabric of existence reinvigorated within the Democratic Republic of the Congo. Their magnanimous commitments and noble sacrifices stand as an empyreal beacon, illuminating the path of collective aspiration, and reaffirming Bangladesh's resplendent constancy in the sacred voyage towards global peace and security.

The United Nations Transitional Authority in Cambodia (UNTAC) occupies a distinguished chapter in the annals of peacekeeping history, having been established by the United Nations Security Council in 1992 to play a seminal role in executing the Paris Peace Agreements, thereby bringing closure to the protracted and devastating Cambodian Civil War. Amid the league of contributing nations, Bangladesh emerged as an eminent force, dispatching over 850 peacekeepers to Cambodia and etching an enduring legacy within UNTAC's framework. These Bangladeshi peacekeepers, characterized by their unyielding commitment and consummate professionalism, embraced multifaceted responsibilities that encompassed disarming and demobilizing former combatants, ensuring electoral security, spearheading infrastructure reconstruction, and delivering vital humanitarian aid. Their pivotal role in shepherding peace and stability into Cambodia earned them widespread acclaim, their indelible imprint significantly altering the nation's trajectory and beckoning acknowledgment. A closer examination of their transformative endeavors under the aegis of UNTAC offers a panoramic view of their remarkable contributions. During their tenure, Bangladeshi peacekeepers undertook the monumental task of disarming and demobilizing over 20,000 former combatants in 1993 alone, a feat that precipitously reduced armed factions and laid the bedrock for enduring tranquility. The same year witnessed their instrumental role in fortifying the security and credibility of Cambodia's first free and equitable elections in three decades, a momentous stride that not only advanced the nation's democratic voyage but also validated its burgeoning governance on the global stage. Moreover, their dedication to Cambodia's infrastructural resurgence was manifest as they tirelessly revitalized roads, bridges, and schools, fostering socio-economic advancement and enriching lives. In concert with these extraordinary feats, Bangladeshi peacekeepers extended indispensable humanitarian aid to the Cambodian populace, facilitating the dispensation of sustenance, potable water, and medical care to assuage the afflictions wrought by conflict. Their valorous pursuits throughout the UNTAC mission crystallize Bangladesh's eminence as a conscientious and adept member of the global community. This dedication to peace has yielded tangible dividends, having ushered enduring stability to Cambodia and ameliorated the human experience. Contemplating the profound legacies of these Bangladeshi peacekeepers evokes a profound appreciation for their instrumental role in shaping Cambodia's narrative of resilience and renewal. Their unwavering resolve, poised at the intersection of duty and compassion, stands as an indomitable testament within the historical tapestry of both nations.

The United Nations Mission in the Central African Republic (MINUSCA), established by the United Nations Security Council in 2013, is a paramount peacekeeping endeavor with a compelling mandate encompassing the protection of civilians, facilitation of the political process, and crucial stabilization efforts. Bangladesh stands out as a significant contributor to MINUSCA, with its peacekeepers playing an instrumental role in safeguarding civilians, disarming and demobilizing combatants, and providing vital humanitarian aid. Renowned for professionalism and unwavering dedication, Bangladeshi peacekeepers have earned commendations for their indispensable contributions, exemplified by their valor during the 2014 Bangui Siege, vital role in disarming combatants in 2015, and humanitarian assistance efforts in 2016. These peacekeepers' unwavering commitment has become a source of immense national pride, demonstrating Bangladesh's accountability and proficiency on the global stage. Their endeavors transcend borders, profoundly impacting the lives of Central African Republic citizens, fostering peace, stability, and improved quality of life.

The United Nations peacekeeping mission in Western Sahara, known as MINURSO, has played a critical role in maintaining stability and facilitating peace between Morocco and the Polisario Front since its establishment in 1991. Among the dedicated peacekeepers from around the world, Bangladesh stands out as a steadfast contributor, consistently sending a significant number of troops to MINURSO. These Bangladeshi peacekeepers have exemplified professionalism and dedication, fulfilling a diverse range of responsibilities including security, humanitarian aid, and support for both the UN and the local population. Notably, their compassion extends to providing essential medical assistance, with over 10,000 patients, including children, treated in 2017 alone, underscoring their commitment to well-being. Their efforts have garnered global recognition, with the UN secretary-general and the Moroccan government acknowledging their exceptional dedication and contribution to advancing the peace process. Despite formidable challenges, from violence threats to adverse conditions, Bangladeshi peacekeepers have resiliently overcome obstacles, making substantial contributions to regional peace and stability. Their accomplishments serve as an inspiration, embodying the transformative potential of individual dedication even in challenging circumstances. Their selfless service, including safeguarding civilians, providing humanitarian aid, infrastructure reconstruction, economic development, training security forces, and promoting reconciliation, resonates throughout Western Sahara, establishing a more peaceful and secure environment for a brighter future. In profound gratitude, we acknowledge the remarkable and unwavering contributions of Bangladeshi peacekeepers to the United Nations peacekeeping mission in Western Sahara, as they exemplify the pursuit of peace and the collective power of selfless service.

The UN peacekeeping mission in Mozambique, known as ONUMOZ, stands as a remarkable testament to international peacekeeping endeavors, achieving its mandate of ending the devastating Mozambican Civil War in a mere four years and leaving an indelible imprint on the nation's journey towards lasting peace and stability. Among the myriad factors that underpinned ONUMOZ's success, the pivotal role played by Bangladeshi UN peacekeepers emerges as a beacon of significance, as their unparalleled dedication and contributions reverberated profoundly. Bangladesh, the largest troop-contributing country, assumed a vital mantle in safeguarding civilians, ceasefires monitoring, and facilitating post-conflict reconstruction, encapsulating a commitment that garnered profound respect from Mozambicans who embraced them as architects of hope. Their unwavering professionalism not only made an immediate impact but also laid the foundation for a brighter, more harmonious Mozambique. Bangladeshi peacekeepers displayed a multifaceted approach, securing civilians, stabilizing ceasefires, rebuilding critical infrastructure, and extending essential humanitarian aid, epitomizing their role as catalysts for recovery and unity. Their legacy endures as Mozambique basks in lasting tranquility, a testament to their unwavering commitment and exceptional professionalism.

This robust participation exemplifies Bangladesh's unwavering dedication to global peace and security, embodying courage and resilience in the face of formidable challenges. Their steadfast pursuit of peace is a testament to national pride, an inspirational emblem for all. As we contemplate ONUMOZ and the pivotal role of Bangladeshi peacekeepers, we celebrate their enduring legacy of harmony and reconciliation, a steadfast reminder that collaborative determination can forge a more peaceful world.

Established in July 2004 subsequent to the overthrow of President Jean-Bertrand Aristide, the United Nations Stabilization Mission in Haiti (MINUSTAH) operates with a multifaceted mandate encompassing civilian protection, human rights promotion, humanitarian aid facilitation, and peace process support within a challenging context. While confronting persistent political instability, gang violence, and a cholera outbreak, MINUSTAH has achieved substantial headway in fulfilling its mission objectives, contributing to national stabilization, violence reduction, and enhanced service delivery. Bangladesh, a significant troop-contributing nation to MINUSTAH, has played a pivotal role through its dedicated and professional peacekeepers, contributing to civilian protection amidst conflict and gang activity, bolstering human rights by investigating abuses and reinforcing the rule of law, facilitating crucial humanitarian assistance, and supporting disarmament and demobilization initiatives. These efforts translate into tangible improvements for Haitian society, forging a more tranquil and secure future while exemplifying a commitment to international peacekeeping. Moreover, Bangladeshi peacekeepers' contributions extend beyond their core tasks, encompassing infrastructure reconstruction, Haitian National Police training, healthcare enhancement, and agricultural and rural development support. This comprehensive engagement underscores the substantive impact of Bangladeshi peacekeepers in stabilizing Haiti, fostering an environment of peace and stability, and laying the groundwork for sustainable development, thus serving as a beacon of inspiration for the broader peacekeeping and international cooperation community.